PHENOMENOLOGY OF LIFE
FROM THE ANIMAL SOUL TO THE HUMAN MIND
BOOK I

ANALECTA HUSSERLIANA

THE YEARBOOK OF PHENOMENOLOGICAL RESEARCH

VOLUME XCIII

Founder and Editor-in-Chief:

ANNA-TERESA TYMIENIECKA

The World Institute for Advanced Phenomenological Research and Learning

Hanover, New Hampshire

For sequel volumes see the end of this volume.

PHENOMENOLOGY OF LIFE FROM THE ANIMAL SOUL TO THE HUMAN MIND

Book I

In Search of Experience

Edited by

ANNA-TERESA TYMIENIECKA

The World Phenomenology Institute, Hanover, NH, U.S.A.

Published under the auspices of

The World Institute for Advanced Phenomenological Research and Learning

A-T. Tymieniecka, President

Springer

Library of Congress Cataloging-in-Publication Data is available

ISBN 978-1-4020-5191-3
ISBN 978-1-4020-5192-0 (eBook)

Published by Springer,
P.O. Box 17, 3300 AA Dordrecht, The Netherlands.

www.springer.com

Printed on acid-free paper

In memory of
the phenomenologist Stephan Strasser
a great scholar and a noble friend
(Catholic University of Nijmegen 1947–1975)

TABLE OF CONTENTS

SECTION II

SOURCES OF HUMANITY

SECTION III

IN SEARCH OF EXPERIENCE

SECTION IV

MORAL ELEMENT OF EXPERIENCE

SECTION V

THE CREATIVE TURN

SECTION VI

CREATIVITY AND AESTHETIC EXPERIENCE

ACKNOWLEDGEMENTS

This volume of studies, like the following one (Volume 94), is dedicated to the memory of Stephan Strasser, great scholar and friend of The World Phenomenology Institute. It gathers papers read at our Fifty-Fifth International Phenomenology Congress, which was held in August 17–20, 2005 at The Radboud University in Nijmegen, The Netherlands. It was at this university, formerly the Catholic University of Nijmegen, that Stephan Strasser taught Phenomenological Psychology.

As one of the Counselors on the Board of The World Phenomenology Institute he has worked with us and supported the Institute from its inception until his death. We owe him innovative ideas and wise counseling, as well as dedicated friendship.

The participants in the conference – authors in the present collection – who have come from various parts of the world, deserve our heartfelt thanks. We would also like to thank Springer for their help with the copy-editing and proof reading of this volume.

A-T.T.

GRAHAME LOCK

OPENING ADDRESS

As the leader of the hosting Political Theory and Philosophy of Management Group at Radboud University – until recently the Catholic University of Nijmegen – I am delighted to welcome the participants to the Fifty-Fifth International Phenomenology Congress.

Its theme is "Phenomenology of Life: From the Animal Soul to the Human Mind". But more particularly, the congress is dedicated to the memory of Stephan Strasser, who from shortly after the Second World War until 1975 taught as a professor of Philosophical Psychology and Anthropology at the Catholic University. At the same time he was also a close collaborator of the World Phenomenology Institute and of its President, Prof. Anna-Teresa Tymieniecka.

I know that our colleague Daniel J. Martino is scheduled to speak on Strasser's philosophical legacy later in this congress. But I wanted to mention a number of things about his biography, his Nijmegen career and say a very few words about philosophical teaching and research in Nijmegen now.[1]

Our great Nijmegen expert on the work of Stephan Strasser is Anton Monshouwer, presently teaching here at the Department of Pedagogy in the field of the Philosophy and History of Education, but who for reasons of ill-health could unfortunately not be with us at this event. Monshouwer has published, among other works, *Ratio Militans. Perspectieven op Stephan Strasser*,[2] as well as a bibliography of Strasser's writings.[3] Readers of Dutch may therefore profitably make a study of those volumes.

The University of Nijmegen was from its inception, in 1923, a Catholic institution, whose foundation was inspired by the general struggle for Catholic emancipation in the Netherlands. It originally contained three faculties: Theology, Law and Humanities, including Philosophy and Pedagogy. Up until the outbreak of the Second World War it was, in the philosophical field, largely inspired by neo-Thomism. A fascinating and controversial figure in its history was Ferdinand Sassen, professor from 1929 of Greek and Roman philosophy and, in his own words, a progressive neo-Thomist. An article by Sassen on Thomas and Hegel, published in 1932, provoked a concerned reaction in Rome, though he was defended by the Dutch bishops, if not

A-T. Tymieniecka (ed.), Analecta Husserliana XCIII, xiii–xv.

entirely whole-heartedly. Sassen left Nijmegen shortly after the end of the Second War, for the Ministry of Education and then Leiden University, a non-confessional institution.

The Catholic University of Nijmegen emerged from the war and the German occupation with its reputation high: it had refused to collaborate with the enemy. It was in this period that Strasser received his appointment, at roughly the same time as Dries van Melsen. Strasser's philosophical attention was to be directed in particular to the students of psychology and of pedagogy, Melsen's to the students of the natural sciences. Thus, in a sense, Strasser worked on the periphery of the rather small faculty. Neither was a priest, though many of the local students of philosophy were, and the Church kept an eye on the Faculty's teaching.

Born in 1905 in Vienna of Hungarian-Jewish parents, Strasser escaped from Austria in 1938, at the time of the *Anschluss*, and found asylum in Belgium, where however, with the Nazi occupation, he soon also had to go into hiding. The Phenomenology Online website notes: "[Herman] van Breda offered him work at the Husserl Archives, where, in the space of 25 months, Stephan Strasser, his wife and mother-in-law transcribed 20,000 pages of Husserl's shorthand into ordinary text. These experiences and his studies with [Alphonse] de Waelhens in 1944 were formative for Strasser's philosophical career."[4]

Strasser's doctorate had not been on a typically philosophical topic: it concerned the novelist and poet Joseph von Eichendorff. His teachers in philosophy were de Waelhens and Albert Dondeyne. My Nijmegen colleague Harm Boukema adds that he later came into contact with Paul Ricoeur, Emmanuel Lévinas, Helmuth Plesner and F.J.J. Buytendijk. For Nijmegen he represented a good point of contact with Louvain as well as, more generally, with phenomenology. But, as Boukema points out, he certainly preferred some phenomenologists to others: Ricoeur, Lévinas and Gabriel Marcel to Merleau-Ponty, and Merleau-Ponty to Sartre.

The above-cited source on phenomenological scholars remarks that for a time, Strasser "became intensely interested in Heidegger, but eventually he moved closer to Merleau-Ponty and in his later years especially to the work of Lévinas. Strasser exercised significant international influence. In North America, his writings provided access to continental thought; in Germany, he helped introduce ... Lévinas; in France, he helped introduce ... Husserl; and in Japan, he helped introduce the human science approach." It concludes: "Throughout his career, it was Strasser's ambition to practice human science without doing violence to what is human. His 1947 inaugural lecture was on the theme 'Objectiviteit en Objectivisme' (Objectivity and Objectivism). In

1950, he introduced the *Husserliana* series by publishing the first volume: *Cartesianische Meditationen und die Pariser Vorträge* (Cartesian Meditations and the Paris Lectures)." He also published *Fenomenologie en empirische menskunde*, translated into English as *Phenomenology and the Human Sciences* in 1963.[5] For Nijmegen, he clearly represented an opening to a wider philosophical world. But the general and the philosophical situation at Nijmegen has changed so radically in the period since his retirement that this contribution is nowadays not always fully recognized.

Any specifically "Catholic" signature of the university as well as of its philosophers is now almost, though not entirely, unrecognizable. Philosophy is by the way not taught only in the Philosophy Faculty itself, but also by philosophers appointed to chairs in other faculties, to teach philosophy of the natural sciences, of the educational sciences or of the management sciences and so on. This phenomenon is itself in part a heritage of the old Catholic University, which wanted to promote philosophical reflection among students of all disciplines.

The Philosophy Faculty presently represents – as indeed do all such Dutch faculties – a more or less eclectic mixture of philosophical positions and tendencies. An interest in phenomenology does exist, if not in the sense in which, in Strasser's time, it was a window through which neo-Thomists were able to take account of the wider philosophical world to which I referred above.

But to conclude, with an eye to the coming days: Nijmegen welcomes the World Phenomenology Institute and its President, as well as Mr Alexander W. Schimmelpenninck, from its Dordrecht publishers, and wishes all participants a very successful congress.

NOTES

[1] In this respect I am entirely indebted to information publicly provided by colleagues from the Nijmegen Philosophy Faculty, especially Dr Harm Boukema and Dr Ad Vennix. This information is contained in Dutch-language lectures held on the occasion of the 80th anniversary of the foundation of the Philosophy Faculty in 2003. But see the remark in the text on Dr Anton Monshouwer.

[2] Bemmel: Uitgeverij Concorde, 1997.

[3] "De filosofische en wijsgerig-pedagogische werken van Stephan Strasser (1905–1991). Een bibliografisch overzicht." *Tijdschrift voor Filosofie*, 56, 1994.

[4] See http://www.phenomenologyonline.com/scholars.

[5] loc. cit.

THEMATIC STUDY

ANNA-TERESA TYMIENIECKA

FROM SENTIENCE TO CONSICOUSNESS

SENTIENT INTENTIONALITY AS THE THREAD OF LIVING CONTINUITY

IN THE ONTOPOIETIC UNFOLDING OF LIFE

Proposing an investigation of living human beings as stretching from the stages of the animal soul to the human creative spirit, we have already indicated the line of the unfolding of life in individual beingness as it has come to unfold. Since the phases of this unfolding pinpoint the essential transformations that have occurred along the trajectory of the unfolding that bring about growth and progress—since "to live" means to grow or to diminish and never to stand still—there has been raised since the beginning of modernity the question of the continuity of this trajectory. In question is the diversity but concurrently the continuity of the sense of life that this unfolding maintains.

It is enough to mention the Cartesian distinction between the seemingly drastically divergent entities that are body and mind, whose functioning introduces and advances innumerable transformatory threads of sense.

I have been proposing and elaborating a radically different approach to classic philosophy at large. Before I come to focus in on the issue at stake, it is necessary to return to the foundation stones of this new conception of philosophy.

Before outlining the new approach to the continuity of the temporal progress of self-individualizing life, let me state that no instrumental/operational rationality may account for the vital, psychic, intellective, and creative metamorphosis that engenders a multiplicity of sense-giving factors. No scientific effort to seek the passage from one category of sense into another in the minutiae of operative and symbiotic mechanisms can bridge the discrete stations of these transformatory moves. Were it not for the foundational role of the logos of life carrying the self-individualizing processes, those moves would not have occurred. *It is the core of the thread of the logos of life, namely,*

A-T. Tymieniecka (ed.), Analecta Husserliana XCIII, xix–xxiv.

sentience, that advances the translacing rationale running through and carrying the life processes in the discrete (to our mind) continuity of their unfolding.

Now let us outline the foundations of our philosophical outlook.

A. THE POINT OF DEPARTURE OF OUR ONTOPOIETIC INQUIRY: THE HUMAN CREATIVE CONDITION WITHIN THE UNITY OF LIFE

We abandon, in fact, the time-honored classification of philosophical problems with their separate realms of inquiry, and in contrast approach their common groundwork, which is life itself at its basic onto-metaphysical level. *Phenomenology/ontopoiesis of life indeed descends intuitively to the level of the pristine logos of life,* wherefrom all scientific and philosophical problems have their common root. It is on this level that problems that through the centuries could not be resolved find their natural setting and solution.

It is upon the basic metaphysical-ontic platform, therefore, that I will attempt to show how the numerous questions that "human development" poses may be treated profitably. Foremostly I will focus on the issue of the discrete continuity of life's unfolding, avoiding the undue reductions that became fashionable with the developments of neurobiology and its associated sciences. Upon the platform of the ontopoiesis of life we will see that no basis presents itself for such "naturalizing."

Let us first of all, then, correct some classic preconceptions of the appropriate point of departure, for the concept of what is "human" cries out for revision. Traditionally, the human being has been specified by its "nature," that is, identified by the salient features that distinguish us from other living beings. With the penetrating progress of natural science, this approach can no longer be upheld.

The biological, chemical, medical sciences of today compel us to move away from this approach. The human being's salient specific character is not a once and forever established entity. On the contrary, we are as a type first and foremost the fruit of a long line of development within the natural unfolding of life. But each individual also unfolds, grows, and declines, remaining dependent throughout on his or her natural milieu with its laws. In agreement with these concrete findings of science, our inquiry into the logos of life reveals that *the human being cannot be defined by its specific nature but by the entire complex of individualizing life,* of which complex it is vitally part and parcel. And so *we speak not of human nature but of the Human Condition-within-the-unity-of-everything-there-is-alive.* We do, however, make one very essential specification by referring to "the HUMAN CREATIVE CONDITION."

It is precisely by the human creative act, instead of by a cognitive act, that we may enter into this innermost line of the workings of the logos of life, which in its basic thread of vital force—*vis viva*– – –brings individualizing life about, promotes its unfolding, and controls its course.

It is the descent into the becoming of beings, living creatures and nonliving objects in their origination, generative ties, existential connectedness, interactive unfolding—and this in the innermost logos that prompts them—that is indispensable to achieving understanding of the intricacies of existence. I will say after Periander that to understand a single thing one has to know all things. Thus the logos of this origination and becoming is the crucial focus of our researches.

The human creative condition as a conception of the human being is already the fruit of such ontic discovery. As a matter of fact insofar as this continuity of the living process (progress/regress) is a basic question of epistemology as well as a major part of our knowledge at large, it is by focusing on the nature and extent of the human creative act that we discover the Human Condition-within-the-unity-of-everything-there-is-alive, discovering an ontologically basic plane of life's generation and becoming. Upon this plane, in contrast to classic ontologies focusing on the structure of things and living beings, we inquire into and may follow *poiein*– – –the making of, the becoming, the very unfolding of those structures in their circumambient context of resources, forces, intergenerative energies, as these are arrayed in basic self-individualization in existence. Upon the ontopoietic level it is confirmed that life consists of constructive *poiein,* becoming, unfolding, development. On the ontopoietic level of the logos of life, we uncover its incipient forces, its directional law, its LOGOS. But the logos of life could not implement its order were it not for its sentient core, which allows all the joints of the instrumentality of ordering to work together.

To understand the Human Condition-within-the-unity-of-everything-that-is-alive, we have to focus on two of its basic features: (1) the discrete continuity displayed through the disruptions in life's unfolding, (2) the ingrownness of individual existence into its circumambient existential network. Ingrownness is a paradoxical way of becoming. In order to establish order, to particularize, to individualize, and to unfold, incipient beingness is necessarily existentially implicated in a radius of external conditions. Each living being is supported on a system of life subtending it in living beingness. No living element may be seen as standing apart, in itself, beyond its circumambient cogeneric radius and outside of its inscription in the network of the earth, its ground, and of the cosmos with its laws. These two spheres of existential dependencies codetermine the autonomy of the living/becoming individual.

B. THE MATRIX OF BECOMING

This twofold inward/outward oriented existential route makes it manifest that in order to grow in selfhood from within, living beingness has to draw on forces and energies from the "external" circumambient sphere. What lies at the center of our inquiry is the question of how the directional devices operating from "within" find their fulfillment in the sphere of their application.

In this respect we have to differentiate the span of life's individualizing becoming two existential matrices of poiein along with a third "in between" matrix:

1) the ontopoietic vital matrix of the generation, unfolding, development of the organic/vital significance of individualizing life;

2) the transformatory climax of the unfolding of living types that goes *in between* the natural/vital and the creative specifically human phase of unfolding life following on the advent of Imaginatio Creatrix;

3) the creative matrix of specifically human activity.

It is of primary importance to differentiate these three matrices, which each have their specific functional systems and yet are inwardly fused together in collaborative variations. There is the differentiation of their guiding principles, on the one hand, and their innermost existential discrete continuity of developmental advance, on the other hand. Here is the key to the understanding of human development with its autonomy, on one side, and its exitential dependencies, on the other.

I. Let us first of all come back to the incipient instance of life. It is in its self-individualization from the prompting *vis viva* of the logos that beingness sets out on its life career, carrying with itself the entire initial endowment concentrated in its ontopoietic sequence. The ontopoietic sequence carries not only life's germinal endowment but simultaneously as well all the indispensable dynamic directional devices enabling it to unfold within favorable rudimentary circumstances in such a way that using them according to how they fit with a living being's own constitution it works out its unfolding, simultaneously transforming appropriately this circumambient groundwork. The ontopoietic sequence prepares as well further propitious conditions for the extended development of its sequential virtualities. It initiates thereby a thread of unfolding that develops the living beingness according to the constructive design that it brings with itself. This constructive encounter of the inner virtualities prompted by logoic forces with fitting material and propitious circumstances constitutes the individualizing beingness and its vital matrix.

However, what is of crucial significance is that in the unfolding of life from its incipient instance, and even before, already at the forming of prelife conditions, each and all of the constructive moves of the individualizing logos proceed along the line of this core—a rationale of coalescence, "fitness," the amalgamation of modulating constructive moves, of operative moves, etc. that is made possible by the sentient core of the constructive line of the logos of life. I speak of a "sentient intentionality" that allows all the innumerable operative connections to come together. To reemphasize, it is owing to the sentient virtuality of the logoic sequence that it may perform its modulating/directing role in the constructive metamorphic generation of new sense that is the mark of each step in life's advance. To put the matter sharply, the order, self-ordering, of the course of individualizing life is not a "neutral," automatic fitting together of matching elements. To the contrary, this ordering—effected by living intentionality, *vis viva*—is a sentient selection, the logos of life of congenial forces and virtualities deploying their capacities in contact with appropriately chimed energies.

This amplification that I bring to my investigation of the logos of life has far-reaching consequences. First of all and critically, it allows us to lay down the foundations for understanding the metamorphic continuity of life's unfolding, the extended history of transformation worked in the growth and decline of individualized life. Secondly, our amplified account of life undercuts any unwarranted "reduction" of one phase of sense attained in this unfolding to any preceding phase. Thirdly, it provides the core explanation of the passage that takes place in the unfolding of individual types from among living agents, whose unfolding is an interior expansion of sentience into selfhood, a progressive advance toward full human consciousness. Lastly, our understanding makes it most manifest how one of the major imports of the Cartesian controversy, a "naturalizing" that still lurks in phenomenology, has no justification.

To summarize, the vitally significant ontopoiesis found in nature progresses and promotes the flow of life, organizing it around the self-individualization that flows out of the coalescing moves of becoming that follow as the intrinsic line of a directional sequence comes to constitute a circumambient radius of elements suited to the continuance of that sequence. The telos of living beingness is projected by its *intrinsic logoic sequence,* which is transformed in an extended course of unfolding into a self-prompting agency to be followed. *This ontopoietic course of self-individualization occurs as a function of the sentient intentionality proper to the logos of life.*

II. The perfecting of the self-prompting agency occurs in its unfolding of sharing-in-life. This is the phase of life's ontopoiesis at which Imaginatio Creatrix will enter into play.

The vitally significant ontopoietic matrix leads in its constructive unfolding to the more developed sharing-in-life matrix of communicative animality. A climactic constructive apex of the inward agency of self-individualizing beingness is its acquisition of the organic basis for higher communicative sharing of life's necessities. Communication enters into the play of becoming.

With the progressive unfolding of receptive/remitting organs in living agencies, the system is readied for the entry into the game of life of the unique force that is Imaginatio Creatrix, which surpassingly transforms the functional system of the ontopoietic/animal matrix of momentum, internal programming, and external selection into a self-directed imaginatively programmed and deliberately selected project, the autonomous creative matrix of the specifically human significance of life.

It is in this intermediary field of essential transformations that the metamorphosis of the animal/vital functions occurs. All the life-significant forces—organic, vital, psychic, and spiritual—meet at this field forming the essential constructive functional links that transform the living agent into a self-conscious selective agency.

This transformed functional field, whose leading orientation comes from creativity, opens a new theater for the logos of life. (To be continued in Analecta Husserliana XCIV.)

DANIEL J. MARTINO

STEPHAN STRASSER'S PHILOSOPHICAL LEGACY AND DUQUESNE UNIVERSITY'S SIMON SILVERMAN PHENOMENOLOGY CENTER

It is truly an honor to have been given the opportunity to present for the 55th International Phenomenology Congress the connection between Stephan Strasser's philosophical legacy and The Simon Silverman Phenomenology Center at Duquesne University, the place he saw as the proper repository for his personal library and scholarly archival materials. Because it is an interesting and important connection for appreciating the growth of phenomenological thought in the 20th century, this author felt a certain onus of responsibility and challenge. On the one hand, Strasser deserves all the due diligence necessary to ensure that his scholarly contributions are preserved as accessible resources to guide and inspire like-minded scholars of every generation. On the other hand, his level of prolificity and degree of involvement in the phenomenological world do not lend themselves to succinct exposition. In striving to achieve a manageable balance between these two daunting demands, I fully resonated with the frustration that he expresses in a piece of his voluminous correspondence. There, as in nearly all his letters to colleagues, the passion for his work is palpable. However, in this instance his zeal has temporarily overwhelmed him, as he writes, "The paper I am busy with deals with Merleau-Ponty and metaphysics. The difficulty I am coping with is that I have so *much* to say while I do not know yet the most adequate manner [of] *how* to say it ... I am pretty sure that I shall overcome this difficulty."[1]

As evidenced by Strasser's lengthy list of publications, he met the challenge over and over again by more than adequately articulating his thoughts on a wide range of issues that drove his philosophy, his "love of wisdom." Likewise, in a very isolated sense, I too was able to resolve the tension associated with the charge to adequately represent the extensive documentation of Strasser's philosophical career in a respectful way. Respect and honor in no way should be seen as a eulogy. This is an exercise that reverentially highlights the record of Strasser's scholarship in order to showcase its timeless dynamism and ongoing source of scholarly inspiration. His intellectual life

A-T. Tymieniecka (ed.), Analecta Husserliana XCIII, xxv–xli.

as preserved in the Strasser Alcove at The Simon Silverman Phenomenology Center is not guarded by docents in a museum as a relic of dead thought nor as a passing reference in downstream philosophies. Instead, Strasser's thought and the materials that supported and sustained that thought (correspondence, manuscripts, notes) are held in the Center as if they were in a garden – a place that constantly nurtures his thought as an ahistorical body of thinking that will continue to pervade and positively influence the understanding of human experience.

1. STEPHAN STRASSER AND DUQUESNE UNIVERSITY: ALLIES IN THE ADVANCEMENT OF PHENOMENOLOGY

Ernst Orth and Thomas Seebohm have included Strasser as part of the first generation of postwar phenomenologists. He, along with Ludwig Landgrebe, Eugen Fink and Walter Biemel, worked at the Husserl Archives in Leuven. This was seen as the center of the new beginning for the phenomenological movement.[2] With 50,000 pages of Edmund Husserl's writings (40,000 of which in Husserl's untranscribed handwriting) safely brought to Leuven by the heroic offices of Herman Leo van Breda, O.F.M., the stage was set for phenomenology's flourishing. Without question Strasser played an instrumental part in the successful launch of this exciting new epoch by his transcription of 20,000 pages of Husserl's *Gabelsberger* beginning in 1942. This experience, however, was not merely a detached chore for which he received only a stipend in return. Deciphering the shorthand meant becoming familiar with Husserl's thought, and it is in this process that something of inestimable worth was given to him – a philosophical awakening that would have an abiding influence on his life's work. Bas Levering and Max van Manen have identified the Husserl Archives' years as a significant period of opportunity and grace for Strasser: "These experiences and his studies with De Waelhens in 1944 were formative for Strasser's philosophical career."[3]

While Strasser's appreciation for the value of Husserl's thinking continued to grow throughout the late 1940s, the emergence of yet another postwar center of phenomenological flourishment was taking place across the Atlantic at Duquesne University. In 1948, the Dutch philosopher Henry Koren, C.S.Sp. came to Duquesne University. He was a member of the Spiritans, the religious order that founded and continues to sponsor the university, and as David L. Smith, C.S.Sp. explains: "It was through the efforts and vision of [Father Koren] that phenomenology first came to Duquesne and took deep root there."[4] When Koren became the chair of the philosophy department in 1953/54 he oversaw a Thomistic program – the prototype for Catholic philosophy departments at that time. While his own philosophical background and orientation were also along Thomistic lines, Smith recounts Koren's sense of disconnect between Thomism and the *Zeitgeist* of the mid 20th century: "while convinced that Aquinas' formidable synthesis offered a satisfactory overall view, he was keenly aware that he left many modern questions untouched."[5] In order to give the philosophy department greater relevance and purchase as it instituted a Ph.D. program under his watch, Koren turned to phenomenology. Most interestingly he arrived at this potential source for philosophical awakening and rejuvenation through Van Breda, the same person who coincidently introduced Strasser to Husserl's phenomenology. In correspondence with André Schuwer, O.F.M., Koren credits Van Breda for influencing the shape of the doctoral program that would become one of the first in the United States to emphasize the concentrated study of phenomenology and contemporary continental philosophy, "Van Breda woke me from my dogmatic slumbers (*pace* Kant) and directed my attention to phenomenology, whose very existence was unknown to me."[6]

The confluence of phenomenological activity in Europe and its nascent development in the States was effectuated by one of Koren's perspicacious programmatic initiatives. Throughout the 1950s he would invite visiting professors to teach at Duquesne University who were committed to serious phenomenological scholarship. These lecturers included Andrew G. van Melsen (University of Nijmegen – now Radboud University), Herman Leo van Breda (Catholic University of Leuven), Remy Kwant (University of Utrecht) and most importantly for the current context, Stephan Strasser (University of Nijmegen) in 1956. At various levels, Koren's strategy would have far-reaching impact for the advancement of phenomenology. Beyond satisfying the short-term need to shore up the credibility of a fledgling graduate program, the promotion of its very distinctiveness would attract and so form interested scholars in North America who might not otherwise have the resources necessary for a European education. Besides this comprehensive seeding of phenomenology in the U.S., a more particular result of Koren's series of visiting professors was the forging of a close relationship between Strasser and Duquesne. In looking back on the beginnings of the relationship that would last nearly four decades until his death in 1991, Strasser credits its causality with reverential regard. It was neither mere chance nor the work of the fates but an act of providence, as expressed in his acknowledgments during the dedication ceremony of the Strasser Alcove at The Simon Silverman Phenomenology Center: "it was providential that in 1956 I started in America as a Fulbright-fellow at the [*sic*] Duquesne University."

But this beginning would not only be viewed as providential from Strasser's perspective. From its vantage point, Duquesne University would also rejoice in its new found friendship with Strasser as a moment of providence. His interaction with Duquesne lent prestige to its efforts to advance phenomenological thought. It is important to understand that the effectiveness of Strasser's influence was felt in a pointed sense by various entities within the university. In other words, it was not merely a generic endorsement to a faceless structure or a narrowly confined boost to the philosophy department. Instead, his reach of influence to promote phenomenology extended to the Duquesne University Press and the psychology department, deemed at one point as "the capital of phenomenological psychology in the New World."[7] From 1957 to 1985, the Duquesne University Press would publish four of Strasser's books. This ensured increased access to his writings for an eager English speaking readership. As indispensable as these press runs were for the greater dissemination of Strasser's thought and the profile enhancement of Duquesne's commitment to phenomenology, the relationship cultivated between the psychology program and Strasser has a significance in its own right and cannot be overlooked.

After his initial visit to Duquesne in 1956 to lecture in the philosophy department, Strasser received repeated invitations to return. He obliged twice, once in 1968 and a second time in 1982. The occasion for the latter visit was to conduct a mini-course for the department of psychology entitled "Understanding and Explanation: Basic Ideas Concerning the Possibility of Human Sciences." Strasser's appearance for the course was the direct result of a key faculty member in the department at the time, Amedeo Giorgi. From 1962 to 1987, Giorgi did all in his power as a scholar and administrator to ensure the flourishing of Duquesne's graduate psychology programs and their internationally renowned exclusive commitment to phenomenology. Giorgi first came to know Strasser as an author. When his book *The Soul in Metaphysics and Empirical Psychology* became available in English in 1957, Giorgi saw this as an appropriate text for his graduate psychology courses in the early 1960s. A more personal relationship developed during a visit to Nijmegen in 1961. After their initial meeting, Giorgi visited Strasser numerous times and the two men maintained an ongoing correspondence. In a recent conversation, Giorgi summed up his meaningful and long standing association with Strasser as follows: "it was a very cordial relationship between two phenomenologists, one a philosopher and the other a psychologist."[8]

By the fall of 1982 as Strasser completed the three-day psychology mini-course, the cordiality between him and Giorgi reached a level of comfortableness. It was during a dinner hosted by Giorgi at his home that he broached a potentially delicate topic – what were Strasser's plans for his personal library and scholarly papers after his death?

Giorgi had the perfect place in mind for Strasser's materials – the newly instituted Simon Silverman Phenomenology Center. Founded in 1980 with the support of the University's Academic Vice President, Edward L. Murray, C.S.Sp., Giorgi was the driving force in the establishment of the Center. When the phenomenological psychiatrist Erwin Straus died in 1975, Giorgi worked closely with Mrs Straus so that her deceased husband's books and scholarly papers would be given to Duquesne. The collection of Straus' materials became the cornerstone of what would evolve into The Simon Silverman Phenomenology Center.

Giorgi and John Sallis, at that time a professor of philosophy at Duquesne, approached Simon Silverman, the president of Humanities Press and asked him if he would be willing to support the idea of a center which would be a repository for world literature on phenomenology. Silverman was a fitting prospect since he had an affinity for Duquesne and its commitment to phenomenology, as Smith explains: "Silverman's relationship with Duquesne University went back to 1972 when Humanities Press became the exclusive distributor for the Duquesne University Press."[9] Silverman was favorably disposed to Giorgi's idea of a hub for phenomenological research and so gave a generous endowment in 1980, and the Phenomenology Center bearing his name came to be. When Straus' *Nachlass* arrived at Duquesne and was placed in the special collections area of the University Library, soon to follow was the personal library and manuscripts of Aron Gurwitsch. Each of these scholars also gave memorabilia and other personal possessions. These items helped to foster a warm and inspiring climate for scholars visiting the Center.

Having seen this space and trustful of Giorgi's stewardship of the Center, it is easy to understand why Strasser quickly agreed to designate The Simon Silverman Phenomenology Center as the optimal place for his scholarly materials.

The official dedication of the Stephan Strasser Alcove at The Simon Silverman Phenomenology Center comprised a two-day celebration on October 15 and 16, 1984. Aside from the ceremonial festivities there were also intellectual segments of high caliber. Strasser's successor at the University of Nijmegen, C.E.M. "Kees" Struyker Boudier, delivered the dedicatory lecture titled "Philosophical Anthropology: Buytendijk and Merleau-Ponty" to which Strasser responded. On the second day there was a round table discussion devoted to "Phenomenology and Philosophical Anthropology." Reflecting on this course of events, Strasser wrote these sentiments to Richard Rojcewicz, the Center's first director: "Gertrude and I want to thank you for the cordial manner we were received by the people of Duquesne University. I had the impression that the opening of the Strasser Alcove and the Round Table conference took place in a dignified manner but also in a warm human atmosphere."[10] In the next section I hope to highlight the more salient aspects of Strasser's materials which enabled the alcove to maintain scholarly dignity and human warmth beyond the initial hope stirred at its dedication.

2. THE ALCOVE OF STEPHAN STRASSER

Because Strasser was an active scholar at the time of the inauguration of his alcove in 1984, it was necessary to ensure that his personal library was at his ready disposal. As a result, only a portion of the total materials that would comprise the alcove at Duquesne was sent prior to the dedication *fête*. In a letter to Rojcewicz, he ruminated on the separation anxiety which afflicts all serious scholars and their love affair with the potentialities of the written word: "I wonder whether the first part of the materials has arrived by now. I emphasize that it constitutes only a third of my library and the less valuable part. It is easy to explain the cause of this disparity: while pondering whether I shall need this book or that series in the coming years I mostly answered 'yes'. Can you understand the ridiculous optimism of mine?"[11] This first installment of materials did safely arrive and proved to be a fitting backdrop for the alcove's formal establishment as well as contributing greatly to the quality and depth of the Silverman Center holdings. Nearly a decade would pass before the final installment would be received in the summer of 1995. While Strasser died in 1991, the directors of the Center were sensitive to Mrs Strasser's request that her husband's materials not leave Nijmegen until after her death. Liesbeth Strasser, the Strassers' second child, diligently oversaw the shipment of her father's library and made it a point to thank the Center's leaders for their sensitivity to her mother's wishes: "In your letter from January 20, 1993 you and the Simon Silverman Phenomenology Center kindly gave permission to our family to postpone the transfer of my father's archives and books till after my mother's death. My mother was very grateful that you took this decision ... Thank you again for your consideration for my mother."[12]

It was therefore a bitter sweet moment when all of Strasser's books and papers were finally installed in his alcove at the Silverman Center. On the one hand their arrival was the direct result of his passing and then the death of Mrs Strasser. However, with the full complement of his materials now in place in their new home a decorous and accessible space was finally established to preserve his legacy and to invite other scholars engaged in phenomenological research who will keep his work alive and meaningful in new ages.

What can scholars expect to find when consulting the resources of the Strasser Alcove? No doubt books, journals and nearly forty archival boxes containing correspondence, manuscripts and off-prints first greet any visitor. However, this initial greeting is given immediate warmth because of a handsome portrait of Strasser that adorns the alcove. The artist, Henry Koerner, was commissioned by the Silverman Center to paint the portrait for the dedication ceremonies. Koerner was a cousin of Mrs Strasser who by coincidence lived in Pittsburgh. The portrait lends inspiration to any research project, since it quickly and clearly contextualizes the books and papers as Strasser's, which at one time were the life blood of his academic achievements. The number of books comprising his personal library is 1,637 volumes. Even the most cursory glance over the stacks will evoke a sense of awe for the diversity of philosophical sources that nurtured his thinking – from the golden age of Plato and Aristotle, to the high middle ages of Thomas Aquinas and of course to the contemporary period which includes every well known phenomenologist and existentialist, and even to postmodernity and post-structuralism with such figures as Derrida and Foucault being represented.

Aside from the books, there is also the sizable collection of personal archival materials. Most noteworthy here is the correspondence. A lively and extensive exchange of letters took place between Strasser and other luminaries who have left a lasting mark on continental thought. Among those with whom he exchanged missives on a regular basis include Paul Ricoeur, Emmanuel Lévinas, Werner Marx, Samuel IJsseling, Rudolf Bernet, M.J. Langeveld, Jacques Taminiaux, Helmuth Plessner, Hermann Lübbe, Adriaan Peperzak, Bernard Waldenfels and of course Anna-Teresa Tymieniecka.

The frequent and intense correspondence between Strasser and Ricoeur as well as between him and Lévinas has a singular significance and is deserving of further study. Ricoeur's recent passing will generate a renewed interest in his thought, while Strasser's acclaim for having first presented Lévinas to Germany's intellectuals will be of lasting import, a point not missed by Anotonino Mazzù. "His more recent book [*Jenseits von* Sein und Zeit: *Eine Einfürung in Emmanuel Lévinas*] may be considered as a classical introduction of Lévinas' thought to the German-speaking public."[13] A more careful examination of this correspondence will be of inestimable scholarly worth, since it will disclose rare insights about the thought behind the thought that reaches the reader only in its final and published form. Because of the unique forum correspondence provides, a perspective of an author is captured that can enhance a reader's comprehension in a more holistic way. Lotte Kohler and Hans Saner made this observation in their edited work on the correspondence between Hannah Arendt and Karl Jaspers: "Because neither, probably, ever thought this correspondence would be published, and because they had absolute trust in each other, there is hardly a trace of self-censorship in these letters. They reveal themselves in a more personal, spontaneous, warmer and, at the same time, more ruthless way than they do in their works."[14]

The more personal and warmer manner in which Strasser reveals himself is especially noticed in the correspondence where he assumes the role of mentor. A long paper trail gives evidence of how numerous younger colleagues sought his counsel on their writing and research. It is possible to trace the sequence of this interaction in different ways. Henk Struyker Boudier, for instance, vetted an article concerning F.J.J. Buytendijk and Erwin Straus through Strasser prior to its publication in the 19th volume of the *Journal of Phenomenological Psychology* (1988). Angela Ales Bello, on the other hand, welcomed Strasser's reactions to her work on Husserl and God after its publication. Unmistakable in this exchange is an appreciation for the revelatory power of correspondence. At that point in time the issue of Husserl and God was also of interest to Strasser. Having presented a paper titled "History, Teleology and God in the Philosophy of Husserl" at the Sixth International Phenomenology Conference,[15] the receipt of Ales Bello's *Husserl Sul Problema di Dio*[16] was a timely opportunity for scholarly symbiosis. By simply noting the very length of Strasser's reactions – a three-page letter – a clear testament is given to the simultaneity of his scholarly rigor and genuine human concern which was brought to bear on the review of her book. Ales Bello's response in turn demonstrates a moment of academic challenge and deepening as well as personal affirmation:

> First of all I am very grateful to you for your review of my book, it is really an honor for me because of your authority in the phenomenological field – I am sorry that you spent so much time reading it – and secondly I thank *you* for your kind remarks upon my interpretation of Husserl's thought. It is true that I did not deal with the question of pantheism and that Husserl's manuscript on teleology (*Intersubjectivität* III, 378–386) may suggest a pantheistic point of view … It is true that from a philosophical point of view there is a deep difference between pantheism and theism, but Husserl's aim was not to discuss the problem of God it [*sic*] itself; for this reason we can discover an ambiguity in his position…In any case I thank you very much for having asked my opinion so that I could think over this argument.[17]

Aside from the invaluable unique glimpse that the correspondence provides of Strasser's intellectual activity as he interacts with others, there are additional alcove materials that complement this view from a different angle. Instead of grasping his thought as it is worked out through correspondence, there are also draft manuscripts and personal reading notes that show the more solitary practices necessary for scholarly growth and refinement. For instance, it is interesting to note the editorial discernment to which Strasser submitted his writing when comparing the rough drafts and final published form of *Welt im Widerspruch: Gedanken zu einer Phänomenologie als ethischer Fundamentalphilosophie*.[18] It is also scintillating to pour over the meticulous notes he took when reading primary philosophical texts. Included here are not only the more obvious authors that would have propelled his thought – Husserl, Heidegger, Merleau-Ponty, Gadamer – but also to be found are notes on the works of Kant, Hegel, Popper, Kuhn and Lacan. This diversity speaks volumes of the multi-colored palette which hued Strasser's intellect.

This intellect expressed itself in Strasser's prolific output. He read much in order to write much. One particular outlet for his writing in which the Silverman Center had direct involvement was the publication associated with the 1984 dedication of his alcove – *Clefts in the World: And Other Essays on Lévinas, Merleau-Ponty, and Buytendijk*. When the Center's leaders at that time, Giorgi and Rojcewicz, proposed the idea of the publication to Strasser, it was met with enthusiasm. The original content plan for book included one chapter by C.E.M. Struyker Boudier and the remaining five as original contributions from Strasser. Upon its final publication, he viewed the work with pride and hoped that it could reach a wide audience. On one occasion he called the work "a tangible proof of my philosophizing,"[19] and at another time saw his yearning for its potential outreach as something to be taken seriously: "you should understand, like every author, I wish that my book be read; and you know the Scholastic adage: "*Desiderium naturale non potest esse inane*."[20] Aquinas used this expression to caution against the facile dismissal of natural human desires for God as illusory.[21]

Indeed the Silverman Center continues to heed Strasser's natural desire to be read and studied. That is the purpose of his alcove, where scholars will always be welcome to avail themselves of the distinctive resources that will stimulate new inroads in phenomenological scholarship as inspired by the seminal work of Strasser. While not restricting the course that these inroads may take, he nonetheless provides some possible directions. These are expressed in a piece that was slated for inclusion in *Clefts in the World*, but was later excluded because of spatial constraints. The "Critical Self-Presentation" is a meditative retrospective of his scholarly life. Though first published in 1981 as volume seven of Peter Lang's series *Philosophers on Their Own Work*, Strasser believed strongly that it would a most fitting inclusion in the publication accompanying the inauguration of his alcove as well as a most effective means to ensure that his writings were more widely read. He asked

Rojcewicz to translate it into English from the German. I am proud to now share with you its first presentation in English. It is therefore in keeping with Professor Strasser's wishes to now reiterate the self-appraisal he made of his philosophical legacy. As he had hoped that it would generate thought and avenues of research at the occasion of the founding of his alcove in 1984, so too is it hoped that his words will serve as a similar catalyst at the 55th International Phenomenological Congress devoted to his memory in 2005. His natural desire to be read and studied is accorded the greatest respect and without doubt there will be innumerable re-readings and reexaminations of his works.

3. STRASSER'S "CRITICAL SELF-PRESENTATION"[22]

A critical self-presentation is incumbent on me. Such an assignment can be interpreted in two senses: the word "critical" can be understood as referring to the "self" or as calling for a taking up of a critical position versus other philosophers and philosophical currents. To satisfy both these requirements I will formulate a few theses at the end of my presentation; they will express certain convictions that I still – after decades of philosophical development – consider valid today. The theses will exhibit the required critical aspects.

In a self-presentation, what is under discussion is the "self." In my case this is all the more necessary, since I came to philosophy only late in life and as a consequence of all sorts of fateful events. How was it that I became conscious of my philosophical vocation so late?

A first circumstance lies no doubt in the fact that Vienna, the city of my birth, was a center of musical and literary culture but not so much an abode of the philosophical spirit. The opera, theater, literature, and politics captured, at the time of my youth, wide segments of the public, while the intellectuals of Vienna were occupied with the battle between Freud and Adler, between psychoanalysts and exponents of individual psychology. As for the philosophical discussions of the "Vienna Circle," they were limited to a few initiates. My own interests were primarily directed toward literature, psychology, and pedagogy. In addition, I felt very happy in my vocation as a teacher (an "assistant master") in a secondary school. The wave of Austrian school reform bore me on; I was successful as a teacher; the pedagogical eros allowed no other ambition to arise in me. I did not think of a scholarly career.

This all changed as a consequence of two world-historical events. The first was the *Anschluss* of Austria in 1938, which led me to emigrate, with my family, to Flanders, the Dutch-speaking part of Belgium. Then, in 1940,

Belgium itself was occupied by the Nazi despots. The director of the Husserl-Archives in Leuven, H. L. van Breda, found himself deprived of his two most important collaborators when Ludwig Landgrebe and Eugen Fink were called into military service. So Van Breda entrusted to me, in 1942, the task of working on the transcription of the posthumous papers of Edmund Husserl, many thousands of pages in *Gabelsberger* shorthand. While living as illegals in Belgium, my wife, her mother, and I steeped ourselves in Husserl's handwritten notes and learned to decipher them. But I also realized I had to familiarize myself with Husserl's philosophy itself, since without an understanding of his thinking a correct transcription was impossible.

In the 25 months that I lived this clandestine existence, I was converted to philosophy. I grasped that philosophy is a decisive matter, because it is the matter of the great decisions. And it was Husserl's genius, and also his unconditional truth-ethos, that fascinated me. The many repetitions to be found in his unpublished papers, the academic style, the long-winded sentences – these did not impair my admiration. For me, aesthetic-literary norms were now laid aside.

It is then understandable that, after the liberation of Belgium, I worked at first as a Husserl scholar and interpreter. Since Husserl's papers were systematically made available in the course of the following decades, and since every volume of the *Husserliana* opened new perspectives, a wide field of activity was offered to me. As a scholarly collaborator of the Husserl-Archives in Leuven, I edited in 1950 the first volume of *Husserliana*, the *Cartesianischen Meditationen und Pariser Vorträge*. My publications on the problem of apodicticity (1946), on the question of God (1954), on social philosophy (1975), and on Husserl's monadology (1978) were contributions to the understanding of this great thinker.

A further event that determined my life in a decisive way was my call to be a professor at the University of Nijmegen in The Netherlands. In 1947 I was appointed professor of philosophical psychology and philosophical anthropology, and in 1949 that was made to include the philosophical foundation of education and the history of education. Thereby areas of instructional, organizational, and administrative activity opened themselves to me. Together with my closest colleague, Joseph Gieler, I managed to call into life, starting from almost nothing, a full pedagogical sub-department. But the situation at Nijmegen also gave rise to new goals for me as a philosopher.

My philosophical interest expanded first of all inasmuch as it was directed to the phenomenological movement as a whole. The interpretation and critique of the great phenomenological thinkers followed as a matter of course and occurred in the form of conferences, lectures, and articles. *Wesen und Grenzen des Schöpferischen* (1952) contained a confrontation with Sartre, *The Concept of Dread* (1957) with Heidegger, and *Betrachtungen über eine phänomenologisch fundierte Sozialphilosophie* (1967) with Merleau-Ponty. General problems of phenomenology, such as that of intentionality (1965), of "Being-for-me" (1968), and of the concept of the world (1977) likewise gave rise to publications.

My professorship included the instruction of a large number of students of psychology. The section "philosophical anthropology," which I directed, was housed in a psychological laboratory of 400 rooms. This resulted in many contacts with psychologists, but also with sociologists and cultural anthropologists. The exchange of ideas with these representatives of the human sciences stimulated me and led me to confront concrete problems of modern methodology and theory of science. I dealt with this problematic in two works. In *Seele and Beseeltes* I posed the question of whether the "soul," still spoken of at that time – 1950 – by introspectively oriented psychologists, was identical with the "psyche" that played a specific role in Aristotelian-Thomistic philosophy. My answer to this question was: No. Already at that time I pleaded for a clear distinction between philosophical and empirical research, a distinction which, to be sure, must not amount to a divorce.

In the meantime, psychology had developed into an experimental science. American behaviorism, with its characteristic methodology, exercised a powerful influence which made itself felt in the other human sciences. In my work *Phänomenologie und Erfahrungswissenschaft vom Menschen* (1964), I opposed this one-sided orientation. The view I developed at that time can best be called "hermeneutical." What I understand by that will be clear at the end of my comments.

A third area which had to be worked on scientifically was the phenomenology of the life of feeling. Scheler, Plessner, Buytendijk, and Sartre had offered promising beginnings, which nevertheless remained only beginnings. The behaviorist methodology, as empirical research, blocked the way to further development. I endeavored in my work *Das Gemüt* (1956) to bring about a synthesis of what had already been achieved and to lead the way to a new philosophical-anthropological reflection on human emotionality. In articles such as *Erziehung der Gefühle* (1950), *Contribution à la phénoménologie du bonheur humain* (1953), and *Feeling as a Basis of Knowing the Other as an Ego* (1979), I treated parts of this very broad problematic.

I had been occupied for years with the philosophical foundations of education and instruction. In the time after the Second World War the task was to blaze new paths without consigning the entire pedagogical tradition to the dust bin. Thus I had to fight various battles on two fronts. While I, even from my Vienna period, sympathized with the modern efforts at reform, I rejected certain doctrines that placed education as such into question. On the one hand, I energetically promoted factual pedagogical research, but on the other hand I opposed the positivistic misunderstanding of the practical character of pedagogy. I thereby had to cross swords often with other theorists. My article *Pädagogische Gesamttheorie als praktische Wissenschaft* (1972) testifies to that.

Above all I was occupied with the question of why the numerous individual discoveries in the domain of education and instruction were not integrated into an educational wisdom. In my book *Erziehungswissenschaft-Erziehungsweisheit* (1965) I tried to answer that question. Another problematic was closely joined to it: the question of the scientific character of pedagogy. I discussed this latter problematic in lecture courses, conferences, and articles. My comments on the *Wesen und Grenzen der erfahrungswissenchaftlichen Forschung auf dem Gebiet der Pädagogik* (1968) is an example.

I took very seriously my professorship in the history of education and instruction. In my own instruction I often employed the method of monographic presentation; I held year-long courses on Rousseau, Pestalozzi, Fröbel, Herbart, and the pedagogical reform movement. Because of my manifold administrative responsibilities, only one of these lecture courses took the form of a book. Together with my assistant Anton Monshouwer, I published a monograph on Herbart (1967).

I already mentioned my interest in the phenomenological movement. The historian of contemporary philosophy knows very well that this movement lacked new impulses in the 1960s. In addition, Heidegger abandoned the method of Dasein analysis, Sartre converted to a dialectics of Marxist coloring, Merleau-Ponty died early, and Paul Ricoeur moved toward analytic philosophy. The effect of all this was confusion. Then in France a new thinker emerged, one who expressly declared his allegiance to phenomenology: Emmanuel Lévinas. With him came forth the great adversary of Heidegger, one that many people had long been waiting for. His program, which can perhaps be characterized by the slogan "Ethics as first philosophy," fascinated me from the start. Yet the understanding of Lévinas' philosophy is very difficult. His conceptual system is novel, his terminology idiosyncratic, his language baroque. To make the philosophical thought of Lévinas available to the German-speaking public seemed to me a difficult though worthwhile task. In my work, *Jenseits vom* Sein und Zeit (1978), I offered a first introduction, simultaneously a commentary and critique. I use the word "first" advisedly; *Jenseits vom* Sein und Zeit is only the beginning of German-language research into Lévinas, not its end. On the contrary, I am convinced that much is still to be accomplished in this field.

The abstract heights of Lévinas' metaphysical thinking do not, however, present the spiritual climate in which I can work autonomously. More concrete philosophical tasks were apparent to me. On the basis of my forty-years' work with students, young men and women, I knew the problems confronting them. The young intellectuals expected from their teachers an "engaged philosophy" and, above all, a philosophical thread of Ariadne which could allow them to find their way through the labyrinth of modern doctrines and ideologies. In my *Betrachtungen über Intimität* (1978), I treated one set of problems facing young people. If I am granted more years of health, of outer and inner freedom, I hope to be able to contribute more to the spiritual struggles of our time.

So much for my self-presentation. I come now to the second part of my assignment, the critical. What have I to say first of all in critique of myself? Anyone who has followed my remarks will already suspect the reproach I must make to myself: I have ventured into too many heterogeneous domains of philosophy. It is clear that I should have remained a Husserl or Lévinas scholar, that modern scientific theory requires a deepening that only specialists can provide, that the spiritual history of pedagogic thinking offers tasks for which one human life is too short. I can, to be sure, refer to my unique professorship as a mitigating circumstance. Such arguments on the part of the accused might lead to a reduction of sentence, but they cannot obviate the judge's verdict of "Guilty!" As I look back on my scholarly career, I must recognize that I have always resisted becoming a one-sided specialist in a restricted domain. If I had accepted such a restriction, I would have made things much easier for myself and my collaborators. But perhaps, through my unconscious resistance to specialization, I remained more faithful to the original sense of philosophy.

As to my critique of other philosophical views, I might formulate two groups of theses: one scientific-theoretical and one with regard to the philosophy of education.

With regards to epistemology, my position is that every science is constructed on a prescientific understanding of its own domain. Negatively expressed, this means that no science is presuppositionless in an absolute sense. Specifically, the claim of the natural sciences to offer the only objective view of reality cannot rightfully be maintained. But even the mathematical-logical disciplines presuppose many things as self-evident, such as the understanding of what constitutes unity and multiplicity, identity and difference, contradiction and non-contradiction. The task of philosophy is precisely to investigate these things taken for granted.

The human sciences add the circumstance that they imply a certain relatedness to value. In other words, the phenomenon of the human is characterized by its accent on values. Prior to all scientific reflection, indeed prior to any explicit axiological judgment, what is human appears as related to values. In this sense, Max Weber, for example, speaks of the "cultural sciences" and calls for a *verstehend* method. The structures and the changes in structures of human groups require this same sort of understanding.

Furthermore, if there are human sciences which are practically oriented, then it must be taken into account that certain very general value-judgments – at least implicit and tacit ones – are there and assumed as valid. Thus it is clear to every doctor that health has a priority over sickness; to the jurist that justice counts as a value while injustice as a non-value; every psychiatrist endeavors to promote normal modes of behavior and to obviate abnormal modes; every pedagogue strives, directly or indirectly, to promote intelligent, appropriate, social behavior and to ameliorate unintelligent, inappropriate, unsocial behavior. The opinions of what in any determinate situation is right, healthy, appropriate, etc., can indeed be very different; yet all such discussions presuppose the validity of those most general value-judgments. For these reasons, I maintain that the human sciences are essentially distinct from the logical-mathematical and the natural sciences, while I take the practically oriented human sciences again for a special group of disciplines.

With the thesis just formulated I already enter the domain of a philosophy of pedagogy. I know that it is now more fashionable to speak of the "educational sciences" instead of "pedagogy." I have no objection to that, as long as one remains conscious of the fact that here the word "science" has a different meaning than it does in a mathematical or physical context. More precisely, two misunderstandings must be obviated. On the one hand, it is senseless to try to deduce concrete educational goals exclusively from moral philosophy or moral theology, as happened often enough in the past. Descriptive pedagogy, i.e., a most exact description of reality under the viewpoint of possible educational activity is indispensable. It is clear that for this description psychology and sociology render service as assisting sciences. But it must not be forgotten that the educator wants to lead the child to a place where the child is not yet – or else he is no educator. His setting of goals is not to occur without the information provided by factual research, but it does not proceed exclusively from that. What is also required is a value-consciousness, a norm-consciousness, which characterizes the practical sciences as such. Therefore the "science of education" is still something other than applied psychology, applied sociology, applied theory of behavior, etc.

The slogan "anti-authoritarian education" must also be used very circumspectly. If this technical term is supposed to mean that no norm ought to be imposed on the child or on the young person which he himself is not in a position to grasp and acknowledge, then most modern pedagogists will concur. But if someone denies that "fate," "nature," or "society" entrusts the small child in the first place to the power of his elders and his educator, such that those elders and educators first make value judgments vicariously for him, then one is denying the very phenomenon of education and arrives at absurd consequences. In this domain it is necessary to steer between Scylla and Charybdis, between a rigid pedagogy, which knows nothing of the facts of the concrete situation, and a slack permissiveness which completely abandons all thoughts of a norm. To produce equilibrium between these two tendencies would be a matter of educational wisdom. Education science might help but can never substitute for it.

From the idea just sketched, there might, I believe, be derived far-reaching consequences touching the domain of practical philosophy in our time. Anyone who believes he can solve modern social problems while ignoring the insights of science, technology, and economics, is indulging in utopian daydreams – utopian in the pejorative sense. Anyone who thinks that the ultimate decisions can be derived exclusively from the sciences, from technology, from economics, or that they can be expected from a computer, can be called a technocrat – in the pejorative sense. Wisdom is at work only when the scientific-technological-economic cognitions are placed in service to a normative setting of goals. Thereby a preeminent role is accorded to the lover of wisdom, the philosopher.

Duquesne University

NOTES

[1] Stephan Strasser, letter to Richard Rojcewicz, 5 April 1984 (emphasis original).

[2] Ernst Wolfgang Ort and Thomas M. Seebohm, "Germany," In Lester Embree et al. (eds.), *Encyclopedia of Phenomenology*, (Dordrecht: Kluwer Academic, 1997). pp. 273–274.

[3] Bas Levering and Max van Manen, "Phenomenological Anthropology in The Netherlands and Flanders," in Anna-Teresa Tymieniecka (ed.), *Phenomenology World-Wide: Foundations – Expanding Dynamics – Life-Engagements: A Guide for Research and Study* (Dordrecht: Kluwer Academic Publishers, 2002), p. 277.

[4] David L. Smith, C.S.Sp., *Fearfully and Wonderfully Made: The History of Duquesne University's Graduate Psychology Programs (1959–1999)* (Pittsburgh: The Simon Silverman Phenomenology Center, 2002), p. 11.

[5] *Ibid.*

[6] Henry J. Koren, C.S.Sp., letter to André Schuwer, O.F.M., 22 April 1976.

[7] Henryk Misiak and Virginia Staudt Sexton, *Phenomenological, Existential, and Humanistic Psychologies: A Historical Survey* (New York: Grune & Stratton, 1973), p. 62.

[8] Amedeo Giorgi, telephone conversation with the author, 19 May 2005.

[9] Smith, pp. 183–184.

[10] Strasser, letter to Richard Rojcewicz, 29 October 1984.

[11] *Ibid.*, 12 May 1984.

[12] Liesbeth Strasser, letter to Richard Rojcewicz, 6 February 1995.

[13] Antonino Mazzù, "The Development of Phenomenology in Belgium and The Netherlands," in Anna-Teresa Tymieniecka (ed.), *Phenomenology World-Wide: Foundations – Expanding Dynamics – Life-Engagements: A Guide for Research and Study* (Dordrecht: Kluwer Academic Publisher, 2002), p. 269.

[14] Lotte Kohler and Hans Saner, (eds.), *Hannah Arendt Karl Jaspers Correspondence: 1926–1969* (New York: Harcourt, Brace, Jovanovich, 1992), p. vii.

[15] Strasser, "History, Teleology, and God in the Philosophy of Husserl," in Anna-Teresa Tymieniecka (ed.), *Analecta Husserliana*, Volume IX: *The Teleologies in Husserlian Phenomenology: The Irreducible Element in Man*, (Dordrecht: D. Reidel, 1979), pp. 317–334.

[16] Angela Ales Bello, *Husserl Sul Problema di Dio* (Rome: Edizioni Studium, 1985).

[17] Ales Bello, letter to Stephan Strasser, 19 September 1985.

[18] Strasser, *Welt im Widerspruch: Gedanken zu einer Phänomenologie als ethischer Fundamentalphilosophie*, Phenomenologica, Vol. 120 (Dordrecht: Kluwer Academic Publishers, 1991).

[19] Strasser, letter to Richard Rojcewicz, 13 December 1986.

[20] *Ibid.*, 26 January 1987.

[21] Thomas Aquinas, *Summa contra Gentiles*, Book II, Chapter 55 (Notre Dame, IN: University of Notre Dame Press, 1975), pp. 162–163.

[22] Strasser, "Kritische Selbsdarstellung," in André Mercier and Maja Silvar (eds.), *Philosophes Critiques d'Eux-mêmes/Philosophers on Their Own Work/Philosohische Selbstbetrachtungen*, Vol. 7 (Berne: Peter Lang, 1981), pp. 183–207. [Translated here by Richard Rojcewicz under the aegis of The Simon Silverman Phenomenology Center with permission to publish the English granted by Peter Lang.]

SECTION I

ANIMALITY AND CONSCIOUSNESS

DANIELA VERDUCCI

NIETZSCHE'S BESTIARY. ANIMAL, MAN, SUPERMAN

1. INTRODUCTION

Surprisingly, most of Nietzsche's references to animals are contained in *Thus Spake Zarathustra*,[1] the work in which Nietzsche announces the coming of the superman. Such an abundant animal presence certainly contrasts with some of the very first interpretations that were given of the superman figure, for instance those of Oskar Ewald in 1903 or of Georg Simmel in 1907, or another, about ten years later, of Julius Reiner.[2] These especially underline the superman's nature as a symbol, as a category of thought or moral will; they bring to light the superman's absolute non-belonging to concrete reality and its wholly psychological function of driving towards a future, meant as a surpassing and a non-continuance of the present, within a non-linear and non-finalistic vision, that is totally opposed to classical evolutionism.[3]

However, in the changed interpretative climate of the 1930s, even Nietzsche's superman doctrine was traced back to an evolutive matrix and, in 1932, Ludwig Haas proposed to "restore a scientific-natural conception of the superman," connected with the cross-breeding (*Züchtung*) idea, by which the superman would belong to a superior biological species, where the development of the spirit would move in parallel with the superior development of physical conformation.[4]

Accepting this hermeneutical context, obviously purged of every concession, however minimal, to deplorable theories of despotic supremacy on the basis of race, the constant references to animals in *Zarathustra* appear not only more plausible, but even inevitable. Indeed, the superiority that the superman expresses can thus manifest a level of life which is both whole and supreme, and hence can always be increased. Here there is also an opening for that biological development which, if it has already started in the plant kingdom, only in the animal dimension does it acquire the impulse and propulsion that drive it as far as man, from whence it can reach beyond man.

Although the thought of the superman must not in any way "necessarily depend on and refer to the evolution theories" as Elisabeth Förster Nietzsche

A-T. Tymieniecka (ed.), Analecta Husserliana XCIII, 3–15.

believed,[5] it is however true that Nietzsche certainly goes along with the Darwinist suggestions of which precisely the bestiary of *Zarathustra* works both as an original transposition and a radical step beyond. The decade between 1876–1886 saw Nietzsche starting his contact with Herbert Spencer's evolutionism and John Stuart Mill's utilitarianism, precisely attested by the presence in his library of densely annotated works by the two thinkers.[6] From this comparison, that extended also to the spencerism of A. Fouillée, E. Littré, A. Espinas and J.-M. Guyau, Nietzsche drew the "substantial cognition" according to the moral genealogy, by which our morality derives from the "herd instinct," from the gregarious instinct, that is, what the species has considered the most suitable to attain its ends of conservation and what it has selected and fixed, together with all the other "unforgotten and unforgettable" experiences, into which flowed the vital sense of what is useful and functional to the species.[7] From here emerged the Nietzschean idea of the rootedness in the depths of the physiological level of the utilitarian-evolutionist morality of modern Europeans, for whom nothing "is older, stronger, more inexorable than the instinct" to "do something that benefits the conservation of the human species."[8] From here also comes out Nietzsche's need to find, in the human physiology, the energy and drive to establish within the current morality of altruism and of the sacrifice of the individual for the benefit of the species, an illuministic counter-movement. This generates "dominating natures,"[9] that, unlike the last men, where the dependency on the gregarious instinct is total, constitute supermen,[10] capable of sweeping away instinctive atavism and of facing existence with untried tools. This is why Nietzsche in speaking about himself in *Ecce Homo*, attributes crucial importance to the moment in which the discovery of "physiology" allowed him finally to rid himself of the "cursed idealism" or "ignorance *in physiologicis*," that had until then, as a good German, characterised his culture.[11]

According to Nietzsche, to make the new emerge in the consolidated and tyrannical physiological conformation of the contemporary European, it was now time, after centuries of indirect and inefficient ethical education "to work, directly influencing the organism," so that it forms "another corporality which will from itself create another soul and other customs."[12] Also Francis Galton,[13] the founder of modern eugenics, hypothesised that if the intention was to promote progress and reinforce the human species, to the processes of natural selection it should be given orientation. As nature tends to eliminate exceptions and favour the stabilisation of "average" types,[14] in order to improve the human, Galton proposed to support nature with the technical intervention of the science of heredity, observing the way in which factors of cultural superiority develop and reproducing them artificially.

In a completely original way, Nietzsche believed, instead, that in order to assist the strengthening of the human, an endogenous method should be used, represented by the socio-cultural contribution of "degenerating natures," which are the only ones capable, in his view, of helping to progress the bio-cultural human type, that strong natures have fixed. Degenerating natures, when they manage not to deteriorate due to their weakness, "soften the stable element of a community and sometimes cause a wound," through which "something new is, so to speak, inoculated into the whole community."[15] Thus within the man dominated by a gregarious instinct of conservation, precisely to the extent to which he is ill, the intimate struggle between instincts breaks out, destroys the old idiosyncracies and promotes the re-orientation and restructuring of capacities, enabling totally new possibilities of life to emerge and the superman to affirm himself. To document this, Nietzsche cites his own experience of life. It was in 1879 when, in his thirty-sixth year, his vitality had touched its lowest point: having left the professorship in Basel, he stayed in St. Moritz and in Naumburg. He wrote *The Wanderer and his Shadow* and then, at his lowest physical ebb, *Daybreak*,[16] where with the dialectic clarity and sangfroid that only illness and physical pain can produce, he set about "sillabizing *la décadence*." Thus he learnt to overturn viewpoints and he realised that, whilst being *décadent* in the detail, he was healthy *in summa*. The sickness was an "energetic stimulus to live and to live-more," to create his philosophy from his will-for-health and for life.[17] He had at last started to pose himself authentic problems, asking himself, for example: "You, how exactly should you feed yourself, to obtain your maximum strength and virtue in the Rennaissance sense, virtue that is free of hypocritical morality?" Furthermore, he concentrated on choosing the most suitable place and climate for his life. He who does not dwell where the air is dry and the sky pure, loses the chance of enjoying a "rapid metabolism," "of continuously attracting to himself big, almost immense, quantities of strength." In him "animal *vigor* will never become so big as to reach that freedom which overflows to the summit of spirituality, where a man can recognise that: "only I can do this."[18] It was sickness that allowed Nietzsche slowly to liberate himself from everything: it gave him the right to completely change habits, it ordered him to forget, it gave him the need to rest, to be idle, to wait, to be patient, in a word, to think. It freed him from the "book." In the end, his deeper self, the animal/vital self, buried beneath the mass of culture, of social and moral habits, awoke and spoke again,[19] and asked what is great in man: *amor fati*.[20]

Certainly, Nietzsche points out, only an individual or a people that altogether is still strong and healthy can bring into its midst the infection of the new and incorporate it to its advantage, giving rise to untried "mutations" that strengthen the human. This is why the "conditions of birth and conservation" of the superman type, albeit different from those of the average man,[21] presuppose the existence of such a man. Indeed it is precisely the species' adaptive movement that, merging all the energies into a *continuum* whose function is directed at the fixing of a type, guarantees *minimum* expenditure and *maximum* capitalisation of energies: the superman draws advantage from such an original accumulation of strength and he uses it to break the evolutive continuity and "to invent his superior form of being."[22]

In this respect Nietzsche takes up Henri Joly's conviction, according to which: "Genius is nothing but long patience,"[23] and he reiterates that "one needs to study how a great man is effectively created" in order to reach something with humanity.[24] And yet, what Nietzsche's superman possesses above all is that "spontaneity of conception" that William James attributes to man and that, finding nothing that corresponds with it in the phenomena of the external world, he traces it back to a cause within the organism. Thus the living man can organise his experience in an unusual and peculiar way.[25] Indeed, Nietzsche's "higher type" is defined with increasing clarity as originating not from simple genetics but from "cultural genetics," that removes it from the casual dynamics of evolution and passes it to the exercise of will.[26] "My problem – says Nietzsche – is not to establish what can take the place of man, but what kind of man should be chosen, wanted, bred as a species of superior value."[27]

A single longing (ZA, *Of Old and New Tablets*, 2) not to live, but to augment life, a single will to power (ZA, *Of Self-Surpassing*) thus seems, according to Nietzsche, to run through the whole living cosmos, that without a solution of continuity proceeds, from the animal level to the threshold of the dimension of the superman, through man, who is a "suspended cord between animal and superman" (ZA, *Preface*, 4) and thus is keeping in a constant, and, therefore, permanently incomplete, creative transevaluation in the newness of the identical that returns. This is the task that Zarathustra must want to undertake, to begin his decline and the coming of the superman that he announces. This is the content of the "chant" that the eagle and the serpent repeat to him. They also exhort him to sing with them, because Zarathustra has "become ill by his own redemption" and is still recovering from the trauma caused from having bitten and spat out the head of the ugly beast that slid into his throat to choke him. He cannot sustain the very weight of the words, that the healthy can utter. Now Zarathustra is only able to sing and to use words cloaked with "all the lying of sounds," that give names to things (ZA, *Of the Convalescent*, 2).

The bestiary that accompanies Zarathustra, the announcer to men of the superman, is thus not to be considered only as an expedient of poetic decoration. Through it, rather, life speaks to us, and opens us to the infinite possibilities of realising the will to power that animates it.

2. THE LIFE THAT THE ANIMALS TRANSMIT TO ZARATHUSTRA

The voice of a bird accompanies the beginning of Zarathustra's decline, the prelude to the coming of the superman. In the last paragraph of the Preface to Nietzsche's *Zarathustra*, the protagonist's attention is attracted by the "shrill cry of a bird": Zarathustra raises his eyes and delights in seeing his animals gliding through the air in wide circles: the eagle, "the proudest of the animals under the sun" and "coiled round his neck" as a friend and not as prey, the serpent, "the wisest animal under the sun" (ZA, *Preface*, 10). They arose in flight "to explore if Zarathustra was still alive." And in reality Zarathustra, who "solitary" came down from the mountain and ventured into the town, in spite of advice to the contrary given by the old wise man he met (ZA, *Preface*, 2), believes to have encountered mortal dangers among men, such that he still wonders: "And in truth, am I still alive?" (ZA, *Preface*, 10).

A quick perusal of the facts that occured seems, however, to show that Zarathustra's life never encountered real danger: in the market square where he had manifested his intention of teaching about the superman, he was confronted by the misunderstanding and derision of the people (ZA, *Preface*, 3); after his first speech the townsfolk had refused to be taught about the superman and they asked him, exultant, to give them, rather, the last man, who invented happiness (ZA, *Preface*, 5). It was the tightrope-walker who in fact lost his life when he fell from his rope stretched between the two towers of the market square, when the clown jumped onto the rope and over his head (ZA, *Preface*, 6). But at dusk Zarathustra was also threatened with death: if he had not left the town, the next day the clown would have jumped over him, passing over him, like "a living man over a corpse" (ZA, *Preface*, 8). So Zarathustra walked all night: only at dawn did he fall asleep when he reached a thick wood, by then far from the town. And when he woke up, in the late morning, he rejoiced, finally catching a glimpse of a new truth: "Zarathustra must not speak to the people, but to companions"; nor should he "become the herd's herdsman and hound" nor a grave-digger, who looks after the dead and can take them with him anywhere he wishes. "I need live companions – says Zarathustra now – that follow me because they want to follow themselves – and precisely to the place I want to go." In his heart he now knows he has to be accompanied to creators, to higher men like him (ZA, *Of Higher Men*, 11): he will show the rainbow to them and "all the steps towards the superman"; he will go with them along his road, leaping "over the loitering and the tardy" (ZA, *Preface*, 9), following his new guides, his animals (ZA, *Preface*, 10).

The animals that live with Zarathustra are wild, and for Nietzsche this means that they are carriers of the biological vitality in the fullness of his autopoietic dynamism, to use the expression coined in the 1970s by the Chilean neurophysiologists H. Maturana and F. Varela.[28] And they can safeguard this autopoietic fullness, as they are devoid of culture (ZA, *Preface*, 5). What is more, what was the force that moved Zarathustra to wish for his decline and to go among men? It was the drive of his unsatisfied vitality that mixed gall into his industriousness and into his best honey (ZA, *The Grave-Song*). After ten years on the mountain enjoying "his own spirit and solitude," he felt saturated with wisdom and, like the "bee who has collected too much honey" (ZA, *Preface*, 1), he had felt the urgent need to "overflow" for "hands that are outstretched."

Even the sun, which for ten years climbed up to his cave, would have wearied of its journey if Zarathustra had not been waiting for it every morning with his eagle and his serpent to remove "the superfluous" and bless it, liberating it of the repetitiveness of its movement (ZA, *Preface*, 1).

Wild animals, especially preying animals like the lion, whilst incapable of creating new values, have the strength to "create freedom for themselves for new creating" (ZA, *Of the Three Metamorphoses*) and thus they represent the ideal company for those, like Zarathustra, who seek to decline in order for the superman to rise. Pet animals are different: it is time to take our distance from them, because the "little men" who made a dog from the wolf, exercising the bedwarfing virtue, have now made "man himself man's best pet" (ZA, *Of the Bedwarfing Virtue*). Peter Sloterdijk agrees with this view having recently instituted the contrast between traditional humanistic anthropotechnique and the incipient genetic anthropotechnique. He identifies as the main crisis factor of the former, precisely the biopolitical complex of home, man and pet, that qualify it.[29] Zarathustra prefers the smell of the eagle and the serpent, mixed with good air outside his cavern, to that of higher men, who, gathered in the cavern (ZA, *The Song of Melancholy*, 1), have not yet learnt to dance and laugh beyond themselves (ZA, *Of The Higher Man*, 20).

When Zarathustra, after having evoked the "abysmal thought" of the eternal return (ZA, *The Convalescent*, 1) collapses onto the ground and lies at length as if dead, it is the animals near his cavern to run to his aid. His animals then constantly watch over him and the eagle, from time to time, flies out to fetch him some food: yellow and red berries, grapes, rosy apples, sweet-smelling herbage, and pine-cones. The eagle even forcibly takes two lambs from their shepherd (ZA, *The Convalescent*, 2): flavoured with spices and sage (ZA, *The Supper*) they are the favourite meal of Zarathustra who loves nourishing himself with innocent things (ZA, *Of The Spirit of Gravity*, 1). Zarathustra cloaks himself with the lamb's fleece, because he is and wants to be a firstling to be sacrificed (ZA, *Of Old and New Tables*, 6), but he is "against all lamb-spirits" and against their sheeplike kindliness (ZA, *The Song of Melancholy*, 3). When, after seven days of illness, Zarathustra recovers, it is again his animals who push him to go out of the cavern: the renewed contact with the world will be therapeutic for his soul, which, struck by "new knowledge, bitter and grievous," swells up beyond its bounds, "like leavened dough." But Zarathustra resists the invitation: he wants to be content with the restoring garden that the talk of his animals open before him. The eagle and the serpent remind him however that he has now matured that relationship with life of unconditional affirmation, that is similar to their unfailing instinctive reaction to life and thus he can do without the filter of beautiful words. The animals say:

Oh Zarathustra, to those who think like us, things all dance themselves: they come and hold out the hand and laugh and flee – and return (ZA, *The Convalescent*, 2).

But the naturally determined flowing of animal life, its *autopoiesis* devoid of the resistance of knowledge and morality, does not represent Zarathustra's condition. Indeed, to the extent that he knows that where there is a living being, there is will to power (ZA, *Of Self-Surpassing*), he is also aware of the fact that in order to live he must want what his human living involves or that he must want an increase of life through the exercise of will for the surpassing of everything that has been completed in time and consolidated in being. In order to fulfil this destiny, however, it is not enough to overturn all values,[30] moreover one must want positively to transvalue them,[31] creating new values, promoters of the *ontopoiesis* of life[32] and once again liberating that enormous intimate excitation, that took hold of primitive men when they discovered measures and measuring, scales and weighing, so much that they wanted to adorn the name of their greatest discovery and so they named themselves "man" that means "he who measures."[33]

3. THE LIFE THAT LEADS ZARATHUSTRA TO THE SUPERMAN

Thus, in order for Zarathustra to carry out his task it is not enough to be in contact with the spontaneous and full vitality that the animals have kept, unlike men. Even the announcer of the superman must await the hour of his decline. So, between "old broken tables of the law" and "new half-written tables," Zarathustra waits for the appearance of "the laughing lion with the flock of doves," the sign that shows the maturity of his will to decline (ZA, *Of Old and New Tables*, 1). It is true that man has become "the best animal of prey," the wisest, the most intelligent, as "he has preyed upon their [animal of prey's] virtues," having had a more difficult existence (ZA, *Of Old and New Tables*, 22), but precisely because of this incessant defending against life, for fear of it (ZA, *Of Science*), he has lost his real vitality, distorting and deviating towards inadequate ends, the will to power that also animates him.

Zarathustra thus has a dual crucial task: on the one hand he urgently needs to make the will to power of life flow again through his veins in its biological-animal dimension, but on the other hand it is essential that he works on re-establishing in himself a level of wanting that is adequate for the strengthening of natural life, required of him by the very announcement of the superman of which he is the carrier.

Indeed, from the very first flight from the market square of men, amongst whom he also wishes to descend, Zarathustra, still uncertain of his vital status, questions himself on his cleverness: he wishes it were like that of his serpent, clever to the core, but he is aware that his cleverness "loves flying away" and that he is thus still not capable of fully pursuing his destiny (ZA, *Preface*, 10). If Zarathustra had already possessed a clever vitality, for instance, he would have been able to take the advice of the old saint he met just after coming down from the mountain to make himself into an animal with the animals. He would have interpreted this in the positive sense of bringing his animals with him on his journey among men, or in the sense of solidly rooting in the secure ground of the autopoietic spontaneity of the life of animals, his wish to surpass, a weak prelude to the will to power of the superman, as well as being erratic, although he is a higher man. He was thus able to avoid the rashness of the solitary man, who "too easily holds out his hand to those he meets" and to offer not his hand but his paw armed with claws (ZA, *The Way of The Creator*). He could have gained strength from the contempt of men, like an animal who even after lowering his head to be sacrificed, repeats with his eye: "My ego is something that must be surpassed, my ego is for me the great contempt of man" (ZA, *The Pale Criminal*).

But Zarathustra did not bring his animals with him. He descended among men alone and, what is more, while he disseminated the message of the superman, he dragged with him a man's corpse (ZA, *Of the Higher Man*, 1), a "dead dog," as the clown called it (ZA, *Preface*, 8), evoking all the misery of the race of dogs, who "immediately lie on their backs" submissive and obsequious (ZA, *Of the Three Evils*, 2) and lick the sweat of man who, worn out, languishes "at a step from the goal" (ZA, *Of the Three Evils*, 18).

This is thus a clarification on what led Zarathustra to doubt he was still alive, after passing through the town of men: he had really run a mortal risk, because the threat that the clown had brought him at dusk was not only directed against his biological life, like for the tightrope-walker who fell from the stretched rope, but it was intended especially to deprive him of the "secret" of life, of what that very life had revealed to Zarathustra, when life told him: "Look, I am what must always surpass itself" (ZA, *Of Self-Surpassing*). The clown would have killed him by jumping beyond him (ZA *Preface*, 8; also *Of New and Old Tables*, 4, but referring to an acrobat) or by taking away the necessary space for achieving his own vital self-surpassing, given that man consists of a "crossing" (ZA, *Preface*, 4). By placing himself with a jump beyond Zarathustra, the clown would have taken Zarathustra's *humanly* vital space, he would have eliminated the condition for all possible self-surpassing and so of his own specifically human life. Zarathustra's death would not have been violent, but its effects would have been lethal for the whole human population, to whom no one would have been able to teach of the new vital space that could be conquered, the new will to self-surpass that could be inspired. No one would have continued to repeat to the "higher men," who do not want to put themselves on the market and hold contempt and feel desperation, but they see themselves as badly made: "Oh, how many things are still possible!" (ZA, *Of the Higher Men*, 15). If Zarathustra had been surpassed by the clown, the destiny of all men along with him would have been that of the *Indios*, described by the Brazilian anthropologist Darcy Ribeiro: wrenched from their habitat in the Amazonian forest, they died not for biological reasons, but because, believing that the conditions for exercising the will to live and for developing their being no longer existed, they ceased to want to live to the point that they abandoned life.[34]

This is why, thinking over what happened, Zarathustra reproaches himself for not having been clever enough: he had committed the foolishness of the hermit, who puts himself onto the market (ZA, *Of The Higher Man*, 1), disregarding the advice of the old wise man (ZA, *Preface*, 10) he met at the foot of the mountain, not to go among men but, rather, to make himself "a bear among bears and a bird among birds." Zarathustra had automatically ascribed the old wise man's words to that perverse mentality that uses identification with the animals to betray the nature of life. The nature of life is self-surpassing becoming, whilst the old saint still believed he could preserve himself, by practicing ascesis, which reduces life[35] rather than "gymnastics of will," that strengthen it.[36] As if his ear had not heard the cry of the madman who at the market had announced: God is dead!,[37] the old saint had not assumed the stature of the higher man, oriented towards the great noonday, that opens to the coming of the superman (ZA, *Of the Higher Man*, 2). He appreciated the bear's a-sociality, without considering at all that sometimes "he unjoyously returns to his cavern" (ZA, *The Stillest Hour*). He did not learn the enmity of the birds for the spirit of gravity and their healthy and wholesome love for themselves (ZA, *Of The Spirit of Gravity*, 1, 2): the old saint loved the god that was closest to him, "his" god by whom his love was totally gratified and thus was also totally exorcised of creative capacity. Even his choice of being with the animals was leading him to the total loss of the vitality of his life, which soon would become manifest nihilism, as the old saint, like every priest at the service of the god-spider,[38] was behaving like a "venemous spider of life."[39]

But Zarathustra had run the mortal risk of becoming enmeshed in all these events when, as he was seeing off the old saint who asked him what gift he could bring for him and for men, Zarathustra answered: "What do I have to give you? Just let me leave now, so I take nothing from you!" (ZA, *Preface*, 2). Thus doing, Zarathustra showed himself to be lacking in cleverness: he did not exercise any surpassing of the situation in which he found himself and he preferred to escape from it, thus betraying the very law of the superman that he was going to teach, as well as the great law of life, the will to power that "subjects to itself and assimilates the external more and more,"[40] nor is it ever an adaptation of internal conditions to external conditions. However, he got away with it and it was his pride that made his foolishness fly far from the danger, like the eagle, who in his reconnoitre flight, looking for Zarathustra, carried the serpent around his neck, as it could not rise up in flight (ZA, *Preface*, 10).

CONCLUSION

In the progressive approach to his decline for the superman, therefore, Zarathustra's constant reference to animals seems to have the precise meaning of underlining that among all living things there is an evolving continuity, whose law is called will to power. The way in which the will to power expresses itself, however, varies: in animals it is spontaneous and unfailing, in men it is filtered by morality and culture, so it can assume distorted and deviated forms, as in fact occured in history with the affirmation of the unique axiological-cognitive Judaeo–Platonic–Christian paradigm. The possibility of the coming of the superman is precisely linked to the fact that man voluntarily accepts to assume the will to power, of which the animal condition of life is the carrier. But uttering these words is terrible for Zarathustra, whose single destiny, of announcer of the superman, represents both "his greatest danger and his greatest illness." This means, as his animals repeat to him, that even he were to die, Zarathustra will return "with this sun, with this earth, with this eagle, with this serpent" and above all that "the little man," the man of whom he is tired, will eternally return (ZA, *The Convalescent*, 2). On hearing such words, Zarathustra's soul, which does not have the innocent vitality of the animals, weeps, laments, sobs, accuses and to make it "overflow" in his song, as the animals advise him, Zarathustra must order it to his soul (ZA, *Of the Great Longing*) and make new lyres for his soul. Only in this way can his soul recover and accept, to the point of actively assuming for himself, his "great destiny": to be and to become master of the eternal return (ZA, *The Convalescent*, 2).

University of Macerata
Translated by Paul Bowley

NOTES

[1] F. Nietzsche, *Thus Spake Zarathustra*, trans. W. Kaufmann (New York: Penguin, 1978) (from here onwards: ZA). Original Title: *Also sprach Zarathustra*, in *Nietzsche Werke. Kritische Gesamtausgabe*, G. Colli and M. Montinari (eds.), V_1 (Berlin-New York: De Gruyter 1968). Cfr.: W. Kaufmann, *Nietzsche, Philosopher, Psychologist, Antichrist* (Princeton: Princeton University Press, 1950). For some of Nietzsche's English editions: *Complete Works*, ed. Oscar Levy (New York: Russell & Russell, 1964); *Complete Works of Friedrich Nietzsche*, ed. Bernd Magnus, (Stanford, CA: Stanford University Press, 1995 onwards).

[2] O. Ewald, *Nietzsches Lehre in ihren Grundbegriffen. Die ewige Wiederkunft des Gleichen und der Sinn des Übermenschen. Eine kritische Untersuchung* (Berlin: Hoffmann 1903); G. Simmel, *Schopenhauer und Nietzsche*, (Leipzig: Duncker & Humblot 1907), now in *Gesamtausgabe*, hrsg. v. Otthein Rammstedt, Bd. 10 (Frankfurt am Main: Suhrkamp 1995); J. Reiner, *Friedrich Nietzsche der Immoralist und Antichrist* (Stuttgart: Franck'sche Verlagshandlung 1916).

[3] Cfr.: M. C. Fornari, *Superuomo ed evoluzione*, in F. Totaro (ed.), *Nietzsche e la provocazione del superuomo. Per un'etica della misura* (Rome: Carocci 2004), p. 45.

[4] It is common knowledge that it was A. Baeumler, *Nietzsche, der Philosoph und Politiker* (Leipzig: Reclam 1931), who first presented the interpretation of Nietzsche as a forerunner of National-Socialism and the theories of race, where, however, an "heroic" and "Germanic" reading of the superman prevails. On the contrary, L. Haas, *Der Darwinismus bei Nietzsche. Inaugural Dissertation zur Erlangung der Doktorwürde bei der philosophische Fakultät der Hessischen Ludwigs-Universität zu Giessen* (Giessen: Kindt 1932), maintains the consideration of the superman as the result of an evolutive progression, regulated according to Darwinian precepts, or rather, following the principles of Darwinian selection, consciously directed and optimised, considering that causal changes for Nietszche are not sufficient and also the time for a natural development for the breeding of a new humanity is too slow (Fornari, *Superuomo ed evoluzione*, p. 31). Furthermore, in 1890, Joseph Diner had already observed a point of accord in Nietzsche's philosophy with the most recent scientific discoveries, demonstrating how the teaching of Ch. Darwin and E. Haeckel eventually led to the superman. Cfr.: J. Diner,

"Friedrich Nietzsche: Ein Dichterphilosoph," in *Freie Buhne*, 1 (1890): 371, cit. in: G. Moore, *Nietzsche Biology and Metaphor* (Cambridge: Cambridge University Press 2002), p. 5 (note).

5 Cfr.: letter of 7 February 1883, from Elisabeth Förster Nietzsche to Ida Overbeck, in Goethe–Schiller-Archiv Weimar, 72/818.

6 On this subject: M. C. Fornari, *La morale evolutiva del "gregge". Nietzsche legge Spencer e Mill* (Pisa: ETS 2004).

7 F. Nietzsche, *Zur Genealogie der Moral*, in: *Nietzsche Werke*, VI_2 1968 I, 3. Cfr. also: *On the Genealogy of Morals*, trans. W. Kaufmann and R.J. Hollingdale (New York: Vintage, 1967).

8 F. Nietzsche, *Die fröliche Wissenschaft*, in: *Nietzsche Werke*, V_2, 1973, aph. 1. Cfr. also: *The Gay Science*, trans. W. Kaufmann (New York: Vintage, 1974).

9 F. Nietzsche, *Nachgelassene Fragmente*, 27 [80], Summer–Autumn 1884, in: *Nietzsche Werke*, VII_2, 1974, p. 295.

10 Nietzsche, *Nachgelassene Fragmente*, 7 [21], Summer–Autumn 1883, in: *Nietzsche Werke*, VII_1, 1977, p. 252.

11 F. Nietzsche, *Ecce Homo. Wie man wird, was man ist*, in *Nietzsche Werke*, VI 1969, *Warum ich so klug bin*, 2. Cfr. also: *Ecce Homo, How One Becomes What One Is*, trans. W. Kaufmann and R.J. Hollingdale (New York: Penguin, 1979).

12 Nietzsche, *Nachgelassene Fragmente*, 7 [97], Spring–Summer 1883, in: *Nietzsche Werke*, VII_1, p. 283.

13 It was J. Paneth who gave Nietzsche the work of F. Galton, *Inquiries into Human Faculty and Its Development* (London: Macmillan, 1883). He had other contacts through reading the journal *Das Ausland* XLIII (1870): 928–931, where the review appeared by Oskar Peschel to F. Galton, *Hereditary Genius: An Inquiry into Its Laws and Consequences* (London: Macmillan, 1869) Cfr.: R. F. Krummel, *Joseph Paneth über seine Begegnung mit Nietzsche in der Zarathustra-Zeit*, in *N-S* XVII (1988): 478–495; M.-L. Haase, *Friedrich Nietzsche liest Francis Galton*, in *N-S* XVIII (1989): 633–58. Already C. Richter, *Nietzsche et les théories biologiques contemporaines* (Paris: Mercure de France, 1911), spoke of possible points of contact between Nietzsche and Galton. Cfr.: Fornari, *Superuomo ed evoluzione*, pp. 49–53.

14 Since 1869, in *Hereditary Genius*, Galton pronounced his theory on the stability of types, where he hypothesises a statistical regression in the heredity of exceptional characters, in favour of the typical at every generation. In: Fornari, *Superuomo ed evoluzione*, p. 52.

15 F. Nietzsche, *Menschliches, Allzumenschliches, I*, in *Nietzsche Werke*, IV_2, 1967, §224. Cfr. also: *Human, All Too Human*, trans. R.J. Hollingdale, 2 Vols. in 1 (Cambridge: Cambridge University Press, 1986).

16 Cfr.: F. Nietzsche *Daybreak: Thoughts on the Prejudices of Morality*, trans. R. J. Hollingdale (Cambridge-New York: Cambridge University Press, 1982). German title: *Morgenröthe*, in *Nietzsche Werke*, V_1, 1971.

17 Nietzsche, *Ecce Homo*, "Warum ich so weise bin", 1–2.

18 *Ibid.*, "Warum ich so klug bin", 1–2.

19 Nietzsche, *Menschliches, Allzumenschliches*, "Vorrede", 4.

20 Nietzsche, *Ecce Homo*, "Warum ich so klug bin", 10.

21 Nietzsche, *Nachgelassene Fragmente*, 10 [17], Autumn 1887, in *Nietzsche Werke*, VIII_2, 1970, pp. 128–129.

22 *Ibid.*, 9[17] e 10 [17]. On this subject cfr.: G. Campioni, *Scienza e filosofia della forza in Nietzsche*, in Id., *Sulla strada di Nietzsche* (Pisa: ETS, 1993), pp. 161–96; C. Richter, *Nietzsche et la biologie* (Paris: PUF, 2001); G. Moore, *Nietzsche, Spencer and the Ethics of Evolution*, in *Journal of Nietzsche Studies*, XXIII, 2002, pp. 1–20.

[23] Nietzsche, *Nachgelassene Fragmente*, 9 [69], Autumn 1887, in: *Nietzsche Werke*, VIII_2, where Nietzsche repeats the motto contained in H. Joly, *Psychologie des grands hommes* (Paris: Hachette, 1883), p. 240.

[24] Nietzsche, *Nachgelassene Fragmente*, 25 [309], Spring 1884, in: *Nietzsche Werke*, VII_2, p. 87.

[25] Cfr.: S. Franzese, *L'uomo indeterminato. Saggio su William James*, Roma: D'Anselmi 2000, pp. 19–63. Nietzsche may have read the essay by W. James, *Great Men and Their Environment* (*Atlantic Monthly* XLVI: 441–59), in French translation on the *Critique philosophique* (Jan.–Feb. 1881). As found in: G. Campioni, *Les lectures françaises de Nietzsche* (Paris: PUF, 2001), p. 32, note 2.

[26] Nietzsche, *Nachgelassene Fragmente*, 9 [7], 9 [84], Autumn 1887, in *Nietzsche Werke*, VII_2, p. 6 and p. 42.

[27] Nietzsche, *Nachgelassene Fragmente*, 9 [153], Autumn 1887; 11 [413], 1887–1888, in: *Nietzsche Werke*, VIII_2, pp. 88–90; p. 433.

[28] Cfr.: H. R. Maturana, F. J. Varela, *Autopoiesis and Cognition. The Realization of the Living* (Dordrecht: Reidel, 1980).

[29] P. Sloterdijk, *Regole per il parco umano. Una replica alla lettera di Heidegger sull'umanismo*, in: *aut-aut*, 301/302, 2001, pp. 127, 132, 137. Cfr. also the interview, conducted by Heik Afheld and Bernd Ulrich, appeared in *Der Tagespiel on line*, 8 March 2001. P. Sloterdijk's well-known titles are published by Suhrkamp-Verlag, Frankfurt a. M.: *Sphären I. Blasen, Mikrosphärologie* (1998); *Sphären II. Globen, Makrosphärologie* (1999); *Regeln für den Menschenpark. Ein Antwortschreiben zu Heideggers* Brief über den Humanismus (1999).

[30] Nietzsche, *Menschliches, Allzumenschliches I*, "Vorrede", 3.

[31] Nietzsche, *Der Antichrist*, in *Nietzsche Werke*, VI_3, 1969, §62. Cfr. also: *The Antichrist*, trans. W. Kaufmann, in *The Portable Nietzsche* (New York: Penguin, 1982). Nietzsche, *Nachgelassene Fragmente*, 11 [415], in *Nietzsche Werke*, VIII_2, pp. 435–436.

[32] Cfr.: M. Konegger and A-T. Tymieniecka (eds.), *Life. The Human Quest for an Ideal*, in: *Analecta Husserliana* XLIX (1996), p. 15.

[33] Nietzsche, *Menschliches, Allzumenschliches II*, 1967, "Der Wanderer und sein Schatten", §21.

[34] Cfr.: Darcy Ribeiro, *Os Índios e a Civilização – A Integração das Populações Indígenas no Brasil Moderno* (Rio: Editora Civilização Brasileira, 1970).

[35] F. Nietzsche, *Jenseits von Gut und Böse*, in *Nietzsche Werke*, VI_2, 1968, a. 51. Cfr. also *Beyond Good and Evil*, trans. W. Kaufmann (New York: Vintage, 1966).

[36] Nietzsche, *Nachgelassene Fragmente*, 9 [93], 1887–1888, in *Nietzsche Werke*, VIII_2.

[37] Nietzsche, *The Gay Science*, aph. 125.

[38] Nietzsche, *The Antichrist*, §17.

[39] *Ibid.*, §38.

[40] Nietzsche, *Nachgelassene Fragmente*, 7 [9], 1885–1887, in *Nietzsche Werke*, VIII_1, 1974.

TED TOADVINE

"STRANGE KINSHIP": MERLEAU-PONTY ON THE HUMAN–ANIMAL RELATION

Of Merleau-Ponty's published works, only his first book, *The Structure of Behavior*, devotes significant place to the life and world of the animal. Responding to both Henri Bergson and Max Scheler, Merleau-Ponty here treats the "vital order" of the animal as one integrative level of the dialectic of "form" or "structure" that he takes as ontologically fundamental. But the situation of animal life in this work remains ambiguous: On the one hand, Merleau-Ponty's description of animal life as a melodic unity aims to respect the originality and irreducibility of the animal level of structure, and his treatment of both animal soul and human mind as levels of structure recognizes an ontological continuity between animal and human orders that is lacking, for instance, in Scheler's distinction of animal soul from human spirit.[1] Consequently, Merleau-Ponty can claim that animals truly *exist* in the sense of "being-in-the-world."[2]

But on the other hand, the perceptual life of animals is never strictly comparable to our own, since the human order integrates perceptual life into a larger dialectic and effectively transforms it. Thus, "Man can never be an animal; his life is always more or less integrated than that of an animal" (SB, 196/181). Furthermore, structure, as ultimate reality, is understood as the intentional correlate of perceptual consciousness, which in the end is treated as one level of human consciousness. The problem of perception, which Merleau-Ponty considers in the last part of *The Structure of Behavior* and which occupies him entirely in *Phenomenology of Perception*, concerns the relation between human perceptual and reflective consciousness. This is why the animal's world is scarcely discussed in its own right in *Phenomenology*, where examples of animal behavior appear alongside those of the child, the "primitive," and the pathological only in order to illuminate the pre-reflective life of human perceptual consciousness. One may conclude, then, that Merleau-Ponty's early work, like Heidegger's *Being and Time*,[3] treats animal life merely as a privative mode of human existence.

Merleau-Ponty's approach to animal life in his final lectures on nature, however, presents a remarkable reversal of this relationship, with important

A-T. Tymieniecka (ed.), Analecta Husserliana XCIII, 17–32.

consequences for the development of an ontology of inter-animality. The goal of these lectures is a positive determination of Being that resolves the ontological ambiguity of the earlier concept of "structure," and animal life is Merleau-Ponty's guiding clue in this reformulation. Rather than starting from human perceptual consciousness and working backward, privatively, to disclose the essential structures of animal life, Merleau-Ponty starts from animal life and its *Umwelt* to demonstrate that Being is constitutively phenomenal. This leads him to recognize an irreducible *Ineinander* of animality and humanity such that "the relation of the animal to the human will not be a simple hierarchy founded on an addition" but rather a "strange kinship" that is perhaps best indicated by "mythical thinking."[4] This "strange kinship" allows us to recognize differences between the content of animal and human worlds while insisting on their shared rootedness in nature; thus, commenting on the work of Jakob von Uexküll, Merleau-Ponty affirms that "there is no break between the planned animal, the animal that plans, and the animal without plan" (N, 231/176). What we find in Merleau-Ponty's final treatment of animals, therefore, has much more in common with Gilles Deleuze and Félix Guattari's concept of "becoming-animal,"[5] emphasizing the bi-directionality of the animal–human exchange, than with Heidegger's insistence on the absolute difference between the world-poverty of animals and humanity's openness to Being.[6]

I. INSTINCT AND INTELLECT IN *THE STRUCTURE OF BEHAVIOR*

When Merleau-Ponty announces, in the opening lines of *The Structure of Behavior*, that his goal is to "understand the relations of consciousness and nature," he is announcing what will become the trajectory of his entire career. But only in this first book does he offer a thematic treatment of the mind–nature relation, providing us with a comprehensive ontology of nature as a series of structural orders – physical, vital, and human – and with a theory of mind's inherence in and emergence from other natural structures. *The Structure of Behavior*, then, is perhaps the earliest attempt to offer a truly "naturalized" phenomenology. In this respect, we can read Merleau-Ponty's first book as a response to Henri Bergson's *Creative Evolution* (1907)[7] and Max Scheler's *Man's Place in Nature* (1928). Each of these works provides an account of the place of human consciousness in the natural world, and both proceed by a contrast of our consciousness with that of non-human animals. Merleau-Ponty's interpretation of the phenomenological reduction is inspired by Bergson's account of "intuition," which brings together instinct and intelligence in a more encompassing account of consciousness that may remain true to the diverging and complementary paths of life. But, as we will see, Bergson fails to appreciate the originality of the human level, namely, the extent to which this structure subsumes the vital orientation toward need within a dialectic that transforms it. Scheler, on the other hand, thematizes the phenomenological reduction precisely as a break with instinct and vital need. But in so doing, he locates the agent of this break, spirit, outside of the dialectic of life as a whole. In seeking an immanent ontology of life and mind, Merleau-Ponty's first work aims to steer between these two extremes.

In *Creative Evolution*, Bergson argues that animal instinct and human intelligence represent two complementary strands in the evolutionary unfolding of the *élan vital*. Instinct, as found paradigmatically in the arthropods and the insect order of *Hymenoptera* in particular, is turned inward toward life, while the consciousness of the vertebrates, culminating in human intelligence, is focused outward toward matter. To grasp life as a whole, Bergson contends, we must widen our understanding beyond the limits of the intellect and regain possession of our latent instinctual nature. By so doing, we arrive at "intuition," animal instinct elevated to self-awareness, which comprehends the entire movement of the whole of life from within. Intuition operates, then, by a kind of epochē, a "bracketing" of the intellect's tendency to schematize all experience into the practical categories of pure space and morcelized time. This bracketing is a becoming-animal of the human consciousness, or, more precisely, a becoming-insect, a taking-up of the insect's undivided sympathy with life into human self-consciousness.

According to Bergson, instinct is oriented to the thing itself, toward what is, while the intellect is oriented toward relations, such as the hypothetical "If x, then y." The insect's world is therefore determined by the particular organic form that either opens or closes it to other living things. Different insects perform the same function – for example, paralyzing without killing caterpillars by a series of precisely delivered stings – but they perform these same functions in different registers or keys:

> We seem ... to be before a *musical theme*, which had first been transposed, the theme as a whole, into a certain number of tones and on which, still the whole theme, different variations had been played, some very simple, others very skillful. As to the original theme, it is everywhere and nowhere. It is in vain that we try to express it in terms of any idea: it must have been, originally, *felt* rather than *thought* (CE, 641/171–2).

The "original theme," the *élan vital* itself, is expressed only through the concrete activities of individual living things, which are different improvisations of this theme. However, while the key and variation of the insect is fixed by its organic equipment, only the human being achieves true freedom, which makes it possible for the human being to recapture the "original theme" of life – not in intellectual concepts, which always fall short of the movement of intuition, but rather through creative works of art. Thus, Bergson's example of the kind of order appropriate to the vital or the willed, in contrast with the automatic order of matter toward which the intellect is oriented, is a symphony of Beethoven (CE, 685/224).

Merleau-Ponty's account of phenomenology is inspired, in part, by this Bergsonian effort to overcome the limitations of the intellect and to think life from within. But while, for Bergson, the positive movement of life is opposed by the inverse, entropic tendency of matter, in Merleau-Ponty's account of structure the physical and vital are treated as two equally positive orders. Nevertheless, Merleau-Ponty appropriates Bergson's metaphor of musical structure to characterize the contrast between animal and human consciousness. Throughout *Structure*, Merleau-Ponty characterizes animal behavior as a "melodic unity" or a "kinetic melody," but only at the symbolic level of behavior characteristic of humanity do we find an orientation toward the theme as such. This orientation toward the theme is demonstrated, for instance, in the ability to play a musical instrument from a score, where we find three independent systems communicating internally: "The character of the melody, the graphic configuration of the musical text, and the unfolding of the gestures participate in a single structure, have a single common nucleus of signification" namely, their common expressive value (SB, 132/121). What characterizes human symbolic behavior is precisely the orientation toward this "structure of structures," the musical essence itself, which makes possible true expression and creativity, for instance in improvisation: "It is this possibility of varied expressions of a same theme, this 'multiplicity of perspective," which is lacking in animal behavior," since, for them, the possible themes are fixed by the "*a priori* of the species" (SB, 133/122). The "structure of structures" cutting across the different symbolic systems can be reached only by recognizing the theme as "the same" in different keys and different symbolic representations; thus, it requires an awareness of identity across a "multiplicity of perspectives." What separates human symbolic behavior from animal instinct, then, is an orientation toward ideality; spontaneity and creativity presuppose such an eidetics, so that human behavior no longer *has* a signification but *is* itself signification (SB, 133/122).

But it is at this point that Merleau-Ponty's account breaks with Bergson's, because the latter fails to adequately thematize the structural differences between instinct and intelligence. For Bergson, both are solutions to the problem life poses to the organism, while Merleau-Ponty sees in the human dialectic of "perceived situation-work" an original development that transcends the orientation toward vital needs. The Hegelian term "work" here signifies, for Merleau-Ponty, the "ensemble of activities by which man transforms physical and living nature," the more encompassing dialectic that surpasses the fixed *a prioris* of the species oriented toward need (SB, 175–6/162–3). That the understanding of human consciousness requires a break with the orientation toward vital need echoes Scheler's conception of the relation between life and spirit, the influence of which is also apparent throughout Merleau-Ponty's account.

For Scheler, "man is more a problem to himself at the present moment than ever before in all recorded history" (MPN, 10/4). This apparently negative judgment actually marks a progress, since it is precisely this self-problematization that is characteristic of "man": man becomes man only through an alienation from nature or life that is simultaneously the dawn of philosophy. Thus, Scheler notes that the word "man" has a deceptive ambiguity: on the one hand, "it signifies the particular morphological characteristics of man as a subclass of the vertebrates and mammals," so that man is subordinate to the concept "animal" and in truth "occupies a relatively small corner of the animal kingdom" (MPN, 12/6) But this sense of "man" as merely one animal among others is not the essential meaning of the term, for, according to its second and essential sense, "man" signifies a "set of characteristics which must be sharply distinguished from the concept 'animal' – including all mammals and vertebrates" (MPN, 13/7). In other words, man-as-animal and man-as-such have entirely different meanings and origins.

But the difference between humans and animals does not rest on the contrast between instinct and intelligence, for Scheler, since the animal participates in all four stages of psycho-physical life: the "vital feeling" or impulse common to all organic beings; the purposive but innate and species-specific behavior of instinct; the habitual behavior characteristic of association and conditioned reflex; and practical intelligence, defined as the "capacity of responding, without trial and error, to a new situation meaningfully" (MPN, 33/29). This capacity requires "insight into a set of affairs on the basis of a structure of relations whose basic elements are partly given in experience, partly completed in anticipatory representation" (MPN, 34/30). Thus, practical intelligence is productive rather than reproductive, and involves a kind of anticipatory prevision of a situation that is new for the individual. Drawing on Wolfgang Köhler's work with chimpanzees – the same studies that will be central for Merleau-Ponty's account of animal behavior – Scheler calls the insightful grasp of a new solution an "Aha experience" (MPN, 34/30). For Scheler, therefore, it is true that animal and human intelligence exist along a continuum, and that, from this perspective, no sharp ontological distinction is admitted between animal and "man." But even though the animal demonstrates genuine intelligence, this intelligence remains "practical," that is, oriented toward obtaining a goal set by the vital drives. While the animal has the capacity to choose between different means of satisfying this drive, and may even postpone immediate for greater satisfaction in the future, what separates the animal from "man" is precisely the capacity to choose values as such, and in particular to transcend the values prescribed by life.

The principle that defines human existence, then, is not intelligence but spirit, which "is opposed to life as such, even to life in man" (MPN, 38–39/36). While the animal remains immersed "ecstatically" in its environment, spiritual beings are free from the environment; that is, they have a "world" and are open to the objective nature of things. Animals lack a true world because they encounter only those "centers of resistance" within the environment that interest their instincts or drives; in other words, they fail to constitute true "objects" that remain identical across multiple perspectives. The animal cannot accede to objectivity as such, nor to self-consciousness, since both require a decentering of the focus of vital need. What is essential to "man," on the other hand, is precisely the capacity for what Husserl had termed the "phenomenological reduction" (MPN, 53/52), although in Scheler's interpretation this is not a suspension of judgement about the world's existence so much as a suspension of the primordial encounter with resistance that gives experience its existential index. The paradigmatic act of spirit, for Scheler, is therefore ideation: through a "tentative experimental suspension of reality," the "essence is peeled off, as it were, from the concrete sensory object" (MPN, 53/51). Our distinctively human capacity, the ability to distinguish between essence and existence – or, in other words, the ability to wonder and engage in philosophy – necessitates our break with natural existence:

> when man, becoming himself, broke with the methods of preceding life to adjust or to be adjusted to the environment and embarked upon the opposite direction of adapting the revealed world to himself and to his own life of organic stability, when man separated himself from nature and transformed it into an object subject to domination and to the control of symbolic manipulation – at this moment man was also driven to anchor his own central being in something beyond this world. He who had placed himself so boldly above this world could no longer regard himself merely as a "member" or "part" of this world (MPN, 88/90).

Spirit, consequently, cannot be a part of the natural world; in order to provide "man" with a true picture of the world, it must be situated outside of this world (MPN, 48/47).

Scheler's account anticipates Merleau-Ponty's own by identifying ideation and the capacity for withdrawal from vital need as indicators of specifically human consciousness. Merleau-Ponty's critique of Bergson focuses on the latter point. While Bergson was correct to criticize earlier accounts for treating perception as immediately contemplative, he does not carry this idea far enough: "To do it justice completely it would have been necessary first of all to stop defining consciousness by knowledge of self and to introduce the notion of a life of consciousness which goes beyond its explicit knowledge of self" (SB, 178/164). Bergson attempts to bring consciousness closer to action, but he narrowly conceives of action in strictly vital terms. A richer response requires two movements: first, consciousness must be understood primarily in its pre-reflective, perceptual mode, of which the refined form of self-conscious awareness is only a variation; but, secondly, this perceptual consciousness cannot be immediately associated with the consciousness of the animal, since human consciousness participates in a more encompassing dialectic, a level of structure that is specifically human. By taking up the immanent meaning of specifically human action and its internal structure, we can avoid Bergson's fault of treating instinct and intelligence on the same plane, that is, of conflating animal and human life. Therefore, for Merleau-Ponty, "the word 'life' does not have the same meaning in animality and humanity" (SB, 188/174).

Human life, for Merleau-Ponty, is oriented toward the virtual, the iterative, the transposable theme that links the melodies of behavior across different modalities. This orientation toward the virtual – evident in language, revolutionary action, and suicide, which Merleau-Ponty says have the same signification (SB, 190/175) – situates our being outside of "biological existence," since it presupposes our capacity for "rejecting the given milieu and of searching for equilibrium beyond any milieu" (SB, 190n/245n97). Human being is characterized, then, as Scheler had said of spirit, by its capacity for saying "no" to life. It is this gap between humans and life, our ability to break with the milieu, that allows us to stand back from it, to treat it as a spectacle, and to "take possession of it mentally by means of knowledge properly so called" (SB, 188/174).

Merleau-Ponty seems quite close, therefore, to Scheler's sharp division between the animal and the human. But his approach is set off sharply from Scheler's on two points that reverberate through his ontology and his appropriation of the phenomenological method: first, Merleau-Ponty reintegrates the "reality function" or existential index into the phenomenal dimension of perceptual consciousness, while Scheler had treated spirit (and the phenomenological reduction) precisely as the break with this existential index, which he had located on the side of life. Although consciousness, for Merleau-Ponty, breaks with the orientation toward vital need, it does not break with the orientation toward action, as Bergson had correctly intuited (SB, 189/175). This means, secondly, that the two perspectives on consciousness – that of life, on the one hand, oriented toward action and the real, and that of reason or spirit, on the other hand, oriented toward an objective and true universe – cannot simply be juxtaposed as Scheler had attempted to do. The problem of the relation between these two perspectives on consciousness – consciousness as life and as spirit – is precisely what Merleau-Ponty terms the "problem of perception" (SB, 191/176).

The relation between the animal and the human is thus a duality inhcrcnt within human consciousness itself, the relation between the psychological and the spiritual. For Scheler, the psychological situates us firmly within the ontology of life and establishes our kinship with animality, while the spiritual is a definitive break, even a rejection, of this continuity with life. But, for Merleau-Ponty, both terms of this relation must be rejected: our psycho-physical life is not that of the animal, since this life has been taken up and transformed by its participation in an original unity of greater complexity. Therefore, "Man is not a rational animal" and "can never be an animal," since "his life is always more or less integrated than that of an animal" (SB, 196/181). But, on the other side of this relation, the spiritual is not a break with existence or a retreat to another world but is precisely the more encompassing form into which these relations are subsumed:

> neither the psychological with respect to the vital nor the spiritual with respect to the psychological can be treated as substances or as new worlds. The relation of each order to the higher order is that of the partial to the total.... The advent of higher orders, to the extent that they are accomplished, eliminate the autonomy of the lower orders and give a new signification to the steps which constitute them (SB, 195/180).

Now, what is original in Merleau-Ponty's account of the relation between life and mind is that he can maintain a sharp distinction between the form of animal life and human reason without drawing any sharp ontological boundaries between the two. This seems to be an advance over the metaphysical dualism espoused by Scheler, in which humans, by their very essence – by the very essence that makes philosophy possible – are opposed to nature and life. But this ontological kinship with animals comes at a cost: there is no longer a level or layer of our lives that we may truly be said to share with other animals, since the incorporation of life into our specifically human dialectic yet again separates us absolutely. Although Merleau-Ponty, unlike Scheler, was willing to accord to animals a certain unique manner of "being-in-the-world," in the end he insists that this world can have nothing in common with our own.

Merleau-Ponty focuses his attention, in the conclusion of *The Structure of Behavior* and in its sequel, *Phenomenology of Perception*, on the relationship between "perceptual" and "objective" consciousness, both taken at the specifically human level. This relation is the key, as he has told us, to the "problem of perception," the problem to which all of the investigations of *Structure* are reducible (SB, 240/224). He can close his first work, then, by reiterating this key question: "What are the relations of this naturized consciousness and the pure consciousness of self? Can one conceptualize perceptual consciousness without eliminating it as an original mode; can one maintain its specificity without rendering inconceivable its relation to intellectual consciousness?" (SC, 241/224). These questions set the stage for the investigation of *Phenomenology of Perception*, but they are notable in what they leave out of consideration. Merleau-Ponty, at this juncture, is no longer concerned with the world of the animal as such, but rather with the relation between human perception and human reason.

This ambiguity of Merleau-Ponty's treatment of animals in his early work is once again explicit in his 1948 radio lectures, recently published as *Causeries* [*The World of Perception*], the fourth lecture of which is entitled "Exploring the Perceived World: Animality."[8] Merleau-Ponty argues here that the rediscovery of the perceptual layer of experience demands a revalorization of the "extreme or aberrant forms of life and consciousness" found in "animals, children, primitive peoples and madmen" (WP, 34/70). The animal is neither reducible to the collection of "wheels, levers, and springs" that Descartes saw in it, nor to a privative modification of normal, human consciousness. By reopening the investigation of these "irrational roots" of consciousness, contemporary philosophy and art find in them "the movement by which all living things, ourselves included, endeavor to give shape to a world" (WP, 37/73, 40/76). Animals, then, despite the suggestions of mechanistic biology, have their own originary interiority and "proceed to trace in their environment, by the way they act or behave, their very own vision of things" (WP 39/75). But the animal's struggle to give shape to the world reminds us, above all, Merleau-Ponty says, "of our failures and limitations" (WP, 40/76). The "healthy, civilized, adult human being" strives for a kind of coherence that is lacking in these aberrant forms of consciousness, and they serve mainly as a reminder of our failure to ever achieve a complete coherence. Thus, normal human thought takes precedence over that that of the child, the primitive, or the animal on the condition that it "not masquerade as divine law, but rather … measure[s] itself more honestly against the darkness and difficulty of human life and without losing sight of the irrational roots of this life"

(WP, 37/73). So, despite the originality of its world, the animal still mirrors back to us a wavering and unsteady reflection of our own faces.

II. "STRANGE KINSHIP" IN MERLEAU-PONTY'S LECTURES ON NATURE

Merleau-Ponty revisits the issue of animal–human relations in his lectures on "The Concept of Nature," especially in the second and third courses from 1957 to 1960, and it seems clear that he intended to include the conclusions of these investigations in the complete text of *The Visible and the Invisible*. For instance, in the last working note included with the published text, dated March, 1961, Merleau-Ponty indicates that the second part of the complete work would be devoted to Nature: "not Nature in itself, a philosophy of Nature, but a description of the man–animality *intertwining*."[9] Although no treatment of this "man–animal intertwining" appears in the published portion of *The Visible and the Invisible*, we can gain some sense of Merleau-Ponty's intentions from his course notes.

In Merleau-Ponty's 1957–58 course on "Animality, The Human Body, and the Passage to Culture," he takes up again the musical figure of animal behavior, this time by way of a reading of ethologist Jakob von Uexküll. Already in *The Structure of Behavior*, Merleau-Ponty had noted Uexküll's proposal that "Every organism … is a melody which sings itself," but there Merleau-Ponty had only been quoting a passage cited without reference by Frederik Buytendijk (SB, 172/159). Merleau-Ponty's only investigation of Uexküll's work on its own terms occurs in this lecture course two decades later. Melody is again at the center of Merleau-Ponty's interests, and he concludes that "it is the theme of the melody … that best expresses the intuition of the animal according to Uexküll" (N, 233/178). What the notion of melody elucidates is the ontological status of the animal's *Umwelt*, which, according to Uexküll, can be explained neither through physical processes alone nor through an explicit plan in the consciousness of the animal. In fact, the *Umwelt* must be evoked as an explanatory principle both for the animal's physical development and for its behavior. But the *Umwelt* itself cannot be reduced to subject or object; it is, instead, a kind of self-organization at the level of life. Thus, it is a kind of "melody that is singing itself":

> When we invent a melody, the melody sings in us much more than we sing it; it goes down the throat of the singer, as Proust says. … We think naturally that the past secretes the future ahead of it. But this notion of time is refuted by the melody. At the moment when the melody begins, the last note is there, in its own manner. … It is in this way that things happen in the construction of a living being. There is no priority of effect over cause. … It is impossible to

distinguish the means and the end, the essence and the existence in it. From a center of physical matter surges an ensemble of principles of discernment at a given moment, which means that in this region of the world, there will be a vital event (N, 228/173-74).

Whereas Merleau-Ponty's earlier use of the musical metaphor had emphasized the fixity of the organism's melody by the *a priori* structures of vital need, here the accent is on the ecological relationships formed between the organism, other creatures, and their milieu. In fact, the animal's melody enters into a kind of counterpoint with its milieu that remains causally inexplicable: "Each action of the milieu is conditioned by the action of the animal; the animal's behavior arouses responses from the milieu .... In brief, the exterior and the interior, the situation and the movement are not in a simple relation of causality" (N, 229/175). The relation between organism and milieu is therefore one of meaning, and the *Umwelt* as a "natural plan" is already the introduction of symbolism and a "beginning of culture." But there is no subject to which the meaning of this contrapuntal melody can be attributed, since the "unfurling of the animal is like a pure wake that is related to no boat" (N, 231/176).

Merleau-Ponty draws similar conclusions from Robert Hardouin's account of animal mimicry and Adolf Portmann's examination of animal appearances, both of which resist explanation either in terms of adaptation for survival or as a mere projection on the part of the outside observer. Animal mimicry points to an intrinsic relation or indivision of the animal from its surroundings, a mutual "contamination" that seems to entail a kind of "natural magic" (N, 242–43/185–86). But the indivision of mimicry is no more magical, Merleau-Ponty suggests, than the miraculous emergence of sense organs, which, without violating the constraints of causality, succeeds in making the "ordinary and the extraordinary communicate" (N, 243/186). Portmann's study of the forms of animal appearance underscores the fundamentally expressive character of the animal's relation with its milieu and the internal relation or indistinction that exists between them. The animal's form presents an "existential value of manifestation" that arises neither from the pursuit of survival nor any intentional purpose but in a "perceptual relation" between animal and milieu (N, 246/188).

Furthermore, we cannot understand the relations between animals within a species as a sum of individuals exterior to one another: "What exists are not separated animals, but an interanimality," wherein each animal is the mirror of the others (N, 247/189). This aspect of animal life finds development in Konrad Lorenz and Nikolaas Tinbergen's descriptions of animal communication, which demonstrate the emergence of symbolism from natural signs and the "reciprocal expression" manifest in ritualistic behaviors. "In short,"

Merleau-Ponty concludes, “we can speak in a valid way of an animal culture” (N, 258/198).

But Merleau-Ponty’s effort here is not to raise the animal to the level of symbolic behavior or reflective activity he had earlier reserved for human consciousness. His conclusions are, instead, ontological: “behavior can be defined only by a perceptual relation and ... Being cannot be defined outside of perceived being” (N, 247/189). This remark illustrates the reversal of method that Merleau-Ponty’s studies of nature and animality follow in his lecture courses, as compared with the earlier focus on the problem of perception posed in terms of the relation between human perceptual and reflective consciousness. In *The Structure of Behavior*, the concept of structure attempted to hold together two incompatible approaches to nature: on the one hand, Merleau-Ponty saw in vital behavior an immanent and self-organizing intelligibility. But on the other hand, his commitment to phenomenological principles required him to treat this immanent intelligibility as an object for human consciousness. This problematic relationship between the Concept sunk into Nature and then made explicit in self-conscious awareness is precisely the “problem of perception” around which Merleau-Ponty’s first two books are oriented. But in the nature lectures, Merleau-Ponty takes nature as the non-instituted, as what precedes the intentional activity of consciousness, and this allows him to develop his earlier insights into the melodic unity of animal life into the articulation of an ontology of perceived being. Rather than returning to being by way of perceptual consciousness, these studies of animal life indicate the constitutive reference to perception within Being itself. Renaud Barbaras expresses this methodological reversal as follows:

> In [*Phenomenology of Perception*], the natural world ... was reduced to the advantage of an incarnated subject, and the constitutive reference of this subject to a perceived world was made to appear. In the later works, Merleau-Ponty suspends subjectivity and becomes interested in only natural being, at the heart of which he discovers a constitutive reference to perception.[10]

In this reversal of his approach, Merleau-Ponty is no longer thinking nature, and the animal in particular, in accordance with what Giorgio Agamben has called the “anthropological machine,” that is, with the aim of marking an internal schism in “man.”[11] Approaching both the animal and the human by way of “natural being” allows him to recognize an irreducible *Ineinander*, an intertwining, of animality and humanity. The investigation of this *Ineinander* is found primarily in the third course on nature, “Nature and Logos: The Human Body,” from 1959–60.

The aim of this course, Merleau-Ponty indicates, is to consider the human “at his point of emergence in Nature,” and it is at this point that the interwining with the animal comes to the fore: “Just as there is an *Ineinander* of life and physicochemistry, i.e., the realization of life as a fold or a singularity of physicochemistry – or structure, so too is the human to be taken in the *Ineinander* with animality and Nature" (N, 269/208). The human being is neither reducible to the animal, then, nor to be understood by mere addition as the "rational animal." The relation is not hierarchical, but lateral, "an overcoming that does not abolish kinship" (N, 335/268).

The best clue that we have for unraveling this "strange kinship," Merleau-Ponty suggests, is "mythical thinking." By way of example, he refers to a catalog for a contemporaneous exhibit of Eskimo masks at the Musée Guimet:

> Masks of the Inuit type recall the original double nature. The animal and its human double, the Inuit, are inscribed on the same side, presented either simultaneously or alternatively, thanks to a dispositive of mobile flaps opening and cutting in on each other. In this way is restituted the primitive state, when the envelope was a mask that we willfully separated in order to appear as man or as animal, changing appearance but not essence .... It recalls a time ... when the separation was not yet effected.[12]

Commenting on this passage, Merleau-Ponty emphasizes the transition from a "primordial indivision and metamorphoses," an origin in which the human doubles the animal and the animal doubles the human, to the present state of separation. But the caricatures and strange becomings that mark the lateral relationship between humans and animals are not always so unidirectional, as Merleau-Ponty had recognized in 1948, when he noted the role that animals continue to play "in the secret reveries of our inner life" (WP, 40/76). Perhaps, then, the separation marked by Inuit masks may be matched, in an opposite direction, by the sorts of metamorphoses that Franz Kafka describes, which would bring Merleau-Ponty close to what Gilles Deleuze and Félix Guattari had described with their concept of "becoming-animal."

Merleau-Ponty's own interest, however, lies in what this primordial indivision and bi-directional metamorphosis may teach us about ourselves and about mind's emergence within nature. The "extraordinary representation of the animal as variant of humanity and of humanity as variant of animality" in the Inuit masks demonstrates that "there must be a vital foundation of man and of spirit," namely, the human body (N, 277a/307n11). This "strange kinship" reveals to us our intertwining with "sensible Being," so that even mind or spirit is "incredibly penetrated by its corporeal structure" (N, 339/271, 335/268). While *The Structure of Behavior* had placed the accent on spirit or mind as an integrative dialectic of form that subsumes life, Merleau-Ponty here treats mind as emergent from life by way of the "figural form" of reflection in the human body (N 335/268). What distinguishes mind is precisely its orientation toward the invisible dimension of Being.

Merleau-Ponty says of this invisible dimension, which he equates with mind or spirit, that it is "not another positivity: it is the inverse or the other side of the visible," a "brute and savage mind" that underlies our language and cultural acquisitions (N, 274/212). We can only understand this invisible dimension of being starting from the visible, and this process of reflection is the "coming-to-self of Being ... without the notion of a subject" (N, 335/268). Although our kinship with animals is grounded within the visible, our difference stems from the human contact with the invisible dimension as well:

> Animality and human being are given only together, within a whole of Being that would have been visible ahead of time in the first animal had there been someone to read it. Now this *visible and invisible* Being, the sensible, our *Ineinander* in the sensible, with the animals, are permanent attestations, even though visible being is not the whole of being, because [Being] already has its other invisible side (N, 338/217).

The traditional definition of humanity by reflection, as one finds it, for instance, in Teilhard de Chardin, must be rejected, since it eliminates our relation with life. The ontology of the visible and the invisible aims precisely to avoid this schism between life and mind.

More than once while working through this relationship between visible and invisible, Merleau-Ponty speaks of the need to conjoin "eye and mind." For instance, he writes "Even mind is incredibly penetrated by its corporeal structure: eye and mind" (N, 335/268). This recalls, of course, the title of his last published essay, written during this same period, which examines the ontological relationship between vision and painting. In this essay, Merleau-Ponty prioritizes vision, and consequently painting, for the investigation of ontology since they make possible a certain suspension of the world that the writer, for instance, cannot achieve. In this context, he notes that "Music, at the other extreme, is too far on the hither side of the world and the designatable to depict anything but certain schemata of Being – its ebb and flow, its growth, its upheavals, its turbulence."[13] Merleau-Ponty does not elaborate on this remark, but it suggests a shift in his understanding of mind and reflection. Rather than seeking the theme of life in the melody that cuts across all keys and variations, mind directs itself primarily toward the invisibility that only the painting, paradoxically, can disclose. The shift toward invisibility is Merleau-Ponty's last formulation of the withdrawal or separation from life – the "suspension of the world" – that he continues to find necessary for any philosophical ontology. Thus, at the conclusion of "Eye and Mind," Merleau-Ponty reaffirms that what the painter seeks is "on the hither side of means and goals, commanding and overseeing all our *useful* activity."[14]

But perhaps an ontology of life cannot avoid listening more carefully to the upheavals and turbulence of Being, to the contrapuntal refrains that constitute each organism's characteristic style of "singing the world,"as Merleau-Ponty had described corporeal expression in *Phenomenology of Perception*.[15] Just a few pages after coining this phrase, Merleau-Ponty differentiates the musician from the writer by his comparative freedom from the constraints of tradition: "every composer starts his task at the beginning, having a new world to deliver."[16] It is nature, on this description, that would be the consummate musician, having, as Merleau-Ponty will describe it at the beginning of his second lecture course, "an originary productivity," there "from the first day," that "continues beneath the artificial creations of man" (N, 169/125). Perhaps we can locate the distinction between music and painting, between ear and eye, along the division Merleau-Ponty has drawn between the visible and the invisible. Just as the invisible is, paradoxically, what the painter reveals to the eye, perhaps it is the very being of the visible – its pulsations and vibrations, the intense rhythms of its own peculiar duration – that music is best able to release. While an ontology of the invisible mind should rightly turn its eye toward painting, then, an ontology of nature should listen attentively to the ebbs and flows, the upheavals and dissonances, of the sensible itself.

University of Oregon

NOTES

1 Scheler, *Die Stellung des Menschen im Kosmos* (München: Nymphenburger Verlagshandlung, 1949); *Man's Place in Nature*, trans. Hans Meyerhoff (New York: Noonday Press, 1962). Hereafter, cited textually as MPN, with German preceding English pagination.

2 *La Structure du comportement* (Paris: PUF, 1942), pp. 136–37; *The Structure of Behavior*, trans. Alden Fisher (Boston: Beacon Press, 1963), pp. 125–26. Hereafter, cited textually as SB, with French preceding English pagination.

3 Heidegger, *Sein und Zeit* (Tübingen: Max Niemeyer Verlag, 1963), §10 and 12; *Being and Time*, trans. John Macquarrie and Edward Robinson (San Francisco: Harper & Row, 1962), §10 and 12.

4 *La Nature, notes, cours du Collège de France* (Paris: Seuil, 1995), p. 277; *Nature: Course Notes from the Collège de France*, trans. Robert Vallier (Evanston: Northwestern University Press, 2003), p. 214. Hereafter, cited textually as N, with French preceding English pagination.

5 Deleuze and Guattari, "1730 – Devenir-intense, devenir-animal, devenir-imperceptible," in *Mille Plateaux* (Paris: Minuit, 1980); "1730: Becoming-Intense, Becoming-Animal, Becoming-Imperceptible," in *A Thousand Plateaus*, trans. Brian Massumi (Minneapolis: University of Minnesota Press, 1980).

6 Heidegger, *Die Grundbegriffe der Metaphysik* (Frankfurt: Klostermann, 1983), esp. §62 and 63; *The Fundamental Concepts of Metaphysics*, trans. William McNeill and Nicholas Walker (Bloomington: Indiana University Press, 1995), pp. 268–273.

7 Bergson, *L'Évolution créatrice* [1907] in *Oeuvres*, centenary edition (Paris: PUF, 1959); *Creative Evolution*, trans. Arthur Mitchell (New York: Henry Holt, 1911). Hereafter, cited textually as CE with French preceding English pagination.

[8] Merleau-Ponty, *Causeries 1948* (Paris: Seuil, 2002); *The World of Perception*, trans. Oliver Davis (London: Routledge, 2004). Hereafter, cited textually as WP, with French preceding English pagination.

[9] *Le Visible et l'invisible* (Paris: Gallimard, 1964), p. 328; *The Visible and the Invisible*, trans. Alphonso Lingis (Evanston: Northwestern University Press, 1968), p. 274. See also Merleau-Ponty's outlines for the completion of *The Visible and the Invisible* included in the Editorial Note by Claude Lefort, pp. 10–11/xxxv–xxxvi; the working notes at pp. 222/168, 226/172, 230/176–77; and the Translator's Introduction to *Nature*, pp. xiii–xx.

[10] Renaud Barbaras, "Merleau-Ponty and Nature," trans. by Paul Milan. *Research in Phenomenology* 31 (2001): 37.

[11] Giorgio Agamben, *The Open: Man and Animal*, trans. Kevin Attell (Stanford: Stanford University Press, 2004), pp. 13–16 and *passim.*

[12] *Les Masques Esquimaux*, exhibit catalog to *Le Masque*, Musée Guimet, Paris, December 1959–May 1960. Merleau-Ponty cites the text by Evelyne Lot-Falck (Department of the Archives, Musée de l'Homme), p. 9. This text is reproduced in *Nature*, p. 277a/307n11.

[13] *L'Œil et l'esprit* (Paris: Gallimard, 1964), p. 14; "Eye and Mind," trans. Michael Smith, in Galen A. Johnson (ed.), *The Merleau-Ponty Aesthetics Reader: Philosophy and Painting*. (Evanston: Northwestern University Press, p. 123.

[14] *L'Œil et l'esprit*, 90; "Eye and Mind," p. 149.

[15] I refer here, of course, to Merleau-Ponty's famous characterization of embodied expression in *Phénoménologie de la perception* (Paris: Gallimard, 1945), p. 218; *Phenomenology of Perception*, trans. Colin Smith (London: Routledge and Kegan Paul, 1962; rev. 1981), p. 187.

[16] *Phénoménologie de la perception*, p. 221; *Phenomenology of Perception*, p. 190.

MARY TRACHSEL

HUSSERL'S INTERSUBJECTIVITY AND THE POSSIBILITY OF LIVING WITH NONHUMAN PERSONS

> As much as animalia and vegetation exhibit autonomous natures comparable to that of humans, they do not evidence any intrinsic dependence on humans to exist. In their specific universal natures, they appear to stand in relationship above all to climate, just as humans ultimately do. They in fact share with man the basic conditions of life within the unity of all beingness. And so I revise the Pythagorean maxim and declare "Life is the measure of all things." It is life that maintains the unity of all beingness.
>
> Anna-Teresa Tymieniecka, "The Pragmatic Text of the Ontopoeisis of Life."

In explaining the methodology and field of study that constitute "pure phenomenology," Edmund Husserl describes a "science of consciousness" that is wholly distinct from the objective sciences. Likening phenomenology to the pure mathematical sciences because it seeks to comprehend "essential laws" that govern consciousness, just as mathematical laws govern spatial reality, Husserl prescribes phenomenological methods – intuition and reflection – that are introspective rather than empirical. While objective sciences look outward to the transcendent realm of Nature as their given field of study, phenomenology turns inward to contemplate the *cogito*, the conscious subject in the act of knowing.

With an exclusive focus on the structures and operations of consciousness as an active force in and of itself, pure phenomenology assumes the radical separability of Mind and Nature, and for the individual phenomenological subject this means accepting, as well, the radical separability of mind and body. To apprehend pure consciousness undistorted by received knowledge and preconceptions about the self and the world, Husserl calls for "phenomenological reduction," excluding from consideration all beliefs derived from objective experience. Through this radical suspension of empirical belief, the phenomenological field is cleared of all but consciousness

A-T. Tymieniecka (ed.), Analecta Husserliana XCIII, 33–51.

and what it is conscious of, and the phenomenological subject is accordingly reduced to disembodied consciousness. As Husserl explains, "the actuality of all material Nature is therefore kept out of action and that of all corporeality along with it, including the actuality of my body, the body of the cognizing subject" (Husserl 1981/2002, 130). This reduction of being to the subjective experience of the mind paradoxically opens up a limitless field of inquiry, consisting of "the endless multiplicity of manners of being conscious, on the one hand, and on the other, the infinity of intentional correlates" (Husserl 1981/2002, 131).

Husserl acknowledged that despite its vast terrain, phenomenology runs the risk of solipsism, since the radical suspension of Nature excludes from consideration all other cognizing subjects, leaving each individual subject with genuine experience of only "his own flowing phenomena" (Husserl 1981/2002, 131). Anticipating criticism on this point, Husserl maintained that phenomenology promises to reveal the concrete grounding for the empirical science of psychology by laying bare the essential laws that undergird and animate all manifestations of consciousness. As universals, these essential laws not only describe individual subjectivities – ego-subjects – they also explain what Husserl called "intersubjectivities" – we-subjects, who collectively enact the cultural consciousness that constructs the "life-world" (Husserl 1970, 109). Whether this shared phenomenological construction of Nature, the life-world, is a receptive accomplishment of *recognition* or a productive accomplishment of *attribution*, or whether it is some combination of these, Husserl does not really say, asserting only that the life-world is a "universal mental acquisition" (Husserl 1970, 113).

For the most part, Husserl seems to present the lifeworld as a *human* acquisition and accomplishment, and to view *human* subjectivity as the proper field of phenomenological investigation, but his attention to human mind is by no means absolute. His essay on pure phenomenology suggests that the essential laws governing the workings of consciousness encompass all manner of minds, concluding with the words, "If all consciousness is subject to essential laws in a manner similar to that in which spatial reality is subject to mathematical laws, then these essential laws will be of the most fertile significance in investigating facts of the conscious life of human and brute animals" (Husserl 1981/2002, 132). In Husserl's scheme, then, the conscious lives of nonhuman animals are legitimate objects of phenomenological inquiry, and it is at least theoretically possible that intersubjectivity, the formation of "we-subjects" united in shared subjective accomplishment of the life-world, may be undeterred by species boundaries that are, after all, mere products of empirical science.

PHENOMENOLOGICAL INQUIRY INTO NONHUMAN CONSCIOUSNESS

Certainly one of the most famous instances of philosophical inquiry into the conscious life of "brute animals" is Thomas Nagel's 1974 essay, "What is it like to be a Bat?" In this essay, Nagel, like Husserl, assumes that conscious experience is common to "many levels of animal life" (Nagel, 436), but he rejects Husserl's call for the separation of mind and body. Although he concedes the dearth of evidence for the consciousness of nonhuman animals, Nagel nevertheless adopts as his starting point the premise that "bats have experience" (Nagel, 438) and that there is a particular, species-specific, subjective point of view occupied by bats. In short, Nagel accepts as given that "there is something it is like to *be*" a bat, "something it is like *for*" the bat and for no other creature (Nagel, 436). Nagel concludes that the subjective experience of "batness" is inaccessible to humans because bats are so differently embodied from human beings, relying on forms of perception, notably echolocation, that are outside the range of human experience.

Counter to Husserl's claims that consciousness can be examined in isolation from its embodied manifestation in the natural world, then, Nagel maintains the position, now increasingly supported by the discoveries of modern neuroscience, that mind is necessarily conditioned if not wholly produced by the body it inhabits. He points out, for instance, that although human beings might attempt to hang upside down as bats do, differences in the size and density of human and bat bodies make the experiences very dissimilar for the two species. While he acknowledges that humans have access to the phenomenological experiences of others of their own kind, explaining that "there is a sense in which phenomenological facts are perfectly objective: one person can know or say of another what the quality of the other's experience is" (Nagel, 442), he insists that the possibility of ascribing subjective experience to another being is limited by the kinds and degrees of difference between oneself and the subject undergoing the experience. In the case of humans and bats, Nagel concludes that the species divide is simply too great for phenomenological inquiry to span. "If mental processes are indeed physical processes," he explains, "then there is something it is like, intrinsically, to undergo certain physical processes" (Nagel, 445–46), and the physical processes that constitute bat consciousness are simply beyond our reach. Phenomenological inquiry may provide us access to the subjective experiences of other members of our own species, but the minds of bats and other nonhuman animals remain closed to "the mentalistic ideas that we apply unproblematically to ourselves and other human beings" (Nagel, 438).

Subsequent critiques of Nagel's argument that there is "something" it is like to be a bat, but that this "something" is inaccessible to the human investigator, have tended to emphasize the essential inseparability of mind and body and to fault Nagel for failing to pursue more rigorously the connections he only briefly describes between mental and physical processes. In a 2002 article titled, "Is there Anything it is like to be a Bat?" P.M.S. Hacker denies that Nagel's question even makes sense. His critique first and foremost declares invalid the phenomenological concept of the "quale," designating the "qualitative character" or the "qualitative feel" of subjective experience. Citing Ned Block, who defines qualia as "the experiential properties of sensations, feelings, perceptions and ... thoughts and desires as well," Hacker discounts qualia as mere "figments of the philosophical imagination" (Hacker, 157). In particular, he rejects the phenomenological assertion that perceptions and experiences are accompanied by unique "feelings." He acknowledges, with John Searle, that experiences such as tasting beer and hearing Beethoven's Ninth are different in kind, but he recognizes these differences only in terms of their sensory modality and their objects ("what they are experiences of"). "To claim that each of the several experiences has a unique, distinctive *feel*," he insists, "is a different and altogether questionable claim" (Hacker, 163).

Hacker's second objection to Nagel's argument targets the claim that "an organism has conscious mental states if and only if there is something that it is like to *be* that organism – something it is like *for* that organism" (Nagel, 436). Hacker claims that this formula is "misconceived" as a means of characterizing the conscious experience of a creature, human or nonhuman, because it leads directly to the solipsism that Husserl acknowledged as a primary risk of phenomenological inquiry. "It makes no sense to ask what it is like for me to be me," he explains, "for no one else could be me and I could be no one other than myself. 'I am me' says nothing, so 'There is something it is like to be me' likewise says nothing" (Hacker, 170).

Hacker finally charges that Nagel mystifies consciousness when he claims that our only means of knowing what it is like to be a bat is sympathetic imagination. Hacker insists that in fact consciousness is not at all mysterious; it is merely shrouded by philosophy in "empirical ignorance and conceptual mystification" (Hacker, 174). His prescription for the future of philosophical inquiry into mind and consciousness completely reverses Husserl's demand that consciousness be examined in isolation from its embodiment and apart from empirical science. Hacker calls for less attention to "felt experience" and more attention to the physiological underpinnings of mental processes. "Disentangling one of the roots of the conceptual confusions that conjure qualia into being," he insists, "is a first step towards the demystification of consciousness" (Hacker, 174).

Hacker's position on phenomenological inquiry into consciousness finds support from cognitive philosopher Daniel Dennett, who in 1998 declared the current philosophical understanding of nonhuman animal consciousness to be "a mess" (Dennett, 337). Dennett claims that Nagel's "What is it like to be a Bat?" exemplifies how the field of philosophical inquiry has abdicated responsibility for discovering reasons for our *moral attitudes* toward members of other species. Nagel's conclusion that humans have no way to access the subjective experience of "batness," according to Dennett, illustrates a philosophical tendency to obscure those issues that "promise or threaten" to illuminate the grounds that justify our moral regard for other creatures. The flaw in Nagel's argument, Dennett maintains, is his "one-sided use" of empirical evidence. Nagel invokes scientific knowledge about bats' anatomy and echolocation system to support his assertion that bats are conscious creatures and that there is, in fact, something it is like to be a bat. At the same time, however, he denies that further third-person objective investigations can bring us closer to understanding what bat consciousness might be like. Concluding that it is impossible for us to gain a phenomenological understanding of bats' experience, Nagel begs the question by deferring to the non-empirical methodology of human imagination. It is on this point that Dennett most strenuously objects to Nagel's argument. If we accept Nagel's assumption that we can be morally certain other humans are conscious beings, he argues, we should seek similar moral certainty about the conscious status of other species. For Dennett, as for Hacker, the place to seek such certainty is modern neuroscience and its discovery of the neural mechanisms that enable consciousness and determine the qualities of subjective experience.

CONSCIOUSNESS AND PERSONHOOD

In classifying consciousness as a *moral* concern and not only a topic of metaphysical investigation, Dennett raises the question of how we are to evaluate forms of consciousness once we gain an understanding of what those forms of consciousness are "like." His argument implicitly introduces the concept of "personhood" that figures importantly in philosophical discussions of how to regard the moral status of different forms of consciousness. Tom L. Beauchamp notes that metaphysical definitions of personhood strive to identify the requisite psychological properties that characterize what might be labeled "personal" consciousness, usually including such capacities as intentionality, self-awareness, free will, language acquisition, pain reception and emotionality (Beauchamp, 309–310). Moral definitions of personhood go beyond these metaphysical properties of consciousness, adding to them such capacities as moral agency and moral motivation that presumably govern the interactions of persons – that is, of moral subjects – with others of their kind (Beauchamp, 310).

Like Nagel's attempt to apprehend what it is like to be a bat, Norton Nelkin's inquiry into the category of subjective experience labeled personhood, "What is it like to be a Person?" arrives at no conclusive answer. Nelkin sets out to demonstrate that all attempts to analyze or understand personhood as a particular type of consciousness have hitherto met with failure. His discussion of the matter presents the shortcomings of three prominent attempts to define person-consciousness: the first equating personhood with awareness of consciousness (metaconsciousness), the second equating personhood with language use, and the third concluding that personhood is a unique subjective experience and like all other forms of consciousness is ineffable and unanalyzable (what Norton terms "Nagel-consciousness"). Nelkin argues that none of these approaches to defining person-consciousness is sufficient because all of them fail to capture the moral dimension of personhood, and this he sees as the primary reason for the historical exclusion of most if not all nonhuman animals from the realm of personal consciousness and for withholding from them the status of "persons."

While acknowledging that many nonhuman creatures no doubt have a first-order consciousness of their surroundings and of others in their surroundings, and perhaps even of themselves (what he terms "C1 creatures"), Nelkin denies such creatures the status of personhood because, lacking the ability to objectively view their own subjectivity and to conceive of themselves as unified selves with pasts and futures, they are not members of the same moral community as human persons. The critical element in personhood, the quality that distinguishes moral subjects from mere moral patients, he concludes, is something akin to free will or intentionality, the capacity to exercise conscious choice in determining one's own actions. Accordingly, whether or not a creature has Nagel consciousness is irrelevant to the question of personhood, because what really matter are "determinism-indeterminism" considerations that underlie the concept of moral responsibility. Nelkin explains:

> I doubt very much whether a C1 creature could be a moral agent. I do not see how, if one were not aware of what one's thoughts and feelings were, one could be held responsible either for them or for the actions based on them. I suspect it is partly to do with such matters that we are loath to ascribe moral agency to nonhuman animals. I doubt if that is the *only* reason, but I think it is often *a* reason. Moreover, I strongly suspect that, while a C1 creature might think and feel, a C1 creature would lack many of the kinds of thoughts and feelings that make human beings into persons (Nelkin, 239).

While Nelkin apparently accepts "personhood" as a separate and distinct category of consciousness, he stops short of restricting this category to human subjects, acknowledging "our prejudice that anything that is a person has to be *exactly* like us" (Nelkin, 234) and allowing the existence of "nonhuman animals ... to which we ascribe near-personhood, if not personhood" (Nelkin, 234). His position is somewhere on the way to that of Stephen Clark whose examination of "non-personal minds" denies that "personhood" is a natural category of being but sees it rather as a human construction that privileges our own subjectivities and those we recognize as being sufficiently "like us" to merit our regard. Excluding other species from this category has empirically "honorable roots" in a skeptical reluctance to "assume too much about creatures radically like ourselves," Clark observes. But the practice also has "dishonorable roots" in an anthropocentric desire "to think well of ourselves while causing, as it might otherwise seem, enormous distress to other animals" (Clark, 193). Ultimately, Clark calls into question an anthropocentric value system that automatically privileges a construction of "personal" mindfulness over all other kinds:

> Being a person is like being royal: a collective hallucination having little to do with anyone's ordinary attributes. Non-personal mindfulness is the commoner and maybe the better state ... Mindfulness is as present in the non-personal as in ourselves, and what makes us *persons* is a distraction. Platonizing philosophers have often identified their core self with Intellect, and concluded that, insofar as we think *rightly*, we think the same – *and are the same*. Persons (that is, creatures that think of themselves as isolated subjects and 'captains of their soul') may be in error, and those who think and feel 'unself-consciously' and in accord with truth, may not be persons (Clark, 209).

Others, however, resist giving up the long-established doctrine of human uniqueness and superiority that reaches back through Locke and Descartes to Aristotle. Their reluctance is often tied to the desire to preserve humanity's right to stand as the moral and intellectual measure of all things. As David Cockburn suggests, the argument for human uniqueness frequently hinges on our special status as inventers and users of language. In philosophical thought, language is often posited as the dividing line between human and animal minds, not only because language is a crucial medium for representational thought and complex social coordination, but also because verbalization promises access to the subjective content of other minds, offering "proof" of other human subjectivities comparable to our own. Cockburn notes that this understanding of language values it primarily as a medium of informational exchange, an "outward manifestation" of the "inner mirroring" of the outside world, in short, as the manifestation of the mental phenomena we label "beliefs" (Cockburn, 152).

Roger Fellows' 2000 article "Animal Belief," heavily influenced by the work of Donald Davidson, exemplifies the position Cockburn describes as "the philosophically regimented conception of language and belief" (Cockburn, 155). Investigating the question of whether non-language-using animals can have thoughts, Fellows concludes that indeed they cannot. He further asserts that the absence of thought excludes non-language-using animals from the moral category of "persons," because they lack conscious minds directed and animated by beliefs. As Fellows explains his reasoning:

> A creature which possesses the concept of belief can distinguish between true and false belief. This requirement in turn rests upon a capacity to distinguish changes in the world from changes in the mind. Language-learning, which is community-based, provides for the fixation of belief, in so far as language enables a creature's beliefs to be made manifest and hence held up to scrutiny by itself and others of its kind. Here, what is crucial is learning and training, features which are conspicuously lacking in the signaling systems of language-less creatures (Fellows, 597).

DOES LANGUAGE TURN BRUTE ANIMALS INTO PERSONS?

Fellows' conclusion that "non-language-using animals lack minds because they lack beliefs" (Fellows, 598) upholds a long philosophical tradition attributing human uniqueness to language. This popular construction of human uniqueness is summarily captured by Descartes' contention that "the word is the sole sign and the only certain mark of the presence of thought hidden and wrapped up in the body" (letter to Henry Moore, cited in Fellows, 589). Despite its long and venerable lineage and its continued acceptance today in many and diverse fields of study, however, the claim that language is the mark of human uniqueness and ultimately a reason for the moral elevation of human minds over those of "brute animals" does not go unchallenged.

The challenge has emerged with particular force from a relatively new, interdisciplinary field of study known as ape-language research. Following the U.S. Space Program's decision to man NASA's first space flights with chimpanzees trained to operate the controls of space capsules orbiting the earth, young chimpanzees captured in Africa were brought to the U.S. in unprecedented numbers. Because chimpanzees' lifespans are comparable to those of human beings, and because NASA's eventual transition to space exploration manned by humans eliminated the agency's need for chimpanzee trainees, a large and long-lived population of captive chimpanzees in the U.S. made chimpanzee subjects available to researchers who sought to test scientifically the folk hypothesis that the minds of chimpanzees and other species of great apes are sufficiently human-like to render them capable of language.

When the first reports of researchers' success in teaching rudimentary forms of human language to chimpanzees appeared in the late 1960s and early 1970s (Gardner and Gardner, 1969, 1971; Premack, 1972; Rumbaugh et al. 1973), the news generated intense and mixed responses from the general public as well as within the scientific community. Throughout the 1970s and 1980s, ape language research flourished, branching out from initial studies of chimpanzees to other species of great apes: gorillas, orang-utans, and bonobos. Because humans alone among the great apes have the vocal apparatus to produce speech, the apes who served as subjects in this research were taught to sign with their hands or to use invented languages encoded as lexigrams on keyboards. From early on, the field was fraught with controversy, as researchers argued over the linguistic integrity of the various languages apes were purportedly using. A crucial question posed by skeptics was the challenge, "Is it really language?"

Beneath this question lay an even deeper concern about whether the allegedly language-enabled apes were using their so-called language to express genuine intentions or whether they were merely performing elaborate circus tricks, demonstrating the results of sophisticated operant conditioning rather than the deliberate utterances of conscious minds. At issue was the question of whether ape language studies could provide a window into the minds of nonhuman primates as human language was presumed to do for human beings. Offering the tantalizing prospect of giving humans access to the structures and contents of nonhuman minds, ape language studies have functioned in part as a modern-day empirical test of Husserl's assertion that mind, regardless of its embodiment in human or brute form, is governed by the same "essential laws" of consciousness.

Indeed, most researchers who conduct ape language studies justify their endeavors by claiming that ultimately these studies will serve, through comparison, to illuminate human consciousness. The sharpest critics of ape language studies, however, deny both the accessibility of nonhuman minds and the qualitative continuity of these minds with the minds of human beings. They argue that language remains an obvious dividing line between human and nonhuman varieties of consciousness – that without language there can be no thoughts or beliefs or intentions, and that when nonhuman creatures appear to be learning and using human language, it is not really language. Often these claims rest upon the apparent failure of apes to exploit the creative capacity of complex human language syntax to produce novel utterances expressing original thoughts and meanings. Maintaining that the syntax of mature human language is the essence of linguistic competence and the hallmark of human consciousness, some critics have asserted that the "primitive" syntax of apes' linguistic productions proves the disjuncture between human and other minds.

Barbara J. King (1994) has characterized the debates over the findings of ape language studies as clashes between evolutionist and essentialist theorists. While the former maintain that the human capacity for language can be traced to its origins in our shared biological ancestry with the other great apes, the latter support the position, most famously articulated by Noam Chomsky, that language is "qualitatively unique to and originated with humans" (King, 2). King herself obviously favors the evolutionist perspective, but she faults both groups for their shared tendency to work within a narrow framework that requires the application of a human standard of analysis to nonhuman communication. By this she refers, for instance, to the importance ascribed to the syntax of human language as a definitive indicator of a desirable quality of mind, signifying the kind of conscious awareness that is prerequisite for "authentic" language capacity and that entitles man alone to stand as the measure of all things. "The species-specific human form of communication, vocal or otherwise, is said to contain the criteria of choice," King observes. "Apes or monkeys are then evaluated against that standard in order to discover whether they can do what humans do (or what we assume humans do)" (King, 5). Although King faults the essentialist camp for imposing anthropocentric demands upon the nonhuman subjects of ape language studies, she similarly criticizes the evolutionist camp for responding defensively to these demands by redoubling their efforts to prove that the minds of nonhuman apes are in fact sufficiently like ours that their use of *our* language opens a window onto a conscious and intentional mindscape that is recognizably like our own.

In the end, King contends that the field of ape language studies is affected by deeply rooted human fears over the erosion of the human/animal, person/non-person boundary, and she compares it to paleoanthropology, a field that similarly suffers from its own "tacit objective" of "policing and maintaining" the divide between humans and other animals (King, 7). Citing the archaeologist Adam Cartmill's 1990 essay, she notes that in the service of the field's unspoken objective, paleoanthropologists have continually redefined the distinguishing traits of human identity (e.g. upright posture, large brains, technology and language) in terms that apply uniquely to modern *homo sapiens* "whenever it was necessary to protect their axiomatic human uniqueness from the threat of empirical reality" (King, 7).

But if human fears attend the prospect of an eroding divide between ourselves and other species, so too do human hopes that we are not alone in our conscious habitation of the universe. The cultural force of these hopes, as well as their cultural containment as naïve and sentimental longings, can be seen in the historical popularity of children's stories that feature animals who talk among themselves and communicate their intentions unambiguously to humans. Deemed suitable for children, but largely unacceptable for mature adults, the *Dr. Doolittle* fantasy of talking to animals is strictly policed in the realm of empirical science; accordingly, those researchers who have claimed that their ape subjects have learned to use human language as a medium of intentional communication face charges of anthropomorphism, the projection of human subjectivity onto nonhuman animals and the attribution of consciousness to minds that are in fact qualitatively unlike our own.

Researchers who endeavor to learn whether and how much nonhuman apes are capable of learning human language are well aware that their findings are subject to charges of inadequate objectivity and overinterpretation. Herbert Terrace (1983), Jean Umiker-Sebeok and Thomas A. Sebeok (1980), for instance, all leveled charges that the early ape-language projects were flawed by "expectancy effects" – that the researchers' desire for their subjects to demonstrate humanlike language comprehension and production leads them to make unconsciously inaccurate observations and recordings of the apes' linguistic behaviors or to unintentionally modify those behaviors in the direction of the hoped-for results (Hillix and Rumbaugh, 52). And Seidenberg and Petitto's (1979) critical review of research on signing behaviors in apes raised similar concerns, pointing out, in addition, that the data of ape language studies are difficult for skeptics to replicate because apes are such expensive research subjects to train and support. To counteract such charges, ape language researchers have sought to show that their findings are scientifically sound and to demonstrate the empirical unassailability of their research designs and data collection techniques, employing, for instance, double-blind observational methods, high standards for determining when an ape has "learned" a linguistic sign, and video-taped records of the tests they have used to measure the ape's linguistic accomplishments.

At the same time, ape language researchers complain that science applies a double-standard to language studies of human and nonhuman subjects. Sue Savage-Rumbaugh points out that human child-language research *assumes* that the child subject is endowed with intentional consciousness and that the child's utterances are therefore meaningful and qualify as language – or as language-in-the-making, despite the absence of syntax or the imperfect imitation of adult-language forms. Ape-language research, on the other hand, must start from the assumption that the subjects do *not* possess intentional consciousness, and the burden of proof must be taken on by the research study itself. T. J. Taylor (1994) describes the double standard as a "rhetorical dilemma," explaining that different *systems* of measurement are employed in comparative studies of human and nonhuman communication, even though the *units* of measurement are the same (e.g. reference, intentionality, meaning). As Taylor explains: "It must be proven, according to the canons of scientific proof, that any application of one of those terms to a nonhuman animal is justified; but for humans such an application is taken for granted to be justified" (Taylor, 126). Savage-Rumbaugh argues that if we were to apply the research standards of ape-language studies to our studies of human language development, we would have to enact a "reduction" of field even more extreme that Husserl's. That is, we could not even assume the presence of consciousness in human subjects. While acknowledging the problems inherent in attributing intentional consciousness to other species, Savage-Rumbaugh maintains that "the only other comparative option is to treat ourselves as mindless creatures and apply to our own behavior the standards we currently reserve for animals" (Savage-Rumbaugh 1998, 166).

In the scientific deployment of this double-standard, what Savage-Rumbaugh has labeled "schizophrenia," we see the currency of Nagel's assertion that the conscious experience of others of our own species is easily accessible to us through "the mentalistic ideas that we apply unproblematically to ourselves and other human beings" (Nagel, 438). Our ready acceptance of human children's language, however primitive, as a demonstration of intentional consciousness, while withholding this same interpretation from the language acts of nonhumans, functions as a preconception, a construct of the conceptual life-world that Husserl describes as a collective accomplishment. What, then, are we to make of those language researchers whose long and close acquaintanceship with nonhuman apes leads them to insist, often in very unscientific terms and in popular venues, that their research subjects, although they are not human, nevertheless belong to the category of "persons"?

In an essay titled "A Case for the Personhood of Gorillas," Francine Patterson and Wendy Gordon make this appeal to a general audience, using their interpretations of the signed utterances of the gorilla, Koko, as evidence to support their case. Their essay presents Koko as an individual who creates meaningful, novel combinations of signs, who recalls the past and anticipates the future, who expresses love, grief and other emotions, and who even contemplates death. Koko's reported combinations of signs often deviate considerably from the syntactic forms of standard English (e.g, when asked where gorillas go when they die, Koko reportedly signed "COMFORTABLE HOLE BYE" [Patterson and Gordon, 136]), but the authors nevertheless construct Koko on the page as a character who embodies all the traits we commonly ascribe to the category of persons. The same is true of Roger Fouts' best-seller, *Next of Kin*, in which Fouts tells the story of his life with Washoe, one of the very first chimpanzees to be taught to sign with her hands. Fouts presents Washoe as a "chimpanzee girl," a "sister," a "friend," and as a "person." He justifies his willful participation in "the heinous sin of anthropomorphism" by explaining, "My college professors neglected to tell me that apes behave anthropomorphically because they are anthropoids" (Fouts, 58).

Similarly, Lyn Miles, trainer and companion of the language-enabled orangutan, Chantek, makes the case for Chantek's personhood on the basis of his intellectual capacity (measured at the level of a 2–4-year-old human child), his self awareness (demonstrated by recognition of his own reflection in a mirror), his abilities to imitate and deceive, and his capacity to use sign language to communicate with humans. Miles' argument finds some support from primatologists W. A. Hillix and Duane Rumbaugh, who conclude after observing a number of the nonhuman primates who have gained a degree of celebrity status in the field of ape-language research, that Chantek's abilities make him, like other language-trained apes, "appear to be almost human" (Hillix and Rumbaugh, 199). Indeed, the theme of nonhuman personhood runs throughout Hillix and Rumbaugh's comprehensive survey of animal language research. Their closing words affirm the subjective perception, shared by many who study animal language, that their research subjects are in some sense "persons": "Every animal language researcher recognizes the different personalities of his or her subjects. What a great word PERSONality is in this context! Communication with an animal encourages, or forces observers to recognize the personhood of animal subjects" (Hillix and Rumbaugh, 273)

EMPIRICAL SCIENCE AND THE PHENOMENOLOGY OF RELATIONAL MINDS

In an age when Husserl's desire to establish a science of mind uncontaminated by empirical studies seems a quaint and antiquated fantasy, current efforts to understand consciousness have abandoned the ideal of pure phenomenology, traversing empirical fields as far flung as linguistics and neuroscience, evolutionary biology and artificial intelligence. The study of consciousness today is as vast in its scope as Husserl predicted it would become, but it has strayed very far from his prescribed methodologies of introspective reflection and subjective intuition carried out in disembodied isolation from cultural prejudices and the "alien interferences" of empirical science (Husserl 1970, 133).

Daniel Dennett, who seeks models of artificial intelligence in neuroscience, voices the contemporary view that the study of consciousness is and should be an empirical science that takes the brain and central nervous system as its field of inquiry. Proclaiming this the most empirically sound and therefore the most morally responsible path to understanding consciousness however embodied, Dennett dismisses "intuitive" methods of understanding as the products of romantic delusion. He traces the contest between responsible, empirical scientific studies and uncritical intuitions about nonhuman consciousness to the skeptical, empirical stance of Descartes, "an honest scientist who was apparently the first victim of the lunatic fringe of the animal rights movement," and the erroneous "intuition" of Montaigne, "a gullible romantic of breathtaking ignorance" who accepted unquestioningly folktales about animal intelligence (Dennett, 338). Present-day heirs of Montaigne, according to Dennett, include popular writers like Elizabeth Marshall Thomas, whose books about canine social life contaminate keen empirical observations of animal behavior with "romantic declarations" about their significance. Dennett complains that instead of seriously addressing important questions about whether and how dogs are conscious creatures, Thomas appeals to the anthropomorphic preconceptions of readers who "want to believe in the consciousness of dogs" (Dennett, 338).

Himself a designer of artificial intelligence systems, Dennett maintains that intuition is an unreliable way to perceive the inner lives of nonhuman animals, pointing to the reactions of human observers who "intuitively sense" conscious awareness in robots, like Cog, an automaton programmed to mimic human movements and the human gaze. Although Dennett is well aware that the robot is a machine, he reports that he nevertheless finds himself gripped by "an almost overwhelming sense of being in the presence of another conscious observer" when Cog's robotic eyes automatically follow human hand gestures. "When you acknowledge the power of such elegant, lifelike motions to charm you into an illusion," Dennett concludes, you must also acknowledge that you might be similarly "charmed" by your cherished family pets (Dennett, 340).

When ape-language researchers who experience long and intense relationships with nonhuman primates report their intuitive perceptions of personal consciousness in the minds of their research subjects, are they, as Dennett suggests, merely victims of romantic illusion? Are they reverting to childish fantasies of a social reality constructed and maintained by humans in cooperative harmony with other kinds of animals? If they are simply "charmed" into this belief by their nonhuman subjects, we must at least question why the delusion is shared by so many. Hillix and Rumbaugh report that most animal-language researchers, including those who work with dolphins and parrots, "become as attached to their subjects as most people do to their children, in an extreme version of what happens to most owners of dogs and cats" (Hillix and

Rumbaugh, 22). They report that many of these researchers are couples who eventually dedicate all or a significant portion of their lives to the care of their nonhuman research subjects. Often, this dedication extends beyond individual subjects to the entire species, and in many cases the researchers come to advocate some degree of "human rights" for the species. Hillix and Rumbaugh report that this advocacy derives from the researchers' "new awareness that animals share more of our intellect and feelings that we previously believed" (Hillix and Rumbaugh, 23).

The "new awareness" that Hillix and Rumbaugh refer to seems to derive from the state of "living together" that Husserl identifies as a fundamental condition of conscious subjectivity attempting to make sense of the world:

> In whatever way we may be conscious of the world as universal horizon, as coherent universe of existing objects, we, each "I-the-man" and all of us together belong to the world as living with one another in the world; and the world is our world, valid for our consciousness as existing precisely through this "living together" (Husserl, 1970, 108).

The social consciousness that results from living together and collaboratively producing the grand phenomenon of the lifeworld is intersubjectivity, the formation of we-subjects. In his quest to abstract from the lifeworld to its universal subjective essence, Husserl regards intersubjectivity as the key, explaining that "the world of straightforward intersubjective experiences" is a "world nucleus" to which the concrete manifestation of the lifeworld can be distilled (Husserl, 1970, 133). Universal consciousness, in other words, is a collective subjectivity of which all individual manifestations of consciousness partake. While Husserl acknowledges the participation of even "brute animals" in the universal, ultimately functioning subjectivity that animates the world and gives it meaning, he defines the subjective essence of being in human terms, asking, "Is it not in the end our human being, and the life of consciousness belonging to it, with its most profound world problematics, which is the place where all problems of inner living being and external exhibition are to be decided?" (Husserl, 1970, 114).

But what if at least some of the minds with which an individual consciousness lives and co-constructs an interpretation of the world are not the minds of other human beings, but rather of our closest relatives just across the species divide? What if intersubjectivity admits a blending of the subjective experiences of humans and apes? This in fact seems to be the experience reported by many who have worked and lived closely with the research subjects of ape-language studies. Francine Patterson and Wendy Gordon, for instance, describe Koko the gorilla and her companion, Michael , as "persons" who live in a "multi-species family." As two human members of that family,

Patterson and Gordon maintain that the intersubjective experience of interspecies family life has shaped their own subjectivities in significant ways:

> Perhaps our most interesting findings relate to how astonishingly like us gorillas are – or how like them we are. Through what they have taught us about gorillas, Koko and Michael are helping to change the way we view the world. They force us to reexamine the ways we think about other animals. With an emotional and expressive range far greater than previously believed, they have revealed a lively and sure awareness of themselves as individuals (Patterson and Gordon, 141).

Koko's human companions describe her acquisition of language skills as a "bridge" across the species divide, and as a "window" into the feelings and thought processes of gorillas. Because language functions so importantly for humans in forging intersubjective bonds, it is no surprise that the possibility of linguistic communication with apes would encourage the human participants in this research to *feel* as though they have heightened ways of "knowing" or "connecting with" the minds of their nonhuman companions. But Savage-Rumbaugh, reflecting on her years of "living with" Kanzi and other members of her own multi-species family, insists that though language assists her in establishing intersubjective bonds with nonhumans, it is not the only means of doing so. Like Patterson and Gordon, she asserts that her work with chimpanzees and bonobos has persuaded her that the minds of humans and apes are more similar than she previously believed, but this realization is not entirely due to the apes' abilities to articulate what is on their minds in terms understandable to humans. "It is possible, even without words, to perceive much of what they are thinking," she writes, and goes on to explain that the essential elements of intersubjective knowing are intuition in combination with close observation and receptivity to the communicative intent of the apes. She describes the resultant intersubjective experience – "our common knowledge of the surrounding events" as "a sort of joint awareness that leads to joint perception and joint knowing" (Savage-Rumbaugh 1994, 260).

Although ape-language researchers pursue the science of consciousness well into the terrain of empirical science (primarily behavioral science, with increasing appeals to evolutionary biology and neuroscience), their protests against the constraints of empiricism belie their tacit or explicit assent to Husserl's contention that consciousness exceeds the grasp of science and that the study of consciousness requires additional ways of knowing. While modern neuroscience has made dramatic strides toward disclosing connections between neural structures and mental experiences, such advances in our understanding of consciousness apply primarily to the individual organism. Science has provided considerably less enlightenment when it comes to social cognition and the ways that minds work together to produce shared subjective experiences.

One promising development is research into "theory of mind," an organism's recognition (or attribution) not only of itself as a locus of intentionality and subjective experience but also its recognition (or attribution) of others as similar sites of consciousness. Theory of mind, generally considered a prerequisite for the development of communicative abilities, including language, supplements self-awareness with other-awareness, thereby making possible communication and other forms of social cooperation. Research with human children suggests that theory of mind is a developmental accomplishment, and behavioral research with apes suggests that it is a mental capacity we share with our closest genetic relatives. While neurological studies of humans suffering from communication disorders such as autism have identified the frontal neo-cortex as the physical location of theory-of-mind activity, this understanding of the physiological mechanisms of consciousness does little to address the pheonomenological question of what it is *like* to be a creature who does or does not theorize other minds. Accordingly, though science may tell us that apes, like humans, exhibit this social and communicative capacity, and though apes who have learned human language skills may tell us something about what it is like to be an ape, ape-language researchers repeatedly tell us that empirical knowledge and linguistic reports still fall short of telling us who these creatures are, and what it is like to be one of them. For this kind of understanding, ape-language research suggests, intersubjective knowledge alone reveals the consciousness that unites us with other forms of life.

University of Iowa

REFERENCES

Anna-Teresa Tymieniecka, "The Pragmatic Test of the Ontopoeisis of Life," in *Analecta Husserliana* LXXXIV (2005) pp. xxxvii–xl.

Barbara J. King, "Evolutionism, Essentialism, and an Evolutionary Perspective on Language: Moving Beyond a Human Standard," *Language and Communication* 14:1 (1994): 1–13.

Beatrix Gardner and R.A. Gardner, "Teaching Sign Language to a Chimpanzee," *Science* 165 (1969): 664–672.

Beatrix Gardner and R.A. Gardner, "Two-way Communication with an Infant Chimpanzee," in A. Schrier and F. Stollnitz (eds.), *Behavior of Nonhuman Primates* (New York: Academic Press, 1971), pp. 117–183.

Daniel Dennett, "Animal Consciousness: What Matters and Why," in *Brain-Children: Essays on Designing Minds* (MIT Press, 1998), pp. 337–352.

David Cockburn, "Language, Belief and Human Being," in Anthony O'Hear (ed.), *Minds and Persons*. Royal Institute of Philosophy Supplement 53. (Cambridge University Press, 2003), pp. 91–105.

David Premack, "Language in Chimpanzee?", *Science* 172 (1971): 808–822.
David Premack, "Teaching Language to an Ape," *Scientific American* 227 (1972): 92–99.
David Premack and G. Woodruff, "Does the Chimpanzee Have a Theory of Mind?", *Behavioral Brain Science* 4 (1978): 515–526.
Duane Rumbaugh, Timothy V. Gill and Ernst von Glasersfeld, "Reading and Sentence Completion by a Chimpanzee (*Pan*)," *Science* 182 (1973): 731–733.
Edmund Husserl, "The Way into Phenomenological Transcendental Philosophy by Inquiring Back from the Pregiven Life-world," from *The Crisis of European Sciences and Transcendental Phenomenology: An Introduction to Phenomenological Philosophy*, trans. David Carr (Evanston: Northwestern University Press, 1970), pp. 103–137.
Edmund Husserl, "Pure Phenomenology, its Method, and its Field of Investigation," in Peter McCormick and Frederick Elliston (eds.), trans. Robert Welsh Jordan, *Shorter Works*, 1981. (University of Notre Dame Press), pp. 10–17. Reprinted in *The Phenomenology Reader*, Dermot Moran and Timothy Mooney (eds.), (London: Routledge, 2002), pp. 124–133.
Francine Patterson and Wendy Gordon, "A Case for the Personhood of Gorillas," in Linda Hogan, Deena Metzger and Brenda Peterson (eds.), *Intimate Nature: The Bond Between Women and Animals*. (New York: Ballantine, 1998), pp. 132–141.
P.M.S. Hacker, "Is there Anything it is like to be a Bat?", *Philosophy* 77(300) (2002): 157–174.
Herbert S. Terrace, "Apes Who 'Talk': Language or Projection of Language by their Teachers?", in J. de Luce and H. T. Wilder (eds.), *Language in Primates: Perspectives and Implications* (New York: Springer-Verlag, 1983), pp. 19–42.
W.A. Hillix and Duane Rumbaugh, *Animal Bodies, Human Minds: Ape, Dolphin and Parrot Language Skills* (New York: Plenum, 2004).
D. Jean Umiker-Sebeok and Thomas A. Sebeok, "Introduction: Questioning Apes," in Thomas A. Sebeok and D. Jean Umiker-Sebeok (eds.), *Speaking of Apes: A Critical Anthology of Two-Way Communication with Man* (New York: Plenum Press, 1980), pp. 1–59.
Jesse M. Bering, "A Critical Review of the 'Enculturation Hypothesis': The Effects of Human Rearing on Great Ape Social Cognition," *Animal Cognition* 7 (2004): 201–212.
John Harris, "The Concept of the Person and the Value of Life," *Kennedy Institute of Ethics Journal* 9(4) (1999): 293–308.
Mark Seidenberg and Laura Petitto, "Signing Behavior in Apes: A Critical Review," *Cognition* 7 (1979): 177–215.
Michael Tomasello and H. Rakoczy, "What Makes Human Cognition Unique? From Individual to Shared to Collective Intentionality," *Mind and Language* 18: 2 (2003): 121–147.
Michael Tomasello and Josep Call, "The Role of Humans in the Cognitive Development of Apes Revisited," *Animal Cognition* 7 (2004): 213–215.
Norton Nelkin, "What is it like to be a Person?", *Mind and Language* 2:3 (1987): 220–241.
Roger Fellows, "Animal Belief," *Philosophy* 75 (2000): 587–598.
Roger Fouts, with S. Mills, *Next of Kin* (New York: William Morrow, 1997).
Rose A. Sevcik and Sue Savage-Rumbaugh, "Language Comprehension and Use by Great Apes," *Language and Communication* 14(1) (1994): 37–58.
E.S. Savage-Rumbaugh, *Ape Language: From Conditioned Response to Symbol* (New York: Columbia University Press, 1986).
E.S. Savage-Rumbaugh, "Nonhuman Species: Implications for the Innateness Debate," *Developmental Psychobiology* 23 (1990): 599–620.
E.S. Savage-Rumbaugh, with Roger Lewin, *Kanzi: The Ape at the Brink of the Human Mind* (New York: Wiley, 1994).

E.S. Savage-Rumbaugh, "Scientific Schizophrenia With Regard to the Language Act," in Jonas Langer and Melanie Killen (eds.), *Piaget, Evolution and Development* (Mahwah, NJ: Lawrence Erlbaum, 1998), pp. 145–169.

Stephen R.L. Clark, "Non-Personal Minds." in Anthony O'Hear (ed.), *Minds and Persons*, Royal Institute of Philosophy, Supplement 53. (Cambridge University Press, 2003), pp. 185–209.

Stuart Shanker, "Ape Language in a New Light," *Language and Communication* 14(1) (1994): 59–85.

T.J. Taylor, "The Anthropomorphic and the Sceptical," *Language and Communication* 14(1) (1994): 115–127.

Thomas Nagel, "What is it like to be a Bat?", *Philosophical Review* 83 (1974): 435–450.

Tom L. Beauchamp, "The Failure of Theories of Personhood," *Kennedy Institute of Ethics Journal* 9(4) (1999): 309–324.

Uta Frith and Francesca Happe, "Theory of Mind and Self-Consciousness: What is it like to be Autistic?", *Mind and Language* 14(1) (1999): 1–22.

MARTIN HOLT

VERTIGO AND THE BEETLE OUT OF THE BOX. ON THE REPRESENTATION OF INNER MENTAL STATES

For a long time I have wondered what happens when inner mental states are represented in works of art, particularly in films. One of my favourite film directors, Hitchcock provides some of the most vivid examples of this in the film *Vertigo*, examples I shall examine at the end of the paper. But first I want to provide an account of these states and of how they can become public, and be represented in language or in art, and I also want to examine the scope and limitations of this representation. I will concentrate on the representation of experiential states – states arguably with a "raw feel" or subjective properties. First, let me explain what I mean by "inner mental states": any mental states qualify as inner mental states if they have no components normally publicly observable by the subjects of them or by anyone else, by which they can be identified.[1] In contrast, states that are identifiable from their normally publicly observable components, I shall call public states, and I shall assume that some public mental states, if not all, have normally unobservable components.[2] So on this scheme, after-image states qualify as inner states, and perceptual states qualify as public. In perception the publicly observable components are the object of perception, and the discrimination behaviour, which can both be used to identify the state. But perception has brain state components, and these qualify as inner states, on my definition; perception might also include qualia, or internal functional states, and these are also inner states.

I should add that I am not only an externalist when it comes to perception, but also when it comes to meaning and language, as well, for reasons I will not rehearse here, but many of which can be found in the works of Wittgenstein and his followers. In this paper I want to examine the challenge presented to those of us that believe meaning is public in some strong sense, but experience has private components – "inner" components in the sense just outlined. However I do not intend to meet Wittgenstein on his own ground, and argue with him in his own terms – as you will see, in arguing for my own account, where I do not agree with Wittgenstein, I more or less successfully, cut the Gordian knot.

A-T. Tymieniecka (ed.), Analecta Husserliana XCIII, 53–73.

Wittgenstein has a famous allegory, the "beetle in the box", that illustrates how problematic it is for inner mental states to become public, and also illustrates how hard it is for these states to enter into the language.

Suppose everyone had a box with something in it: we call it a "beetle." No one can look into anyone else's box, and everyone says he knows what a beetle is only by looking at his beetle. – Here it would be quite possible for everyone to have something quite different in his box. One might even imagine such a thing constantly changing. – But suppose the word "beetle" had a use in these people's language? – If so it would not be used as the name of a thing. The thing in the box has no place in the language game at all; not even as a something for the box might even be empty. – No one can 'divide through' by the thing in the box; it cancels out, whatever it is (PI 1 293).[3]

Although Wittgenstein is clearly talking about what we now call qualia, just before he makes them vanish, it is important to note that this allegory applies to many inner states. Brain states, for example, are not private in principle in the same way that qualia are, but they certainly are in practice. If sensations and the like can be reduced to brain states or internal functional states or instantiations, this allegory still illustrates a problem about how we can gain concepts for these states, and come to know and talk about them.

Whereas Wittgenstein is interested in showing that such inner states cannot get into the language, my approach is to show that we can become aware of these states – or at least discriminate them – and then refer to them in language, even if only obliquely. What I would like to stress though, is that while it will become obvious in what follows that I do not agree with some of the conclusions of the private language argument, of which this allegory is a fragment, I am persuaded by considerations Wittgenstein and his followers raise, that qualia are extremely problematic, and not the certain self-evident items of consciousness that the post Cartesian consensus would have us believe, and I will refer to Wittgenstein, and use some thought experiments of my own, to illustrate some of the difficulties.

My revision of the allegory, which I shall call "the beetle out of the box," illustrates a possible method by which inner mental states can become public, a method that I shall flesh out in the rest of the paper. Imagine each person has a box with a beetle only he can look at, as in Wittgenstein's allegory, but he can also point at similar beetles others can look at, and he can make comparisons between the inner and outer beetles. If a person can find a public beetle like his private one, then if he gets others to look at it, they will know what his private beetle is like. My allegory suggests that we can compare one private experience to another public one, and sometimes at least, match the inner and outer states, and via this discrimination of something inner with something public, we can, though in an oblique way, share the experience with others.

But the revised allegory is also used to illustrate the difference between 'seeing' an orange patch – an after image say – and seeing an orange. 'Seeing' an orange patch is like having a beetle in the box, which you make sense of, and compare to seeing the public beetle – let us add that the box makes it more difficult to scrutinize the inner beetle, than the outer. So once the matching of the two has been done, the outer beetle can be used to talk about, and indeed think about, the inner beetle. The inner beetle enters the language through the outer beetle, and the concepts and predicates that apply to the outer are applied to the inner. So I agree with Wittgenstein that there is a problem of referring to inner states, and of how these states can get into the language; by extension, I also think there is a problem about how inner states can be represented at all. But, contra Wittgenstein, I do think these difficulties can be overcome, though the route via which I think any of these things can happen is oblique, problematic, and limited, as we shall see.

In the new allegory a person can describe the qualities and structure of his private beetle. He can do this by comparing it directly to the public beetles all can see, as we have already noted, or more abstractly, he can use the language he would apply to public beetles: "it has six legs, wings, an oval shape, the body is divided into three" and so on. In a parallel way, with inner mental states, one can say: "it looks like an orange," or more directly "it is a round orange colour patch" to which one might also add "it flickers and floats – it must be internal, because it moves with my eye." A sceptic's common reply is that nothing is round, orange, moving, or flickering, and so on.

Even if one accepts the sceptic's point, which I do, this structural and qualitative language used to describe inner states seems to be informative, and I will give a fuller account of how it may work later on. Meanwhile we can note that my allegory rather hides a complexity – as, indeed, does the original – namely that the beetle in the box, whatever it is, has to be a very different kind of thing from any real beetle, in order to really parallel the difference between inner and public states. And then the problem arises for my allegory, how can one thing be used to represent and talk about a very different kind of thing?

While Wittgenstein's allegory is used to illustrate the unimportance to language and meaning of inner mental states, my revision is used to illustrate and defend a particular way of referring to and talking about inner states, and I want to approach this issue from a different perspective to Wittgenstein. I am interested initially in giving a description of human beings as discrimination machines, rather than language users.

My general justification for taking this approach is that the ability to make outer and inner discriminations is a necessary component of all public sensory states in complex animals, and is more basic than the ability to use language. So, the ability to discriminate does not require concepts, or the use of language, and it does not require consciousness either, which arguably language does. Computers with neural nets can now discriminate between faces, but this does not mean they understand the concept "face," or "nose" and so on; insects can discriminate all kinds of things, from prey, to mates, to food; I doubt they are conscious nor do they employ concepts. Higher mammals can make extremely complex and fine discriminations. A lion will attack the neck of what it identifies as a prey animal; a monkey will eat some kinds of fruit and not others, and only when they are ripe; a dog will recognise other individual animals by scent alone, and so on.

How does this sorting work? One way to understand it is via an abstract description of the minimal conditions animals and other things that make discriminations have to fulfil. I will call anything that fulfils these conditions a discrimination machine. I think there are two types of discrimination machines, simple and complex. A simple machine sorts things without any intermediate states; it goes from inputs, created by objects, properties, or events, straight to outputs: sortings of objects, properties, or events. Take, for example, a sieve sorting grains of sand into different sizes. Note that such a simple machine cannot learn to sort, its ability to sort is built in. A complex machine can learn how to sort, which it does according to intermediate states; it goes from inputs created by objects, properties, or events, to intermediate states created by inputs, together with background conditions in the machine, to outputs created by the intermediate states together with background conditions in the machine. The outputs are sortings of objects, properties, or events. Note that a complex machine may also have the ability to sort some things built in, for example, by evolution or a programmer.

In order to learn, a complex discrimination machine has to be able to monitor intermediate states, monitoring at least all the intermediate states to be used in making discriminations. This monitoring may be provided by any means of telling when two states are or are not of the same sort or similar, and when a state repeats. The machine requires some kind of memory to fulfil this last condition, which also allows the machine to judge if any past and present states are of the same sort or are similar. The matching or otherwise of intermediate states are directly used to recognise and sort objects, events and properties. But in a sense we have an inner kind of discrimination going on to produce this outer sorting. The key question, which I shall return to, is how can we become aware of this sorting of inner states?

These conditions for a discrimination machine are extremely abstract and far from sufficient, if our goal were to try to understand in detail how such machines work, or indeed, to build such machines. They say nothing about how the inputs are produced by objects, properties or events, or about how the intermediate states are related to the inputs, or how the inputs together with the background conditions in the machine, produce the intermediate states, or how intermediate states are memorized, or compared as similar or not, or how these states produce the discriminations. State monitoring need not be anything like consciousness, and could be software based, as in computers, or entirely mechanical, as in drinks machines or beetles. An object, property or event could, together with the machine background conditions, produce an intermediate state entirely unlike itself in all properties, which could then be used to pick it out. Two entirely different kinds of thing, together with machine background conditions, could create the same intermediate state in the same machine, and so be sorted together, or two different intermediate states may even be sorted together because of a bias in the machine.

Thus while sorting may match objective categories, it may not. But even when it does not, on my scheme there is a objective test for sorting: the same machine must be able to make the same discriminations in the same conditions of the same stuff on different occasions, and one example of a type of machine has to be capable of sorting the same stuff, under the same conditions, in the same way as another of the type. (I will allow there may be only one example of the type.) If a machine cannot do either of these things, we may still call it a categorizer, but it is not a discriminator. Note though, that what is sorted together may in fact be heterogeneous, and the sorting may cut across real categories.

My claim is that anything that can learn to discriminate fulfils these minimal conditions, and while this claim is hard to prove, I have just argued for its *a priori* plausibility, and it is open to confirmation and counterexample. What happens, for example, when a dog can recognise its master from smell? Most if not all modern psychological theories, even if they are behaviourist, will agree that what happens involves the dog getting an input from his master's scent, and then something happening in its brain; then the behaviourist will argue that we can cut out this middle man, but not that it is not there. The dog then reacts in a certain kind of way, a reaction that includes behaviour, and in some theories, the dog is conscious, and has a smell sensation, and perhaps expectations to the effect that it will see its master soon.

The sensation, or brain state, are prime candidates for the intermediate state used in it making its fine-grained smell discriminations. But how can either of these states be used in making smell discriminations if there is no state monitoring, to tell when the intermediate state matches other intermediate states, and when it repeats? I cannot prove state monitoring, but I can challenge anyone to produce a better or even a different account that enables learned discriminations and fits the basic known facts about animal biology. In order to recognise its master, does the dog not at a minimum have to have a memory of an intermediate state produced by the smell of its master, and match that state, via memory to the present intermediate state? If not, how does it do it?

Human beings provide a problem for this picture, since we use concepts to make some discriminations, and, arguably, these concepts tincture through all our perceptions, rather as Thomas Kuhn, amongst others, argues that in science all observation language is theory laden.[4] I remain neutral on this issue, but have allowed "background conditions" in a discrimination machine to interact with the intermediate states to produce an intermediate state to serve as the basis of the discrimination; I see no reason why these background conditions cannot include concepts. I think it is even possible that the intermediate states themselves may be concept laden. I still want to maintain that the automatic processes of state monitoring, if they exist in all the higher animals, must exist in us, and that state monitoring is still essential to us in making perceptual discriminations, even if they are concept laden.

In all higher animals, except perhaps squid and octopuses, intermediate states are in fact inner states – that is states which cannot normally be observed by the subject of them or a third party – so in what follows I shall simply refer to "inner states" when talking about these intermediate states.

What are the problems for a human discrimination machine sorting inner states from the point of view of materialism, functionalism, and dualism? If materialism is correct, then when we make discriminations between objects, properties and events, we do this in part on the basis of brain states, and two questions arise. How can we become aware of our brain states in any sense? And how can we tell one brain state from another? While I have noted that a discrimination machine does not require that the inner state monitoring is done consciously, or even that it relays its information to consciousness – and indeed there are many examples of where this does not happen in machines, computers, and in other animals – if we are to allow that humans are conscious of inner states at all, then we have to defend one of these two possibilities.[5]

If we defend the rather natural idea that the state monitoring is done consciously for experiential states, the problem is that it seems quite clear that one may have no idea of what brain states one is in, even when, on the face of it, one is consciously sorting inner states; so when in I am in pain, and distinguishing it as "sharp," or "dull," I may have no idea that my somatosensory cortex, limbic system, and frontal cortex are all being stimulated, and when I "see" an afterimage, and "see" it as a certain colour, I may have no idea even that anything is happening in my eye, leave alone in my occipital lobe. How then can I be aware of these inner states or discriminate one from another?

I will return to this problem later, but there is one answer we can look at right now, the functionalist answer. We can know we are in a functional state either by knowing directly that the causal role of the state is instantiated, which in our case entails knowing we are in a brain state with all the problems we noted above, or we can know we are in the state by its typical causes and effects. If the causes and effects fit the behaviourist stimulus/response model, then we do have an external and publicly observable route to our inner mental state; but often the functional roles that are candidates for mental states have at least some of their causes and effects that are internal, as mental states are often at least in part caused by mental states, and cause other mental states. And unlike the behaviourist, the functionalist can make sense of the brain in the vat case: the inputs and outputs are atypical, but the brain states typically instantiating normal external functional states remain, for example the states for smell, or vision. Hence the brain in the vat's internal experience remains the same. But this rather suggests that to the subject of these states, the important thing to the inner experiential character of these states is the state that instantiates the functional role, whether or not it is playing the functional role – in the case of human beings, this means the brain state.[6]

Anyway, using the typical external causes and effects of brain states – by which I mean stimuli and behaviour – to talk about inner states, and indeed to become aware of them, provides one kind of answer for talking about and discriminating between inner states, even for the materialist – though for the materialist this is a rather indirect sort of knowledge. Take, as an example, the anxious man who is initially oblivious to his anxiety. He is told he is anxious and eventually realizes he must be, because he is on a plane, and planes typically cause many people anxiety, and because his breathing is shallow and rapid, and he keeps checking his passport – typical anxiety behaviour, given the context. He might even notice that he is saying to others in a snappy way, 'no, I'm not anxious', something he knows people typically do when they are anxious. In other words, he eventually recognizes the typical causes and effects of his anxiety, and so its functional role, though, as we noted, he may even initially deny the anxiety. For the functionalist, in a way he has then understood his mental state, even if he does not know how it is instantiated in his brain, but even for the materialist, he has found out something important, perhaps even identifying, about his mental and thus brain state, namely its causes and effects.[7] One should note also, that this route of getting to inner states via their external function, can also be exploited to talk about qualia, if only very indirectly.

Which brings us to the question, how can qualia be used in making discriminations and how do they get into the language? According to the Cartesian view, they are transparent to consciousness and incorrigible. I shall characterize this view as follows: incorrigible states are such that if the subject of them believes something of them, it is always true; transparent states are such that if something is true of them, then the subject of them believes it. So if colour sensations form the basis of our colour discriminations, then according to this view of them we cannot fail to know when we have colour sensations, or fail in knowing if they are "red" or "green" sensations, and so on, and no one can correct us if we say, "I see a green after image."

I think qualia can be inner states used as a basis for discriminations; I even think this is what happens with colour sensations, but I disagree with the Cartesian view of them. I think these butterflies of the mind, if they exist at all, are exceptionally hard to pin down, as Wittgenstein and his followers so persuasively argue. I have already rehearsed problems in locating and describing them, problems we can distil into two basic questions: where are these qualia? What are their properties? The beetle out of the box allegory is in part devoted to trying to pin qualia down a little bit more, as is what follows, but at least I want to show how we can learn about inner states, talk about them, and share them with others even if these states are just brain states or internal functional states.

There are a number of different proposals to make sense of the descriptive language of inner mental states that we seem to have and use, apart from Wittgenstein's sceptical public meaning account of them, which we have so far left unchallenged. Ryle, in a famous passage from *The Concept of Mind*, says that in the case of "inner" visual imagery, it is not that we in fact have some inner state that resembles the thing in reality that we imagine, but that we resemble someone seeing in reality what we imagine.[8] In contrast, J.J.C. Smart says when someone says he 'sees' a yellow/red after image, he is in fact saying "what is going on in me is like what is going on in someone who is looking at an orange"; Smart elaborates that in fact we are very vague about what we are talking about, and he dubs these descriptions "topic neutral," and makes an analogy with existentially quantified sentences where the description of the quantified thing is very vague, as in "someone is in the room," inferred from the sound of footsteps in the room below.[9] Smart's view seems to me to be an advance on Ryle's, in that he allows that something internal is the point of resemblance to the present inner experience, but both agree that to understand what is happening with pure visual imagery, you have to make use of a perceptual case, in order to understand the pure case.

On Smart's topic neutral account the beetle out of the box allegory works, since it allows that you can realize when inner states are similar to the inner component of a public perceptual state, however, you know little about the inner part of the state. Topic neutral description is not only topic neutral, but vague. This account of mental state description seems rather thin. The inner 'orange patch' shines, shimmers, is of a particular shade, size, the inner 'sound' has a rhythm, a repetitive white noise quality, is crisp or dull. There is, on the face of it, a detail and precision in these descriptions that is ignored by Smart.

Let us assume for a moment what Smart argues for, the identity of mind and brain. Now perhaps a subject has what he describes as a 'blue, circular' after image on one occasion and a 'red, circular' one on another. Smart as an identity theorist, could maintain that the two states are in fact two different brain states with nothing in common, and in particular, nothing in common apt for picking out circularity. But how plausible is this?

Consider the perceptual case used as the comparison with the after image state. Suppose that the brain state component of the perception changes utterly, if the colour of the object viewed changes or its shape changes. If for each object picked out, we need an entirely different brain state to pick it out, then not only would this require a great number of entirely different brain states, it also would not provide anything in the brain states that allowed you to pick out the similarities we in fact make between things of the same shape or colour and so on.

Aspects of brain states with the powers to pick out shapes and colours, aspects which could be repeated across different brain states, would allow these similarities to be made, and would allow discriminations made on the basis of brain states to work on compositional principles; a perceptual brain state or its after-image counterpart, would be composed of parts that fix and determine what discriminations they are apt for.

Now Smart is not committed to this compositional view, as an identity theorist, but for the reasons given, and others I have no time to go into here, I think it is more plausible than the first view I discussed. If it is right, then one can make the case for a very tight connection between at least some of the properties obliquely ascribed to after image states in accurate topic neutral descriptions and those properties of brain states apt for picking these properties out in perception. So if the orange patch of an after image appears circular, then if we assume compositionality, we have good reason to believe something is happening in the brain which happens when we see something circular; if it appears orange, pitted, and so on, then we have good reason to believe the same thing is happening in the brain, as when we perceive things with those qualities.

Smart's point is that you do not have to assume that there must be something in the head that has these properties, and that the inner state of seeing an after image could be a brain process – a possibility allowed by the topic neutral first person mental state description. I want to add, that given all the detail in these reports, you have indirect evidence of some precision about the state of your brain, for all this detail is actually identifying, if you are a materialist – different people accurately describing themselves using the same mental state descriptions, will not only be in the same mental states, but in the same brain states too. If compositionality is right, the connection between brain states and accurate topic neutral sentences can be very tight, and at an abstract, admittedly functional level, we can know a lot about the brain state from these descriptions. We not only know what a state is apt for picking out, but also we can make comparisons between different brain states according to shared elements, and we have a basis for making some kind of ordering of brain states in terms of similarity.

So while I agree with the topic neutral account, that inner state descriptions do not describe real structure in the brain or mind, I would argue that they are not vague as topic neutral statements are meant to be. What this descriptive language of inner mental states does is describes in detail the properties that a real object or scene would have to have to cause a similar effect in a subject via normal perception – it describes the private beetle obliquely, by saying what the public beetle would have to be like to cause a similar experience. But in doing this, I think it also identifies the inner state, whether or not this is a brain state; it does this by reference to a public state, which has this inner state as a necessary component.

There is something rather familiar about this suggestion. We are almost back to the strange philosophical language of being appeared to, or seeing an appearance, with all those curious phrases such as "I am appeared to redly" or less oddly, "I see what appears to be a red patch." But there is a clear implication in this language that something of a particular sort is happening to consciousness in both cases; in the one, consciousness is modified, and the modification is being described, in the other the subject becomes aware of some inner item of consciousness, which is being directly described. My slightly different proposal, building on Smart's account, has the merit that it does not have to get directly tangled up with consciousness. I may be talking about my brain state, and with some oblique precision.

But if we want to say a little more about the language descriptive of these mental states, in order to do justice to it, and to the mental states themselves, and certainly if we want to try out qualia seriously, as the candidate for these inner experiential states, then we have to get tangled up with consciousness a bit.

One way to think of 'seeing' an after image, is that you are having a visual experience, which itself is a detailed description of a possible arrangement of objects. As a description, it does not have to have the form or qualities that it describes; your visual red colour sensations are not themselves red, on this view, anymore than the word "red" is red, and the round patch you apparently see, is not round, but a description of roundness, and so on. When we have colour sensations, these sensations are not colours, and so we cannot tell from having red colour sensations, what red is. Rather these sensations are the means we use to see an object; we do not see the sensations; we see through them. An eye cannot see itself. The eye enables us to see the object, and to discover properties of it we could not otherwise know. In the same way colour sensations tell us about real qualities of objects – the light waves they reflect or transmit.

We are still left wondering how these mental colour descriptions differ from verbal descriptions, and one interesting proposal, inspired by psychologists interested in advancing mental representation theory, is to take Nelson Goodman's distinction between notational symbols and non-notational symbols – an example of the one would be natural language, an example of the other would be painting – and apply some of his characterisation of the latter to these mental descriptions.[10] So mental descriptions function, I think, in some respects, and at a very abstract level, as paintings do. They are *semantically dense*: which means that the referents of them are distinguished by fine differences in certain respects; they are *relatively replete*, which means comparatively many aspects of these descriptions are significant, and one needs to pay attention to an indefinitely large number of aspects of them; they involve *multiple and complex reference*, which means the descriptions perform a number of integrated and interacting referential functions and that there is no readily accessible and unambiguous content.

So our candidates for inner states involved in perception and after image type states, are complicated mental descriptions of the structure and other qualities of objects and events. At this point we can add that the descriptions may be provided by qualia, although they may also be provided by other inner states. When we see a shimmering surface we have an inner state that gives us a very complicated description of the patterns of light on the surface, and just as a picture is worth a thousand words, this description of the surface is worth countless verbal descriptions.

We have the mental state of "seeing" an orange patch, for example, which is to be thought of as a description, possibly with nothing in common with an actual orange, and we have another description, a verbal one that could apply equally to the orange or to the mental state. For the one it *is* a true description, for the other, it is at least consistent *as* a description. Not only do we have two consistent descriptions, but also the verbal description gives us some of the content of the non-verbal one. Compare this to having a painting of a real event, and a description that can apply equally well to the painting or to the event; and also compare this to having a painting of a fictional event and a description of the fictional event. In both the real and fictional picture the description, if it is used as a description of the painting, would have be oblique, in order to parallel what happens in mental descriptions; it would have to be in terms of the content of the painting, that is, the real or fictional event depicted, and in terms of what one might call the depicted form of the painting, such as the depicted perspective, and not in terms of the paint or drawing, style or canvass. This is, by the way, quite a natural way of describing paintings. It is quite natural to talk about Washington crossing the Delaware, rather than talking about the brushstrokes or paints used in the painting.

Instead of a verbal description, someone could provide a model for this mental description; this model would be intended to show the content and structure of the real scene or objects that could be components of a perceptual experience with the same or similar inner components, and not to describe the inner experiences directly – after all, we only see through the inner states, on this view, we do not see them directly.[11] (I think we can extend the scope of this definition a bit, but more of that later.) It may also be possible for someone to externally recreate an inner experience, via a perception of a real scene or objects that contain an inner component which matches the inner experience. I think that both modelling, where this is understood as falling short of recreation, and recreation itself, are two common ways that inner mental states are represented in visual works of art; oblique verbal description of these states is common in works of literature.

Before looking at the film *Vertigo*, to see how this theory applies, there is one final argument from Wittgenstein against becoming aware of inner states that I would like to consider, "The Sensation S Game." My Beetle out of the Box allegory depends on inner states being systematically correlated with outer states, and on us being able to notice this correlation; Wittgenstein's thought experiment on the face of it is against just this idea.

> Let us imagine the following case. I want to keep a diary about the recurrence of a certain sensation. To this end I associate it with the sign "S" and write this sign in a calendar for every day on which I have the sensation – I will remark first of all that a definition of the sign cannot be formulated. – But still I can give myself a kind of ostensive definition … I speak or write the sign down, and at the same time I concentrate my attention on the sensation – and so as it were concentrate on it inwardly … (PI 1 258).[12]

Let me just point out in passing, that making a mark in the sensation S game is like tying a knot in a handkerchief to remember something. It is not informative as to the content of the experience to be recalled, as a note, or picture might be, but it does say that there is something to remember; indeed it is a public sign of this! One might say, such a sign serves as a public prod to memory, and if memory is reliable, the prod may aid the recall of the experience.

Anyway, after dismissing a couple of defences of this sort of definition, Wittgenstein finally considers the idea that we can get a purchase on private sensations when they are correlated with public states, in this case a manometer reading blood pressure.

Let us now imagine a use for the entry of the sign "S" in my diary. I discover that whenever I have a particular sensation a manometer shows that my blood pressure rises. So I shall be able to say that my blood pressure is rising without using any apparatus. This is a useful result. And now it seems quite indifferent whether I have recognized the sensation *right* or not. Let us suppose I regularly identify it as wrong, it does not matter in the least. And that alone shows that the hypothesis that I make a mistake is mere show. (We as it were turned a knob of the machine, but it was a mere ornament, not connected with the mechanism at all.) … (PII 270).[13]

If Wittgenstein's analysis of this thought experiment is right, then it is hard to see how inner states of any sort can get a purchase on the language, or indeed, our consciousness. I must add, however, that though some see this experiment as the heart of his private language argument, I do not; and although I am convinced by much of what Wittgenstein says about the public nature of language and meaning, I find what he has to say here rather unconvincing.

Before I rehearse my criticisms, let me introduce a couple of real life case where such a correlation appears to exist. According to the Mayo Clinic website, in temporal lobe epilepsy the seizures "…often are preceded by a warning sensation – called an aura – that can cause a wide range of thoughts and emotions, including déjà vu, anxiety, panic, joy or overwhelming, strange sensations that are impossible to describe."[14] This last almost seems like a shallow caricature of what Wittgenstein was criticising. After all if a sensation is impossible to describe, how can we name it, or pick it out again, so the name has a use? If it really is like nothing else, and has no properties we can talk about, how can we even notice it?

Let us return to this point in a moment, but first note that there is a name for this sensation although it shares it with all the describable sensations; it is called an "aura." Now let us suppose, as Wittgenstein argues, the warning sensation is constantly changing; if it changes from joy, say, to panic, we would surely notice. But Wittgenstein would just point out that these sensations or moods already have names, set by public criteria, and that they are describable; what about the indescribable sensation? Could this not indeed be changing all the time? How would we know, since we have no purchase on it, no properties with which to describe it, or tell when it repeats? And, Wittgenstein would add, if we could not actually know any of this, does the possibility that it is repeating really make any difference? If it is still an aura then it will warn us.

But how does something become an aura? The obvious reply is that you notice it because it is a sign of a seizure; a natural sign in the way that smoke is recognized as a sign of fire. But you can only recognise something as a natural sign after it repeats in correlation to what it is a sign for. So you must first be able to identify the sensation in some way, just in order to be able to recognize when it repeats. And a reliable sign of epilepsy cannot be a changing sensation. What sense can we give to knowing that the "epilepsy" sensation is recurring if it is constantly changing? Let us try, "any unknown sensation means epilepsy." But this sounds as though we are saying we recognize something as unrecognisable! Perhaps less paradoxically, we can say we *identify* something as unrecognisable.

The problem is, then, how can we identify an indescribable aura? Suppose we do identify it negatively, as that sensation that is unlike any other. Even if we could make this kind of brute comparison, it would be useless for the purposes of recognizing the same state recurring. How would we know if it is that state we previously identified as unlike any other sensations, or another sensation unlike our describable sensations? What can we use to pick out that it is the previous indescribable sensation recurring? If we cannot even pick out the same indescribable state recurring, how could it serve as a warning? Well, given a particular sensation is correlated with epilepsy, and that we have never had other indescribable sensations – I think we can agree these are at least unusual! – then perhaps we can rely on the fact that any indescribable sensation is of the same kind. We are still left with no direct way of recognizing the sensation.

But there is a way of recognizing the indescribable sensation, a way that relies on memory, and inner comparisons of past to present states – what I have called state monitoring. State monitoring does not require that the discrimination machine knows how the inner matching is done, or can describe in what respects the two states are similar. If they are matched, that is enough. State monitoring can be unconscious or preconscious, and the information that the state match can be relayed to consciousness.

At this point I should sketch my own position on transparency, since it is directly relevant to the issue at hand. I do not believe that inner mental states are entirely transparent as described earlier, in my thumbnail version of the Cartesian view. I think that we can be more or less aware of the properties of inner states by my beetle out of the box methods, and the issue becomes one of the relative transparency of these states. Just as we may not be able to count, or even be aware of all the speckles on speckled hen, to use

Wittgenstein's famous example, I think we can make a parallel point about a 'speckled' after image. I have already provided another sort of example of an occurrent mental state that is not transparent – in this case it is entirely opaque – that of the anxious man in the airport, oblivious to his anxiety, and note that unlike the after image case, he can be corrected by outside observers, so incorrigibility fails as well as transparency.

What makes a state more or less transparent? I think the degree to which an experience can be verbally described in terms of public states with matching inner components, or indeed represented by recreation or modelling, is probably closely allied to the degree to which it is transparent. The indescribable aura is naturally recreated, but cannot be represented by an external recreation, nor can it be modelled, or described. I think that the repeatability under certain public conditions probably allows the experience to "stick" and become more transparent to consciousness and accessible to memory. And I suspect the subject of these states can describe something about these states, for example, the duration, intensity, location, even if vague, the rhythm of the feelings, and so on.

But if nothing is describable of the content of it, by our methods of representation, I find it hard to see how we could consciously be aware of the same sensation recurring, apart from intuiting this was so because of state monitoring. I think that if the experience is relatively indescribable, then at best we can only be vaguely aware of it, however vivid, intense, and overwhelming it is – it is like a car crash, only even harder to grasp. But if the experience is truly incomparable, perhaps we cannot be aware of it at all; it becomes too smooth to get a grip on. But then it is no longer an experience. Perhaps in order to be conscious of inner states, state monitoring has to be done by consciousness; we have at least to be able to consciously compare and contrast the experience, to see what it is like and unlike, but then, of course, it is describable.

There is another kind of warning aura, one that precedes migraine headaches of a certain sort. This is typically a visual disturbance, and is often reported by sufferers as visual colours and shapes projected onto normal visual perception. This aura, unlike the indescribable one, is classifiable as a kind of sensation; it is visual, and, moreover, is quite verbally describable. There is even a version of this sort of migraine where you get the aura without the headache, which is the sort I have. I can describe what happens to me as "seeing coloured zigzag lines, which come across the left side of the scene I am looking at, which are in a hook shape, and which to begin with are very transparent, but become increasingly solid, and end up dominating all that I see." Now, if I am right in my theory of these inner states, none of this description is literally true. There are no lines I am seeing, or colours, and no hook shape; I still think that if you want to know how what is happening to me 'looks,' and therefore what my experience is like, this description is quite a useful tool. A description, as it were, of the strange coloured zigzag plastic filaments you would have to see to be in a state like I am. I can do better though.

I was looking on a website devoted to migraine, and found a section devoted to migraine art. Interested to see if I would see an image that matched my own experience, but rather sceptical that I would, I looked at a number of pictures. There are a wide variety of pictures on the site, and descriptions to accompany them, and they were all far from my own experience, until I found a picture from the digital artist Conny K. Jonsson, where I felt immediate recognition. And I cannot help but feel that if I could show you the picture, this would recreate in you something very like what is happening in me when I have an aura.[15] Notice this migraine aura without the headache is a case of a sensation without an objective correlation – but even this aura can be talked about, and represented, as I have described, and the Jonsson picture of it is a prime example of recreation.

At last we return to the problem that started us off on this journey, the representation of vertigo in the Hitchcock's film of the same name. Hitchcock remembered an incident when he was very drunk: "I had the sensation" he says in an interview with Truffaut, "that everything was going very far away from me."[16] He goes on to say that he then spent fifteen years thinking about how to represent this sensation in film. Eventually he devised a method using the then new technology of the zoom, combined with a tracking shot, so that the viewpoint is fixed, and thus the object filmed remains in the same position in the frame and roughly the same size, but the perspective changes. In effect what happens is that the perspective elongates backwards, separating the foreground from the background.

Is the vertigo effect a model for vertigo, or a recreation of the sensation? Well, if it is a recreation of vertigo, and therefore similar to the experience, then it is only visually similar; the key feeling of dizziness cannot be recreated, or at best only mildly, in a film, and thus neither can the way this infects the visual perception of someone with vertigo. What we have represented is the feeling that the background is going back, separating from the foreground; depth is increasing dramatically, and yet we are at the same time visually aware nothing is really moving. All this is shown in a vivid easy-to-read way. Not only is the increasing depth represented, but also the timing and the rhythm of this feeling. Moment by moment we know what state the central character is in. A close up of Scottie, the subject of these states, by itself would not indicate his vertigo, just his terror. A very wide angle point of view shot, contrasting with the "normal" lens camera shot, would just indicate his feeling of depth, not the stomach churning apparent movement of the background.

But after some reflection, and having seen the effect many times, I now think this is not enough like the visual experience of vertigo to call it recreation. Vertigo, from my admittedly scant and mild experience of it, does not create a visual sensation of perspective elongated backwards, and I am not sure the feeling of depth in real vertigo can be recreated visually. There is the common example where depth perception does not function normally: when we look out of planes and the distant houses and cars look like toys. Here, everything seems the same visually, as if we saw it in depth, but the houses and cars do not feel as though they are distant. If this is right, and if the feeling that is lacking in this example is exaggerated in vertigo, then the feeling cannot simply be visually recreated in film, because film cannot recreate this kind of depth perception, leave alone the inner vertigo sensation. But it can still be modelled, because increasing depth can be visually represented.

Perhaps the film does recreate vertigo, I may be wrong – I think fleeting and rare inner states such as this are difficult to analyse and far from transparent, as should now be clear. However, I also think that good modelling or recreation of inner states aids transparency, even if only in making clear, as in this case, that exactly what happens is unclear, and thus dispelling Cartesian illusions. Anyway, I think this model is useful. For the paper I was easily able to re-see the vertigo sequences a number of times. The sequences gave me a tool for investigation, something to examine, and to compare with my vertigo memories.

Initially I thought the vertigo effect recreated vertigo; after study, and reflecting on my own memory of vertigo, I now think it is probably best to see the vertigo effect as an example of modelling. But how do we know if a representation is a model or recreation if we have not experienced the inner state before? I do not think we can always know, if, as in the vertigo example, the model is perceived by the audience using the same sense or senses, as would be used in a recreation, and closely matches the oblique description of the represented sensation. Moreover, vertigo is modelled using visual means that are not abstract, as a diagram would be, but very closely matching to subjective descriptions of vertigo such as Hitchcock gave of his experience.

Does it really matter if this is a recreation or a model? Well, perhaps a model is a more analytic tool, and liable to make us more aware of the structure of a state, but I am not too sure it does matter in this example. Although if I take this image to be a recreation, I will be seduced into thinking the separation of foreground from background by this elongation of perspective can recreate real vertigo.[17] (I am now assuming this wrong, for the purposes of argument.) But the aim of this sequence is not to investigate the nature of vertigo, but rather to help us identify with the central character, and to represent what he is undergoing in a vivid and powerful way, and in this context, it works very well. Modelling can be a more analytic tool than recreation, as I have said, but it is also more flexible, and it need not always appeal to the same sense or senses in the representation as would be used in a recreation; and if we broaden our definition of modelling a bit, models may even be made of what it is impossible, at least in practical terms, to recreate; this allows the representation in films of inner mental states that are not totally or even at all visual or aural.

Let us take another example from *Vertigo*. It occurs in the scene where the central character Scottie kisses his new girlfriend after he has made her over in the image of his dead love, Madeleine. As he kisses her he mentally revisits the stable where he last kissed Madeleine – perhaps "revisits" is misleading: he *feels* the place. And this feeling is shown by the slightest flick of Scottie's eyes, as the camera circles the kissing couple, and we see her hotel room turn into the stable behind them (not a matt shot, but an ingenious double set). The flick of his eyes serves to show us that the image of the stable behind them is relevant to him, not to her; but at no point does he look at this image. The length of time we see the stable indicates the length of time of the feeling, and the concrete nature of the image indicates the intensity of the sensation, for the character, and its vividness.

We do not normally, if ever, have a detailed visual image when we mentally revisit a place. So it would be totally wrong to put the stable in front of Scottie, and have him flick his eyes to look at it; this would suggest that the image he has of the place is like the image we see of it. The ingenious thing about the sequence is that we are shown the subtle nature of what Scottie is feeling, the evanescent yet vivid impression of the stable that he has. It is like that public feeling of place that we all have when in a place, and seeing and hearing it, and perhaps more perspicuously, it is like that feeling of atmosphere of a place we derive from our perceptions of it; but here only the feeling of the place and its atmosphere and perhaps some vague images of it remains. Images like this are usually much less visually detailed, as it were, "at the back of the mind." In other words, our eyes can be open, and we do not see the images projected onto our visual field. (Now we have a new oblique descriptive language for inner mental states because of film, and can say, "it is not like a superimposition in a p.o.v shot".) One might even argue these feelings are not really visual at all, at least not for most of us.

Is there a common inner experience most of us can refer to, to check the accuracy of this representation? Well, I think we have a prosaic version of the same thing when we suddenly feel the ghost of our bedroom, momentarily, when we wake up in a hotel, even though we may remember we are in the hotel. Our oblique descriptions this time suggest the mental imagery is not visual; I might say "I do not 'see' my bedroom in the hotel room," and add "but I feel it is there."

In a way, then, Hitchcock's model for this feeling of place is partial, modelling the experience by referring to it, showing us its content, and structure, and showing us what it is not – it either does not involve a sort of inner sensory recall of place, or this recall is very vague, and 'in the back of the mind.' However, given this feeling of place is part of our normal perception of a place, there is an element from the inner part of a public perception we can use to understand this 'pure version' that is modelled.

What happens in recreation and modelling? Recreation lets the spectator experience something like the experience represented, and so understand it from the inside. A public mental state – seeing a picture say – has as one of its necessary components an inner state. This inner state matches the inner state we want to represent, for example, my migraine aura. In this way the picture recreates the inner state, and thus the experience. In contrast, modelling shows the spectator the content and structure of the experience obliquely, and does not recreate it. This modelling is oblique because it is a model of a public state that could be used to recreate the experience, and not a model of the experience itself. I think we can extend this notion of modelling a little bit, as I suggested above; perhaps modelling can also occur for states that cannot, at least in practical terms, be recreated via perception. In this case a model consistent with an oblique verbal description can be created.

Instead of perching spectators at great heights to reproduce in them feelings of dizziness, and great and increasing depth, we use a film sequence that models the feeling of great depth; perhaps this feeling cannot be recreated via visual perception alone, in those that never suffer from vertigo. The second example from *Vertigo* is another example of modelling where recreation is problematic, if not impossible. The model makes use of a public component of a public state, the perceived stable, with an inner component, the feeling of place, that is at best only partly like the inner state modelled.

Recreation, and oblique description and modelling, are, I think, major ways in which mental states become transparent. And we should note that the recreation we have discussed is what one might call "external recreation," and that these three ways of representing inner states tell the spectator more about any inner state than a pill recreating the state would do. If we only had pills to recreate experiences, or to refer to in describing or modelling them, then perhaps we would be almost be back to the land of Wittgenstein's "beetle in the box."

Without external recreation, or oblique description and modelling, we are left with an indescribable sensation, and surely such a sensation only has at best a kind of half-life; to steal a couple of lines from Wittgenstein, it is "not a something, and not a nothing either," and "a something about which nothing can said." But with these methods of representation at work, we have a means of investigating that inner, experiential part of our minds, a means which does not involve us in a private language, or private meanings, for we have followed the schema 'from the outside in' and so public meaning, and communication, is retained. If I am right, and experience is not entirely transparent, then these ways of representing inner states are an indispensable aid to transparency, but they also guarantee these states are not entirely private.

The uses to which these three ways of representation can be put in art and everyday life are legion, and not confined to becoming aware of our own mental states or to making them public. A patient may want to show a doctor what is happening to him or her, but an artist may want to engage our sympathies and interest, or aid in the mysterious process known as identification with a character, as much as teach us something about how the mind works – though this may indeed be a valuable spin-off. I believe this kind of representation is grounded in more than mere fancy, indeed, that it is vital to our mental life. I hope I have succeeded in providing a sketch of how representations in art and language can make our inner mental states public: understandable to others and transparent to ourselves.

City University, London

NOTES

[1] Often, when philosophers use "state" they mean "a relatively stable condition in which a thing is"; I am using it in the everyday sense of "condition in which a thing is", and so when I say "inner state," I am referring to processes, properties, and events, as well as relatively stable conditions. For example, seeing is a mental event, rather than a state in this last sense, but I still refer to it as a state, in this paper.

[2] For the first and best statement of the component approach to mental states, see Irving Thalberg, *Perception, Emotion and Action* (New Haven: Yale University Press, 1977).

[3] Ludwig Wittgenstein, *Philosophical Investigations*, trans. G.E.M. Anscombe (Blackwell, 1976), p. 100. In the rest of the paper I will use the standard formula for referring to the *Investigations*, and mention the part number followed by the section number; here, for example, PI 1, 293 refers to *Investigations* part 1, section 293.

[4] Thomas Kuhn T. S., *The Structure of Scientific Revolutions*, 2nd ed. Chicago: (University of Chicago Press, 1962/1970).

[5] An alternative is just to claim that we are never conscious of any inner states while conscious, or, more heroically, that there is no such thing as consciousness.

[6] For a discussion of these issues by a functionalist, see D. K. Lewis, "Mad Pain and Martian Pain," in Ned Block (ed.), *Reading in Philosophy of Psychology*, Volume One, (Harvard University Press, 1980), pp. 216–222.

[7] This points to a deficiency in the functional analysis, since it cannot explain the difference between someone who knows he or she is anxious because he or she feels anxious, and someone who knows this by seeing causes and effects of the state.

[8] Gilbert Ryle, *The Concept of Mind* (London: Peregrine Books, 1976) p. 234.

[9] J.J.C. Smart, "Sensations and Brains Processes," in David M. Rosenthal (ed.), *Materialism and the Mind-Body Problem* (Prentice Hall Inc, 1971) pp. 61–62.

[10] Nelson Goodman, *Languages of Art* 2nd ed. (Brighton: Harvester Press Limited, 1976) pp. 130–156. I have left out syntactic density, where the finest differences in form may constitute a difference between symbols, because this requires these descriptions have a form, and they only have an apparent form – at least only the apparent form is accessible to us. (Even if so, we could still talk about apparent, or oblique syntactic density.) I have also left out exemplification – symbolization by converse denotation – where something serves as a sample of properties it literally possesses.

[11] "Model" is, admittedly, a vague, if useful notion, rather like similarity. I am not using the word in a special sense here, and will assume the reader has a grasp of the everyday usage. A good account of "model" is the subject matter of a paper at least, if not a book. For now I will be content to say that models allow one to see in them something of the structure and content of the original they model; they do this via reference and similarity to the original. They may model using very objectively similar structures and contents to the original, as some scale models do; if they do this then they may be used in what I call external recreations, which is why I do not have these kinds of models in mind. Other examples of models are maps, diagrams, hydraulic models of the economy, models used to plot the disposition of forces in war, idealized drawings of the anatomy, weather models in computers. Models usually involve a great deal of simplification and distortion to make then easier to interpret and to pick out some aspect of what is modelled. For a more detailed discussion see Max Black, *Models and Metaphors* (Cornell University, 1962), pp. 219–243 and Nelson Goodman, *Languages of Art*, 2nd ed. (Brighton: Harvester Press Limited, 1976) pp. 170–173.

[12] Ludwig Wittgenstein, *Philosophical Investigations*, trans. G.E.M. Anscombe (Blackwell, 1976) PI 1 258.

[13] *Ibid* PI 1 270.

[14] By Mayo Clinic staff, *epilepsy section*, April 27, 2005, www.mayoclinic.com,

[15] http://www.migraine-aura.org/EN/Conny_K_Jonsson.html.

[16] François Truffaut, *Hitchcock* (Paladin Books, 1978) p. 308.

[17] Fiction sometimes purports to recreate what it cannot. At the end of the poisoning scene in *Notorious*, for example, we see subjective shots from the poisoned woman's point of view full of extreme distortions, and in this case surely we are meant to see, more or less, not only what she "sees," but also *how* she "sees" it. Yet the sequence in *Notorious* is a work of the imagination, since Hitchcock could only guess at what the heroine's inner state was like, having never experienced it himself; research into descriptions of such a rare, extreme and confusing experiences would, I suspect, be of limited use, but I doubt he bothered. He was not after veracity, but vividness, and so probably invented an inner state for us to share with the heroine.

MARY JEANNE LARRABEE

BODIES AND MORE BODIES: TRYING TO FIND EXPERIENCE

The last decade has seen an avalanche of books on "The Body": absent bodies, deviant, imaginary, indeterminate, and leaky bodies, rejected bodies, and volatile bodies, among others. A lot of this work was initiated by postmodernist and feminist critiques of the Cartesian self which attempt to unravel Descartes' split of mind from body. Recently, writers have claimed that the experiential side of embodiment or corporeality – of fleshiness – has been somewhat overlooked in the primarily postmodernist shift to viewing "the body" as socially, culturally or discursively constructed. Robyn Longhurst notes: "The fluid, volatile flesh of bodies, however, tends not to be discussed … [that is,] the runny, gaseous, flowing, watery nature of bodies" (Longhurst, 22). Or Lynda Birke adds: "What was awkward for those who insisted on social construction was the sheer impossibility of dealing with pain or bleeding, with the general messiness of bodies" (Birke, 199) – the messiness of how real bodies transgress the boundaries of our concepts of them, of our interpretive nexi which may be satisfying to our reason but are constantly jolted by our further experiences of ourselves as embodied.

This call to the "experienced" body, however, cannot be a simple return to the nature/culture dichotomy, now emphasizing again the "natural" body as somehow existing completely separated from cultural (conceptual) constructions of it. Some postmodernist moves have collapsed both "the body" and "nature" into "culture/discourse" and such shifts make sense to some extent. For example, I might be experiencing myself as having a mechanized body; here a postmodernist interpretation might state it is clear that I am socially constructed, as a life-long U.S. resident, to take my body as that sort of thing, to consider an ill body as similar to a broken automobile and to seek the expert for that type of body, usually the allopathic medical physician. I agree that I experience my body through the habituated meanings for bodies that I have encountered, heard and read about, seen in the images and representations of popular culture – and those bodies for me are mostly the bodies of Western medicine and thus in a sense of Eurocentric philosophy. However, in this paper I look at how *this*

A-T. Tymieniecka (ed.), Analecta Husserliana XCIII, 75–95.

"body" – the one of Western medical science, one for which I will use the label "biomedical body" – is taken as paradigm in much of contemporary United States culture (by implication, then, by many contemporary philosophical accounts of "the body," even postmodernist ones – a point I will leave for another time to prove). I describe the paradigm and then look at contestations of that paradigm from persons with disabilities and accounts of bodies in other cultures. My aim in this essay is not so much to dispute this paradigm as a viable one, since within the cultural context of modernist Euro-Western society this paradigm rests on the successes of modern medical science to achieve its goals of reducing disease. Rather contestations of the paradigm point to the greater complexities of real lived lives and the realities in which they exist. The paradigm is a theoretical construct, yet it does construct people's bodies. Contestations, however, show how these constructions may be operable alongside persons' understandings deriving from alternative ontologies and personal constructs of embodiment from outside biomedicine, and also they clarify the nature and the extent of the paradigm's influence so as to open freedom from the way it is used to some extent hegemonically to judge normality from one narrow (albeit successful) perspective.

Paradigms function most strongly, but often most insidiously, when they are apparently fully in one's face, consciously visible. It is thus possible, following upon the strong postmodernist critique of the Cartesian paradigm in Eurocentric everyday culture and its underlying theoretical/philosophical nexus, that this paradigm and its cooptation by modern medical science and much of medical practice nonetheless has a fixity that belies the postmodernist threats that nevertheless allow an ongoing dislodgement of the paradigm.

First, let us briefly investigate the major features of the paradigm of the biomedical body. As noted already, it functions as a theoretical construct within the medical disciplines, but also in the broader culture when people both think and deeply experience their bodies according to the paradigm. This does not mean, however, that it is the only type of "body" operative within the *practice* of these disciplines – it primarily comes from the *science*, not the *practice* of Western medicine. The paradigm prescribes the norm for bodies, it legislates normality: any diseased body is abnormal, a result of an invasion of pathology-causing entities, or abnormal genetics, external or internal violence causing harm (the auto accident, the wrongly ingested liquid) or dysfunction. This body is mechanical, and the object of medicine is to exert a control over this body that will normalize it. The bodies that medicine tries to normalize, however, are actual lived bodies, the embodied expression of individual human persons. As individuals they are both subject to the influence and pressures from outside forces such as cultural and discursive construction and free to exist to some extent how they will.

One way to frame the situation that arises from within this concatenation is to draw on Husserl's phenomenology, here his distinctions about essences. Taken as an invariant, with distinctions along the lines of genus and species, any essence has types ending at the "bottom" point with the individual essences; for materially filled essences such as those of consciousness this would be the singularity. Husserl states: "A This-here, the material essence of which is a concretum, is called an individuum" (Husserl, 1983, 29), and "a concretum is obviously an eidetic singularity" (Husserl, 1983, 30). I will take this as applicable to human individuals, so that each of us have an eidetic singularity as our material essence (or part of) that is distinctively our own. My codicil to the Husserlian notion of essence that might not totally reflect his position is the fallibility of the source of claimed essences found within the phenomenological undertaking. Thus, any essence claimed as an invariant is that in the moment, given one's experience thus far, and until future eidetic seeing sees further. In other words, either the individual's eidetic seeing has missed something or the something has not appeared in its totality – always a problem with morphological essences, which include all those relative to consciousness, for these are the ones with fuzzy edges, ever waiting for further filling out. (Husserl discusses this point about morphological essences in *Ideas I*: they are not exact essences and can only "'approach' more or less closely without ever reaching" the ideal limits reachable by exact essences' ideation [Husserl, 1983, 167].)

So, on the one hand, we have the essence of the individuum – for our discussion, of this distinctive individual human self – and, on the other, "essences" or the labels with their underlying social meanings that indicate various social identities that can be ascribed to a person by an other – What theorists call social or discursive or cultural constructions (for the moment I will leave out of the picture the extent to which the individual may "buy into" these constructions). Tensions between the former and the latter, both for the individual labeled and those doing the labeling, can arise in many cases where labels are applied by persons outside an identity group who draw on the constitutional concretions (sometimes viewed as an "essence") of social identity-formation; for instance, a Kenyan American ascribes characteristics supposedly true of most Brazilian Americans to one individual Brazilian American. These "constitutional concretions" I would understand as constituted by what Husserl calls "personalities of a higher level" (Husserl, 1960, 132) than individual persons, yielding through intersubjective or interpersonal constitution a wide variety of communities, "spiritual objectivity," a "cultural world," among others (Husserl, 1960, 132–133). These communities or social groups are distinct from the individual person (even from those in the group), and from the primary type of "community" established in the constitution of "this world" for us, that is, for myself and the other apperceptively constituted as an other person who is in some fundamental ways like myself as a conscious person (Husserl, 1960, 129–30)]. Thus, the labels and their meanings that come out of social identity-formation can be established by a culture, which I consider a personality of a higher level. Unlike a lot of the individual's constituting activities, culturally constituted senses can tend heavily towards immobility, a factor Mikhail Bakhtin noted in his discussion of the phenomenon of the centripetal forces of speech communication (Bakhtin, 271–272). Murray Edelman also notes this point, showing how the rhetorical structures used within social groups are, while changing and changeable over time, marked by the power variation between the defining in contrast with the defined groups. Edelman notes the power implicit in professional terminologies: "Any categorization scheme that consigns people to niches according to their actual or potential accomplishments or behavior is bound to be political, no matter what its scientific function is" (Edelman, 115). Since power is teleologically inclined to maintain itself as stable, such social categories, the language and meanings that are their expressions, remain in place readily. The individual's awareness of an own-self, a self constituted from a cross-section of history, interests, etc., yields a self-identity with some force as the eidetic singularity. Given this experience, it makes sense that individuals experience incongruence between their "selves" and social identities ascribed by others as are apparent in the expectations shown by these others.

In this paper I am primarily concerned with the labels or identities relative to health conditions, which are often assigned by the medical professions to what are called "patients," and the connection of these to one's experiences as embodied. And of particular concern is the use of the terms "normal" and "abnormal" in the perception of bodies that are diseased or disabled. My paper will use a framework from Husserlian phenomenology for discussions of essences, individual and social or cultural identities, embodiments and bodies, and the perception and attribution of normality/abnormality to the latter. I will also draw on medical narratives to destabilize some of the more common theoretical discussion of these topics.

Medical literature lately has distinguished "disease" as object of the medical scientific gaze from the experiential side of bodily abnormalities called "illness" (Kleinman, 1988). This distinction does fall along the dualist lines of nature (disease) and culture (illness), particularly when used by anthropologists and sociologists of medicine, as well as the dualism of objectivity and subjectivity, but I hope to problematize these dualities in their application to the experiential realities. Not all writers employ this distinction, but I will use it in this essay. We will see that allowing for the full experiential side of illness to find appropriate expression might move beyond the "body" as locale of illness, allowing the experiential side of "physical" disease to have both behavioral and emotive-cognitive repercussions that might also not be completely separable from the illness viewed as "disease", the biomedical entity. [For me, this allows a contestation of "mental" disorders as wholly other than "physical" ones.] Arthur Kleinman indicates how "illness means in social senses. There is the taken-for-granted, ostensible meaning of the symptom ... a pain, deformity, disfigurement, disability [which are]... shared meanings ... more or less standardized in a local culture." Further, there are the "cultural significances certain disorders carry in particular societies" and finally "peculiar meanings" from personal and interpersonal arenas (Kleinman, 1986, 145).

In narratives of illness, in distinction from disease, what comes to the surface is a person's embodiment, the experiential side of living the lived body. This point indicates a central distinction in dealing with the body phenomenologically: Husserl names the lived or experiential body the *Leib*, and the body as object-body *Körper*, which includes the body usually spoken about by physicians – the body when dead and when viewed by the most narrow lens of Western biomedicine, which treats disease and the body in a specifically scientized way. Elizabeth Behnke (66) summarizes this distinction, noting that what Husserl calls *Leib* "involves the turn to experiential evidence." This assignment of distinction, however, may have some cultural pecularities. Thomas Kasulis (305) notes that East Asian medical practice, while it used autopsy for a short period, discontinued the practice because the dead body was seen as so different from the living body as to make autopsies useless in learning how to handle illness. For them, then, only dead bodies were *Körper* in the true sense, while living bodies were something else.

The biomedical construction of the body – the paradigm of the biomedical body – has been touched on above. Husserl's discussions of the mathematization of the natural world within Eurocentric cultures (Husserl 1970, section 9) fit here: the medical profession as developed in Europe and other Eurocentric cultures treats disease and the body in a specifically scientized way. Zbigniew Szawarski describes it thus: "the subject of medical activity is a damaged or malfunctioning biological mechanism – the human body" (Szawarski, 41).

[He points to one alternative operative among some Eurocentric health practitioners, the systems approach model or the axiomedical one, that sees health as "a state of dynamic equilibrium of many systems that comprise biological, psychic and social levels of an individual's functioning" (Szawarski, 45).]

Whatever the interpretation of the body there is a corresponding understanding of the nature of disease and of embodiment. Richard Baron (M.D.) comments on the way the paradigm of the biomedical body functions in medical practice; he notes: "In general, modern medicine takes disease to be an anatonicopathologic fact. [That is,] we [physicians] tend to see illness as an objective entity that is located somewhere anatomically or that perturbs a defined physiologic process" – a viewpoint that takes "illness from the universe of experience ... moving it to a location in the physical world" (Baron, 606). The problem with this, Baron finds, is that "by believing that the real world is somehow 'out there', we see ourselves as anatomic and physiologic entities, and we inevitably devalue the world of our experience, drastically limiting our ability to understand ourselves and others as people." Baron is addressing his fellow physicians in the "our" here, but I would claim that anyone who comes under the influence of Westernized medicine might incorporate such attitudes themselves in relation to their embodiment.

This paradigm has been challenged by various calls to a return to experience, especially as those with marginalized experiences contest the hegemony of either the biomedicalized body or the postmodernist "body." One avenue of attack on the biomedical body has been one termed "phenomenological" in the sense of drawing upon the narratives of persons with illness, especially in their confrontation with medical practitioners. For example, Kay Toombs's discussions in *The Meaning of Illness* address the gap between the two worlds of "physician" and "patient," resulting from the differences between the physician's perspective and that of the person receiving medical treatment (Toombs, 12). This gap seems wide in Eurocentric medicine – it narrows where the "patient" or medical client constitutes embodiment and its abnormality in ways congruent with the physician's understanding of body.

There are other contestations of hegemonic readings of the body than phenomenological accounts of illness experience. In the rest of this essay I will look briefly at one from critical race theory and then more extensively at two others – one from disability narratives and the other from cultures other than modern Eurocentric culture. Critical race theories, in describing some possible experiences of the "raced" self, have noted the doubling or multiplying of "the body" under racialization, in particular the body experienced by the "racialized" person versus the "body" that racists see as "raced", in a racial category that is considered inferior in some ways. Franz Fanon recounts his racialization by a white boy who began staring at him on a train, finally pointing Fanon out to his mother: "And then the occasion arose when I had to meet the white man's eyes. An unfamiliar weight burdened me. The real world challenged my claims. In the white world the man of color encounters difficulties in the development of his bodily schema.... Then, assailed at various points, the corporeal schema crumbled, its place taken by a racial epidermal schema. In the train it was no longer a question of being aware of my body in the third person but in a triple person" (Fanon, 110–111). For Fanon this experience of a tripling of his body included his recognition of the white boy giving a particular type of racialized meaning to "the black man," as well as his more "normal" senses of himself as embodied.

Our second source of contestation of the paradigm of the biomedical body comes from the experiences and challenges posed by persons with disability, who have challenged not just biomedicine for its singular notion of normality as deficiency of any sort to the body, but the influence biomedicine has exerted on the culture within which there is widespread penchant to consider anyone with perceivable and chronic medical problems, even when stabilized, as "broken." John Hockenberry describes an incident from the end of the first Gulf War, when he followed Kurdish refugees fleeing from Iraq. He took a donkey to get up the high passes of the mountains on the border between Turkey a nd Iraq in order to file a first-hand account for National Public Radio. While on the donkey he was seen as an individual from the United States, but after he was removed from the donkey in order to sit and take interviews, something else was seen – when the refugees realized he was "different": paralyzed and unable to walk. There occurred a shift that Hockenberry had experienced often, including in the Americas and Europe: he was no longer a person, he was a fractured body, an object of pity (Hockenberry, 367), despite the fact that he held one of the most active and world-engaged occupations, war correspondent.

The situations experienced by these persons illustrate the double-sidedness of embodiment, its shifting meanings, its momentary concatenations of body-for-me and body-for-others, of the self-constituted sense of my-self-embodied and of the other- or culture-constituted senses of body-seen-by-an-other. It is in this concatenation that we find fissures of meaning where discordance and resistance can occur. Resistance narratives offer some degree of rejection of globalizing theories from the dominant cultural strand. Some also question whether rejecting a globalizing theory necessarily yields "experience" as a pure category. Toombs speaks to this point from the position of a person receiving treatment: "The patient's experiencing of illness is influenced by the theoretical understandings that are embedded in the life-world" (Toombs, n. 34). Neverthelesss we need to look at those experiences that are considered "outside" the range taken as "normal" in our biomedically informed culture.

Before clarifying the potential impact of the resistance narratives of those with disabilities, I think that Husserl's writings on the dyad normality/abnormality can help illustrate part of the underlying tensions of challenges to biomedicine's paradigm concerning the "normal" body. I will look at three definitions that can be found in Husserl's post-early writings on genetic phenomenology (see Steinbock, 129–130). The first is where there is a lack of conflict among presentations, e.g., within perception; when conflict of perceptive presentations occurs, there is abnormality. For example, when I walk into my office, I [passively] do see most of my things where they were left the day before – my computer on the left of the desk, a pile of papers to the right. If vision presents these in the expected "order," then my perceiving is normal and the perceived is normal. If I walk into my office and things start looking blurry or the computer is on the right and no papers are anywhere, then I take my perceiving, in the first instance, and the office, in the second instance, to be abnormal. In the latter instance, if I did not use a key to walk into my office, I might check the office number and find I entered the "wrong" office after all (see, for example, Husserl 1973: Part I, p. 363). This first understanding of the dyad normal/abnormal might function in some arenas of medical practice or experienced embodiment, but I will not use it here.

The second understanding of the dyad is based on the notion of concordance. It draws on the processes whereby a field with differentiated givennesses comes into a synthetic unity, a concordance (for example, with what is expected in a perceptual experience) and this concordance is understood as normality. In the case of my office, if I take a severe blow on the head in it and then cannot focus so that things look all alike, this would be a lack of concordance, compared to how I "normally" see the office – here it is not a matter of some things in the office not being where expected, rather the whole field of vision is not "coming together" and therefore is not as expected. We can look at a medical instance – I may feel my self embodied as "normal" in regards to comfort in relation to heat/cold, but when, despite the coat I am wearing, I get a sudden chill when a wind whips up, that is discordance, an interruption or break has occurred (Steinbock, 133). Husserl's examples include how an injury (a bone fracture) would be experienced as "abnormal," an anomaly in relation to my previous days of bodily integrity, or how an illness can be an inhibition of my usual embodied capacities (see, for example, Husserl 1973, Part II, 123). Of importance to notions of health as normal and injury/disease as abnormal, under this interpretation of normality, is that the abnormal or anomalous must go away – be healed or cured – in order for normality to be restored, for much of biomedicine is "restoration" in aim.

This account applied to ill health fits the biomedical model well and corresponds to what current writers have to say. Baron (609) states: "The experience of illness seems to be intimately related to a sense of disorder, of loss of control, of things not being right with the world. ... Illness can be described as a split between body and self where the usual effortless, unselfconscious unity of the body and the self is disrupted, making one pay explicit attention to the body as suddenly problematic..., a disturbance in a person's ability to relate to and function in the world" [his suggestion is a "phenomenologically informed medicine"] (Baron, 610). In connection with the physician's perspective, Jan Hartman sees one definition of disease as "a fault of the bodily mechanism – a malfunction" (Hartman, 87). Thus, Husserl's notions of normality as concordance and abnormality as discordance fit this version of the biomedical paradigm of body well – physicians' goals are always restoration of "normality" taken as the healthy body and the healthy body is one operating well, a confluence of appropriate processes and parts. The account will also help us understand the problems that come with putting all the following into one category of "abnormal" (medicalized) bodies: those with acute disease (pneumonia) and with chronic conditions (diabetes), those with acute injuries subject to healing and with injuries resulting in chronic loss (of body part or process – a broken leg versus a paralyzed one) – in other words, the diseased and the disabled. The account of normality as concordance makes clear how physicians and medical clients can often be "on the same page" in regard to a medical condition; but it also can help us understand that a physician and a person with a physician-defined abnormality can disagree about its central meaning, since there can be other senses of normality at issue, not just the one about "repairing" the medical condition. Lynda Birke (218) illustrates this difference, given the various social/cultural meanings that attach to "the heart": "Having severe pains in the chest is a terrible experience.... This experience is not helped by talk of problems with plumbing, for such talk fails to address the embodied experience of people for whom the heart carries much more symbolic power than 'simply a pump'." Frances Chaput Waksler notes (80): "Medical practitioners and those who are deaf may have dramatically different perspectives on the medical status of deafness... [Some] in the deaf community... argue that deafness is simply a way of being in the world, not a medical problem" and hence resist medicalizing the condition. Maureen Connelly (178) puts the point concerning the different worlds of physician and the ill/impaired person rather strongly, claiming there is a "profound incompatibility between traditional or typical biomedical approaches to the body, and lived body approaches to the body," the experiential living out of embodiment. Thus, she suggests, "these approaches cannot co-exist,... the thrust of most clinical encounters is in the direction of eliminating – by various means – the lived body of the person seeking medical care," a claim that echoes that of Baron above.

The third account of the dyad, normality/abnormality, in Husserl can capture some of what Waksler's claim is pointing to. It focuses on optimality, providing an interesting departure from the second account described above, based on concordance. This third understanding of the dyad allows for a different mode of experiential shift – as Steinbock puts it: "the anomalous qua discordant can become normal" (Steinbock, 144), a contrast with the concordance understanding for which the discordant is always abnormal and must be done away with. In the optimality account, for instance, what in my experience of myself as embodied might begin as abnormal can gradually shift in its meaning for me (or sometimes also for the culture) into the normal for me (or for the culture) *without* its being fixed or going away; the abnormal remains but is normalized. Husserl speaks of this as the "originary normality" being broken through, so that a new system of "if-so" becomes instituted and the idea of an "originary optimal" – in the case of the body, health – is broadened in light of the changed capacities of the body (Husserl, 1973, Part II, p. 121). Thus, Steinbock (145) puts it, it is the case "both that the lived-body is normal in transcending norms when it is optimalizing or creating new norms, and that anomaly can function innovatively in the institution of norms." In this case of the normal as optimal, we are dealing with a process over time whereby the ongoing existing with the abnormality allows it to be integrated in the day-to-day experience of oneself as embodied *in this way*, a way which has no prognosis for the return to "normality as concordant" but nevertheless a way that the individual can find appropriate adaptations so that much of the rest of his existence can become coordinated with this way of embodiment and therefore this way of being with the rest of his world and with others. The new sense of oneself as "whole in this way" becomes stronger than the old sense of "this way is broken."

This discussion of normality becomes complicated when we add contrasts between the self as lived body and social perceptions of an "abnormal" body, because one person might have shifted the meaning of her disability or chronic disorder from "abnormal" in Husserl's second and socially/medically accepted sense to "normal" in the sense of optimized for herself, but without others taking up this new sense. Steinbock's discussions of normality in Husserl deal only with the dimension of individual genetic phenomenology and not with the intersubjective and thus social and cultural dimensions of the genesis of meanings. Husserl, however, does recognize the possibility of a shift from a condition being understood as abnormal to being optimalized into a new "norm" happening both subjectively and intersubjectively (Husserl, 1973: Part II, 122). This is important, since, as we have noted earlier in discussing social and discursive construction, these arenas of meaning cannot be so readily separated in regard to the experience of illness and disease. One aspect of this matter is the accrued professional language that is needed and has developed over time – of relevance to our discussion here, that professional language is that of the medical profession. Husserl would probably see in the dynamism of social groups the interactive constitution of practices upon their world of interest, the language that describes that world, and the world as found by them, their practices and their language; all three of these are constituted by the subjects and those intersubjectivities, or social groups, which for our discussion include, among others, physicians, a hospital, the American Medical Association, a medical school (etc.). Mikhail Bakhtin (270) explains the dynamisms out of which such intersubjectivities become constituted, noting how "ideological discourses" of social groups can form due to centripetal forces and "serve to unify and centralize the verbal-ideological world" and thus yield "a common unitary language" for that group. The social identities applied to ill and impaired people emerge out of these habituated practices/languages. For example, in our culture it is assumed that people with a chronic disorder like diabetes must be sickly: "You can't expect them to do a lot of physical things" is the way Margot Joan Fromer (12) describes the usual expectation by persons unfamiliar with many diabetics, pointing out that for diabetics with their disease under control, however, this is not the case – they may be "healthier" than many "healthy" but unfit persons. Susan Wendell (2001, 17–18) reminds us that there are "healthy disabled." John Hockenberry is an example of such; he speaks forcefully of the impact such social attitudes can have upon the individual: "Inside me is the engine that thrashes about never stopping, always mindful that someday those same forces [that put my uncle in a home for the chronically impaired] could decide my fate, claim that I am really helpless, that my life is not worth living, give me a label, and send me away to a place for those like me" (Hockenberry, 337). In his life Hockenberry in some sense races away from the social meaning ascribed to paraplegics and others with "abnormal" bodies in order to avoid the fate so many of them have endured.

Husserl's third understanding of the dyad normality/abnormality, where normality is understood as optimality, reflects these individual experiences, in particular those with chronic illness or impairments: In the ongoing living through of their embodiments as marked by these, they gradually become accustomed to the chronicity of the condition and find new optimalities which for them give new norms. The social interpretations of their condition according to "old" norms can invite a resistance to others' labeling that is based on stable social norms, since the latter now contrast with the newly created individual norms. Yet from this creation of new norms for themselves individually can come a progressive creation of new norms first for smaller and then larger social groups, possibly finally for the entire culture (although not necessarily for all individuals within it). The political movement behind the U.S. Disabilities Act would have instituted a number of such creations. Husserl probably had in mind the situation of normalization as an individual "becoming accustomed" to a new norm, with the exemplar being a shift from an experienced abnormality. But I am claiming that resistance narratives like Hockenberry's add a type of creative normalization which can function both on an individual level and within a social group or culture. Jenny Morris proclaims the motivation for pursuing such normalizations: "If we [with disabilities] don't impose our own definitions and perspectives [on disability], then the non-disabled world will continue to do it for us in ways which alienate and disempower us" (Morris, 9) – "alienate" because of the incongruence between the impaired/disabled experience of their embodiment and the understandings of disability at the social level.

It might help to remind ourselves that it is less than a century since the "modernity" of the Eurocentric human was supported by public displays of "freaks," those with anomalous bodies, which, while challenging "the accepted distinctions that ordered the separation between human and animal; civilized and primitive, normal and pathological..." (Rosemarie Garland-Thomson, 131), nonetheless pointed to both recent achievements and a certain glorious future for the newly scientized biomedicine of the twentieth century. Now at the beginning of the twenty-first century, Richard Zaner speaks to us on the "new" threshold of biomedicine, genetics-based nuclear medicine: "Unlike in restorative medicine, the new paradigm – like most other [Eurocentric] cultural projects – is centrally concerned to conquer, replace and transcend nature. Thus, on the new agenda is not merely curing but nipping heritable diseases in their genetic bud, correcting otherwise debilitating anomalies, and altering the human condition itself...." (Zaner, 133–134). So again we see how the paradigm of biomedical science is geared toward the abolition of abnormality, with control as a second best. This feature of biomedicine and the telos of its scientificality operate in Eurocentric cultures, as we have noted, to render abnormal all bodies that have not been cured or repaired by it (as well as those not yet "perfectable" – brought to the exactness of the paradigmatic body) and, as abnormal, often also marginalized (mistakes belong in closets).

What happens when we do listen to the experience of the disabled? The disabled, many of whom do not experience the vulnerability and need for immediate medical attention suffered by the acutely ill, have in the last half decade demanded that they be treated differently. John Hockenberry describes recent changes in the way people in the U.S. culture talk about people with disabilities. These changes have a history, since it was, according to Douard, by the early twentieth century, that "a disability came to be understood as a dysfunctional, and therefore an abnormal, physical or psychological condition" (Douard, 156). A distinction was made by Kenneth Hamilton in 1950, that between disability and handicap, the latter being "a set of obstacles disable persons must confront" (Hamilton, 1995, 147). Douard states one source of ambiguity, that the underlying notions of "normality" were defined by the "normality" of the nondisabled: "historically disability policy has been linked, on the one hand, to the concept of a 'normal' range of employment opportunities and, on the other hand, to the concept of a 'normal' range of physiological or psychological functional capacities" (Douard, 159–160). Since many disabled were not hired simply because of their medical disability and not due to a lack of other relevant abilities, this understanding of "normal" maintained the disabled as abnormal in the eyes of much of society.

Susan Wendell points to one resolution of this ambiguity, citing another shift embodied in the United Nations declaration on disability in 1983, which distinguished "impairment" from "disability" and "handicap". Thus, "impairment" is "any loss or abnormality of psychological, physiological, or anatomical structures or function" – we can see that this definition could be taken as a purely "physical body" one. "Disability" is "any restriction or lack (resulting from an impairment) of ability to perform an activity in the manner and within the range considered normal for a human being" (note the "normal" here), and "handicap" is "a disadvantage for a given individual, resulting from an impairment or disability... [occurring] when they encounter cultural, physical or social barriers ..." (Wendell 1996, 11, citing United Nations Decade of Disabled Persons, 1983–1992). Here "disability" is not the impairment itself, but results from the impairment, referring to ability for activities; the impairment will be what the physician diagnoses and treats as part of the physical body; the disability will be dealt with by the individual, family, or other caretakers, for instance, occupational therapists. The implication is that disability is also objectively verifiable and thus at least in part treatable; the definition of "handicap", however, seems to tilt us into the "subjective" – "for a given individual."

These distinctions may still be insufficient to express the experience of the disabled, who in their recent political movement have insisted on defining themselves, declaring the terms "disabled" as stemming from the viewpoints of the non-disabled. Jenny Morris distinguishes "disabled persons" from "persons with disabilities", claiming the first group are how they are, not just due to having a disability, but being limited by "disabling barriers of prejudice, discrimination, and social exclusion" (Morris, 2001, 1–2), from mainstream society and its activities. In other words, the "disability" defined by the United Nations is in part due to the impairment and in part due to social attitudes, etc., and the "handicap" is not just due to something the individual (but no one else) takes as a barrier. According to Alexa Schriempf (59), this "social model" of disability sees disability as "located in society and not in the individual's impairment" and uncovers the presupposition of the biomedical model under whose authority disability becomes "a system of values (ableism) that oppress disabled people." Schriempf names the oppression, and its source is others.

Yet we know that a clear separation of "subjective" and "objective," of body as experienced and body as looked at and labeled by others is not possible, particularly if we take seriously the interweavings of constitutings across the personal and the interpersonal dimensions of an individual (see Carr 269ff.). We saw that Hockenberry's impaired embodiment as partially immobilized (in terms of the "I can" of his person as paraplegic) became the normal for him, although for others it remains saddled with the social identity of abnormality, a factor that stills bears notice by him: "For almost all my life in a wheelchair ... I have been stared at" (Hockenberry, 367). The other continues to matter. We are reminded again of the interconstitutionality of the personal and the intersubjective and cultural, and the gap between them that allows for resistance. And we can understand to some extent this gap as experienced by those with "abnormal bodies" and their insistence on moving away from "a series of labels and their signifiers derived from scientific positivism," as Mairian Corker puts it (35).

The interpretation of bodies in Western biomedicine also faces criticism from the bodies of other cultures. As Frances Chaput Waksler (76) reminds us: "Using Western medicine as an absolute standard inevitably leads to the denigration of alternative healing practices ... and especially a denigration of alternative explanations," that is, ontologies – whether these are Asian, African, Aboriginal or 'traditional' – usually understood by biomedicine as "premodern" and thus not scientific. (Charles Finch refutes this characterization when applied to African medicine, both in ancient Egypt and in current traditional African societies.) Thus, my final approach to abnormal bodies in relation to the paradigm of the biomedical body and medical care is to look at cultures operating, at least to some degree, outside of the biomedically influenced one, in order to appreciate the experiences of embodiment in such cultures. If the point made earlier about the influence from one's life-world and its embedded theory upon one's experience, then it will help to look at other cultures, where both medicine and the understanding of the body might differ from those of Eurocentric cultures. Another reason to look at cultures is given by Kleinman (1980, 381), reminding us that cross-cultural investigation shows us that biomedicine – and psychiatry – are "not what they appear (and claim) to be. Cultural analysis calls in question their purpose and scope... [revealing the inadequacy of] the biomedical reductionism and technological 'fixes'."

Western biomedicine has moved into cultures foreign to its presumptions and, for the well-being of their inhabitants, shaped their bodies to fit the biomedical paradigm; but such a move can often ignore underlying local ontologies and their value networks, bringing the negative effects of Eurocentric colonization, which includes contemporary economic globalization. Mary-Ellen Kelm traces one such encounter in Canada with Aboriginal First Nations, noting: "bodies are made not just in flesh but also in words" (Keln, xviii). In a comparative study of the body as understood in ancient Greece and China, Shigehisi Kuriyama (13) reminds us that "differing ways of touching and seeing the body were bound up with different ways of *being* bodies." He details the history in these two concerning concepts of blood and breath in relation to the "divergence of embodied experience in China and in Europe" given distinctive developments of medical knowledge and practice (Kuriyama, 237). Kuriyama speaks of the different ways that wind as a causal agent in illness was understood in Greece and China: the former took the warmth/coolness and dampness/dryness of winds to affect aspects of the body leading to disease, but for the Chinese wind "swept straight into the body's interior and harmed by intrusion" (Kuriyama, 251). Kuriyama concludes: "the discourse of winds once gave voice to an embodied experience of space and time, a physical feel for local airs, seasonal atmosphere, shifting moods, contingency. Personal breaths could harmonize with cosmic breath, and habitually the two might be reasonably in phase" (Kuriyama, 257). Kuriyama (8–9) questions the universal assumption of the biomedical paradigm: "The true structure and workings of the human body are, we casually assume, everywhere the same, a universal reality. But when we look into history, and our sense of reality wavers ..., accounts of the body in diverse medical traditions frequently appear to describe mutually alien, almost unrelated worlds"; he then notes the way culture can limit what we perceive to be the reality of the body: "muscularity was a peculiarly Western preoccupation.... Even now in China, complaints that English speakers express as 'sore', or 'tense', or 'sprained muscles' are habitually experienced in other ways" (Kuriyama, 12). What is not in our theory is not in our experience. Kuriyama notes an example: "the tracts and points of acupuncture entirely escaped the West's anatomical vision of reality" (Kuriyama, 9). Thomas Kasulis (306) captures the difference in the bodies of Eurocentric biomedicine and South Asian traditional medicines further: "While we Westerners studied the body as hydraulic and electrical systems, the East Asians studied it as a system in which flows a matter-energy or life-force called *ch'i* in Chinese or *ki* in Japanese." Some Asian Indian philosophies claim that we have multiple bodies: a gross body that is the physical body of waking consciousness, a subtle body of dream-consciousness, and the causal body which is the body that yields another incarnation if it holds the causal remains of our immoral deeds (Dissanayake, 41). Such fundamentally different ontologies also affect persons' experience who live within cultures infused with them.

Orona Oroka notes how the Isoko people of Africa are dualistic, a feature similar to many Eurocentric peoples, and consider human beings a composite of body and soul, such that the latter survive the death and decomposition of the former. But Oroka also indicates that for many Africans the soul as "mystical power or force is not easy to classify in African ontology" – it may indeed in some sense permeate the entire universe (Oroka, 7–9), in a way similar to the energy, *ch'i*. Placide Tempels gives us a parallel in Banto philosophy – the vital force, which each entity has in some way; he distinguishes this fundamental ontological category from eurocentric philosophy's "being": "There is not an idea among Bantus of 'being' divorced from the idea of 'force'" or "vital power" (Tempels, 430), so that for Bantus "force" yields a "dynamic" conception of being, in contrast to the static one of Eurocentric culture. Such a conception will have bearing on how embodiment and its abnormalities are taken by the Bantu culture and bodies experienced by Bantus. Charles Finch (122) notes how this understanding of "force" works out in terms of medical practices, those of both ancient Egyptian and traditional African cultures: there is a "complete interpenetration of 'magico-spiritual' and 'rational' elements" and thus of the traditional and the "scientific," even in ancient Egypt where much anatomy was understood and sophisticated surgery was practiced, comparable to early Eurocentric biomedicine. Finch goes on: "Thus in this setting, moral, social, or spiritual transgressions are likely to lead to illness because they create both individual and communal disharmony. Without the psycho-spiritual cure – without reestablishing this sensitive harmony – the medicinal cure is considered useless" (Finch, 129).

We have actually been speaking as though we each have only one body. The point arising from Hindu philosophy reminds us that we can experience ourselves with more than one. In fact, some of us find certain physicians' approach to our bodies uncanny in that it seems the physician is seeing a body we do not ourselves see (which to a certain extent is true) and, in fact, do not want to have. Ailbhe Smyth (19) speaks from her experience: "'My body' is no more one or abstract than 'I'... Continuous plurality, non-sequential simultaneity. There is no story of my body, only the daily, momentary mediations between experience, sense, memory, desire, understanding – mediations between my bodies, and beyond." Sheba Chhachhi reminds us also of variations not just between individual human bodies, but between bodies of different cultures: "Any attempt to engage with women's health in India ... must, of necessity, incorporate an examination of the simultaneous co-existence of many layers of bodily meaning" (Chhachhi, 103) that come from both "traditional" and "modern" sources, yielding for Indian women multiple embodiments, a "multilayered polyvalent way in which contemporary Indian women look at their bodies ... [including] overlapping, often contradictory codes which constitute body and self for postcolonial Indian women." (Chhachhi, 102).

These forays into other cultures give us some food for thought concerning the Western and modern biomedical paradigm for the body, asking us to imagine the embodiment experiences of those who live within cultures with such different ideas of the body. Reading multiple cultural meanings for our bodies brings me to my original point concerning paradigms. Janet Price and Margrit Shildrick (224) remind us of this paradigmatic body for Eurocentric biomedicine: "It is the young, healthy, well-formed, preferable white male body that is the most acceptable; the 'normal', tractable body, that is, that in drawing no attention to itself can be forgotten," and thus is the paradigm for health as normality. This paradigm makes more abnormal the bodies "othered" by biomedical culture, those that do not fit well its definitions of health and ill health. It operates even in contemporary disability studies, where the paradigm is a match, except the male body is in a wheelchair and well capable of moving it – the fit paraplegic.

One way to understand this paradigm is to remember the characterizations of Eurocentric biomedicine, especially its need for control. Genetic nuclear medicine illustrates this issue of control over nature, the resounding motif of modern sciences in general; and control over nature requires first that it be made to fit the paradigm of exactness embedded in what Husserl

(1970, section 9) calls the process of the mathematization of the sensuous plenum, our originary access to the world. Granted biomedicine is not the only source enforcing this paradigm in our culture, but it probably flows into Western popular culture's infatuation with the body beautiful, perfected, fit and youthful. Susan Wendell links the types of bodies we need to have with the culture more broadly, including its expressions in media: "In the commercial-media-soaked societies of North America, the body is idealized and objectified to a high degree; these cultural practices foster demands to control our bodies and to attempt to perfect them, which in turn create rejection, shame, and fear in relation to both failures to control the body and deviations from body ideals" (Wendell, 85). It is thus interesting that specific physicians will service both necessary and elective procedures to correct perceived "deviations from body ideals" – the anomalous bodies of those born with, for example, hair lips and the rejected bodies of the well-heeled with too much fat tissue or the drooping cheeks from growing older. Wendell recognizes the pervasiveness of what the telos of the paradigm of biomedicine can "exact" of us, making us want perfect specimens of "human body": there are many "disciplinary practices of physical normality... [which are] internalized by most of us, and [are] socially pervasive ... [requiring] us to meet physical standards, to objectify our bodies, and to control them" (Wendell, 88). Kathryn Pauly Morgan (327) even speaks of a "normalization" of elective cosmetic surgery, linking this to a continuing dominance of the "Cartesian machine-modeling of the body" (Morgan, 343, n. 4).

It would seem the resistances of persons with disabilities and of others with "abnormal" bodies might bring new norms, creative normalizations in the sense of optimality, that could affect all of us. For it is not that this paradigm of the biomedical body needs abandoning, as noted earlier, but that its own leakiness and fuzzy borders must be addressed, particularly in the practice of medicine. We need to look to what Drew Leder calls the "absent body." Hence, there must be awareness of the experiences of those persons constructed too strongly by the paradigm and with a disregard for their lived experiences of their own embodiment, together with a recognition of the normalizations to which their experiences have led them. These creative optimizing normalizations, as well as alternative ontologies and ambiguities resisting codification by the paradigm indicate a continuing need to keep the paradigm under a reflective observation, until such time that it itself will be shifted through new social and cultural constructions due to the variety of experiences, practices, and persons that it attempts to encompass but cannot completely explain.

DePaul University

REFERENCES

Ailbhe Smyth, "Loving the Bones: Medi(t)ing My Bodies," in Margrit Shildrick and Janet Price (eds.), *Vital Signs: Feminist Reconfigurations of the Bio/Logical Body* (Edinburgh: Edinburgh University Press, 1998), pp. 18–44.

Alexa Schriempf, "(Re)fusing the Amputated Body: An Interactionist Bridge for Feminism and Disability," *Hypatia* 16(4) (2001): 53–79.

Anthony Steinbock, *Home and Beyond* (Northwestern University, 1995).

Arthur Kleinman, *Social Origins of Distress and Disease* (Hartford: Yale University Press, 1986).

Arthur Kleinman, *The Illness Narratives: Suffering, Healing and Human Experience* (New York: Basic Books, 1988).

Arthur Kleinman, *Patients and Healers in the Context of Culture: An Exploration of the Borderland between Anthropology, Medicine, and Psychiatry* (Berkeley: University of California Press, 1980).

Charles Finch, *The African Background to Medical Science: Essays on African History, Science and Civilizations* (London: Karnak House, 1990).

David Carr, *Interpreting Husserl: Critical and Comparative Studies* (The Hague: Martinus Nijhoff, 1987).

Drew Leder, *The Absent Body* (Chicago: University of Chicago Press, 1990).

Edmund Husserl, *Cartesian Meditations: An Introduction to Phenomenology*, trans. Dorian Cairns (The Hague: Martinus Nijhoff, 1960).

Edmund Husserl, *The Crisis of European Science and Transcendental Phenomenology*, trans. David Carr (Evanston: Northwestern University Press, 1970).

Edmund Husserl, *Zur Phaenomenologie der Intersubjectivitaet: Texte aus dem Nachlass*. Part I: 1905–1920, Part II: 1921–1928, Iso Kern (ed.) (The Hague: Martinus Nijhoff, 1973).

Edmund Husserl, *Ideas Pertaining to a Pure Phenomenology and to a Phenomenological Philosophy*. First Book: General Introduction to a Pure Phenomenology, trans. F. Kersten (The Hague: Martinus Nijhoff, 1983).

Elizabeth Behnke, "Body," in Lester Embree et al. (eds.), *Encyclopedia of Phenomenology* (Boston: Kluwer Academic Publishers, 1997).

Frances Chaput Waksler, "Medicine and the Phenomenological Method," in S. Kay Toombs (ed.), *Handbook of Phenomenology and Medicine* (Boston: Kluwer Academic Publishers, 2001), pp. 67–86.

Franz Fanon, *Black Skin, White Masks*, trans. C. L. Markmann (New York: Grove Press, 1967).

Jan Hartman, "Illness as Evil and Illness as Good," in *Analecta Husserliana* 64 (2000), pp. 85–90.

Janet Price and Margrit Shildrick, "Uncertain Thoughts on the Dis/abled Body," in Margrit Shildrick and Janet Price (eds.), *Vital Signs: Feminist Reconfigurations of the Bio/Logical Body* (Edinburgh: Edinburgh University Press, 1998), pp. 224–249.

Jenny Morris, "Impairment and Disability: Constructing an Ethics of Care That Promotes Human Rights," *Hypatia* 16(4) (2001): 1–16.

John Hockenberry, *Moving Violations: A Memoir – War Zones, Wheelchairs, and Declarations of Independence* (New York: Hyperion, 1995).

John Douard, "Disability and the Persistence of the 'Normal'," in S. Kay Toombs, David Barnard and Ronald Carson (eds.), *Chronic Illness: From Experience to Policy* (Bloomington: Indiana University Press, 1995), pp. 154–175.

Kathryn Pauly Morgan, "Women and the Knife: Cosmetic Surgery and the Colonization of Women's Bodies," *Hypatia* 6(3) (1991): 25–53. Reprinted in D. Welton (ed.), *Body and Flesh* (New York: Blackwell, 1998), pp. 325–347.

Kenneth Hamilton, *Counseling the Handicapped* (New York: The Ronald Press Company, 1950).

Lynda Birke, "The Broken Heart," in Margrit Shildrick and Janet Price (eds.), *Vital Signs: Feminist Reconfigurations of the Bio/Logical Body* (Edinburgh: Edinburgh University Press, 1998), pp. 197–223.

Mairian Corker, "Sensing Disability," *Hypatia* 16(4) (2001): 34–51.

Margot Joan Fromer, *Healthy Living with Diabetes* (Oakland: New Harbinger, 1998).

Mary-Ellen Kelm, *Colonizing Bodies: Aboriginal Health and Healing in British Columbia, 1900–50* (Vancouver: UBC Press, 1998).

Maureen Connelly, "Female Embodiment and Clinical Practice," in S. Kay Toombs (ed.), *Handbook of Phenomenology and Medicine* (Boston: Kluwer Academic Publishers, 2001), pp. 177–196.

Mikhail Bakhtin, *The Dialogic Imagination: Four Essays* M. Holquist (ed.), trans. C. Emerson and M. Holquist (Austin: University of Texas Press, 1981).

Murray Edelman, "The Political Language of the Helping Professions," *Politics and Society* 4 (1974): 295–310.

Orona Oroka, *Philosophy of the Isoko People* (Port Harcourt: Abe Publications, 1998).

Placide Tempels, "Bantu Ontology," in Emmanuel Chukwudi Eze (ed.), *African Philosophy: An Anthology* Orig. *Bantu Philosophy* (1959) (Oxford: Blackwell, 1998), pp. 429–434.

Richard Baron, "An introduction to medical phenomenology: I can't hear you while I'm listening," *Annals of Internal Medicine* 103 (1985): 606–611.

Richard Zaner, "Thinking about Medicine," In S. Kay Toombs (ed.), *Handbook of Phenomenology and Medicine* (Boston: Kluwer Academic Publishers, 2001), pp. 127–144.

Robyn Longhurst, *Bodies: Exploring Fluid Boundaries* (New York: Routledge, 2001).

Rosemarie Garland-Thomson, "Making Freaks: Visual Rhetorics and the Spectacle of Julia Pastrana," In J. Cohen and G. Weiss (eds.), *Thinking the Limits of the Body* (Albany: State University of New York Press, 2003), pp. 129–143.

Sheba Chhachhi, "Raktpushp (Blood Flower)," in Margrit Shildrick and Janet Price (eds.), *Vital Signs: Feminist Reconfigurations of the Bio/Logical Body* (Edinburgh: Edinburgh University Press, 1998), pp. 102–131.

Shigenori Nagatomo, *Attunement through the Body* (Albany: State University of New York Press, 1992).

Shigehisi Kuriyama, *The Expressiveness of the Body and the Divergence of Greek and Chinese Medicine* (Zone Books, 1999).

Susan Wendell, *The Rejected Body: Feminist Philosophical Reflections on Disability* (New York: Routledge, 1996).

Susan Wendell, "Unhealthy Disabled: Treating Chronic Illness as Disabilities," *Hypatia* 16(4) (2001): 17–33.

Thomas Kasulis, "Introduction: The Body – Japanese Style," in T. Kasulis (ed.) with R. Ames and W. Dissanayake, *Self as Body in Asian Theory and Practice* (Albany: State University of New York Press, 1993), pp. i–xx, 299–319.

Kay S. Toombs, *The Meaning of Illness: A Phenomenological Account of the Different Perspectives of Physician and Patient* (Dordrecht: Kluwer, 1980).

United Nations Decade of Disabled Persons, 1983–1992, *World Programme of Action Concerning Disabled Persons* (New York: United Nations, 1983).

Wimal Dissanayake, "The Meaning of Body in Classical Chinese Philosophy," in T. Kasulis (ed.) with R. Ames and W. Dissanayake, *Self as Body in Asian Theory and Practice* (Albany: State University of New York Press).
Zbigniew Szawarski, "Two Models of Medical Knowledge," in *Analecta Husserliana* 64 (2000), pp. 39–47.

SECTION II

SOURCES OF HUMANITY

LESZEK PYRA

NATURE AND MEN. THE COMMON DESTINY

One can speak about different aspects of time and, correspondingly, about different histories. One can distinguish astronomical (cosmic) time, geological time, evolutionary time, social (human) time, subjective time, etc. As far as the natural history of life on Earth is concerned of philosophical interest are two points. The first is the origin of life in general, the second the origin of human life and, as result, the origin of human culture superimposing itself upon the evolutionary history and radically changing the face of Earth. People are changing the natural history of our planet, threatening both the future of life in general and of human life in particular. Science has made people increasingly competent in knowledge and technological power but at the same time man has become less certain about what can be done, and what cannot, in regard to nature out of which he evolved and in which he is wholly immersed. For the first time in the evolutionary natural history there have appeared beings with conscience, *Homo sapiens*, the wise species that should take care of both human and non-human life on Earth. There have developed different concepts of nature within the history of human ideas and different practical directives followed them as regards biotic and abiotic nature. Some of such concepts are sketched below and then the role of people in natural history is shortly presented, while special attention is paid to human moral obligations in regard to the community of life on Earth, especially nowadays, when man has turned to the third millennium and when certainly the time has come to cope with different ecological problems.

1. THE CONCEPT OF NATURE – A DIACHRONIC VIEW

The importance of the category of nature, both within traditional and contemporary philosophy, is unquestionable. When one looks at the history of philosophy one notices immediately how the concept of nature changes, how it evolves in time. The ancient philosophers were inclined to search for the ultimate ground of all reality, for the so-called *arché*, and finally found very different things. They assumed that through philosophical reflection the nature

A-T. Tymieniecka (ed.), Analecta Husserliana XCIII, 99–107.

of the world would be revealed to them. The early Greek thinkers, such as Thales, Heraclitus, Anaxagoras, Pythagoras and others, treated philosophy as we now treat contemporary science, and pointed to different categories which, according to them, constituted the deepest reality of the world. Such categories may roughly be divided into two groups: material and ideal. And so, as it is well known, some of them claimed that the essence of reality was formed by four material elements: water, air, earth and fire.

Thales, for example, invented an ingenious hypothesis according to which all objects were variations of one basic ingredient – water. For water, he argued, if heated, becomes steam, and thus all entities which are gaseous can be perceived as rarefied water; water in its natural state is liquid, and appears in all other sort of liquids; and then, if water is cooled, it becomes a solid, ice. Water is needed for life: without water a seed does not germinate. Eventually, it seemed plausible, that all things consisted of water, one way or the other.

The other kind of *arche* was of an ideal character, e.g. *apeiron*, numbers, or of a semi-ideal character, e.g. atoms. As far as atoms are concerned, it must be remembered that Democritus was able to work out the atomic theory in ancient times when no careful investigation could produce any empirical confirmation of such a theory. Among the ancient views on nature one can also easily discover those combining two spheres: spiritual and material. This is well seen in the case of Plato's views, which I call semi-dualistic, using the prefix "semi" because the two components of reality: Ideas (sometimes called Forms or Universals) and Matter, were not treated on equal terms – the ideal component certainly prevailed. According to Plato the material component of reality was in the best condition just after it had been built by Demiurge, the god-like builder of the world, and when the time elapsed, the condition of the material world slowly but constantly deteriorated. Next, his pupil and successor, Aristotle, continued to propagate *de facto* the dualistic philosophy (some claim that a pluralistic one), best expressed by the categories of matter and form, the philosophy much more resembling the traditional classical dualism of modern times than Plato's views. In the medieval philosophy nature was thought of as the emanation of God; as a consequence, when compared with God, she was eventually treated as much less important. Perhaps, with a certain amount of oversimplification, it may be stated that the medieval philosophy was greatly dependent upon the ancient one. It is probably best seen when one compares the views of Aristotle with those of St. Thomas of Aquinas. In the case of the latter one can easily notice that the material reality, when well examined, could supply proofs for the existence of God; such proofs seem to be a necessary ingredient for any systematic philosophy in the Middle Ages.

In the so-called classical dualism we find the concept of nature resembling that of a mechanism. Also in the field of philosophical anthropology some attempts were undertaken to present man as resembling some kind of a mechanism (however complicated). In the same line of thought as Galileo Galilei and Descartes remained the theory of nature as propagated by Isaac Newton at the end of the 17th century. During the 18th century, in the Age of Enlightenment, the Newtonian paradigm prevailed and all living beings were thought of as mechanisms of some kind. But in the legacy of Newton especially the biological sciences were too mechanistic, too reductionistic and therefore unable to explain different creative phenomena, especially the origin and the continuation of life in natural history.

The situation changed in the first half of the 20th century when ecologists adapted, it seems from the classical physics, the theory of the "balance of nature". Accordingly, ecosystems were thought of as having a certain optimised point to which they were inclined to return if disturbed by some external factors (flood, fire, volcano eruptions, etc.) But if shaken too much, ecosystems were likely to break down, being unable to come back to the steady-state again. Man's activities, especially logging, agriculture, mining and the like, when performed on a large scale, were able to threaten whole ecosystems and, if uncontrolled, were likely to destroy many of them irreversibly. In addition to this, man has become a global species, present and very active in every part of the world. According to Anna-Teresa Tymieniecka: "We are, with all living beings existentially symbiotic, which entails the basic life fact that we are in various, more or less developed degrees 'attuned' to *everything-there-is-alive* (...) But the intellectual advance of humanity in the last centuries might have weakened this attunement" (Tymieniecka 2005, XXXIII).

Then the theory of evolution introduced a new paradigm into the thinking about nature, and especially in regard to man, presenting him as evolution's highest achievement – on the one hand, but also stressing man's animaliness – on the other. Such a theory also suggested that there was no unsurmountable discrepancy between man and nature as such, that man was a part of nature (however extraordinary) participating with other beings in the odyssey of evolution. The contemporary model of nature appears to be different from the mechanistic (Newtonian model) because it stresses the importance of two phenomena absent in the mechanistic model, such as the view that all the components of nature are, to a lesser or greater degree, dependent on each other, and that nature changes constantly, that she evolves all the time. Therefore, the concept of nature nowadays is much better conveyed by the comparison to a living organism ("open system") than to a mechanism

(Prigogine and Stengers, IX). It includes a picture of closely connected parts and processes sustained by the co-operations but also competitions that appear among them. When we look closely at the natural history we get a mixture of life persisting in the midst of its perpetual perishing (Pyra 2004). The concept of nature as semi-organism succeeds also in abolishing the sharp dualism between man and nature. Especially the relativity and quantum theories prove that the act of knowing the world cannot be wholly separated from acting upon it. Therefore, in the light of the above formulations it is rather easy to accept the suggestion that the holistic concept of nature should be introduced. In this context Anna-Teresa Tymieniecka notices: "Indeed, we may infer an ontopoietic dictum that only reference to the whole provides us understanding of its vital fragments!" (2005, XXIII) The holistic approach to nature is certainly worthy of much consideration, although it does not necessarily require the Gaia hypothesis (Pyra 2001, 2003).

At the end of the 1980s there appeared some opinions suggesting "the end of nature" and also the end of the concept of nature, at least in its traditional meaning. Although formerly man could count on natural givens, he cannot count on them any longer. Holmes Rolston III writes: "In this century, humans have stressed these natural systems to the breaking point. The water is polluted; the soil is degraded; wildlife is gone or going; forests are cut down; deserts advance on overgrazed lands. Humans are upsetting, irreversibly, even the climate; the change will be disastrous because it will be so rapid that natural systems cannot track it. In the twenty-first century, there will only be nature that has been tampered with, no more spontaneous nature" (Rolston III 1994, 197). According to McKibben contemporary man lives in a "postnatural world" in which there is no such thing as nature any more. The same author also claims that nature has definitely achieved her end. McKibben writes: "Having lost its separateness, it loses its special power. Instead of being a category like God – something beyond our control – it is now a category like the defense budget or the minimum wage, a problem we must work out. This in itself changes its meaning completely…" (McKibben 1989, 210).

One cannot quite agree with McKibben, however. Although there really seems to be no more (or at least scarcely any) pristine nature, because the anthropogenic activities have changed the world almost globally, there still remain in the world some quite wild areas in which we should let nature take her course. And, according to me, it is enough to think about such areas using the following definition of wilderness: "a wilderness, in contrast with those areas where man and his own works dominate the landscape, is hereby recognised as an area where the earth and her community of life are untrammelled by man, where man himself is a visitor who does not remain" (Nash 1974, 5).

2. THE EVOLUTIONARY HISTORY OF MAN

For more than a hundred thousand years the history of man has been closely connected with nature surrounding him. The evolution of man has always been based upon interactions with the natural environment, which in turn determined man's cognitive structures and the content of his consciousness. Man always was, on the one hand a part of nature, but on the other he became more and more alienated from her. The notion "ecology" slowly lost its purely biological character and began to be used to describe man's different relations with the environment. The ecological problems, as understood by John Passmore, began to be the objects of interest of different sciences, and also of the humanities, which started to examine the relation man-nature, especially in its ethical dimension, in order to limit man's destructive activity in reference to natural environment (Passmore 1974). Of course all this was done within the actual, prevailing paradigm of culture and had as its primary aim defence of man's different interests, it was supposed to guarantee the survival of man on the Earth.

The theory of evolution shows that during long periods different organisms evolved, leading ultimately to the appearance of creatures possessing self-consciousness and intelligence. In the case of man evolution "invested", so to say, in intelligence appeared for the first time (as far as we know) in the Universe. One should also notice, however, that within culture man developed something as unique as intelligence, namely moral sense, which warns him that he should stop destroying carelessly the biosphere. Something appeared on Earth that gave rise to a global sense of obligation. It seems that morality appeared rather lately in the history of culture, functioning as its instrument of adaptation to life on Earth. The Darwinian paradigm – random mutations and natural selection of the fittest – was especially well suited to biological sciences, but the story did not stop in biology as such because it continued into culture. Charles Darwin in *The Descent of Man* explained how the mechanisms of natural selection could produce "moral sense", despite the assumption of the ubiquitous "struggle for existence". According to Darwin young mammals required a long maternal care in order to survive. Such care was usually given to them by mothers. Motherly love, together with parental care of the fathers, made it possible to form small families or clans which shared so-called "social sentiments" among their particular members. As a consequence, claimed Darwin, such emotions "will have been increased by natural selection; for those communities which included the greatest number of the most sympathetic members would flourish best, and rear the greatest number of offspring" (Darwin 1904, 107) Then the families (clans) connected in order to successfully exploit resources which usually were scarce. Co-operation prevailed instead of competition and rivalry. There appeared a stable tendency to form larger and larger social groups, then tribes, and ultimately nations. The feeling of sympathy (altruism) spread on a large scale. Eventually moral norms are not be accounted for as products determined by the processes of nature, but by cultural traditions (most often religions) and principles that are products of human history.

Following Darwin one can say that in the 20th century nations have a tendency to form still larger communities, e.g. a United Europe. The category of altruism, under the cover of "human rights", spreads throughout the world. And, eventually, the process of transition of certain norms from generation to generation, becomes relegated to the sphere of culture and, as a subject matter of education, takes an organised and orderly form, first of all in schools. When on compares the natural history with human cultural history one should strongly underline that the *category of time* becomes essential. What I mean is that cultural changes appear more quickly than evolutionary changes. The genetics of the people living in the last two centuries has not changed much, whereas cultural views definitely have. Consider, for example, the changes in ethical ideas. In the 19th century opinions about slavery changed radically. In the 20th century ideals of freedom, human rights and, in the last decades especially, negative opinions about capital punishment were widely spread. In such ways human culture transcends biological nature. From the point of view of biology and ecology organisms can exist only within species and ecosystems, whereas from the point of view of social life humans can exist only within their heritages and social institutions of different kinds (Pyra 2003, 24).

In the 20th century also, because of the environmental degradation of nature, which really no one questions, man comes to the conclusion that he cannot do nature anything he wants. There appear some voices questioning man's superiority in the world, at least superiority traditionally conceived as dominion and tyranny. The traditional world's view implied that we, humans, could do anything we wanted on the planet, and that abiotic nature, plants and animals, which preceded us in the odyssey of evolution, would accommodate themselves to our wishes. Nowadays we humans, under the influence of the theory of evolution and ecology, must realise at last that we are within life, not atop of it, and that life as such is a system of interconnections, which we not yet understand well. A very good advice comes from Aldo Leopold: when one has doubts, one should preserve the health (integrity) of the ecosystem (Leopold 1990, 261).

3. PLACING ETHICS IN THE PERSPECTIVE OF NATURAL HISTORY

In the light of the above considerations the necessity of developing ethics concerning fauna, flora, species, ecosystem, landscapes and the like has been suggested with an intent to apply them in different spheres of human activity. It appears, however, that at the bottom decisive is the private ethic practised by particular individuals immersed in their particular, local places. Ethic is not just only theory, given only to those most educated, ethic is first of all convictions concerning, in this particular case, the relation man–natural environment. What is looked for in this case is the increased quality of life in a given habitat, which concerns not only men but also the other forms of life. The life of human beings is very special in this respect because only men live fully (consciously) in nature (biology) and culture at the same time. Thanks to nature human beings gain some dominion over nature which, however, requires some wise restriction. Man certainly is the most free of all beings but it should be remembered that – in culture at least – freedom is necessarily connected with responsibility, and therefore with duties. Man should remember this in order not to commit the sin of pride (*hubris*). The old traditional, anthropocentric ethics made only humans the *loci* of value, but the new environmental ethics must change this, extending the sphere of values in such a way that it would overwhelm also the other beings of nature. According to Holmes Rolston III: “Among the remarkable developments on Earth with which we have to reckon, there is the longstanding ingenuity of the myriads of species that compose natural history; there is the recent, explosive human development; and there ought to be, and is, a developing environmental ethic that optimizes natural values” (Rolston III 1994, 235). One can only add that such an ethic is, so to say, still *in statu nascendi*.

Historia magistra vitae est, says the old Latine saying. Humans should study their history in order to understand what they are, why they are as they are, etc., but apart from studying the history of human species they should also carefully study the history of nature. Some theoreticians disagree with such an opinion because, according to them, nature has no history. They are probably right in the sense that plants and animals do not know their histories.

But they seem to forget that the history of all living entities is written in their DNA, which can be decoded and interpreted, although bioscience cannot explain why some events happened in the past, and is unable to predict what will happen in future. It seems that nature moves towards complexity and individuality, and as a result the Earth story eventuates in *Homo sapiens*, in humans who then must be moral overseers of Earth. In this context Anna-Teresa Tymieniecka pays attention to the so called "symbiotic empathy" with which the gardener, like the farmer, feels his/her beasts for their well being, pain or sadness, and also participates with his or her entire sentient being in the sentient circuit of plants. Therefore she writes: "Hence it is symbiotic empathy that offers us the sought for *filum Ariadne* the all-pervading although differentiating thread running through the entire ladder of life's deployment. What else could we consider to be the key to establishing unity among all the divers attitudes, interests, and perspectives if not life itself? The ontopoietic stream beginning with the most elementary sequence of life and running to the highest sphere of complexity in the Human Creative Condition is the sole philosophy ... " (Tymienicka 2005, XXXIV) Concluding, one must admit that the challenge referring to studies of natural and social history, studies of environment and ecology is of utmost importance. Such studies face humanity as a never-ending task. One should also be constantly aware that with increasing scientific knowledge comes increasing duty.

So far I have referred to the past and to the present time. A remark concerning the future would be in place. Although nowadays we are in the midst of extinction on a global scale, we certainly should not be afraid of the total disappearance of life on the prolific Earth. Even in the so called "normal epochs" the usual result for species is extinction. But it seems that the same factors that bring extinction at the same time contribute to respeciation, bring new species. As natural history shows there were often mass extinctions with dramatic resettings of evolutionary history. For example, according to David M. Raup (1979) probably as many as 96 percent of all species on Earth vanished at the Permo–Triassic period. Also some other mass extinctions happened in the past but in result some new forms of life appeared. In the light of what we know of extinctions it seems obvious that man is, *de facto*, unable to destroy life in general. If *Homo sapiens* would succeed to destroy life on Earth it probably originates again. It might come back in different forms, perhaps in millions of years. But as far as the interplay between extinction and evolution is considered man should support evolution, not extinction.

Krakow, Poland

REFERENCES

Aldo Leopold, *A Sand County Almanac. With Essays on Conservation from Round River* (New York: Ballantine Books, 1990).

Anna-Teresa Tymieniecka, (ed.), "*Introduction*," in *Analecta Husserliana* LXXXIV (Dordrecht/Boston/London: Springer, 2005), pp. xiii–xxxvii.

Bill Mckibben, *The End of Nature* (New York: Random House, 1989).

Charles Darwin, *The Descent of Man and Selection in Relation to Sex* (New York: J. A. Hill, 1904).

David M. Raup, "Size of the Permo–Triassic Bottleneck and its Evolutionary Implications," Science 206 (1979): 217–218.

Holmes Rolston III, *Conserving Natural Value* (New York: Columbia University Press, 1994).

Ilya Prigogine and Isabelle Stengers, *Order out of Chaos. Man's New Dialogue with Nature* (New York: Random House, 1984).

John Passmore, *Man's Responsibility for Nature. Ecological Problems and Western Traditions* (London: Gerald Duckworth, 1974).

Leszek Pyra, "Environmental Holistic Ethics: Leopold and Callicot," in *Contributions of the Austrian Ludwig Wittgenstein Society*, Vol. IX (2) (Austria: Kirchberg am Wechsel, 2001), pp. 202–207.

Leszek S. Pyra, *Environment and Values. Holmes Rolston III's Environmental Philosophy* (Kraków: Zeszyty Naukowe Akademii Rolniczej im. H. Kołłątaja, 2003).

Leszek Pyra, "The Metaphysical Foundations of Environmental Philosophy," in *Analecta Husserliana* LXXIX (Kluwer Academic Publishers, 2004), pp. 749–754.

Roderick Nash, *Wilderness and the American Mind* (New Haven – London: Yale University Press, 1974).

BRONISŁAW BOMBAŁA

IN SEARCH OF THE SOURCES OF HUMANITY

Man will only get to know himself once he realises that he lives with his soul, not his body.

Leo Tolstoy

The most good is done out of love, not out of the logics of concepts.

Bogdan Jański

1. IN SEARCH OF THE STARTING POINT OF THE ANTHROPOGENESIS

All attempts to grasp the phenomenon of the human being must go back to his beginnings and his history, as any non-historical approach would always be subjective and incomplete. This is indicated by the definitions formulated from various research perspectives, in which man may be:

– biological (*animal*),
– social (*homo socialis*),
– political (*homo politicus*),
– economic (*homo oeconomicus*),
– a creator of civilisation and culture (*homo faber et artifax*),
– religious (*homo religiosus*),
– of entertainment (*homo ludens*).

The idea of human beginnings, i.e. the starting point for the anthropogenesis, has a crucial impact on understanding the sense of being of man himself and the world. In contemporary science, the prevailing conviction is that man has developed over the long course of evolution. In speaking of evolution, scientists mean ever more developed forms of life that have emerged. Development is understood as "a biological progress upwards" (Wojciechowski 1985, 53), which leads to realising new possibilities of the living systems and to the creation of ever-higher forms of life (Paturi 1984, 17). This approach assumes that the stages of the human genesis followed a continuous linear pattern: Ramapithecus, Australopithecus, Pithecanthropus,

A-T. Tymieniecka (ed.), Analecta Husserliana XCIII, 109–126.

Neanderthal, Homo sapiens. Each of these stages was connected with a gradual formation of the neuro-cerebral system, predominantly related to the hands, which served as the beginning of the higher functions of reception and reflex, sensation and instinct, information and work.

In the 20th century, another view emerged, according to which man originates from a core of life in a radiantly manner, with the core of biogenesis being shaped like an onion (Teilhard de Chardin 1984, 61). This indicates that man emerges directly from the mystery of the cosmos, while each particular form of "human-like beings" are shells and scales between which there is no direct transformation from one morphological form to another. The continuation of life always takes place through the core. In this view, the first human being ontogenetically and philogenetically was not born directly from the "pre-human" beings, which makes the "beginning" of man as a particular event inaccessible to empirical knowledge. One may only suppose that the roots of the "thinking" man go back to the anthropoids from Oligocene, while apes and homonidae were only lateral shells of the anthropogenesis. The same refers to the subspecies of homonidae, which were also lateral shells, replaced by *hominis sapientis*.

The exceptional nature of the anthropoidal being manifests itself through his unique nervous system, which serves not only the sphere of reception, but also processing of information, intelligence, reactivity, aspirations, activity. The human brain, or more precisely, the mind, is constructed in such a way that it is capable of grasping the entire structure of the universe: the microcosm and the macrocosm. Therefore, there is a significant likelihood that man can grasp the essence of things. The human mind is considerably more powerful than would be required by the needs of the biological being, and is very much focused on learning about the world. Moreover, curiously, it has numerous tools for learning: the ability to perceive and store information, to recall, to think discursively and heuristically, etc. Perhaps the entire structure of the world has already been given to our minds, and we are merely confronting the state of our minds with the actual nature of the world with the use of our senses. Because of the (abstract) reconstruction of the world through scientific discovery, our mind allows for its physical reconstruction (work, culture).

At this point, a question arises: is it possible that the special software known as the human mind has matured on its own as the result of the evolution? It is not easy to answer this question. One thing, though, is certain: drawing conclusions based on analogy, with respect to the application of the studies conducted by zoologists to anthropology, which is the only method available to us nowadays, is to a large extend prone to be unreliable. This is so, because, on the one hand, there is this amazing continuity of man against the world of animals, and on the other non-continuity in the sense of existence of the psychical and spiritual plane. From the time perspective, a human being apparently emerges from the material plane and transcends the entire reality. At the same time, a dialectic tension grows between the physical and psychical/spiritual planes, as well as the realisation of our own subjectivity.

In view of the above, human origins remain a mystery. It is hard to indicate, who was actually the first human being, and where is the border between the zoological unit and a person, i.e. a being with its own subjectivity, a free being, which acts in various ways and transcends the material world. As the reflexive cognition is thought to be the essential determiner of the human being, in order to answer the questions posed, it is sufficient to determine in which period of history this reflexive thinking first appeared.

According to L. von Bertalanffy, the crucial moment in the hominisation was the development of symbolic behaviour (von Bertalanffy 1968, 136). It is only the human being that may be referred to as the *animal symbolicum*. The symbolism covers all that is called culture created by man. Symbolic behaviour brought about the following:

– the replacement of the philogenetic evolution, based on genetic changes, with history based on tradition and behaviour, which resulted in accelerated development,
– the replacement of manipulation with objects with manipulation with symbols, thanks to which man was able to go beyond objects,
– the transition from general biological propriety to conscious human propriety,
– the creation of the world of culture and language, as well as achieving autonomy by man,
– in subsequent stages, the replacement of a word with an algorithm,
– "autonomy" of the symbolic universum and gaining control over man in the form of economic and political ideology.

According to numerous authors, the special criterion that justifies the unique position of man in the world appears to be speech. No other being than man has the ability to convey meanings and interconnect them. However, animals also use speech as a means of communication understood in its broadest sense, as they communicate with the use of gestures, as well as chemical, acoustic and tactile signals.

2. IN SEARCH OF *SACRUM*

However, the most important argument for the existence of man's "higher dimension" is religiousness (Zdybicka 1984, 21–23). The ability to feel and the urge to seek some higher purpose values, together with the capability of reflecting upon philosophical and religious issues, are those criteria of humanity which elevate humanity from the zoological (a biological being) up to the anthropological plane (homo sapiens) and prosopology (a person). We assume that there exists a unique, inseparable bond between humanity and religion.[1] It is, then, seeking for sacrum, both in one's surroundings and within oneself, that is the mark of humanity.

Research conducted by alaeontologists offers evidence that even the first hominids had an organised and definite religion, indicated by the cult of the dead, the first human graves, and hunting rituals (Eliade 1988, 4–8). More complete evidence for religiousness is provided by findings and discoveries related to *homo sapiens*. Most importantly, what appeared in that period is art related to tools, technology, painting, and sculpture; as well as a well-developed cult of ancestors, a ritual oblation, a hunting ritual, and the cult of a goddess-mother. The fact that art appeared on such a large scale (beginning with Europe through Africa to Siberia) speaks volumes of man's full intelligence. As soon as spoken and written language appeared, human self-awareness appeared to be obvious.

In the holy books of India, as early as the beginning of the second millennium before Christ, anthropological reflection occurred. In the oldest, Vedist, terminology, the *individuum* was referred to with the use of the reflexive pronoun *atman* (one who is turned to himself). *Atman* was interpreted as the spirit, the soul, the self, which remains unchanged in the course of *sansary* (the journey of souls). Over time, *atman* also came to stand for the human insight: the complex of life, the complex of waft of life, the set of senses and intellectual functions. *Upaniszady* refers to man with the term *purusza*. In *sankkhja* and *joga* religions, this term stands for a deep self, which is a purely spiritual element, deprived of cognitive, emotional and volitive contents. Here, *Purusza* serves as a complementation of *prakriti* (nature, primary matter). *Prakritia* is certain passivity, non-awareness, senselessness; it has no cognitive, differentiating or awareness capabilities. On the other hand, *Purusza* is cognitive (*jna*), conscious (*cetana*), differentiating (*vivekin*), yet it does not have *gun* (features, properties), i.e. active factors, typical to matter (activity is a feature of the matter, not the spirit). *Purusza* is non-active, non-creative, non-transforming, non-evolving. It is only a cognitive, conscious, and enlightening contemplator of nature, spiritual light, someone who feels. As can be seen, in this system, the human being expresses himself in "spiritual non-activity" rather than in material activity. Man is considered a passive participant of reality. It may be claimed that in Hindu mind frame, man is not a reserved being, but it is rather a playground for hierarchical organised material – spiritual as well as psychical – cosmic functions dominated by the divine Energy-Awareness, which constitutes complete existence and everlasting happiness. Man may only be himself thanks to his relationship with the deity (cf. Brockington 1990).

A new understanding of man began to form in Mediterranean culture. As early as in the second millennium before Christ, there was in Greek culture the term of *prosopon*, which initially stood for face and insight, which expressed the human "self," and which later came to mean being in a relationship with another, social status, a persona, a representative of a community, a mask in the theatre (Bartnik 1995, 66). Its Latin equivalent, *persona*, initially also meant a mask. In Cicero's time, this term was used to determine someone's status and role within the public life. Over time, the term *persona* entered the legislation and defined the subject of the law, the dignity of man and citizen, as opposed to *non persona* or *res* (a thing). A slave was, for instance, a member of *res*.

The stoics, who believed in the theory that the world is a giant stage, erected by the deity, upon which each human being has a role to play, made a major contribution to the understanding of man. This role is in fact his true nature, character and fate, which makes him what he is. Hence, for the stoics, *prosopon* is not an empty personification, invented by man, but rather his essence, which is not yet fully discovered and still mysterious. It is only revealed at the end (*kairos*) in acting as the deity decides. The stoic concept of person is religious, holistic and dynamic, and yet, at the same time, it is subject to the necessary plans from above and it lacks freedom. In this view, man would become a person when he played his role correctly by harmonising drives and behaviour with the common and reasonable natural order. The person was, then, supposed to be the realisation of the cosmic, biological, social and ethical order (Bartnik 1995, 68).

However, it was Christianity that brought the proper understanding of man as a person. The central point of the Christian vision of reality is the figure of Jesus Christ, who appears to be the identity of man and God. He perfectly combines both human and divine features. The theology of early Christianity (patristics), while trying to describe this "most perfect reality", introduced the concept of "a non-created person." The fact of describing God as *hipostasis, prosopon, persona* allowed also to construct new anthropology, which would become personalism (prosopology). The term "person" began to describe man the most accurately. Man became "the mystery of a person" and, in a sense, "a minor God." This new definition of man, compared to the pre- and non-Christian approaches, may be presented in the following points: (Bartnik 1995, 76–78):

- contrary to classical Greek philosophy, according to which man was a tiny part of the cosmos, patristics sees the human being as one originating from

other nature, indefinitely beyond the cosmic order, being "the movement" from the matter towards the spirit,
- Greek philosophy was, to a great extent, deterministic, while the patristic approach presented man as a person conditioned by freedom,
- to the fathers of the Church, an individual human being is not a combination of various elements (matter, energy, ideas), but an integral entirety heading for the Absolute Unity,
- in the Christian sense, man is not determined by "necessity," or "the heaven of ideas," nor is he a stoical "creator of the drama of being"; he is rather determined by God, who created man out of love, and who gave man the internal participation in his own life, by over "divinize it."

As a result, a person gains an unlimited and indefinitely deep internal plane, which, however, does not detach him from the world and other people: the person is individual, but still exists within mankind. Patristics also formulated the idea of a person as a "substantial relation," which is a being, whose entire structure is turned towards someone. In this way, the person appears to be a relative being open to others. In medieval times, the understanding of a person was penetrated even further. However, scholastics, drawing heavily on ancient Greek rationalism, moved away from the formulations of the Church fathers.

3. SOCIAL SCIENCES IN SEARCH FOR THE SENSE OF HUMAN EXISTENCE

Following the medieval fascination with the idea of a person, came a critical period in the study of the human phenomenon. The period was initiated by W. Ockham (Gilson 1987, 443), who claimed that reason cannot grasp nor prove the existence of soul, spiritualism, and everlasting life. All the elevated theses of the human being remain solely a question of faith. The person is, then, a creation of faith. Faith, in turn, is, at the most, compliant with intuition. The logical constructions of "being," "nature," "subsistence," and "species," for him do not possess features of reality. They are merely linguistic figures and may, at the most, have a psychological meaning. The person is a psychic "ego," rather than an ontic individuum. In this way, anthropology began to be replaced by psychology and linguistics. As the result, what appeared were specific sciences, which, while conducting studies of narrow fragments of the human being, claimed the right to make farfetched generalisations. This kind of scientific approach is now known as scientism, or, in short, "ism" (cf. Bombała 2001).

In the case of physiologism, which sees a live (human) being as a kind of robot, driven by reflexes, the study of man becomes the study of a mammal, with anthropology being a mere addition to zoology. Similarly, psychologism, which brings man down to a psychic device, and misses his spiritual dimension, becomes the study of an animated machine. The psychic life, in this view, seems to be a game based on drive mechanisms. Sociologism, in turn, sees man as nothing more than a small element in a game of, this time, social powers. Finally, economism perceives man as a puppet in the game of the so-called market mechanisms. From economism, other "isms" emerge, such as technocratism and bureaucratism.

Contemporary psychology highlights "the adjustment" rather than "the good", and so it is far from ethics and inclined towards ethical relativism. Psychologists tend to forget that man often makes decisions based on evaluation judgements. E. Fromm claims, "The divorcement of psychology from ethics is of a comparatively recent date. The great humanistic ethical thinkers of the past (...), were philosophers and psychologists; they believed that the understanding of man's nature and the understanding of values and norms for his life were interdependent. Freud, and his school, on the other hand, though making an invaluable contribution to the progress of ethical thought by the debunking of irrational value judgements, took a relativistic position with regard to values, a position which had a negative effect not only upon the development of ethical theorie but also upon the progress of psychology itself" (Fromm 1947, vi). Naturally, there have been psychologists, who noticed the evident connection between psychology, and psychotherapy in particular, with man's philosophical and moral problems. This relationship was recognised by C.K. Jung, while E. Fromm went as far as to claim that "psychology cannot be divorced from philosophy and ethics nor from sociology and economics." In view of this, Fromm goes further in recognising reality being man, he sees man's multidimensional nature, man's existence in individual, social, economic and philosophical (leaning towards transcendence) dimensions. According to Maslov, this recognition will result in the theory of needs. The higher the needs are placed in hierarchy, the less of needs they become and the more of the tasks that man faces. Psychology is aware of its own limited character, because in the attempt to reach for the sense of life, it becomes philosophy. To sum up, Fromm claimed, "Psychoanalysis, in an attempt to establish psychology as a natural science, made the mistake of divorcing psychology from problems of philosophy and ethics. It ignored the fact that human personality cannot be understood unless we look at man in his totality, which includes his need to find an answer to the question of the meaning of his existence and to discover norms according to which he ought to live. Freud's 'homo psychologicus' is just as much unrealistic construction as was the 'homo economicus' of classical economics. It is impossible to understand man and his emotional and mental disturbances without understanding the nature of value and moral conflicts" (Fromm 1947, 16).

Let us now move to the theory of economism, which not only claims the right to explicate the purpose of human existence, but also imposes on people the rules of conduct and personal fulfilment. This ideology replaced such ideas as happiness, friendship, right, family, etc., with ideas-tasks: success, career, profit, professionalism, etc. Kurt Dopfer claims that since economics was separated from social philosophy, it has always been in the state of crisis. Hence the question: "Does the current paradigm create the basis to formulate the right kinds of questions, does it allow for formulating important theories, does it provide guidance for adequate empirical verification, and finally, does it suggest significant solutions to the most urgent problems of the future?" (Dopfer 1982, 12). Gunnard Myrdal replies that it is necessary to return to the works of the fathers of economics, who believed that economics must have rational foundations. "In order to do that, they had to derive knowledge from a wider range of sciences, and especially from the assumptions of social philosophy. In the sense, which was already explained by John Stuart Mill in his early works, research on economic problems was 'a moral science' " (Myrdal 1982, 144).

This formulation of economics is based on the belief that there are objective values, which may be recognised as facts, analysed and used as a basis for studying economic facts. This evaluating basis for economic studies is provided by moral philosophy. Representatives of "ethical economics" wish to be successors of that old tradition, seeing economics as one of the moral sciences.

In each of these "isms," the human being is stripped of his sense, becomes a caricature, a puppet, moved regardless of its will by internal and external forces. All theories constructed in this way, form a distorted, rather than a true, image of man, which should not be used in any practical applications (in education, management, politics, or art), as it leads to pathology (e.g. the pathology of organisation, which manifests itself in the so-called vicious circle of bureaucracy and in trained incompetence). In order to realise this state of things, a revision is required of both the paradigm ruling in science and the practical activities based on theories burdened with "ism."

Science, when misunderstood, may try to replace religion and may create "half-truths," by obscuring certain theories with something decent (the term "scientific"), in order to win over undecided minds and justify evil (these are creations such as "the scientific philosophy of life," "homo economicus," "techno," etc.). There are plenty of such half-truths. Among them, there are those who have an individual meaning, as well as those with social and historic meaning. Succumbing to them is brought about by simple laziness, which tempts people to be satisfied with the little offered, to make do in the sphere of mind with what is easy and pleasant, and not with what requires pain and effort. As long as this remains nothing more than a weakness, it is forgivable. Yet, it is often the case that a temptation appears to justify this mental and spiritual inability with an easy, generally accessible half-truth or theory, which requires no mental effort. Once that happens, then there is only one step from assuming that one should not only free oneself from serious independent mental efforts to solve the most important life problems, but one should also disturb others in their "harmful pursuits."

4. "INTEGRAL KNOWLEDGE" – A SYNTHESIS OF SCIENCE, PHILOSOPHY AND THEOLOGY

In the modern world, which is ever more dominated by science and models of thinking emerging on its basis, as well as by technology and the lifestyle created by it (technopoly), both philosophical discourse and religious contemplation seem to have lost their convincing powers. God and science seem to belong to such different worlds that any combination of theology and science is regarded as a threat to the foundations and methods of science. However, the discoveries in modern physics indicate that the world around us is becoming ever less material. It can be compared to a "thought" rather than to a giant mechanism. Yet, when a collapse can be observed of Laplace's hypothesis of the world as a mechanism, what follows is a gradual disintegration of the enormous set of materialistic models.

We are now facing a new revolution in thinking, which was initiated by the quantum theory. It leads to the synthesis of two major contradictory trends in thinking: spiritualism and materialism. This is, in fact, an attempt to cross the border between the spirit and the matter. (metarealism). Metarealistic thinking goes beyond classical logics (metalogics). Metalogics goes beyond language, beyond categories of reasoning: the uncertainty principle in mathematics (which shows that it is impossible to prove whether a given statement is true or false), the complementarity principle in physics, or finally the so-called Planck constant, which discovers a physical barrier of cognition. The complementarity principle claims that the basic elements of reality, such as electrons, are beings with two faces. At one time, they appear as particles of the matter, at others as non-material waves. These are two opposing descriptions, but they are equally correct and complementary. On this basis, Heisenberg concluded that complementarity of a particle and a wave brings to an end the Cartesian dualism of spirit and matter: these are two complementary elements of the same reality. In this way, the basic difference between the matter and the spirit is eliminated. The materialistic vision of the world, enforced over the centuries by theoretical and empirical physics, is falling apart. The new cosmology shows a new, totally different way of perceiving reality: quantum physics, reaching beyond the superficial order and the reality of phenomena, goes as far as transcendence. Both quantum theory and cosmology are pushing the borders of cognition ever further, reaching as far as the greatest mystery faced by man, the source and sense of the universe.

Quantum physics forces us to abandon the images of time and space as we know them. The world, from atoms to stars, is characterised by a total and indivisible order. Each particle contains an entirety; each object is a reflection of all others. In this way, physics, which had its contribution into setting aside metaphysics and theology, today is itself becoming much of "metaphysics and theology." Farfetched as this statement may seem, it helps us understand that human knowledge goes far beyond scientific knowledge. What is proven to exist after all, then, is the old thesis of Russian philosophers, which says that true knowledge must be based on three pillars: revelation, philosophy and science (*wsiejedinstwo*). Thus, the integral knowledge (*theosophy*) comes to being, which opens the gate to "integral creativity" (*theurgy*), which in turn enables to build an "integral society" (*theocracy*).

Therefore, there are three possible areas in which one should search for answers to the question of sources of humanity, i.e. religion, philosophy and science. Following this path, by combining a philosophical reflection with theology and psychology, let us reach the sources of humanity. Man hopes for everlasting life, he wants to live forever, but the rules governing the world of nature do not offer eternal life. Following his reason and conscience, man is looking for the truth, but he has not got sufficient powers to discover it. The two great desires of everlasting life and truth are opposed by two great facts – the unavoidable domination of death over body and the unshaken supremacy of evil over each soul. Man has the desire to rise above his own nature, but death makes him equal with animals, while the inclination for the evil makes him even worse than them.

Let us assume, after Soloviev, that human nature is characterised by three groups of needs: animal needs, mental needs and spiritual needs (in Maslov's need hierarchy, mental and spiritual needs are related to the needs for progress). Animal needs are reduced to two major needs, for nutrition – to temporarily maintain individual life, and for breeding – to maintain the life of the species. The basis for animal life is nutrition, its purpose – breeding. An animal's purpose in life is fulfilled once its offspring is born and raised: its entire existence is simply a means for achieving this goal. That leads to a conclusion that: "Each generation, and each individual within it, only exists to produce offspring, and its offspring also exists solely to bring to the world the next generation. This means that each generation only makes sense when there is the next generation. In other words, the life of each separate generation does not make sense; but if the life of each of them does not make sense, then the life of the entire species does not make sense" (Soloviev 1988, 16).

If each generation only exists to die when a new one is born, which, in turn, will itself face the same annihilation, and if a species only lives in such constantly dying generations, then life of the species is a constant death. In view of that, it may be concluded that life, in fact, does not have any purpose or sense. The domination of death, which hangs over man's animal existence, which turns it into a sudden "fruitless" effort, is nothing exceptional. Science shows that death rules the world. Death reaches not only individual specimens, but entire species of animals and plants. The same is true for stars, planets and asteroids.

Scientific research, then, shows nothing more than the helplessness of life. While looking for a dignified life, man notices that satisfying his animal needs and drives is not his ultimate purpose and does not lend sense to his life. "Man-animal" gives in to this fate even though it is against his will, but he cannot ultimately accept it, because he has an innate promise for a different life. While an animal simply follows his natural life impulse (instinct), man is capable of judging whether or not he should succumb to those impulses. When an animal only tries to live, man has a will to live properly. Therefore, while acting, man asks himself a question: is this the right thing to do? In this way, human activity is limited by the internal call of his conscience.

However, it is the animal nature of the human being that makes him prone to acting against his conscience, to following sensual stimuli. If man only had his drive, the drive itself would be neither good nor bad, but similar to animals, it would be a simple fact of nature. On the other hand, if man only had his moral aspirations, those aspirations without any internal obstacle would work as an ordinary innate strength. There would be no moral problem. Yet, when two opposing forces collide, a moral problem arises. In that case, the voice of conscience, in the face of nature that opposes it, calls itself the law, while the drive, as being against the law, becomes lawlessness (or, theologically, a sin).

By judging natural drives, the law-conscience does not replace them with anything else and does not deprive them of their original power. The law only refers to the external manifestations of evil drives, without removing their causes. The law only makes it possible for us to fully comprehend our evil deeds. It indicates what must not be done, but it does not offer clear guidance as to what should be done. In the positive sense, doing the right thing remains within man's creative activity. Nevertheless, it is neither easy nor simple to choose to do the right thing. The awareness of the moral duty, which emerges in man, removes him from the routine of animal life and abandons him helpless. Conscience has passed its judgement of nature, it revealed the right and the wrong, but it did not provide sufficient strength to change nature, to make sure the right wins and that the wrong is defeated. Therefore, "the natural suffering of the deadly nature is complemented by a moral suffering – an internal dualism, which psychologists refer to as a dignity dissonance."

The fact itself of being aware of a duty does not offer sufficient strength to fulfil it, which constitutes man's moral dilemma. Moreover, since man's nature appears to be criminal (often knowing what is right, man chooses to do wrong), the fact of realising that does not change his nature. Theologically speaking, man's nature is contaminated by sin. And, in order to be able to truly repair this sinful nature, some other, real and inspiring source of life must emerge. Man cannot create the source of this new better life (he cannot invent it); and it must exist regardless of his will. If it is to be real – it must be granted. This new, good life is granted to man from the One who is over and beyond the world. Hence, the new, better life is called *grace*: "Grace is the good, of which man does not think, but which he really receives. And because our nature in itself is not good, and the moral law of our reason (...), has no strength to actually give us the good, all we can do is (...) admit, that it exists regardless of our nature or our reason (...). The true Good, i.e., the being that has in itself both the full grace and its source, is God" (Soloviev 1988, 22).

Nonetheless, in order for man to truly find himself in the path of grace, his knowledge based on reason will not suffice; he needs a deed, which is an internal movement of the will: man should stimulate himself internally to accept grace. Consequently, man looking for the good, need not create anything new; all he needs to do is open himself to grace and get rid of obstacles that separate him from it. The crucial obstacle standing in the way to the good and happiness remains within himself, as man may follow the wrong principles, which originate in his own free will.

To believe in God is to recognise that the good of which our conscience testifies, and which we seek in our lives, exists regardless of our nature and our reason. Without this faith, we would have to assume that the good is merely an illusion or a figment of the human mind, i.e., that, in fact, it does not exist. For moral reasons, we must not make this assumption, because we ourselves as moral beings, as well as our lives, only make sense through faith in the true good. We must believe that it exists in itself and that it is that real truth: we must believe in God.

In order to live according to God's principles, we must live truthfully. Man, who contemplates the form of the world, the divine power forming the world appears to be the world's light eternal. God's omnipotence appears to be an idea that shines on everything. A religious man must not stop at uniting with God through prayer. Having risen to the top of religion, having established a spiritual bond with God, enlightened and renewed (godhumanity), he establishes a new religious unity with other people. The commandment for this new covenant is perfect love: "I give you a new commandment, and that is that you must love each other." Man in the lower stages of religion would bring God sacrifice, as he thought that this was what God required. In this view, the deity not only gives us life, but it also lives at our expense. This particular perception was challenged by the words: "It is love that I yearn, not sacrifice." These words mark a turning point in approaching religion, and when man in the higher stage of religion offers God his greatest spiritual sacrifice of his will, and unites it with God's will in the decision "Let your will prevail", he also accepts God's words: "It is love that I yearn, not sacrifice," i.e., he does not wish to live at the expense of other people, but he wishes to give others. Man who is spiritually united with God should relate to his neighbour as instructed by God: "As you received it for free, give it for free; give your neighbour more than he is entitled to; refer to your neighbour better than he deserves" (Soloviev 1988, 48). The manner in which people should relate to each other, which is derived from this commandment, is superior to all other principles of social interaction. Yet, its implementation calls for an internal transformation, as, "We rebel against God, move away from our neighbours, succumb to the power of the body, while real life, as it should be, calls for something quite opposite: surrendering voluntarily to God, reaching a mutual consent (solidarity) and dominating over nature" (Soloviev 1988, 11).

5. ART – THE FORGOTTEN SOURCE OF COGNITION

Art comprises intuitive cognition with philosophical reflection and, at this highest level, also with mystical cognition. The work of art is an emanation of Truth, Good and Beauty. That is why in a work of art we can discover what is elusive to science and hard to express in discursive language (philosophy). Poetry is undoubtedly among such varieties of art, in which questions about the meaning of human existence appear most often. At the same time, poetry is a source from which one can derive answers to the big questions. This is because in one stanza, a poet can render the beauty of the entire world as well as the fragility of our existence: the meaning of life, suffering and death. This truth about man can be discovered in the poetry of Krzysztof D. Szatrawski. "In Szatrawski's lyrics, says Zdzisław Łączkowski: there is this unusual wisdom, which the poet yearns to share. Wisdom, knowledge, which serve constructive purposes. This wisdom is an experience that often comes from entire centuries, from knowing man with his traumas and downfalls, but also with his happiness. Szatrawski combines knowledge with intuition. And that produces great results in art. Krzysztof Szatrawski's artistic achievements prove that beauty, good, anything that is dignified and noble does not die, but it lasts, so that man who is looking for values, despite the many obstacles on his path, ultimately wins. Victorious is also the art which 'until the last breath' carries the gift of love."

In one of his unpublished poems, "Before The Supreme Judge," which is in fact a philosophical treaty, Krzysztof D. Szatrawski stands "beyond time" and says:

they come before dawn
with the reminder of life in their hands tied with prayer
with a smile of suffering glowing on their blistered lips
and they stand with the silent question about the truth, the good and the beauty
which the monumental mason's hammer is already singing about while engraving inscriptions
and they stand with their eyes full of renaissance imageries
and they strain their eyes looking for the judge and the jury
for archangels armed with fiery swords
but each of them has the decorations that he deserved, and so they stand speechless
wrapped in sheets stained with their own blood
still connected with life with the last string of drip (...)
and in mad persistence, they repeat memorised formulas
or put on their worn out masks
and trembling with cold they offer their cards, recite their titles
and just won't understand that they are still alive,
but their time has come, and they won't understand
that they have never wanted to give in to freedom
to which they are destined, this way or another

and they arrange their sentences in their minds, they weigh the severity of their punishment
and still curse their imagines foes, or they apologise
as if there was anything that could win them favours of the judges
or at least distract their attention
from the unsuccessful reality
all the missed opportunities
lost hopes, envy and mean tricks
to which they surrendered with a simple shrug of their shoulders
and yet they still feel pity
for a sparrow, a mouse and a run-over dog
a butterfly caught in a spider's cobweb
and if their pity could be replaced with love, they'd be salvaged
(...) here it was forgotten of all their honours they achieved as poets and philosophers
winner's diplomas, and award winners, medals, high pensions
have all lost their value
through all the days, that have not passed and will never pass
they usurped the right to poetry and love
loudly speaking of truth, their love of good and beauty, and so
the method became the goal, and they
mincing concepts into dust, suppressing fears
until the truth peeled off their bones
until the good solidified in their hearts
until the beauty put out their eyes

In this way, art becomes the quintessence of theosophy and teurgy, and not "art for the sake of art." Such perception of *eidos* of creation can be found in Szatrawski's poetry. These are not new thoughts, but they are still forgotten. Let us recall the visions of Cyprian K. Norwid, who saw the route to man's reintegration in the process of elevating work, stripping it of its degrading stigma of automatism and coercion, making the subject of work the subject of love. The crucial means to redeem work by love is art, which combines physical effort with free expression and the joy of creating. Work is transformed into art, and art is to descend from its heights to join the "practicality" and enlighten everyday life with its brilliance. Norwid perceived art as a flag on the tower of human work. The idea "art-work" was for him an inspiration for spiritual coupling of working class with elites and the formation of "a church of work" (cf. Bombała 2000, 113).

6. SEARCHING FOR CLOSURE – ROMAN INGARDEN'S HOLISTIC APPROACH TO MAN

Roman Ingarden's concept of man seems to be a good summary of our investigations: it allows for a reintegration of man, who is diminished by the scientific approach. The subject of Ingarden's considerations is man as an active subject and a bodily, psychical and spiritual unity. Ingarden emphasises that man constitutes a consistent unity of what refers to the body and the spirit – a unity whose ingredients remaining in close relations with each other: "It seems that this stream of experiences, the subject, the soul and the personality of man are nothing else than certain elements or sides of one consistently-built conscious being, often called a monad. (...) With the entire diversity of the elements and moments distinguished and distinguishing in it ('normal,' 'healthy') it constitutes an entirety. One entirety – and that is why it may be called 'a monad' (Ingarden 1987a, 194).

Ingarden's approach to man contradicts the radical dualism of body and soul, and at the same time it emphasises the superiority of spiritual elements. There is a certain adequacy of man and the structure of reality, which surrounds him. Reality consists of both the world of nature, to which the biological man is subordinate, and the quasi-reality, otherwise called culture, created by himself to confirm his own humanity: "Man differs from animals in that he not only assumes control over nature to much greater an extent than animals, and he even transforms it and adapts it to his own needs and requirements, but additionally – and this is his significant trait – he creates a certain totally new reality or, what someone could call quasi-reality. Once created, it constitutes a characteristic feature of the reality around him" (Ingarden 1987b, 29–30).

The fact that man has a body affects his selfish behaviour: the need to transform nature for utilitarian purposes. Yet the concept of *homo faber* does not suffice, in Ingarden's view, to determine human nature. What makes man essentially different from animals, what makes subjective self-fulfilment possible, consists in specifying extra-utilitarian values. Says Ingarden: "Without direct and intuitive contact with values and without the joy he derives from that contact, man is deeply unhappy. What makes him happy is the implementation of these values and he succumbs to their special appeal. And this is not (...) about relative values related to his living conditions (such as food) nor to his pleasures (such as good health or joy), but it is about values that are absolute in their immanent quality, even though their fulfilment depends on man's creative potential, in a word: moral and aesthetic values" (Ingarden 1987b, 23).

The development of human objectivity, the search for the meaning and identity, the discovery of the mystery of human life is a fulfilment of the highest values, which consists in forming man by means of true deeds, voluntary and responsible, and, most of all, by his own volition. In his analyses of man, Ingarden takes into account the ontological, axiological and anthropological aspects, and reaches the concept of man who fulfils himself through values. It must be said, however, that human activities centred on man himself are equal to losing freedom. Therefore, what is the most important, and the ultimate purpose of the human being, is not self-fulfilment, but the quest to achieve the ideal of humanity, which consists in substantiating the values of goodness, truth and beauty (Ingarden 1987b, 68–69).

The University of Warmia and Mazury in Olsztyn, Poland

Translated by Joanna Jensen

NOTE

[1] Religion confirms the unique value and moral solidarity of people as spiritual beings and their unstoppable progress to achieve the common ultimate goal – the *absolute*. Cf. Znaniecki (1921, pp. 25–26), Brzozowski (1990, pp. 447–449) and Weil (1985, pp. 57–59).

REFERENCES

C. Bartnik, *Personalism* (Lublin: Of. Wyd. Czas, 1995) (in Polish).

L. von Bertalanffy, "Symbolismus und Anthropogenese," in *Handgebrauch und Verständigung bei Affen und Frühmenschen* (Bern/Stuttgart, 1968).

B. Bombała, "Creativity of a Person – Discovering theSacrum," in Z. Kunicki (ed.), *Is Sacrum Still Sacred?* (Olsztyn: Hosianum, 2000) (in Polish).

B. Bombała, "The Panorama of Nihilistic Changes in Technopoly," in Z. Kunicki (ed.), *The Faces of Nihilism*, (Olsztyn: University of Warmia and Mazury, 2001) (in Polish).

J.L. Brockington, *The Holy Thread of Hinduism* (Warsaw: IW PAX, 1990) (Polish trans.).

S. Brzozowski, *Ideas: Alfred Loisy and Considerations of Catholic Modernism* (Cracow: Wydawnictwo, Literachie 1990) (in Polish).

K. Dopfer, "Introduction: Towards the New Paradigm," in *Economics in the Future* (Warszawa: PWN, 1982) (Polish trans.).

M. Eliade, *The History of Religious Beliefs and Ideas*, Vol. 1 (Warsaw: IW PAX, 1988) (Polish trans.).

E. Fromm, *Man for Himself. An Inquiry into the Psychology of Ethics*, (Greenwich, CT: Fawcett Publications, 1947).

E. Gilson, *The History of Christian Philosophy in Middle Ages* (Warsaw: IW PAX, 1987) (Polish trans.).

R. Ingarden, *The Controversy over the Existence of the World*, Vol. II (Warsaw: PWN, 1987a) (Polish trans.).

R. Ingarden, *A Little Book About Man* (Cracow: Wydawnictwo Literackie, 1987b) (Polish trans.).

G. Myrdal, "The Meaning and Value of Institutional Economics," in: *Economics in the Future* (Warsaw: PWN, 1982) (Polish trans.).

F.R. Paturi, *Evolution or Construction* (Warsaw: Wiedza Powszechna, 1984) (Polish trans.).

W. Soloviev, *Selected Writings: Spiritual Foundations of Life* (Poznań: Wydawnictwo W drodze, 1988) (Polish trans.).

P. Teilhard de Chardin, *Man and Other Writings* (Warsaw: IW PAX, 1984) (Polish trans.).

S. Weil, *Thoughts* (Warsaw: IW PAX, 1985) (Polish trans.).
T. Wojciechowski, *Selected Problems Of Philosophical Anthropology* (Cracow: Polskie Towarzystwo Teologiczne, 1985) (in Polish).
Z.J. Zdybicka, *Man and Religion* (Lublin: Redakcja Wyd. KUL, 1984) (in Polish).
F. Znaniecki, *The Fall of Western Civilisation* (Poznań: Komitet Obrony Narodowej, 1921) (in Polish).

KONRAD ROKSTAD

THE HISTORICITY OF BODY AND SOUL

Body and Soul – and the relation between the two – are, of course, a major issue in the modern philosophical tradition. It entails both decisive ontological and epistemological questions: What is the Body? What are the Soul, the psyche and the Human Mind? And how are they interrelated? How do you move from the Animal Soul to the Human Mind? And how do we obtain the knowledge needed to answering such questions? etc. There are very many questions that might be asked, and in regard to the last two, we might focus on the problem of explaining how the senses are related to and affect the mind, discussing various hypotheses regarding causal relations between the animal soul and the human mind. In this article I will not follow that kind of lead. Rather, I will look for a more primary "access" to the issue and "explain" in a genuinely phenomenological manner that primarily is based on the analyses made by Edmund Husserl in his *Ideas II*[1].

But, then, the title of my article, "The Historicity of Body and Soul", might seem a bit odd and might also call for some explanation: what is meant by it, and how is it at all possible for me to use the term 'historicity' to characterize the Body, especially? Is it not rather a contradiction of terms – since the Body is something that belongs to nature and is primarily something physiological – how can it thus be characterized by "historicity"? And one may, of course, further ask – what about historicity in some connection with the *Ideas II*? Does it make sense at all to link this concept with the philosophy of Husserl in the period of the Ideas-complex? In his last work, the *Crisis*[2], the concept of "historicity" is, of course, of great significance, but not before! At least, that seems to be a commonly held view. Thus, it seems we have (at least!) two problems even before we have actually started, the one of characterizing the body by historicity, and the other of speaking about historicity in the context of the *Ideas*. How is that to be dealt with?

I will not begin by explaining in detail what historicity is and by advocating its relevance in the context of the *Ideas*. I will rather comment briefly on what was meant when I (above) spoke about a more primary "access" in regard to the issue of how the animal soul and the human mind are (inter)related. In this

A-T. Tymieniecka (ed.), Analecta Husserliana XCIII, 127–159.

way I will also present what I will be doing, as I am not primarily focusing on what might be causal explanations for inter-relations. And thus I also hope to make the concept of historicity concretely understandable and highly relevant in regard to both the phenomena of the Body and to the Ideas-context.

In a phenomenological perspective, such as ours, the basic principle will have to be intentionality, and, then, as a first interpretation, say that this principle simply implies that you take things, or the phenomena, as they are given. This also designates, very simply and quite generally, the phenomenological strategy for acquiring both knowledge and understanding of the world and of ourselves living in this (historical) world, thus enabling thinking and reason in a way that should characterize the human mind. The "greater historical phenomena" is as Husserl puts it in his *Crisis*, "humanity struggling to understand itself" (*Crisis* p. 14). And the perhaps most unique and basic feature of the human mind is its ability to constitute *identity and intelligibility of what is* – including present, past and future, thus also creating preconditions for that good and rational living that human life appears to be headed for and aims at. But could the mind accomplish this in its capacity of mind alone? What are the other "aspects" to be working with – within, so to speak, the "underground" that enables the mind's ability and capacities for meaningfully having such aims and creatively reaching for such achievements?

Perhaps the most obvious "other" is the body – and, of course, all that which constitute the body as a living functioning organism, "animated" with its "soul" living in a world, a nature, in space and time, surrounded by different kinds of things and processes, together with other bodies, within situations, and perhaps even embodying a history, as well. Those sets of phenomena constitute the "environment" and embody preconditions for the body's ability to function in a manner yielding preconditions for the functioning of the mind. The human mind with its capacity for both creativity and rationality is essentially interwoven and interrelated with the Body. This, however, is not to be understood in such a way that you could not have a functioning body (as brutes have) without a rational mind. But the crucial question is: *could you have a human rational mind without a functioning body?* And, then, what are the criteria for thus being a living functioning body enabling the life of a rational mind? But *this* question could easily be asked the other way around, too, namely, could you sensibly decide on this without exhibiting some rationality of the mind? Keeping those questions in mind, let us first, then, look into some overall structures or tendencies that seem to be at work within the *Ideas II*, thus providing clues for an exposition.

Quite generally described, there are, it seems two tendencies or movements going on in the text at the same time, and moving, so to speak, in opposite directions. On the one hand, we may follow (1) the progression of the text, first constitutively grounding that "material nature" containing real things, governed by causality, etc., on which both "the animal" and "the spiritual" at first might seem to be grounded, along with, however, in due succession their constitutional grounding. On the other hand, we ultimately learn (2) that it is the spiritual which has priority when it comes to the ontological determinations of the whole issue. The spirit, with motivation as its governing, fundamental law, is first – at last. This is how it appears in an overall-perspective that thus seems a bit puzzling as it exposes tendencies that might contradict each other. And, of course, the context for this is both the *Ideas I* and *Ideas III*,[3] involving transcendental reduction, the pure or transcendental ego that executes those functions that actually do the constitutive work, also involving relationships with different sciences, etc. We shall not, however, enter that rather large field of interesting and difficult issues now.

Rather, we will return to what we previously said regarding intentionality being the basic principle. Then one has to recur to things themselves (in a wider definition, including the "spiritual") such as they appear to people experiencing them in lived experience, life-worldly and quite naturally, to find a "most basic" source of the intellective, understanding and reason. And one has to conduct constitutional analyses exposing concrete structures and functions at work in our experiencing and conceptualization of actuality, showing how basic aspects are interwoven and, then, analytically, through descriptions, how they might be distinguished. And it is not causality, but *motivation* – being the basic relationship of our life-world (and the spiritual world) – that *predelineates the point of view of method*.[4] Thus, by analyzing aspects (regarding both material and animal nature, and the spiritual world) of the rather concrete analyses given by Husserl in *Ideas II*, this presentation, we might quite generally say, aims at developing an argument sustaining the following thesis[5]: The identity and intelligibility of what is, are available in things (that appear, such as they appear), and we ourselves are the ones to whom such identities and intelligibilities are given. That way, we escape aporias resulting from dualism, materialism, idealism etc. and phenomenology, itself highly "intellective", might quite concretely demonstrate itself as reason's self-discovery in the presence of intelligible objects.[6]

First, then, let us start by looking at one aspect of what *we ourselves* – to whom identities and intelligibilities are given – are. In the world in which we live, our "life-world", we are, in quite natural terms, humans, *living* subjects and *personal egos living in a world*. And in the perspective of the *Ideas II* there also is a *pure ego* (actually there are more) executing constitutive functions (this essentially has to do also with the historicity of ourselves living in a scientific-philosophical culture) thus yielding the ultimate grounding for cognition and understanding. And, then, as Husserl says, "[the] course of the lived experience [.... of] pure Ego must assume the apperceptive form of the personal Ego, hence must become the nucleus of all sorts of intentions which would find their demonstration or their fulfillment in series of experience [....]" (II, p. 263). One important point, however, is that there is some constitution of the personal Ego prior to reflection. Even though the I in reflection [...] always finds itself as a personal Ego, it is originally constituted in the genesis pervading the flux of lived experience. Husserl himself states all this in the Third Section (§58) of his *Ideas II*, and he further states asking, the "crucial question here is this: is the personal Ego constituted on the basis of Ego-reflection, hence completely originally on the basis of pure self-perception and self-experience?" And he further asks in an even more elaborated manner:

So the question is whether, [.......], apperception in general can develop and, in particular, those of the personal ego in its regulated comportment relative to subjective circumstances, in such a way that reflection on cogitations play no privileged role there, or whether precisely the reflection has here a special and altogether essential constitutive function. Is it necessary that I, in reflective experience, run through my modes of comportment in order for the *personal Ego to be able to come to consciousness as the unity of those modes, or can it already be "conscious" in pre-givenness, before it was given originally through such series of identifying and realizing experiences, which as reflection on the cogitations, focus on the comportment in relation to circumstances*? (II, p. 263–264, my italics).

By reading this, we might, perhaps, get the impression that we are dealing with two mutually excluded options: it is *either* reflection and self-experience that constitute the personal ego, *or* it is already there in some pre-givenness that does not need any reflection. But this is not what Husserl actually states. In the continuation, as he obviously accepts some pre-givenness, he further asks:

But in that case what is it which is organized in the pre-reflective sphere? Surely, "associations" are formed, references back and forth develop, just as they do with the unnoticed sensuous and thingly "backgrounds". Thus *the content is already there*, and in the subsequent reflection, in the remembering, I can and I must *encounter something already formed*. This is *the presupposition for the "explication", for the "fully conscious" exhibition*, of the "if" and "then", the associative nexuses, *the Ego constituted in reflection refers back to another one* (II, p. 264, my italics).

Thus there is "something" pre-given[7] and maybe even pre-constituted, and it is comparable to how there are unnoticed aspects in regard to sensuous and thingly backgrounds. And this might be the object of further reflection, and may even display itself as a presupposition for explication and the "fully conscious" exhibition. How is that possible and understandable? Husserl is very explicit about this as he further states:

Properly spoken, I am originally not a unity composed of associative and active experience [...]. I am the subject of my life, and *the subject develops by living*: what it primarily experiences is not itself, but *things for work*. The *Ego does not originally arise out of experience – in the sense of associative apperception in which are constituted unities of manifolds of a nexus – but out of life* (it is what it is not for the Ego, but *it is itself the Ego*) (II, pp. 263/264, my italics).

Even though reflection and experience have decisive significance, it is life that has priority in regard to the constitution of the personal ego. It is by living that the subject develops itself, by experiencing "things for work" and not primarily itself. Thus we might say that life with an active [in regard to intentionality this includes passivity as well] and creative living – perhaps the whole life-world, enter the field of constitution, and the intentionality which actually is "first" appears to be the one "only" functioning, the one named by the concept of *functioning intentionality*. It constitutes a "core" that basically "connects" and, as we previously quoted, the personal ego is *originally constituted* in the *genesis pervading the flux of lived experience* and, we might add, within the actual *historical* Life-world. But then, we have to realize the following as well:

The Ego can be *more and can be other than The Ego as apperceptive unity*. It can have latent capacities (dispositions) which have not yet appeared, have not yet been apperceptively Objectified, just as a thing has properties that have not yet been drawn into thing-apperception (II, p. 264, my italics).

And there is nothing odd or mysterious about this. It is similar to what happens in regard to thing-apperception, and, as Husserl further says: "We make all these distinctions even in ordinary, personal way of considering a man and consequently they are there in the human-scientific (e.g., historical) approach, in ordinary experience." One implication is thus that "[n]o one "knows" himself or has "knowledge" of what he is, without learning to know himself." And some other implications might be the following:

Self-experience, self-apperception, is constantly expanding. The "learning to know oneself" is *one with the development of self-apperception*, with the constitution of the "self", and this development is carried out *in unity with the development of the subject itself* (II, pp. 264–265, my italics).

How can we make sense of those statements, interpreting them in a manner that (also) connects back to what was first quoted from Husserl? Are we now able to conceive of those two options (with or without reflection, i.e. by living) so that they do not exclude each other, but instead, on the contrary, rather are conditioning each other within the constitution of the personal ego? I would say so, and explain this by first asking *why is it that self-apperception is constantly expanding?* It is because we are facing a concretely living human being who at the same time is able to experience both itself and the world it is living in. It is a subject *for* the world – even with the potential for constituting it. But it is or, rather, can *become* this *only* if it itself is "constituted" by (the living in) the same world (with "things for work") in which it is living. In this "pre-constitution" by the living in which the ego finds itself pre-given, we find the key. The development of self-apperception, with the constitution of the "self," is something "immanent" and displays some "internal," perhaps even transcendental historicity[8]. But as it is carried out *in unity with the development of the subject itself*, it belongs as well to the actual historical world. And, thus, this "key" is about what the later Husserl (in the *Crisis*) explicitly speaks of in terms of *Historicity and the Life-world*. Thus and now quite explicitly, learning to know oneself can become one with the development of self-apperception *since* it is carried out in unity with the development of the subject itself, which, as we previously learned, *originally develops by living*. The pre-givenness (of the personal ego) with the content that is already there and that reflection must encounter as something already formed, displays something which *in its essence and own constitution is historical*. In regard to the person and the primal constitution of her/his ego (in which s/he encounters what is real as formed), that ego is not "personal" only. It belongs to a historical world and is formed by other humans as well – both actually and previously living human beings, both those who are anonymous and those well known. And it is formed in and by the personal life, living incarnated in situations, in a nature, having a body that I live "through" etc. The Body is perhaps the primal aspect of a personal ego's identity, not only its appearance – as looked at, but its comportment as well, how it behaves and (re)acts, quite generally how it functions, senses and feels etc., has its animal nature and embodies psyche and soul, lives as incarnated person in situations, takes part in a spiritual (cultural) world, actually, as a whole, in the historical Life-world.

Given this – in a perspective predelineated by the personal ego reflected on the basis of that historicity which originally constitute[9] – what will a phenomenological analysis of "animal nature," body and soul look like? In the Second Section of his *Ideas II* Husserl has given his quite substantial answer to that question by conducting constitutional analyses into that region called "animal nature". He does not, however, explicitly reflect on those matters we now have been occupied by with regard to personal ego, life and possible historicity. But this does not mean that they are quite absent or without any relevance. Of course, as we have previously commented on, the spiritual has ontological priority over nature, and this together with the life-world might even predelinate the point of view of the method. Let us therefore now look into some of the main aspects of the constitutional analysis in the Second Section, step "onto" its train of thought and try to revitalize the evidences that might be given. How is this "train" driven and conducted, what are the "stations" passed by, how are the steps (and "stops") of the analysis carried out? What is the first step?

It starts by simply making it perfectly clear that we have to "[...] leave aside all vague experiential lore [....]." And this includes also the learning we might get from the science of psychology: "What we are seeking does not lie in the consequences of theoretical, mediate thinking but in its beginnings; we are looking for the *most original presuppositions*." And then, "[...] *no theory can overthrow the meaning of the soul prescribed to us by the perfect intuition of the psychical*" (II, p. 96, my italics). This, of course, reflects "The principle of all principles" (*Ideas I*, p. 44) and has nothing to do with "theoretical, mediate thinking," but rather essentially concerns the "most original presuppositions." And accordingly, "the task is 'to draw out of experience' the authentic concept of the psychic" which, after having contrasted naïve straightforward experiences, is given a more precise formulation in this manner:

> The task is [....] to examine, in eidetic intuition, the essence of the experienced in general and as such, precisely as it is made explicit in any experience, whether carried out actually or imaginatively [......] in order then to grasp intuitively [....] the sense of the experienced as such [....] (II, p. 97).

This represents some general methodological directory for the constitutional analysis, and after having stated this, Husserl first interrogates "the ordinary talk about the "psychic" (§20), and also "Soul as natural Object" (§19), and then examines the sense which quite commonly might be determined. His analysis, then, quite generally described, follows a "strategy" that moves from that "ordinary talk" that also includes an analysis of "I as man," (§21) thus laying a *quite life-worldly grounding* for, so to speak, framing those next moves and moments to come. Within this "framework" of the life-world, constituting the horizon of the whole analysis, an exposition of the pure Ego (Chapter 2) starts the more concrete constitutional analysis. Even in regard to the pure Ego itself, the analysis moves through moments highly colored by historicity and the life-world. And thereafter, the constitution of the psychic reality itself (Chapter 3) takes place. Concreteness is steadily growing and increases as new aspects are examined, but they do not lose their "pre-constituted" origin and givenness from the life-worldly horizon. The whole analysis, then, culminates by concretely analyzing two aspects that are essential in the constitution of psychic reality, namely the constitution (a) through the Body and (b) in Empathy (Chapter 4 and 5). Both the functioning Body and Empathy are, of course, phenomena that belong to the life-world. We shall examine essential aspects of Husserl's exposition and try to reactivate evidences such as they are interwoven and mutually interact. But before doing this we shall first recall some aspects of what he is dealing with in his "The Origin of Geometry,"[10] the text in which Husserl maybe gives his most pregnant exposition of the concept of historicity. This might now yield interesting clues[11] for our main exposition of the constitution of animal nature.

In "The Origin of Geometry," an analysis of the genesis of geometry, Husserl, of course, says that the geometry belongs to a tradition and that it is historical – even though it is a science that deals with idealities. Asking the genetic question about origin this immediately leads into some conception of the historical. But it also leads into what Husserl calls "superficial commonplaces" – that he further describes, "such as: that everything traditional has arisen out of human activity, that accordingly past men and human civilizations have existed, and among them their first inventors, who shaped the new out of materials at hand, raw or already spiritually shaped." (*Crisis*, p. 355) Obviously we do not need specific (objective) historical knowledge to get those "specifications." They might as well emerge from our everyday, common experience and understanding in life, since they are situated in our actual life-world. The historical knowledge in the objective sense is thus really not the issue; the real issue is rather the *origin and the genesis* in regard to *validity and evidence* in the constitution of geometrical ideality. The historical that Husserl really calls for (as he gets in the "depth") is that which he describes as follows: "[...] history is from the start nothing other than the vital movement of the coexistence and the interweaving of original formations and sedimentations of meaning." (*Crisis*, p. 371) His conception of history thus entails a genuine epistemic aspect that in this manner is not (and cannot be) abstracted from the historical – in precisely this sense of historicity – meaning that the historical entails a genuine epistemic aspect that Husserl labels "the universal a priori of history" (thus also "transcendental historicity").[12]

Given this, the concrete analysis begins with that "first inventor" – who really has no need for being first in the strict historical sense (and of whom we do not have to know much specifically). What counts is the evidence that the "first" the original geometrical formation is constituted with – how it could (and can) be established in a genuine phenomenological manner. Then, we face a problem (of the beginning) that might be described this way: we have to realize that any successfully realized geometrical formation, necessarily involves that self-evidence constituting the formation, but as this evidence will precisely be the grasping of an entity with the consciousness of its original being-itself-there, it seems as though one (already) has to have got the entity (pre-given at hand) as a precondition for exactly this successfully being able to fulfill that self-evidence (then) "producing" the entity successfully. This, of course, seems to entail a somewhat "self-evident" aporia. How is that dealt with? What does Husserl actually do?

He says that what takes place in the beginning occurs "purely in the subjectivity of the inventor [...] within his mental space." Thus the "first" geometrical formation is not yet really constituted. What does Husserl then do to "confront" this problem? He does not really confront it; he rather (actually by implicitly conducting reduction) reveals so that it might be brought into the total class of the spiritual products of the actual (and historical) cultural world as it relates to the subjective. And as this also constitute his genuine concept of tradition, he thus makes the *horizon of historicity* (and not that of history in the objective sense) *the concrete field of inquiry* into the question of the "first" objectivity of geometry.

And, then, what more specifically become issues that have to be considered? It is quite superficially described, such as memory and the ability to repeat evidence; it is language – both oral and written (he also speaks of the "Sprachleib"), communication and community (intersubjectivity), and not least, *language's already established and functioning correlation with things*, its meaningful use and function in our historical world. All this is already established and it is given to the investigator – thus (also) yielding the horizon for the phenomenological (genetic, constitutional) analysis.

And if we now recall what was said in regard to the constitution of the personal ego, it was also "given" – we always find ourselves already "constituted" or pre-constituted by living preoccupied by "things for work" – living in the historical life-world. That field for lived experience and practice constitute the precondition and the starting point for reflection and phenomenological analysis that leads us into the "depths" of that historicity which is the primary and real original precondition for it all. How, then, might this perspective of historicity yield clues for the concrete understanding by reactivating the evidences given in the constitutional analysis on animal nature in *Ideas II*?

Returning, then, to that starting point with the "ordinary talk about the psychic", "I as Man" etc., what is it that should first be notified? In what might be characterized as the ordinary conception of the life-world, it is, according to Husserl, in connection with what is material that the psychic is given to us. There are "merely" material things, but also "things" that have the rank of "Bodies" which display a connection with a new stratum of being, the psychic stratum. Experience, then, first discloses to us here a stream, having no beginning or end, of "lived experiences" which are well known to us from inner perception, and in which each of us grasps his own lived experiences in their originality (II, p. 98). Husserl also underscores the fact that the stream of lived experience, in which are included sensations, perceptions, rememberings, feelings, affects, etc., have some internal connection, and they are not united with one another only through a common phenomenal link to the body. Instead, he says:

[.....] they are one by means of their very essence; they are bound and interwoven together, they flow into one another in layers and are *possible only in this unity of a stream.* Nothing can be torn away from this stream; nothing can be separated off, as, so to say, a thing for itself" (II, p. 98, my italics).

In the plain "ordinary" experience of the psychic there is an essential living unity with the experienced constituting not primarily things but some kind of a (living) field. Nevertheless, there are in this phenomenological original field of the psychic, some unities that crystallize and are also interwoven with the stream, and we are thus brought towards the Ego concepts which may be grasped in different senses, also relating to what might be the proper concept of the soul. Then, Husserl – by having worked through this life-worldly "material" (i.e., "I as man," "ordinary talk about the psychic") – can present his "program" for his constitutional analysis of the animal nature with psyche and soul:

[...] the first question to be raised here concerns the unity of the pure (transcendental) Ego and then that of the real psychic Ego, hence of the empirical subject affiliated with the soul, whereby the soul is constituted as a reality bound together with bodily reality or interwoven with it (II, p. 98).

And given this, the Body has to be given special attention. It is not only some material thing, rather it is "the bearer of a new, extra-material, constitutive stratum which is to be designated as a psychic stratum in the pregnant sense" (II, p. 98). Related to this we furthermore (once again) encounter "I as a man" as another unity that has to be taken into consideration. It represents an Ego-concept which is the ordinary one and is especially rich in content. It concerns what everyone grasps in "self-perception" as himself – also related to other men. And it entails his acts and states, such as "I perceive, I judge, feel, will" and "I am of such a kind," have personal qualities, character, abilities, etc. And this is further related to psycho-physical capacities, as for example being a good dancer, skier, driver, etc. And if I burn my finger, I say I have been burnt, even though, strictly speaking, it is my bodily finger that has been burnt. "Hence in normal saying of 'I' [....], the expression 'I' encompasses the 'whole' of man, Body and soul. It can therefore very well be said: I *am* not my Body, but I *have* my Body; I *am* not a soul, but I *have* a soul" (II, p. 99). All this – i.e., how there are I's that by living unifies aspects – is concerned with quite life-worldly phenomena, and it indicates something that is decisively important. On the one hand, how Body and soul in the ordinary life (and talk) of men constitute a natural functioning unity, which, on the other hand, might as well be distinguished not implying metaphysical dualism of any kind.

Thus Body and soul are not two realities that are externally linked with one another but they are most intimately interwoven and mutually penetrating, and therefore states and properties of either of them, can count as ones of the whole, of the "I as man" itself (II, p. 100). The psychic, however, has a priority and this is what essentially determines the concept of the Ego. Husserl explains this by pointing to the fact that when the soul leaves the body, what is left is only dead matter. And he further discusses it by pointing to the fact that the body cannot leave what is to be designated "I as a man" – not even in the case of a ghost because, as he says, it has to have its ghostly body too. And it is experience that teaches us that real spirituality is connected with material Bodies (only) and not somewhat with mere [...] spatial phantoms. It also tells us that the material Body and soul necessarily belong together in the idea of an actual man. At this point the analysis is, however, in a way modified, and a psychic subject without a *material* Body is admitted as thinkable, but *not without a Body* of some kind [e.g., "ghostly"] (II, p. 100). And why is that so?

It is because, for psychic being to be, to have *Objective* existence, the *conditions of possibility of intersubjective givenness* must be fulfilled. And this presupposes "empathy" which again presupposes an intersubjectivily experienceable Body that can be understood by the one who just enacted the empathy as the body of the corresponding psychic being (II, p. 101). Thus, functioning intersubjectivity with empathy-experience that both presents and appresents, enters this field for the constitution of Body and soul. The Body, in a way, now stands as the "crux of the matter." Why, and what is this "crux"? It is about this "double" character of the Body, material and psychic in one – gliding into each other by living. In this regard it is unique, being one because it is, quite life-worldly, double-functioning in its very special manner. So the Body, such as it is given, requires a comprehension of the psychic and can then be exhibited in further experience. It is, says Husserl, purely as a consequence of this that there arises a priority of the psychic or, if you will [as he continues], of the spiritual over against the Body, namely due to the fact that the Body will not withstand separation [it becomes only dead matter] (II, p. 101). The Body-object is thus a concretely living "double"; it is quite literally a pulsating object with both its material and its psychic/spiritual aspect(s) intertwined, life-worldly, by living "constituting" and conditioning each other:

In order to be *experienceable Objectively, the spirit must be animating an Objective Body* (though a priori precisely not just a material Body), whereas, conversely, the Objective experienceability of a spatial phantom or a material thing requires no animation. Thus if we consider more closely what the anima and animation are and also *what is presupposed for the possibility of an Objective knowledge of them,* we are struck by the fact *that no mere simple connection and perhaps no absolutely simultaneous connection can exist here,* and in fact there is none. The Body is not only in general a thing but is indeed expression of the spirit and is at once organ of the spirit. [.....W]e already recognize that everything properly "subjective" and ego-like lies on the side of the spirit (this side comes to expression in the Body), whereas the Body is called "Ego-like" only in virtue of this animation, and its states and qualities are only thereby called "my" qualities, subjective, of the Ego. It is the *special quality of the animation which accounts for the fact that what is Bodily and ultimately everything Bodily from no matter what point of view can assume psychic significance,* therefore even where at the outset it is not phenomenally the bearer of a soul (II, pp. 101–102, my italics).

Here the Body – in regard to and "interweaving" objectivity – thus the crux of the matter, the original unity, first now based on "I as a Man" (quite naturally, life-worldly functioning by living), is "thing" but at the same time it is the expression and organ for spirit and psyche. And it is exactly *this animation by psyche and spirit that constitutes its identity.* This is what makes the Body "Ego-like", makes it a living subject – thus providing the conditions for the possibility of spiritual and psychic objectivity. In regard to the identity of this animation, what it is "objectively," we are (says Husserl, as quoted above) struck by the fact that no mere simple connection and perhaps no absolutely simultaneous connection can exist here, and in fact there is none. Nevertheless, there is "connection", or rather, *there are connections constituted in the historicity of human life.* Based on the historicity of Man within the (historical) life-world there are networks of connection(s) functioning quite naturally. They are not only "mere simple" or "absolutely simultaneous", but rather *functioning by living* – stretching out a *field* of a subject's life thus exposing something historical as well.

And there is nothing irrational or mysterious about this; bringing it into the world and down to earth, now Husserl can say:

[...] since, *in the unity of the total apperception, "man"*, the psychic, which is attributed to the Body "in empathy", is *apprehended as really united within the Body,* then it is understandable *that the Bodily events are apprehended as properly belonging to this human subject,* i.e., are apprehended as "mine" (II, p. 102, my italics).

In quotation marks Husserl here speaks about empathy in regard to the psychic side of the *body-subject's own connection with what is its own.* Why so? Is not empathy the experience you need in regard to the psychic of the *other* body-subject, not your own in which you should immediately "encounter" your own psychic life? But this is not quite how it is. In life, *living your concrete life*, the own psychic life is both immediately presented *and* appresented; it is (self-)experienced in its presence that at the same time entails an essential *absence*. Even your own psychic life has "depths" and hidden (and, in the wake of individual history, forgotten) "spots" (for yourself, just as the body has too), and is not in all its aspects your "own" in the strict sense, and this involves both spiritual and bodily aspects. In an interesting manner you see (realize) that you do not see and at the same time you (nevertheless) have a pre-conception of what you should or could have seen. Therefore the concept of empathy can very adequately describe how it is in regard to this "total apperception 'man'," how the psychic is apprehended as really united – in an essentially open not closed unity – within the Body, the whole as "mine" (II, p. 102).

And given this (actually the horizon of historicity) it is, of course, the most natural thing to notice that in historical human life, in the actual life of "I as Man," there are extra-Bodily things which, through their relation to man, have likewise assumed Ego-meanings. Husserl points at things like words, goods, aesthetic values, objects to be used, etc. And, then, he says, "[...] the situation is somewhat different in regard to them – because they have indeed 'meaning' but no soul. The meaning does not point to a psychic subject really connected to them, *connected into a single founded reality*" (II, p. 102, my italics). Thus, the unique character of Body and Soul is stated by distinguishing them from those other types of things that have meaning. But these "other things" do indeed belong to our life-world and they are constitutive aspects in the whole historicity of Man. Not least in regard to words and language, which, of course, depends on someone being able to use them (meaningfully). This is decisively important.

In regard to this, let us now look into something Husserl states as he discusses how to distinguish between a ghost and the real incarnation of a subjectivity with her/his Ego. Husserl says that there is something essentially missing. What has not been considered is "[...] the fundamental essential role of the *vocalization* of one's own self-produced voice, related to one's own, original given kinestheses of the vocal muscles." And more explicitly, what is this missing issue? "It seems, from my observation," says Husserl:

> [...] that in the child the self-produced voice, and then, analogously, the heard voice, serves as the *first bridge for the Objectification of the Ego or for the formation of the "alter,"* i.e., before the child already has or can have a sensory analogy between his visual Body and that of

the "other" and, a fortiori, before he can acknowledge to the other a tactual Body and a Body incarnating the will (II, footnote, p. 101, my italics).

This is greatly significant in our perspective founded in the life-world, because (a) it *localizes* the "first bridge" for the Objectification of the Ego (and of the "alter"), (b) it shows the relevance and even the necessity of empathy not only in regard to the other but also to oneself, and (c) as it is language (or rather some kind of pre-language) incarnated in the own vocalization that thus bridges, this brings us fairly close to what has previously been said regarding how historicity is "at work" in the constitution of the ideality of geometry. This might even indicate the "deeper bridging" which Husserl does not go into in his analysis in regard to geometry – thus making the perspective of historicity loaded with this that now is taken directly from the *Ideas II*. And, then, we should also notice another statement Husserl makes, namely: "The Ego-concept we have discussed up to now, Ego, the man, leads back, according to all that has preceded, to a purely psychic Ego" (II, p. 102). From the common ordinary belonging in the life-world we are thus carried over into the pure Ego. And in the continuation of our exposition we now will try to disclose the historicity that functions within the "pure" as well.

Thus, after now having laid open this field for the constitution of Animal Nature quite generally, with its elements taken from naturally lived experiences, the next field of interrogation is the pure Ego itself – as Ego-pole. In the context of *Ideas I* this should not surprise anyone, but given the context now established by what has so far been presented in *Ideas II*, it might perhaps seem a bit surprising. Why could not the analysis continue directly into the constitution of the psychic reality (itself) which is now the proper object of interrogation? One reason might be that the context of *Ideas I* is not "enough"; it is, so to speak, another "version" of the pure Ego (not primarily directed towards the epistemic issues, but rather those pertaining to living subjects) that now is needed, a "version" that is more "adapted" to the concrete field of study. Keeping this in mind, how does the concrete analysis appear. How is it carried through?

First, the Ego is viewed as abstracted from the body, which, nevertheless, involves ourselves as spiritual ego(s) related to the stream of lived experiences and, thus, "spiritual" in the more general sense. More specifically it is about the one identical subject through different cogitos or acts of thinking: The focus is on the ego in which the "I think" reigns, and purely as what reigns therein. According to Husserl, this can be brought to givenness in the a priori possible view of fixating reflection, and there is nothing mysterious or abstract about this:

I take myself as the pure ego insofar as I take myself as purely as that which, in *perception, is directed to the perceived, in knowing to the known, in phantasizing [...] in logical thinking [...] valuing [...], willing* [all of them quite ordinary happenings] [...] In the *accomplishment of each act there lies a ray of directedness* I cannot describe otherwise than saying *it takes its point of departure in the ego*, which evidently thereby *remains undivided and numerically identical while it lives in these manifold acts, spontaneously takes an active part in them, and by ever new rays goes through these acts toward what is objective in their sense* (II, p. 104, my italics).

And the specter of different types of acts is rather broad, including such acts as *loving* – "I find myself inclined to the beloved, attracted by the one I love, perhaps wholly surrendered to and lost in the beloved. In hate, on the contrary [...]." Actions, different motivating feelings, holding back or *I am "into" the things in a practical way*, this and more, which are quite natural and genuinely life-worldly phenomena, are what Husserl examines while concretizing (the acts of) his pure Ego. And this also leads up to and includes the categorical and logical level.

Given this, there are some important comments that have to be made in regard to the Ego itself in relation to all those acts. "On the one hand," Husserl says, "we must definitely here distinguish the pure Ego from the acts themselves, as that which functions in them and which, through them, relates to objects; on the other hand, this distinction can only be an abstract one" (II, p. 105). How should this be understood? Is not this about the concretization of the pure Ego, and how could Husserl do that while making a distinction that he himself describes as abstract? The first part of an answer is that "it is abstract to the extent that the *Ego cannot [even] be thought of as something separated* from these lived experiences, from its 'life,' just as, conversely, *the lived experiences are not thinkable except* as the medium of the life of the Ego." And the second part, which introduces new aspects of the Ego that still are grounded in quite naturally lived experience, is the following:

In this connection it is important to notice that *the pure Ego is not only an accomplishing one*, such as we up to now have considered it exclusively: that is, in acts in the specific sense, those of the "cogito." Just as the cogito of the moment *sinks down into inactuality, in a certain sense the pure Ego also sinks down into inactuality*. It withdraws from the act in question, it is no longer in this act as accomplishing Ego, and perhaps it is not carrying out any act at all. But in that case it is *not something separate from all the lived experiences*, as if now the consciousness that is *not accomplishing any act and the pure Ego would be without any connection between themselves*. Rather, the distinction between actuality and inactuality points to a distinction in the *essential structure of intentional lived experiences* and consequently at once to a distinction, inseparable from them, in the *"how," in the manner in which the Ego has lived experiences* (II, pp. 105–106, my italics).

What is revealed by this quotation is, put simply, the time-dimension primarily experienced in *lived experience* that embody both the pure Ego and its whole activity-field incorporating both possible activity and inactivity. In this field the Ego, similar to the cogito it at a moment carries out, might sink down into inactuality, not actually functioning in its capacity as pure Ego. But since the pure Ego (in its inactuality) is not something separate from all the lived experiences it might have (not acting as pure, but as a concretely living human being in the natural attitude), it can, of course, be conscious, perceive and think thoughts anyway, without (however) acting in its capacity of pure Ego. This is obviously what most of us do most of the time. And perhaps this might even be compared to the difference between, on the one hand, perceiving and judging straitforwardly without actually (seriously) posing the question of truth, and, on the other, actually posing it (seriously). Obviously, we might shift from the one to the other and we might even ask, *philosophically*, the "final" (e.g. transcendental) question about evidence and truth. And similarly the pure Ego might shift actuality into inactuality (and back again). In the *Cartesian Meditations* Husserl also says: "Obviously it can be said that, as an Ego in the natural attitude, I am likewise and at all times a transcendental Ego, but that I know about this only by executing phenomenological reduction."[13] Or as he further says in *Ideas II*:

The ego can never disappear, it is *always present in its acts but in different ways*, depending: if they are or if they become acts in actu, then *the Ego steps forth, as if it were, in them; it comes to the light of day, it exercises an active living function, it directs itself, in an active ray, toward what is objective*. But if it is [...] a latent ego, then it does not cast an active regard onto something; it does not actively experience or have an effect on, or undergo, anything. Of course, there *can be no question here of arbitrary possibilities of sudden coming into existence and stepping forth from nowhere in the accomplishment of acts*; instead it is a matter of *a phenomenological alteration of the pure ego-relatedness that is always extant* (II, p. 106, my italics).

Thus the Ego steps forth and comes to the light of day, and it might step back again, then as a latent Ego, etc., and it is *the same Ego the entire time*. And this is so because there can be no question here of the *arbitrary possibilities* of sudden coming into existence and stepping forth *from nowhere*, but a matter of *a phenomenological alteration of the pure Ego-relatedness that is always extant*. Of course, this alteration (strictly methodically) is the phenomenological reduction itself. But what is it this pure ego-relatedness that is always extant – if it is not that historicity, even that *transcendental historicity* that Husserl explicitly refers to in the *Crisis*. Let us look closer at this, and start by reading the following:

To be sure, something new arises thereby, a changed phenomenon, which here could be called an active self-directing of the Ego, e.g., casting a glance of focused notice onto something, etc., but in such a way that *already in the former phenomenon, in the one of latency, an Ego-structure is extant. This Ego-structure precisely allows us and requires us to say that the Ego in the stage of the specifically "unconscious," of latency, is not nothing or the empty potentiality of the alteration of the phenomena into ones of Ego-actuality but instead is a moment of their structure.* Thus the images of stepping forth of the Ego, of the being actively directed onto something, and of the stepping back into latency have a real [reell] meaning. We know all this, however, in reflection, in which we *grasp in retrospection* precisely *not only individual background lived experiences but also whole stretches of the stream of consciousness which would lack all Ego-activity.* In spite of all the obscurity and all the confusion which adheres to the proper character of such stretches, we *can grasp insightfully the most general essential* properties which have been indicated (II, pp. 106–107, my italics).

Husserl elaborates on his characterization of the pure Ego by clarifying more precisely the points he has already made. And what he has said in regard to the Ego in the state of "unconsciousness", of latency – namely that it is not nothing or the empty potentiality of the alteration of the phenomena into ones of Ego-actuality, but is instead an aspect of their structure – is rather important – leading us towards – if not right into the field of historicity. Let us then take a closer look at what he says in the *Crisis*. The following should be noticed:

In psychology the natural, naïve attitude has the result that the *human self-objectifications of transcendental intersubjectivity, which belong with essential necessity to the makeup of the constituted world pregiven to me and to us, inevitably have a horizon of transcendentally functioning intentionalities which are not accessible to reflection, not even psychological-scientific reflection.* "I, this man," and likewise "other men" – these signify, respectively, a self-apperception and an apperception of others which are *transcendental acquisitions involving everything psychic that belongs to them, acquitions which flowtingly change in their particularity through transcendental functions which are hidden from the naïve attitude. We can inquire back into the transcendental historical dimension [Geschichtlichkeit, historicity], from which the meaning and validity-accomplishment of these apperceptions ultimately stems, only by breaking with naiveté through the method of transcendental reduction.* In the unbroken naiveté in which all psychology, all humanistic disciplines, all human history persists, I, the psychologist, like everyone else, am constantly involved in the performance of self-apperceptions and apperceptions of others (*Crisis*, pp. 208–209, my italics).

And then the next issue (§23) is to examine the possibility of grasping the pure Ego, more specifically the Ego-pole itself. How might that possibility be given to us? It is given in what Husserl now calls "Ego-reflection" (i.e. reduction), in which every cogito might be grasped in another cogito. The pure Ego can therefore become an object, not in the sense of a mundane real object, but in another sense which is broader and keeps the Ego quite distinct from real objects, in a way opposed to them. And, as Husserl further says, we:

[..] can in fact say: everything that is objective, in the broadest sense of the word, is *thinkable only as correlate of a possible consciousness* or, more precisely, as correlate of possible "I think" and consequently as referable to a pure Ego. This holds for the pure Ego itself as well. *The pure Ego can be posited as an object by the pure Ego which is identically one with it.* [and] *The essence of the pure Ego thereby includes the possibility of an originary self-grasp, a "self-perception,"* [...] also self-rememberance, self-phantasy etc. (II, pp. 107–108, first two italics by me).

This encompasses "the fact that [the Ego] grasps itself as something *temporally enduring* from the past now to the actual flowing now of the present." And then a distinction must be made between the objectified and the "originally" not objectified pure Ego, e.g., between the perceived and the perceiving pure Ego. This expresses a phenomenological transformation, and yet it is, Husserl says, evident, in virtue of further reflections, on a higher level, *that this and the other pure ego are in truth one and the same* (II, p. 109). As the ego is objective or again not objective, what changes phenomenologically is not the ego itself [...] but the lived experience. And thereafter comes a very important sequence talking about unities, e.g., the identical cogito, [they] are precisely *themselves already unities constituted in consciousness as unities of duration*, and they are constituted *in a more profound, correspondingly manifold "consciousness"* taken in another sense [i.e. the sense of inner time-consciousness], in which all that Husserl up to now has called "consciousness" or "lived experience" does not occur really but *only as a unity of the "immanent time" which is their constitution* (II, p. 109, my italics). Thus we are brought down to "rock-bottom," into the transcendental in the strict phenomenological sense, and the "functioning intentionality" of lived experience that embody unities themselves already constituted in consciousness as unities of duration, might be grasped within inner time-consciousness and historicity (cf. what was just quoted from the *Crisis*, pp. 208–209) in a genuine transcendental sense:

Each and every cogito, along with all its constituents, arises or vanishes in the flux of lived experiences. But the pure subject does not arise or vanish, although in its own way it does "step forth" and once again "steps back". *[.....] What it is, and what it in itself is and does in general, is grasped by us, or, rather, is grasped by itself in the selfperception which is one of its own actions, such a one that grounds the absolute indubitability of the grasp of being.* The absolutely existing Ego – the only way that it is possible for the pure Ego not to encounter itself is for it not to reflect on itself [....] (II, pp. 109–110, my italics).

And given this, we might (§24) discern with evidence the sense in which the pure Ego changes in the changing of its acts. According to Husserl it is changeable in its practices, in its activities and passivities, in its being attracted and being repulsed, etc. But these changes do not change it itself; in itself the pure ego is immutable. Even though it is, in principle, always possible that I, this person (having body and soul) am not at all, the pure Ego is always self-identical itself. And you need only one single experience of one sole and simple cogito to realize this because the pure Ego does not appear or present itself merely from one side; it is a unity which does not present itself by way of adumbrations. Rather it can be grasped adequately in the reflexive shift of focus that goes back to it as a center of functioning. As pure Ego it does not harbor any hidden inner richness; it is absolutely simple and it lies there absolutely clear. All richness lies in the cogito and in the mode of the function which can be adequately grasped therein (my paraphrase of II, p. 111). All this however, can be realized only by conducting phenomenological reduction.

And further, phenomenological experience gives access to (§25) a polarity of acts in the appearance of Ego and Object:

Insofar as every cogito calls for a cogitatum and insofar as this latter is related to the pure Ego in the accomplishment of an act, we find a remarkable polarity in every act: on the one side, the Ego-pole; on the other, the Object as counter-pole. Each is an identity, but of a radical different kind and provenance (II, pp. 111–112).

And there is a difference between (§26) alert and dull consciousness, in which the "being" of the ego as alert/attentive has some delimitation by the dull, which, however, it might always bring into attentiveness (again). Further on (§27) comes an analysis on how "I as man" *am part of the content of the environment of the pure Ego*. "Connected with this," says Husserl, "is still something else. In virtue of the polarity belonging to the essence of the cogito, the alert Ego is intentionally related to what is objective in the accomplished cogitations." There is thus some relatedness to what is objective in the accomplished cogitations, but the whole field that is exposed by this is much broader, because, as he further says:

[...] This holds potentially, however, for the objects of the *unaccomplished noeses dormant in the background* and which, as it were, *makes up the field of the ego's freedom*. Their objects form *the field of the spiritual regard, the ego's field of activity*. [....] What comes into consideration [...] are *not only* the objects of the *dormant noeses [...] also the objects of the possible noeses to which the motivations of consciousness belonging to the content of consciousness lead over. If we proceed from the actually experienced real objects, then it is not only the environing things, the ones actually constituted as background, [....], but what is mine, what is the pure ego's, is the whole world with all it contains that is still unknown to me but is experienceable by me: i.e., things, fellow men [....] I as a man am a part of the content of the real surrounding world of the pure ego, which, as the center of all intentionality, also accomplishes that intentionality by which is constituted precisely I, the man, and I, the person. [......] In principle, the empathized pure Ego [...] is an "other"; consequently if I posit many human beings then I also posit many pure egos, separate in principle, and many streams of consciousness. There are the same number of pure egos as there are real egos. [...]* (II, pp. 116–117, my italics).

This whole field actually reveals how the pure Ego correlates with the life-world nested in historicity, which is transcendental in a manner that constitutes a field for a descriptive analysis shifting between, "interweaving" and linking up the pure with what is concrete and real. Thus (§28) the real Ego is constituted as transcendent Object whereas the pure Ego is given in an immanence that nevertheless correlates with the real. And we might further speak of (§29) some constitution of unities within the sphere of immanence so that persistent opinions might be regarded as sedimentations in the pure Ego, which, then, exposes the historicity of that Ego.

And given this – together, of course, with all those analyses of material nature (Section One), then the next topic will be *Psychic Reality itself.* What are the moments and steps that have to be taken into consideration in the constitutional analysis interrogating that issue? It starts (§30) with the real psychic subject, and as distinguished from the pure ego, we find that it "makes up the substantial-real double being: the animal, man or brute" (II, p. 128). Faithfully sticking to what is given intuitively, the psychic or soul appears as the identical psychic being that is connected in a real way with the human or animal body. The psychic, embodied in psychic lived experiences, is "bound" to the body, but not primarily related to the corporeal body as material thing. It is rather the reverse:

> The psychic subject has a material thing as his body *because it is animated*, i.e., because he has psychic lived experiences which, in the sense of the *apperception of the human, are one with the body in a singularly intimate way* (II, p. 129, my italics).

This is a fairly decisive point – stating some *core-intentionality* in which *psychic lived experiences by the apperception of the human are one with the body in a singular intimate way*. And in a sense this provides clues, and even the transcendental clue, for what is next to be constituted concretely, thus giving that "core-intentionality" nested in the apperception of the human (again nested in historicity), some primacy which enables the constitution of the psychic and the soul through the body. The first to be further commented on is, then, 'the formal-universal concept of reality' (II, p. 133). This concept embodies what is normally designated as "substantial reality", and the point, then, seem to be that "the concepts of *real substance* (concretely understood as a thing in a very broad sense), *real property, real state* (real behavior) and *real causality* are concepts which belong together essentially." And given this, the "fundamental differences between material and psychic reality" might be concretely questioned. How are the material aspects and the psychic aspects of reality both different and related, and especially, what about the reality of the soul? The soul:

> [....] is a *persistent being versus changing circumstances* [.....] is the *bearer of a psychic life together with the subjective possessions of that life*, and as such it is a *unity extending through*

time [.....] It acts upon *physis*, and it undergoes actions coming from there. It exhibits an identity in that, on the whole, it "behaves" in a regulated way in its reactions under given physical circumstances, e.g., it senses and perceives in such a manner, etc. Owing to these regulated modes of behavior, *psychophysical properties* are attributed to it (II, pp. 134–135, three first italics by me).

The field – and this field (in accordance with the italics in the quotations above) embodies historicity – for examining similarities and differences regarding body and soul is thus laid open, and phenomenological descriptions are given leading to observing a difference regarding the distinction between transcendence and immanence, i.e., the material appears in the sphere of transcendence, whereas the psychic states are:

[....] no longer transcendent unities but instead are nothing else than the *immanently perceivable lived experiences* of the *immanent flux of lived experience, of that flux in which all "transcendent" being is manifest*, ultimately through primordial manifestation (II, p. 139, my italics).

And contrasting this with how the "material thing", and further the "substance" in the Kantian (and natural scientific) sense is constituted, then "we must undoubtedly say there is no soul-substance: the soul has no 'in itself' the way 'nature' has, nor does it have a mathematical nature as has the thing of physics, nor a nature like that of the thing of intuition" (II, pp. 139–140). And similarly, if the natural scientific concept of causality is the one governing, then *as regards the soul we cannot speak of causality at all.* Thus:

The *flux of psychic life has its unity in itself*, and if the "soul", *concomitant to a body*, stands toward this thingly body in a *functional connection of reciprocal dependence*, then the soul surely has its *lasting psychic properties*, which are *expressions for certain regulated dependencies of the psychic on the body*. It is a being that is *conditionally related to bodily circumstances, to circumstances in physical nature*. And similarly, the soul is characterized by the fact that psychic events, in regulated fashion, have consequences in physical nature. On the other hand, *the body itself, too, is characterized by this psychophysical nexus and its regulation.* But *neither body nor soul* acquire "nature-properties" in the sense of logico-mathematical nature (II, p. 140, my italics).

And what is more regarding difference – the material thing can remain unchanged, whereas the psychic:

[...] *in principle cannot remain unchanged*; last of all can it abide in an unchanged psychic state. By essential necessity, *psychic life is a flux*; [..... and] *Every lived experience leaves behind itself a wake of dispositions and creates something new as regards psychic reality.* Hence this reality itself is something *constantly changing*. But even though, *together with other souls*, lend itself to a form of constancy (II, p. 140, my italics).

This form of psychic existence (as shown in the quotations above), Husserl further says, consists in the form of the *social community, founded on the form of corporality, as a community of existence unified by means of the bond of reciprocal understanding.* And the full psychic unity as manifold is the current total state of consciousness, whereas the singular is abstracted out (my italics paraphrasing from II, p. 141). Thus, the *psychic existence* that embodies social community founded on corporality and reciprocal understanding – as a unity as manifold being the current total state of consciousness – is surely embedded in historicity. And in that regard the soul [differing from material things], has no places, no pieces. It also is absolutely not a fragmentable unity. It might, however, as Husserl further says, "transform its dispositions, even entire groups and their nexuses, its whole 'character', and so become 'a totally different' ego." (II, p. 141). And as this obviously can be done only in (and through) time, it also involves some personal history that incorporates the historicity of that psychic ego. According to Husserl, there (also) exists, however:

> [...] a certain partition of the soul, namely, a distinction between strata in the soul, corresponding to strata of consciousness. Higher levels can be missing, and then the stratification [...] is quite different, as occurs, in for example constant sleep where no cogito is accomplished. And the soul of the brute animal in which no theoretical thought in the pregnant sense takes place. [.....] In any case, *the most important stratification is indicated by the distinction between the soul and the psychic subject, the latter understood as a reality, but a reality nested in the soul.* Without the soul, it is unable to stand alone; *and yet again, it is a unity which in a certain sense encompasses the soul and which is at the same time so prominent that it dominates the general way of speaking about human and animal subjects* (II, p. 142, my italics).

Thus we are brought into a stratification of the soul that is well known both in ordinary life and in the tradition of philosophy, in the ancient doctrines of the "parts of the soul" and Husserl gives credit to those who first stated those phenomenological problems meaningfully. The distinction between the soul and the psychic subject (nested in the soul) as reality that cannot stand alone – and yet again encompasses the soul etc., all concretizes the field of historicity.

Then, the next to be examined more closely is the psychic ego's *dependence on "circumstances"*, in which we find some stratification as well. Husserl distinguishes between the following three: (1) the psychophysical (or, better, physiopsychic) side (2) the idiopsychic side, and (3) the intersubjective relations of dependence of psychic realities (II, p. 142). Reflecting on the most obvious historicity of our living embedded in the life-world, the psychic depends on the body and thereby on physical nature and all its many relationships, as we all know and might experience. This is about sensations and their concomitant reproductions, all of which display an all encompassing dependence relation. In regard to the life of the soul, Husserl says, "there exist 'physiological' dependencies that are very far-reaching, indeed in a certain way penetrating all conscious processes" (II, p. 143). Thus, there is a physiopsychic side to psychic reality. In regard to the second stratification Husserl has distinguished, the psychic presents itself in a dependency relation to itself. The impact of this is the fact that within one and the same soul, the present stock of lived experiences, as a totality, as Husserl says, is dependent on earlier stocks of lived experiences that are sedimented in the soul and are in a way independent of those specific experiences that might have generated them. We talk about association, habit, memory, and also motivated change in sense, motivated change of convictions, of orientations of feeling, of will, etc. We might say that the soul has complexes of dispositions and, thereby real qualities, which manifest themselves in it as having originated from itself out of its own influence rather than from its relation to something external. Thus, the soul, we might say, is nested in the historicity of its own. Husserl does not at this point specifically comment on the third, the intersubjective dependency relation, but in regard to the two other, he says, "that even the states of the soul [...] (and the soul encompassing the whole course of lived experiences) are dependent from both points of view: i.e., either merely through the sensuous substrate or as a totality" (II, p. 144).

Given the exposition above, then a "more precise determination of the concept of reality" (§33) might be provided. Some aspects of this have to be noted. First, common to both material nature and the psychic is that what counts as "things," quite formally, is included in the idea of the "unity of lasting properties in relation to pertinent circumstances" (II, p. 144). But an interesting difference soon appears, namely in regard to time and some dimension of history. Material things, according to Husserl, "are conditioned exclusively from the outside and are not conditioned by their own past; they are *history-less realities*" (II, p. 144). Therefore they might return in cyclical processes to the same identical external circumstances under which they had already stood – at least this is possible in principle. On the other hand, as Husserl also says, "it pertains to the essence of the *psychic reality* that as a matter of principle it *cannot* return to the same total state: psychic realities have precisely a *history*" (II, p. 145). Earlier states are, thus, sedimented within the soul itself, and by this historicity of the soul they functionally determine the following one and, we might perhaps add, the very likely possibility of the actual ones influencing the conception (in memory) of the previous states.

In regard to the concept of reality, Husserl presents three options: (a) the mere natural reality (b) the "supernatural" and (c) mixed realities. The second obtion is, however, only an empty option, and the soul belongs to the last one. It has both a natural and an idiopsychic side, and because of this it, unlike nature, admits of no mathematization (II, p. 145). Further, psychic reality is constituted as reality only through psychophysical dependencies. It has its unity in itself, but here it is considered to be a unity in a nexus (II, p. 146). And finally, the unity of the soul is a *real* unity in that, *as the unity of psychic life, it is joined with the body* as a unity of the bodily stream of being, which for its own part is a member of nature. Thus, then, Husserl concludes, "what we have to oppose to the material nature as a second kind of reality is not the 'soul' but the *concrete unity* of body and soul, the human (or animal) subject" (II, p. 146).

And given this, the constitutional analysis of the psychic reality discloses some necessity of making a distinction between what respectively is called the naturalistic and the personalistic attitudes (§ 34). As soon as you take the human subject, thus embodying this concrete unity of body and soul, acting, reacting etc., then also there appears a human person, having his spiritual individuality, his intellectual and practical abilities and skills, his character, his sensibility etc. All those aspects, of course, have an impact on the psychic constitution of a concretely living human subject, and Husserl includes some analysis of this in his exposition of psychic reality.

After having constituted the psychic reality and disclosed this conception for further examination, Husserl follows two paths for making his analysis even more concrete. The first questions "the constitution of psychic reality through the body" and the second interrogates "the constitution of psychic reality in empathy." Those two paths are actually interwoven and constitute, in a way, preconditions for each other. And, of course, what has previously happened in his whole exposition, both in regard to the constitutional analysis of nature and the pure ego, constitute preconditions for what is next to come. The first move, therefore, has to be a (§35) "transition to the study of the constitution of 'man as nature'." The human subject, the concrete unity of body and soul, and then, "man as nature" as distinct from "man as spirit," raise the question of how it is possible to provide the ultimate distinction between those two, as well as the establishment of their reciprocal relations? According to Husserl, this is possible only when both these objectivities have been subject to constitutive study (II, p. 151). And he also remarks, in regard to the constitution of material nature, that it, with its entire intuitive content, is related to animal subjects. And Husserl further says:

Hence when we approach the constitution of the natural object, "man", we *may not already presuppose his body as a fully constituted material thing* but instead must at first pursue *what is already constituted prior to, or correlative with, material nature, as regards the psychophysical subject*. And here, as before, let us first try to see how far we can advance in a solipsistic consideration (II, p. 151, my italics).

What are the substance and the implications of this? Let us first now think back to what was previously said in regard to the general perspective expressed in *Ideas II*, namely those two opposing tendencies pervading the book (the one from "below", from nature and the other from "above", from spirit). At this point in his analysis, in regard to the constitution of the natural object "man," we cannot presuppose a fully constituted material thing, but must initially pursue what is already constituted prior to [...] material nature, in regard to the psychophysical subject. According to Husserl, there is this primacy as the psychophysical subject is in its way "first." And it is first simply due to the fact that it embodies both aspects as it is functioning quite naturally. The impact of this, of course, is that the constitution of the psychophysical subject (the Body) not only "influences" the constitution of material nature, but is also a phenomenological precondition for it. This primacy does not however, imply the bodily subject's unrestricted ("absolutely free") rule over material nature; rather the "freedom" is embedded in historicity and is a relation of interdependence in which both aspects (of the Body, the material and psychic) concretely show how they naturally function together as unities in the lived experiences of ordinary, natural life. And what is it that first appears to be examined further? In the constitutional analysis of the psychophysical subject we first have to examine more closely the "constitution of the Body as bearer of localized sensations (sensings)" (§36). And quite generally, Husserl begins by stating that in all experience of spatio-thingly objects, the Body "is involved" as a perceptual organ of the perceiving subject (II, p. 152).

This bodily involvement is in a way so obvious that we commonly do not think about it. But is it therefore without significance? In our analysis, in which we are trying to explicate the elements of historicity in Husserl's analysis, it is now pertinent to recall some of what was previously said in regard to "The Origin of Geometry". So called "superficial commonplaces" were given significance – even transcendental significance in the genetic constitution of ideality. The Bodily "involvement" in all experience of spatio-thingly objects is of such a kind. And in regard to the constitution of the Body-object's Corporality Husserl now chooses the special case in which the spatially experienced body, perceived by means of the Body, is the Corporal body itself; the Bodily involvement is perhaps the most obvious aspect imaginable. No one can escape using her/his own corporality while perceiving material aspects of her/his body; in a sense they are one. Nevertheless, according to Husserl, this is perceived from the outside although within certain limits, preventing it from being considered, without qualifications, as a thing like any other in a thingly nexus (II, p. 152). And further on the observation is easily made that there are parts of this body which can indeed be perceived by touch but cannot be seen. These however, are initially disregarded by Husserl, and he begins his analysis with parts that we can both touch and see. We can both look at them and feel them, just like other things. But then there is a distinction between the visual appearances and the tactual regarding, and the example Husserl uses is the perceiving of a hand. His concrete analysis goes like this:

> Touching my left hand, I have touch-appearences, that is to say, I do not just sense, but I perceive and have *appearances of a soft, smooth hand, with such a form.* The indicational sensations of movement and the representational sensations of touch, which are objectified as features of the thing, "left hand", *belong in fact to my right hand. But when I touch the left hand I also find in it, too, series of touch-sensations, which are "localized" in it,* though these are not constitutive of properties (such as roughness or smoothness of the hand, of this physical thing). If I speak of a physical thing, "left hand", then I am *abstracting from these sensations* (a ball of lead has nothing like them and likewise for every "merely" physical thing, every thing that is not my body). *If I do include them, then it is not that the physical thing is now richer, but instead it becomes body, it senses* (II, p. 152, my italics).

This very instructively shows how Husserl, by analyzing the *most obvious experience that any normal human being may have*, distinguishes the physical thing from the Body in its quite decisive function in regard to a "first" appearing (as it becomes Body, it senses) of the psychic. Sensing is not just another feature of the physical thing, but is bodily-psychic as distinguished (even self-distinguished!) from the physical. And now we very easily may observe that "touch"-sensations belong to every appearing Objective spatial position on the touched hand, as it is touched precisely on those places. The hand that is touching, which for its part again appears as a thing, likewise has its touch-sensations at the place on its corporal surface where it is touched (or touched by the other) (II, pp. 152–153). Thus an essential correlation is constituted that encompasses quite distinct fields of phenomena. And the sensations – even not themselves material – are localized, i.e., they are distinguished by the place they appear on the Corporal body, and they belong phenomenally to it. Hence the Body is originally constituted in a double way: it is a physical thing, matter; it has extension, in which are included its real properties, its color, warmth etc. Second, I find on it, and I sense "on" it and "in" it the warmth on the back of the hand, coldness in the feet, sensations of touch in the fingertips. The localized sensations are not, says Husserl, properties of the Body as a physical thing, but they are properties of the Body as a field of sensation, and indeed they are effect-properties. They arise when the Body is touched, [...] where it is touched and only under certain circumstances do they still endure after the touching takes place. The touching might very well be described as a physical event, and even lifeless things can touch one another, but only the touching of the Body provides sensations on it or in it (II, pp. 153–154).

Thus, as indicated by this, as it appears in and through lived experience, the Body constitutes itself as a *functioning unity* of both material and psychic aspects; even the body-object considered "purely" as a material corporal thing has to be constituted by the Body, but not, then, this constituting Body as purely corporal material thing. Even in the analyses of material nature (Section One) the Body was functioning in the constitutional analyses of that region. But it was functioning only "anonymously" using functions that had not yet themselves been constituted, as they were already functioning very well. In the life-world in which we all live, it is by *living we first experience a nature*, and we cannot actually bypass this "first" experience. We can, however, abstract from it – by making the lived experience something that only functions anonymously – even perhaps considering it to be a source of fault. In the modern, natural scientific (objectivistic) conception we do that. We abstract *systematically* from the subjective and lived experience. Due to the success of that science, this has created a general paradigm for science and it has also brought about the scientific conception of the psychic. This tendency is what Husserl's quite concrete analyses, founded in lived experience and the life-world, opposes.[14] And by reactivating the evidences they entail we now try to provide an alternative that builds a foundation for what *truly constitute the origin* for all actual kinds of conceptions.

And having reminded ourselves of this broader context together with this life-worldly grounding and sticking to that, we quite naturally also observe a *difference of the visual and tactual realms* in regard to the Body and the animal nature. This is, according to Husserl, a striking difference between touching and vision we experience. Whereas the touching may provide double apprehension (actually it may even be triple: you apprehend both what you touch with the right hand and the touching within both the right and the left hand), there is no such experience in regard to vision: "An eye does not appear to its own vision, and it is not the case that the colors which would appear visually on the eye as localized sensations would be the same as those attributed to the object of the apprehension of the seen external thing and Objectified in it as features" (II, p. 155). And the comparison may go even further since I do not see myself, my Body, the way I touch myself. What I call the seen Body is not something seeing the seen, the way my Body as touched Body is something touching which is touched. And Husserl includes a footnote about seeing one's eye in the mirror – one does, of course, see (a picture of) the physical thing eye, but not the seeing; "[...] my eye, that which sees qua seeing, I do not perceive. I see something, of which I judge indirectly, by way of 'empathy', that it is identical with my eye as a thing (the one constituted by touch, for example) in the same way that I see the eye of another" (II, p. 156). Thus, in regard to seeing, what we are denied is an analogon to the touch sensation, which is actually grasped along with the touching hand. Actually, the eye, too, Husserl further states, is a field of localization but only for touch sensations, and, like every organ "freely moved" by the subject, it is a field of localized muscle sensations. It is an object of touch for the hand; it belongs originally to the merely touched, and not seen, Objects. And this has to do with a primal group of Objects constituted directly in intuition. The eye can be touched, and it itself provides touch and kinetic sensations; *that is why it is necessarily apperceived as belonging to the Body*. And, according to Husserl, all this is said *from the standpoint of straightforward empirical intuition*. Furthermore, and this is important, the relation of the seen color of the thing to the seeing eye *will not* be confused with the givenness of these sensations in the manner of localized "sensings." And the same applies to hearing. The ear is "involved," but the sensed tone is not localized in the ear. And there are further analyses on this issue that are important to both the constitution of material thinghood and of the Body (paraphrasing from II, pp. 156–157).

The outcome is that "we must be perfectly clear on the fact that *localizations of sensings* are in fact something *in principle different from the extension of all material determinations of a thing*." The sensings do indeed spread out in space and cover in their way, spatial surfaces. But Husserl insists on the fact that spreading out and spreading into are precisely something that *differs essentially* from extension in the sense of all the determinations that characterize the res extensa – such as the roughness of the hand, its color, etc. These real properties of a thing are constituted through a sensuous schema and manifolds of adumbrations. To speak in a similar way about sensings, according to Husserl, would be quite absurd – because the touch-sensings, the sensations which are constantly varying and lie on the surface of the touching finger, are, such as they are lying there spread out over the surface, nothing given through adumbration and schematization. (They have nothing at all to do with the sensuous schema.) What, then, in a way summarizes and concludes this analysis of the sensings is the following:

The touch-sensing is *not* a state of the *material thing, hand, but is precisely the hand itself*, which for us is more than a material thing, and the way in which it is mine entails that I, the "subject of the Body", can say that what *belongs to the material thing is its, not mine*. All sensings pertain to my soul; everything extended to the material thing. On the surface of the hand I sense the sensations of touch, etc. And it is precisely thereby that *this surface manifests itself immediately as my Body* (II, p. 157, my italics).

By experiencing touch-sensing and their reflective identification – *the hand as something more than a material thing is identified as well.* The hand is mine, which entails that I, the "subject of the Body," can say that what belongs to the material thing is its, not mine. On the other hand, on the surface of the hand I sense the sensations of touch, etc. And this is precisely why that this surface *manifests itself immediately as my Body*. This, thus constitutes a decisive point linking in a *strictly descriptive manner* the psychic and the material aspects of actuality, so that they are joined together and at the same time might be distinguished in a most obvious way. There is, of course, a profound and genuine life-worldly foundation for all this, and the touch-sensing has some priority, Husserl even saying that *a subject whose only sense was the sense of vision could not at all have an appearing Body* (II, p. 158).

Thus, having constituted what is a most prominent feature of the Body, other aspects might also be taken into consideration. The Body (as a thing with its stratum of localized sensations) is then analyzed as an *organ of the will* and as a *seat of free movement.* It is "the one and only Object which, for the will of my pure Ego, is moveable immediately and spontaneously [....] Only Bodies are immediately spontaneous ("freely") moveable, and they are so, specifically, by means of the free Ego and its will which belong to them" (II, p. 159). And the Body further has significance for the constitution of higher objectivities, such as "sensous" feelings, the sensations of pleasure and pain, the sense of well-being that permeates and fills the whole Body [...] but also for acts of valuing or for the constitution of values, etc. All these groups of sensations, as sensings, have an immediate Bodily localization. "Thus, for every human being they belong, in a way that is immediately intuitable, to the Body as his particular Body, i.e., as a subjective objectivity distinguished from the Body as a mere material thing by means of this whole stratum of localized sensations" (II, p. 160). And there is an important continuation stating:

The intentional functions, however, are *bound to this stratum*; the *matter receives a spiritual forming*, just as, discussed above, *the primary sensations undergo apprehension, are taken up in perceptions, upon which, then, perceptional judgments are built*, etc. Hence in this way a *human being's total consciousness is in a certain sense, by means of its hyletic substrate, bound to the Body*, though, to be sure, the intentional lived experiences themselves are no longer directly and properly localized; they no longer form a stratum on the Body (II, pp. 160–161).

Thus, this descriptively explains, however not in detail, how sensation is linked with the perceptional leading to judgment and thinking. Even though thinking is not actually localized intuitively in the head, *thinking is essentially related to a Body*, due to the hyletic substrate that human consciousness always has. Further on, Husserl provides (§40) "more precision concerning the localization of the sensings and concerning the non-thingly properties of the body." One of the main points he makes is this:

To the apprehension of Corporality as such belongs not only the apprehension of a thing but also the *co*-apprehension of *the sensation field*, and indeed these are given as belonging, in the mode of localization, to the appearing Corporal Body. "Belonging:" phenomenologically, *this term expresses relations of the phenomenal "if-then"*: if my hand is touched or struck, then I sense it. We do *not here have the hand as physical body and, connected with it, an extra-physical consequence*. From the *very outset it is apperceptively characterized* as a *hand with its field of sensation, with its constantly co-apprehended state of sensation which changes in consequence of the external actions on it*, i.e., as a *physical-aesthesiological unity* (II, p. 163, my italics except the last one).

What Husserl here explicates is exactly the uniting horizontal structure that functions in the apperception originally incorporated in every perception. There is this apperceptively given horizontal structure – that "depth" correlating every sensing. And it actually concerns everything sensible, both the Body itself, (the things of) the world, time, space, etc. If it was not already apperceptively functioning we could not either (afterwards) "repair" it by finding causal explanations for the relation. It would have been too late. This is in a sense the so called "classical problem of body and soul" that in phenomenology is dealt with from the perspective of intentionality. On the one hand (the "deep" aspect), its constitution in inner time-consciousness, and, on the other, by the obviousness of intentionality – now especially in regard to the physical-aesthesiological unity constituted by the functioning Body. We might, of course, loosen up this physical-aesthesiological unity and consider each of its aspects in isolation, but that can be done only in the abstract. It would not adequately grasp the hand as part of the concrete phenomena living and functioning Body:

In the concrete perception, the Body is there as a *new sort of unity of apprehension*. It is constituted as a Objectivity in its own right, which fits under the formal-universal concept of reality, as a *thing that preserves its identical properties over against changing external circumstances* (II, p. 163, my italics).

This "reality" – as the Body is apprehended not only as dependent with respect to the primary stratum of sensation, but also with respect to the fields of that which pertain to it mediately and are not properly localized, thus, e.g., with respect to the field of vision, it constitutes the "link" to the whole field of possible experience:

How the visual field of sensation is filled, which motivations can occur therein, and consequently what in the visual field can be experienced by the subject, and in which modes of appearance it must be exhibited, this *all depends on certain qualities of the Body, especially on those of the eye [......And it is] the capacity to be stimulated in general [that] becomes a universal title for a class of real properties which have quite another source than the properly extensive [......] through this stratum, through this new group of real properties, which display themselves as real insofar as they are constituted through a relation to real circumstances within the real, the material Body is intertwined with the soul* (II, p. 164, my italics).

Thus it might be said that:

Soul and psychic Ego "have" a Body; there exists a material thing, of a certain nature, which is *not merely* a material thing but is a Body, i.e., a material thing which, *as localization field for sensations and for stirrings of feelings, as complex of sense organs, and as phenomenal partner and counter-part of all perceptions of things* [...], makes up a *fundamental component of the real givenness* of the soul and the Ego (II, p. 165, my italics).

And given this we might generally point out what constitutes the Body as material thing *in contrast to* other material things. The first characteristic feature is that (a) the Body is a center of orientations of the Subject. It is the zero point in relation to how you experience nearness, farness, left or right etc. The second is what Husserl labels (b) peculiarities of the manifolds of appearance of the Body. This is about my own relatedness to my Body:

[....] I do *not have the possibility of distancing myself from my Body, or my Body from me*, and accordingly the manifolds of appearance of the Body are restricted in a definite way: certain of my corporal parts can be seen by me only in a peculiar perspectival foreshortening, and others (e.g., the head) are altogether invisible to me. The same Body which serves me as means for all my perception obstructs me in the perception of it itself and is a remarkable imperfectly constituted thing (II, p. 167, my italics).

And, finally, the third reminds us that (c) the Body is an *integral part* of the causal nexus of material nature. Despite its freedom to move itself spontaneously and all the other characteristics which are not material, it also is material and is affected by the causality working within material nature. And actually it is the adaptability of the Body in relation to those external "forces of nature" that constitutes one important aspect in regard to the real freedom of both the Body and the human Subject itself. Another important aspect is the relatedness with other Subjects, its intersubjective relatedness. This has not yet be taken into proper consideration, and this is now the next Husserl comments on, namely the character of the Body as constituted solipsistically (§42). Husserl then points at limitations of his exposition so far, and this brings us into the next main field of interrogation, namely the constitution of Psychic reality in Empathy (Chapter Four, Section Two).

I shall now only mention those issues that are dealt with there. First he deals with how other animalia are given, and then the issue of primal presence and appresence has to be dealt with. Thereafter Husserl interrogates how animalia are primally present as Corporal bodies with appresented interiority, and his exposition closes by discussing the significance of empathy both for the constitution of the reality "I as man" and for the constitution of nature. This, then, of course, sheds light back on his exposition of material nature in Section One of the *Ideas II* as well: the "in-it-self" of material nature also depends on empathy and on intersubjectivity – actually nested in the historicity constituted by the historical practices of the scientific community – and, then, all those other aspects of Body and Soul that have been dealt with in our presentation.

The whole exposition in the *Ideas II*, of course, culminate with the constitutional analyses of the Spiritual World. This might appear to be a field of issues that are even more adapted to the *Historicity of Man* than those issues we have dealt with so far. What we have been doing in our exposition is to re-think and try to reactivate the evidences that are given in the rather concrete analyses of Husserl, specifically in regard to animal nature – thus trying to *demonstrate the historicity embedded within the whole issue descriptively*. Certainly analyses further into the constitution of the Spiritual World, concretely demonstrating the *Historicity of the Spiritual World* as well, would have corroborated and completed what actually has been presented in this exposition. But that has to be a topic for another article.

University of Bergen, Norway

NOTES

[1] Edmund Husserl: *Ideas pertaining to a Pure Phenomenology and to a Phenomenological Philosophy. Second Book. Studies in the Phenomenology of Constitution (Ideas II)* (Dordrecht: Kluwer Academic Publishers 1989/1998). Quotations from this book are referred to in the following manner (II, p. 89) immediately after the quotation in the text (not notes).

[2] Husserl: *The Crisis of European Sciences and Transcendental Phenomenology. An Introduction to Phenomenological Philosophy (Crisis)* (Northwestern University Press Evanston, 1970/1977).

[3] Husserl: *Ideas pertaining to a Pure Phenomenology and to a Phenomenological Philosophy.* Respectively *First Book. General Introduction to a Pure Phenomenology (Ideas I)* (Dordrecht: Kluwer Academic Publishers 1983/1998), *and Third Book Phenomenology and the Foundation of the Sciences (Ideas III)* (The Hague: Martinus Nijhoff 1980).

[4] It is Husserl himself who says this in the Appendix XIII to *Ideas II*, p. 384–385. I have previously published two articles in *Analecta Husserliana*, about this and related issues. One of them is called "Nature, Subjectivity and The Life-world" Vol. LXXVII, and the other "The Historicity of Nature," Vol. XC (Dordrecht: Kluwer Academic Publishers). The present paper represents, in a manner, a continuation.

[5] The thesis, of course, is rather far-reaching and we do not (in this article) pretend to cover it all. But we *aim* at developing the needed argument.

[6] The way I express myself here is inspired by Robert Sokolowski who in his brilliant introductory book *Introduction to Phenomenology* (Cambridge University Press 2000), p. 4 uses statements like this to characterize phenomenology.

[7] This is similar to what Husserl says in regard to the thing of material nature, that objects are already constituted pre-theoretically, cf. II, p. 8.

[8] In the *Crisis* Husserl talks about "die transzendentale Geschichtlishkeit" and we now quote (in English) the whole context in which the expression appears, pp. 208–209: "In psychology the natural, naïve attitude has the result that the human self-objectifications of transcendental intersubjectivity, which belong with essential necessity to the makeup of the constituted world pregiven to me and to us, inevitably have a horizon of transcendentally functioning intentionalities which are not accessible to reflection, not even psychological-scientific reflection. "I, this man," and likewise "other men" – these signify, respectively, a self-apperception and an apperception of others which are transcendental acquisitions involving everything psychic that belongs to them, acquitions which flowtingly change in their particularity through transcendental functions which are hidden from the naïve attitude. We can inquire back into the transcendental historical dimension [Geschichtlichkeit, historicity], from which the meaning and validity-accomplishment of these apperceptions ultimately stems, only by breaking with naiveté through the method of transcendental reduction. In the unbroken naiveté in which all psychology, all humanistic disciplines, all human history persists, I, the psychologist, like everyone else, am constantly involved in the performance of self-apperceptions and apperceptions of others. [........] Also the Cartesian two-substance-doctrine [....]."

[9] This will be in accordance both with what Husserl says in an Appendix, namely that it is the living subject and the basic relationship in this life-world which predelineates the point of view of the method (cf. II, p. 384), and in the *Crisis* (pp. 208–209) in regard to transcendental historicity.

[10] Appendix VI to the *Crisis*, pp. 353–378.

[11] There might, of course, be a problem using clues and the perspective given in the last and most mature stage of phenomenology's development to analyze and interpret what it says in previous stages. There might be breaks etc. that make those different stages in the development in a way incompatible. Husserl himself has a very good and relevant comment in regard to this kind of problem. In a letter to his acquaintance D. Feuling in March 1933, he says the following: "[...] a genuine elucidation of the historical development (in the philosopher) can only be given on the basis of the way its meaning takes shape in its full maturity; only then can one understand the structure of the dynamic in each lower stage" (Husserl: Briefwechel p. 89 – here quoted from Brozina's translation in his *Introduction to Sixth Cartesian Meditation*, p. XXVI).

[12] In his article "The Life-world and the Historicity of Human Existence," in *Research in Phenomenology, Vol. XI*, pp. 111–140, Ludwig Landgrebe argues for a transcendental theory of the life-world and of historicity by suggesting that a phenomenological reflection upon the transcendental ego – once correctly understood, as he says – is the proper procedure for constructing such a theory (p. 112). I myself have written about this in an article called "On the Historicity of Understanding," in A-T. Tymieniecka (ed.), *Analecta Husserliana* LIX, pp. 401–422, 1999.

[13] Husserl: *Cartesian Meditations. An Introduction to Phenomenology* (The Hague: Martinus Nijhoff 1960/1988), p. 37.

[14] Of course, we find a proper context for getting deeper into this both in the *Ideas III* and the *Crisis* – but there surely are other places as well.

SECTION III

IN SEARCH OF EXPERIENCE

ILJA MASO

AN EMPIRICAL PHENOMENOLOGICAL APPROACH TO EXPERIENCES[1]

INTRODUCTION

Discussing the essence of consciousness as the theme of inquiry, Edmund Husserl writes in his Ideas,[2] that phenomenologists will "consider conscious experiences *in the concrete fullness and entirety* with which they figure in their concrete context – the *stream of experience*."[3] As an indication of a conscious experience in its concrete context, Husserl points out his experience, his seeing and touching, of the white paper that, in the dim light, lies in front of him, appearing to him from a particular angle, and also to his background awareness of the books, pencils, ink-well, and so forth, that lie around and about the paper.[4]

Such an experience holds good, according to Husserl, both of the conscious awareness of things "in recollections, in representations similar to recollections, and also in the free play of fancy,"[5] and of all conscious experiences of thought, feeling and will (although the directedness towards these experiences differs from the way *objects* are selected and noticed). All these experiences are embedded in "a medium of dormant actuality," viz., the stream of experience.[6]

Although Husserl only gives hints of how such an experience should look, his emphasis on "the concrete fullness and entirety with which they figure in their concrete context" should have become the most important initial impetus to empirical phenomenological research. Because the results of this kind of research are ultimately based on the description of such contextual experiences, these experiences should be formulated as completely as possible. Only then will it be possible to discover all the correctly composed essential characteristics of these experiences. However, empirical phenomenologists do not take this call for completeness sufficiently to heart.

Another consequence of Husserl's enunciation for empirical phenomenological research, one which researchers do take to heart, is that experiences should be considered in relation to their context: a consequence Husserl himself never acknowledged. One of his projects was to discover essences of

A-T. Tymieniecka (ed.), Analecta Husserliana XCIII, 163–174.

experiences independently from the historical, cultural and situational context in which these experiences occurred. In empirical phenomenology this project is regarded as impossible, as will be shown in this paper.

In changing Husserl's project in this way, empirical phenomenology is, however, not radical enough. To consider experiences in their concrete context, the stream of experiences, suggests, and this is confirmed by our everyday experiences, that experiences have a process character, i.e., they have a beginning, a middle and an end. This process character should be reflected in the eidetic reduction of an experience, but this is neglected in empirical phenomenology.

THE RICHNESS OF EXPERIENCES

17 November 2000. On stage is the Dutch comedian Marc-Marie Huybregts, wearing a dress made out of a burlap sack. I have never seen him before because I hardly ever watch television. He is a rather heavyset but short man. I think he is older than he looks at first glance. Not someone who would catch your eye if you passed him on the street. And he certainly would not stand out as attractive. But as a comedian, he certainly knows how to capitalize on his "weak points," such as his stubbiness, his gay demeanor and his nasal voice, and he does so without mercy. Self-ridicule but without really being cynical. And I admire that about him, just as I admire his singing voice: a brilliant countertenor! But at the same time he gives me the creeps.

He is playing a scruffy, misfit teen whose withdrawn existence consists of wiling away each day writing letters he never sends. His mother committed suicide and his father threw him out because he kept mistaking his new stepmother for a throw cushion and kept trying to sit on her. Wry and hard humor. But his personality keeps it all alive and warm.

However, he also played tricks on the audience. He spoke to individual audience members. And I sensed that he was really trying to see how far he could go. For one woman, he acted out the letter about masturbation that he had written to "the lady next door." A letter which, in a social reality, could never have been written; however, in the reality of the theatre, the letter played the feelings of shame as experienced in Drachten and environs, most residents of which belong to the fairly conservative Dutch Reformed Church, and the feelings of this woman in particular. I am sure that she was well and truly mortified and I was glad I was not in her shoes.

But there was still something in store for me, too, I sensed. Because his trick was the same every time: he made contact with someone in the audience and came back to this person later with an even bigger routine. He had already spoken to me. We laughed hysterically a bit about doctors and homosexuality. I was at the theatre with my regular theatre friends: a gay couple, one of whom is a psychiatrist and ex-colleague of mine. And there was a doctor sitting behind us. We were in the third or fourth row. So we were a really noisy group when he was doing this bit; it was really like listening to our own lives. "So what's your name?", the shrill voice with the southern accent suddenly asked from the stage. I tried to look away. "Greet," I said softly. "What?" he asked again. "Greet!" I said again, a little louder. Laughs. The woman from just before, the one from the letter, was called Gretha. He looked from her to me and made a face. "We're not related!" I shouted. It just came to me. More laughs. "You look like my mother," he said in a rather exaggerated manner, and then more softly: "For a moment you looked like my mother." He continued the show with something else, but I was no longer at ease.

But I was not prepared for what came next. The closing song of the show, which suddenly had a serious tone, was an ode to his mother whom he (or the character he was playing) missed and really longed for. He sang it for me. To me. He stood at stage left, at the front of the stage, right in front of my seat and he looked at me. Kept looking at me. Somehow I even was his mother.

Very strange because I knew, thought, that what he did put him in control on a different level. I could not look away, I was forced to endure it, I no longer was in control of my own body. My heart was pounding, it was as if my whole body inhaled and exhaled. I felt my face turning red and saw that he was blushing, too. I felt very, very warm and confused. What was happening here? This was all theatre, and he was playing a part, but at that moment we were really genuinely connected, my body was responding to his and his to mine. It was as if we were in some kind of tunnel. I was sitting in my seat in the packed house but at the same time, I was in a sort of private room with him. I sensed that we knew each other, understood each other, through and through.

Wim and Norbert felt it too and started fidgeting and giggling out of the awkwardness of the moment.

I do not remember what he sang or how long it took but I do remember a stream of emotions going back and forth: tenderness, dependency, it was very strong. It was a gimmick, and yet it was not. It was theatre, but it had its own authenticity. It felt real.

When the ovation started, his eyes met mine again. The guy in front of me was very tall so he had to take an extra step forward to see me. I smiled at him and he smiled back.

Later, as we were having drinks in the foyer, Marc-Marie passed our table on his way to the obligatory mingling with the theatre director (something he clearly did not feel like). "Did you like it?", he asked. Did I like it? "You really shook me up," I said, "But I found it very special!" "I was glad you were there," he said.

When he was leaving and had shaken the hands of all the big shots in attendance, it became all the more clear that the world had returned to normal. We were back in the world where words play the lead and social conventions are a mask. We moved towards each other. We were not sure whether we should shake hands. I think that in the end, it was the hesitation in my step that refrained us. We left it at a nod. I felt bare and lonely. But also a warm afterglow as I thought back to that extraordinary event.

This is one of the accounts written by a student who was asked to describe a significant experience and, in so doing, took a first, hesitant step down the path of empirical phenomenological research. This could also be an account phenomenological researchers could acquire when they ask their co-researchers to describe, for instance, an experience of shame.

Because this is what empirical phenomenological researchers do: they ask their co-researchers to tell them or to write down a certain experience in as much detail as possible. Among other things, this means that co-researchers are asked not only to focus on "the highlights" of their experiences but also on body sensations, feelings, beliefs, thoughts and convictions that accompany these highlights.[7] To supplement this, co-researchers are sometimes asked to say more about some words, phrases and statements they have used. More than this is generally not attempted.

During my research, I have found that what empirical phenomenological researchers do is not nearly sufficient. To acquire an experience as full and complete as possible, researchers must, based on what co-researchers say, attempt to experience the situation as if it happened to them. To really be capable of doing this, they must bracket their own world view and at the same time refrain from prejudgments, conceptions, value judgments or theories over the other person and the other person's experience.[8] From this perspective they will often have to request clarification on things that co-researchers have a tendency to gloss over or omit.

In order to be able to answer these questions, at that moment or later, co-researchers may have to go back into their experiences, sometimes delve deeper into them or tell more about the situations in which the experiences took place. The important thing for them is to recall the experiences as if they are being experienced anew as much as possible. This way the experiences approach the immediacy of direct experience, that is, experiences that "precede all interpretation, discussion or communication."[9] This helps co-researchers to approach their experiences as openly as possible without considering any prejudices, conceptions, value judgments, explanations or theories that are not an element of those experiences.

Co-researchers usually assume that their experiences are something that happened to them (and in which they hardly had any role at all). This usually means that they focus too little on their own inward and outward behaviour. An important tool to help co-researchers to recall that behaviour is to ask them how they brought about having that experience. By using this method, and by repeatedly checking the experience which has been ultimately related for concealed elements, the richness of the experience is revealed to the maximum extent possible.

This can be exemplified by considering the following extract of the example I have presented above.

I could not look away, I was forced to endure it, I no longer was in control of my own body. My heart was pounding, it was as if my whole body inhaled and exhaled. I felt my face turning red and saw that he was blushing, too. I felt very, very warm and confused. What was happening here? This was all theatre, and he was playing a part, but at that moment we were really genuinely connected, my body was responding to his and his to mine. It was like we were in some kind of tunnel. I was sitting in my seat in the packed house but at the same time, I was in a sort of private room with him. I sensed that we knew each other, understood each other, through and through.

This extract was changed as follows.

I am not able to look away. I can, physically, because my body is not impaired; I know that I can still move my muscles. When I swallow I am very aware of the movement of the muscles in my cheeks and throat, they feel normal, not warm or cold. It is more a sort of not wishing to go away from that feeling and a certain persuasive quality from the other side, an appeal to stay with him. A suction, an invitation to keep playing the game with the glowing ball.[10] He keeps looking, I am forced to endure it, I am no longer in control of my own body. It is drawn to him, as it were, it is directed to him, it is as if all molecules point in his direction in some way like an MRI scanner levels on molecules. My heart is pounding, it is as if my whole body inhales and exhales, a feeling of getting larger and smaller, mostly in the chest, a feeling of being made of a flexible material that has the elasticity of a rubber band but at the same time the strength of oak. I feel my face turning red, a glowing, hot, tingling sensation that creeps from chest-high and from behind my ears over my neck and face and up to my hairline. It goes deeper than the skin, it is as if my whole head is full of a bubbling liquid, and I see that he is blushing, too. I feel very, very warm and confused. I think, "What is happening here?" This is all theatre, and he is playing a part, but at that moment we are really genuinely connected, my body is responding to his and his to mine. It is like we are in some kind of tunnel. I am sitting in my seat in the packed house but at the same time, I am in a sort of private room with him. To be honest I know that I am sitting, but I do not feel like that. I mostly feel my upper body, starting at my belly. My lower body is far away, like the people around me. I know and see that they are there but I feel, experience them as very far away. While Marc-Marie feels much closer than he is. It seems like six or five feet while it must be twenty or twenty-three. I have the image of a transparent, grey tube, along which sparks of yellow light dance around at varying speeds. The tube seems to be made of a heavy material somewhere between liquid and glass. Its walls are four inches thick or more. The tube, tunnel, is light inside and everything outside it seems darker. The tunnel is about three feet around. Marc-Marie is at one end and I am at the other. Somehow, I have no sense of what is behind me and I have the feeling that he does not either. And time seems to stand still, or in any case, it seems to be irrelevant. I sense that we know each other, understand each other, through and through.

When we compare this description with the one from the introduction, we notice that the experience is not so much one of shame but more a kind of trance. Repeatedly recalling an experience anew does not only reveal what has been concealed, but also causes co-researchers to remember parts of their experience that they did not at first acknowledge. The consequence can be that other, more fundamental parts of their experience resurface.

The comparison also shows that the extended description is written in the present tense and, as a result, that it is much more vivid and personal than the first version. When co-researchers recall portions of their experiences that they had initially summarized or had initially escaped them, the experiences are increasingly relived as contemporary, new, living experiences.

Because feelings are more important in the relived experiences, something else important changes. While the co-researchers initially view their experiences at a distance, from the present, they increasingly recall their experiences in the situations in which they were when they happened.[11] So, while the student initially saw the whole experience "from the outside looking in," she slowly began to re-experience each successive event "from the inside looking out," seated in the theatre.

The extended description also shows that, sometimes, it is impossible to stay close to an experience using normal language.[12] In that case, co-researchers seek refuge in abstractions, metaphors, comparisons, poetic imagery and whatever tools there are at hand to convey their experiences as well as possible. In this case, the student uses phrases like "it is as if all molecules point in his direction in some way like an MRI scanner levels on molecules," "it is as if my whole body inhales and exhales," etc.

Another thing that can be noticed is that the extended description is more than three times as long. Because language not only reveals, but also conceals,[13] researchers cannot assume that the entire experience has been uncovered until they have tried their utmost to discover everything that lies buried into the mind of their co-researchers. By trying to co-experience the experiences of their co-researchers, researchers and co-researchers are able to use hermeneutics in the sense of Hans-Georg Gadamer. Gadamer assumes that we can only properly explain experiences if these experiences touch us in one way or another, or if these experiences are in some sense about ourselves.[14] In this type of situation the passion arises to abandon or adapt some of our preconceptions that we use to study experiences,[15] to make the concealed into the unconcealed[16] and thereby change ourselves.[17] Partly because there are no fixed, context-free guidelines for this,[18] researchers and co-researchers are aware that although the things discovered will not contain lies, they will, at the same time, also not be entirely true.[19]

A brief analysis of the difference between the original and the ultimate extended description of the whole experience shows that what has been uncovered can be extremely diverse. It can consist of feelings, images, insights, questions, conjectures and expectations that either precede or give rise to certain behaviours or which follow or are brought on by certain behaviours. It can be constructed from the feelings or images to which certain previous descriptions refer, the significance that elements of the experience had then or the associations that were evoked, from opinions the subject had at the time about certain feelings, behaviours and events, from nuances of previous statements and from elaborations on a number of described emotions.[20] This shows that there are no general rules applicable to arbitrating the struggle of researchers and co-researchers to acquire an experience in its fullness and entirety.

The last thing that can be noticed is that, due to the uncovering, the final text does have a quality akin to what we find in novels. This implies a danger. The focus in empirical phenomenology is not good writing but a good description of the richness of an experience. The urge to make a nice complete story must therefore be bracketed. If this fails, and the narrative takes centre stage, the result may be engaging prose, but the language used will convey more in what it conceals than what it reveals. The richness of the experience is then upstaged by the richness of the language.

THE ESSENCE OF EXPERIENCES

I was talking with a group of girls and I said something that probably came out pretty badly and could be taken a few different ways and well. ... I could feel myself blushing so I just said "geez, I feel myself going red now" and they just laughed and it went away a lot quicker and no one seemed to make any comments about it, cause I got in there first. ... .

This example, that originates from a discussion group on the *internet*, shows that experiences do not come and go in a vacuum. This girl said something silly to other girls, experienced a moment of shame, and her experience ended the moment she dealt with it in a certain way. Likewise, the experience of being scared can happen in a situation in which somebody is caught up on some particular experience or is self-absorbed, whereas the experience of shock could end gradually when this person's heartbeat returns to normal or when he regains his composure.

When researchers and co-researchers expand on these and other experiences, the situation in which the experience began, the continuation and conclusion of that experience cannot be left out, because these stages are bound to it. In this way (co-)researchers reveal the process-oriented nature of every experience.

To discover the essence of an experience, we evoke as many examples of this experience as possible, regardless of whether they actually occurred or were imagined. We compare these and make variations on these, and with each variation ask ourselves whether we are still able to recognize it as a specimen of this experience.[21] If we are not able to do so, we have apparently changed an essential characteristic.

This is the normal technique of eidetic reduction in phenomenology. With it, we reduce the personal, the unique, the coincidental elements of an experience to the core, the essence, the eidos, the structure of that experience. These essences are not found in a separate world or exist independently of the world, but are included in our consciousness as a condition and result of the way in which consciousness and world are (intentionally) connected to each other. As a result of this connection, we are never capable of knowing whether and when the experiences (and variations thereon) evoked in our minds are sufficient for discovering the "universality of the essence" as formulated by Husserl. Have we included all experiences influencing the formulation of the essential characteristics in our consideration? Have we perhaps, from our perspective, from our presuppositions, systematically excluded certain experiences or have we failed to "do these experiences right?" To what extent do experiences resembling ours receive other essential characteristics?[22] Seeing that there are no definitive answers to these questions, there is no basis to draw definitive conclusions on the universality of essences.

This conclusion is affirmed if we consider that, as a result of the simple fact that in this day and age we have more knowledge about all kinds of phenomena, we experience all sorts of natural phenomena in a manner essentially different from our ancestors. Due to their symbolic meaning, a snake or a tree was experienced differently in those days than is common nowadays in our present culture, and still is experienced differently in other cultures.[23] As another example, people with supernatural powers in other eras and places are experienced in a manner that is substantially different from the way we experience them in our present Western culture. Simply consider the position of shamans in other cultures compared to the status of fortune tellers and spiritual mediums in ours.[24] Moreover, it cannot even be ruled out that, when taking a closer look, each experience in different cultures/subcultures and historical periods also has essential differences, along with having essential similarities. Whatever the case may be, we, in any case, cannot assume that research into any experience results in the discovery of essential characteristics that are universal. Empirical phenomenological researchers are therefore well advised to place a caveat by the assumption that the essential characteristics found are universal in nature, and assume for the time being that the essences have a validity that is restricted in time and place.[25]

If this premise is correct, it would appear to be generally ill-advised to look for similarities between certain experiences from cultural and historical contexts that vary to a greater or lesser degree and which have cultural and historical variables. If we nevertheless proceed to do so, we run the risk that the outcome of the eidetic reduction, the essence of the experience, is so remote from the richness of each of these experiences that it does not contribute in any way to the existing knowledge about such an experience.

It would appear to be more appropriate to consider experiences as inextricably bound to the situation in which they occur at a certain point in time and, on this basis, compare similar experiences in similar situations. Apart from the fact that research in this manner remains manageable in scope, there is a reasonable chance that the richness of individual experiences will resonate through the essential characteristics we can discover in this manner. Thus, empirical phenomenological research can contribute substantially to the knowledge about the situational experience in question, and, for the most part, the results will consequently have theoretical and practical relevance.

In practice, this means that we will limit ourselves to comparing an experience such as that of shame, to experiences of shame occurring in similar situations. Suppose, for instance, that we start from the experience of shame as presented at the beginning of this section. In this example, the feeling of shame happens in the presence of others *and* is the result of behaviour that could in principle have been different. Looking for similar situations, perhaps we are able to remember our feeling of shame in one of our dreams because, in the company of others, we seemed to have forgotten to put on our pants and, consequently, were naked from the waist down. Or how we experienced a feeling of shame when we got caught stealing money from mother's purse. Or how Piglet was ashamed at being scared about catching a "Heffalump" when it turned out to be Pooh with a honey jar over his head.[26]

The consequence of the decision to refrain from considering experiences independently from the situation and the time of occurrence and as a result, engage in eidetic reduction of similar experiences in similar situations, is that one or more of the essential characteristics which have been found may have a causal relation to the rest of the characteristics. If the essential characteristics of the experience occurring in one situation differ from those in another situation, it is likely that this difference is the consequence of the difference in the situation.

Suppose that we are investigating the experience of shame in a situation in which someone chose to behave in a way he is or later becomes ashamed of, and the experience of shame in a situation in which the behaviour leading to the feeling of shame was not a conscious choice. Suppose that in the first situation "damage to self-esteem" is found as the essential characteristic, whereas this is not the case in the second situation. In that case, there is a reasonable chance that the first situation can be considered to be the cause of this essential characteristic. If we restrict the research to similar experiences in similar situations, these causal connections are not made explicitly, but they are nonetheless present and this hardly appears or does not appear at all in the eidetic reductions of traditional phenomenology.

In addition, essential characteristics representing the process nature of an experience are rarely found in phenomenology. In the beginning of this section, it is made clear that the course of experiences, from beginning to end, is the main concern of empirical phenomenological research. This can imply that eidetic reduction of this course of experience leads to the discovery of essential characteristics which need to be placed in a particular phase of such an experience. If shame always ceases once we assume that our self-esteem is no longer affected, this is an essential characteristic reflecting the end of that experience, which means that this characteristic has a process nature. These types of characteristics can therefore also be the result of eidetic reduction.

The University for Humanistics, The Netherlands

NOTES

1 This paper is based on I. Maso, G. Andringa and S. Heusèrr, *De rijkdom van ervaringen: Theorie en praktijk van empirisch fenomenologisch onderzoek* (Utrecht: Lemma, 2004) (in Dutch).

2 E.G.A. Husserl, *Ideas: General Introduction to Pure Phenomenology* (1913), trans. W.R. Boyce Gibson, 5th imp. (London: George Allen & Unwin, 1969).

3 *Ibid.*, pp. 116; the italics are Husserl's.

4 *Ibid.*, pp. 116–117.

5 *Ibid.*, pp. 118–119.

6 *Ibid.*, pp. 118–119.

7 D.A. Rahilly, "A Phenomenological Analysis of Authentic Experience," *Journal of Humanistic Psychology* 33(2) (1993): 59; A. Giorgi, "A Phenomenological Perspective on Certain Qualitative Research Methods." *Journal of Phenomenological Psychology* 25(2) (1994): 207; P.R. Heath, "The pk Zone: A Phenomenological Study," *The Journal of Parapsychology* 64 (2000): 56; Ch. Pietersen, "Research as a Learning Experience: A Phenomenological Explication," *The Qualitative Report* 7: 2 (2002); Th. Groenewald, "A Phenomenological Research Design Illustrated," *International Journal of Qualitative Methods* 3: 1 (2004).

8 I. Maso, "Phenomenology and Ethnography," in P.A. Atkinson, A. Coffey, S. Delamont, J.F. Lofland & L. Lofland (eds.), *Handbook of Ethnography* (London: Sage, 2001), pp. 139, 141.

9 H.-G. Gadamer, *Truth and Method* (1960) (London: Sheed and Ward, 1988), p. 55.

10 She felt the changing, inner connection between her and the entertainer as a play of warm balls of fire; balls that fluctuate in size and move through each other.

11 Cf. Nigro and Neisser in D.L. Schacter, *De kunst van het geheugen: De herinnering, de hersenen en de geest* (1996) (Amsterdam: Anthos, 1997), p. 38 (in Dutch).

12 Cf. S. Mulhall, *Heidegger and Being and Time* (London: Routledge, 1996), pp. 809–810.

13 M. Heidegger, *Being and Time* (1927) (Oxford: Basil Blackwell, 1973), p. 197.

14 Gadamer, *Truth and Method*, pp. 89, 111–119; D.G. Smith, "Hermeneutic Inquiry: The Hermeneutic Imagination and the Pedagogic Text," in E.C. Short (ed.), *Forms of Curriculum Inquiry* (New York: State University of New York Press,1991), pp. 200–201.

15 M. Polanyi, *Personal Knowledge: Towards a Post~Critical Philosophy* (1958) (London: Routledge & Kegan Paul, 1973), p. 159.

16 Heidegger, *Being and Time*, pp. 56–57.

17 J. McLeod, *Qualitative Research in Counselling and Psychotherapy* (London: Sage, 2001), p. 22.

18 D.E. Linge, "Editor's Introduction," in D.E. Linge (ed.), H.-G. Gadamer, *Philosophical Hermeneutics* (Berkeley: University of California Press, 1976), p. xiv, xviii; Gadamer, *Truth and Method.*

19 Cf. Gadamer, *Truth and Method*, p. 265.

[20] Apart from this analysis there are, of course, other aspects that can be uncovered through the repeated recollection of a given experience. Some examples are: tactile sensations, smells, tastes, feelings of cold or warmth, pain, other physical sensations (such as hunger, itching, tickling, the need to relieve one's self), self-awareness, dreaming, excitement, fatigue or a combination of all of these; D.J. Chalmers, *The Conscious Mind: In Search of a Fundamental Theory* (New York: Oxford University Press, 1996), pp. 6–11.
[21] Cf. H. Spiegelberg, *The Phenomenological Movement: A Historical Introduction*, 2nd ed., Vol. 2 (The Hague: Martinus Nijhoff, 1971), p. 677.
[22] D.M. Levin, "Induction and Husserl's Theory of Eidetic Variation," *Philosophy and Phenomenological Research* 29 (1968): 8–12.
[23] See, for instance: A. Stevens, *Ariadne's Clue: A Guide to the Symbols of Humankind* (London: Penguin Books, 1988), p. 38.
[24] See, for instance: R.S. Broughton, *Parapsychology: The Controversial Science* (New York: Ballantine Books, 1991), pp. 27–32.
[25] Maso, "Phenomenology and Ethnography," p. 142.
[26] A.A. Milne, *Winnie-The-Pooh* (1926) (London: Mammoth, 1995), pp. 61–64.

NANCY MARDAS

THE ETHICS OF ATTENTION

> I don't say he's a great man. Willy Loman never made a lot of money. His name was never in the paper. He's not the finest character that ever lived. But he's a human being, and a terrible thing is happening to him. So attention must be paid. He's not to be allowed to fall in his grave like an old dog. Attention, attention finally must be paid to such a person.
>
> *Death of a Salesman* (Arthur Miller)

Upon the death of Arthur Miller May 2005, this haunting refrain from the second act of *Death of a Salesman* re-entered our cultural orbit. It was never far away; at least not for phenomenologists. For attention is the essence of the science of phenomenology: attention both to the objects which present themselves to our consciousness, and to that consciousness itself. In this paper, I would like to pay attention to attention itself, as an event of consciousness and as an ethical phenomenon, an event which is also a demand, a blessing, and a delivery, in a dual sense.

Attention to consciousness is the hallmark of the phenomenological method. The event of the *epoche* is the phenomenological event *par excellence*: that movement by which human consciousness accomplishes its self-transcendence and reaches simultaneously into its own essence and into the essence of the object of its attention. For too long, I think, we have looked upon the *epoche* as a matter of epistemological, rather than moral significance.

Yet certainly Anna-Teresa Tymieniecka has been pushing us in this direction for quite some time, especially with her re-valuation of the *epoche* as part of the critique of reason that she has termed "a new Copernican revolution." By looking at the creative imagination not merely as a matter of interpretation of perceptions but rather as the means by which we become fully human, Tymieniecka brings our focus back onto ethical agency. For it is a central tenet of her philosophy that it is the moral sense which establishes the human individual in his or her life-world, in two ways. First, ethical action is seen as "the origin and nature of the uniquely human meaningfulness of ... human existence and the world."[1] Second, she sees ethical action as "the specific meaning-bestowing function of the person as the subject/agent

A-T. Tymieniecka (ed.), Analecta Husserliana XCIII, 175–187.

within the social world."[2] Simply put, in her philosophical system it is ethical action which establishes the world as real to the individual, and simultaneously establishes the individual as real within a social context. Ethical action defines the person, both within the self and in the community of others. Ethical action is the means by which the individual becomes real (to itself and others), makes the world real to itself and others, and the means by which meaning is bestowed, both on the individual and on the world.

The process of human "self-interpretative individualization," as Tymieniecka calls it, has been discussed elsewhere.[3] Briefly recapitulated, Tymieniecka outlines a fivefold process of development of the individual, as s/he moves through experience of the world into action upon the world, and then into full engagement with the world as a meaning-bestowing subject inscribing itself into history through ethical articulations of individuality, through which s/he distinguishes him/herself from the circumambient world. As this process unfolds, automatic or primal processes are gradually replaced by conscious processes, and the person moves forward from generic responses to its environment and basic needs to an increasingly specific satisfaction of meaningful desires and intentions. In this way, all of the potentialities of human being can be explored, and the human person is able to express him/herself in creative self-articulation in the context of the lifeworld.

This activity cannot be undertaken in a vacuum. If the individual is to emerge into full-fledged consciousness, Tymieniecka avers, this must occur by means of the impetus of that context which stimulates "me" to move into an exploration of how I can expand meaning. In this schema, otherness (of the lifeworld, of other persons) is necessary to my development in and of meaning. It is not enough for me to bestow meanings on my actions, to articulate (to myself) a sense of significance. Meaning must also be ratified by the recognition of what is not-me. What is felt is experienced and given significance within the context of that otherness. Otherwise, the event would have no transcendent quality, and no possibility of founding meaning; I would merely spin in concentric, solipsistic circles of the same. Tymieniecka has expressed this well:

> Full consciousness means not only the instinctive sharing of self-interest with other individuals, but also the propensity to *expand one's own individual meaningfulness into transactions with other individuals*. The dominant limitation by the *universal scheme of life* ... is broken down and recedes before the *inventive function* by which the individual devises his own way of existential self-expansion. This expansion may be accomplished only in transaction with others.[4]

This intersubjective schema has been well traced in Western philosophy, from Fichte to Hegel to Levinas. Tymieniecka is not even the first woman philosopher of the 20th century to insist on the importance of attention in intersubjectivity – Simone Weil took the notion of attention into a metaphysical and even mystical direction.

In a posthumously published essay, "Réflexions sur le bon usage des études scolaires en vue de l'amour de Dieu" (published in *Waiting For God*), Weil ... suggests that true attention consists of suspending our thought, leaving it detached, empty, and ready to be penetrated by the object; it means holding in our minds, within reach of this thought, but on a lower level and not in contact with it, the diverse knowledge we have acquired which we are forced to make use of. Our thought should be in relation to all particular and already formulated thoughts, as a man on a mountain who, as he looks forward, sees also below him, without actually looking at them, a great many forests and plains.[5]

This method bears a marked resemblance to the Husserlian *epoche*, especially as understood and elaborated by Tymieniecka. Tymieniecka's "step beyond" is to elaborate an onto-poeitic understanding of this progression. In the process of these "transactions," the individual creates herself, invents meaning and projects that meaning out into the world.[6] As the individual chooses among actions, s/he seeks to justify action intersubjectively. It is in this first moment of choice that the individual begins to take shape, and so, its world.

Tymieniecka see the creative role of consciousness as constitutive of personhood. Yet for her, this function is neither merely cognitive nor communicative. Rather, this function is essentially moral, and grounds the individual as essentially moral.

How is this? By virtue of the fact that reason, in her words, is not neutral. Neither is it merely pragmatic. Rather, she says, reason is always already transcendent and moral, because it is always in the service of the Logos of Life:

... it is not reason – whether that of mathematics, that of the Platonic forms or of the universal patterns of working consciousness – that brings order into life. On the contrary, *it is life which brings forth the multiplicity of ratios, rationales, and reasons in order to unfold its constructive course*. These are only seemingly autonomous. In fact, they are all intertwined within the constructive individualizing progress of life ... they share a common life-promoting project in the primeval schema of life itself.[7]

In the service of life, she suggests, reason is always making judgments – judgments which are not merely vital, but *moral* – about action based on the relationships within which it finds itself, and always already acting within the context of those relationships.[8] If, as Tymieniecka suggests, value is created simultaneously with the person, then the process of self-interpretation-articulated-through-action is creative not only of the human person (though that might seem to be enough!) but of something else as well, what Tymieniecka calls the *novum* – that unique new factor in the history of the world, irreducible and unrepeatable, the distinctive value which that one person brings into being in its consciousness of self in its life-world. For Tymieniecka, the *novum* is moral in that it is the elemental translation of the dynamic of life in human activity.

In Tymieniecka's understanding, morality, far from being a matter of intellective and rational activity [*Handlung*], is exposed as a far more exciting *ontological* process: it is the creation of new realms of understanding and action. The creation of the individual is the creation of a clearing in the forest of *das Seiende*. As Heidegger might have put it, the creation of the individual is the creation of a *Lichtung*. What occurs in this space of creation is the moment between decision and implementation, which is full of virtualities, potentialities. Only one will issue forth in (or as) the new aspect of the self-interpretation in existence of the individual. Only one will unfold within the context of the unity-of-everything-that-is-alive, to take its place in history.[9] This one moment will lead either to the greater harmony of being, or to greater differentiation within that unity.

I would like to unfold Tymieniecka's insight one step further, and suggest that it is in this moment of creation that human freedom is born. (This is, perhaps, an explicitly moral reading of Tymieniecka's view of creativity and the creative forge.) For it seems to me that it is only in the clearing thus created that freedom actually has the possibility of blooming into actuality: both my freedom and the freedom of the other. The moment of freedom is the moment at which an ethics can emerge, as an instance of the *novum* of unique significance. This moment, I am suggesting, can be identified with the moment of attention.

In the moment just prior to action, I am at my most free. At that liminal point, I have not yet entered (though I may be aware of) any causal chain. I may also be aware of difference (in fact, I must be, or what impetus would I have to take action?) Yet the moment of attention is one in which I am not bound by difference, but rather can observe it without feeling caught up into it. (Here again, Weil's understanding of attention comes to mind. In *Waiting for God*, she writes: "Above all our thought should be empty, waiting, not seeking anything, but ready to receive in its naked truth the object that is to penetrate it."[10])

The moment of attention holds within it what Tymieniecka would call the "virtualities" of responsibility, freedom, and agency. The moment is moral, within a Tymienieckaian framework, because it is the moment at which the Logos of Life begins to move from mere being to becoming: in "the generative synergy" by which the self-individualizing principle surges forward. It is "moral" because it is the cradle of intersubjectivity.

For even if I am on a determined or determining path, or find myself somewhere along a causal chain, each time I act, I can change the shape of my moral self in any one of a myriad of ways, and emerge from the moment as a new being. At the same time, it is in "the pause at the crest of the breath," as Rilke puts it, that I can act towards the other in total freedom. At the moment prior to decision, I am (more properly said, I can be) free: of desire, will, reason, the demands of intersubjectivity, the burdens of self and other. I can float freely in the moment. I believe that the clearing thus engendered is the moment which Schelling describes as the movement of differentiation between the ground and the unground, and that that movement is best described in terms of freedom. It is, I want to argue, the moment in which *Dasein* makes the free (and freeing) leap between *das Seiende* and *das Sein*. This is the moment of the *epoche* to which phenomenology has drawn our attention, and which we can put to good use in understanding ethical action.

To defend this claim, I would like to call your attention for a moment to the inaugural lecture given by Emmanuel Levinas to the fourth gathering of the World Phenomenological Institute, the International Phenomenology Conference that took place in Fribourg, Switzerland in January 1975. In his address, entitled *Penseé et Prédication*, he suggests that it is at the moment of the interruption of the self by the other (in the moment of intersubjectivity) that the human is revealed to itself for the first time, ontologically and (always already) ethically. It is here, Levinas claims, that *we* begin, in the moment of awakening into the knowledge of our ontological need for the other, our need to open a space for the other in which we can take our first breath as a subject. This moment is eternally before us: it precedes and makes possible our being, both with others and as a self. The first moment, however, is a holding back, a moment of differentiation.

> In me – in the ego [*le Moi*] that yields, that renounces the right-of-way – life ... holds its breath, its vitality as a "driving force." One might be tempted to liken this moment of rupture – the moment prior to every decision – to a sort of self-repression of *jouissance* in its even rhythm of being (*esse*) ... But the *itself* of this *self*-repression would only be at the very threshold of its [coming into] existence! At the tip produced by the continual revolutions made by the spinning of a drill penetrating into the inter-being of the identical, which [is also that which] calls me forth, voluntarily, as *myself*, non-interchangeably, i.e., as the uniqueness of the measure, or, even more strongly, of that measure which is commensurate with that which is irrecusably dedicated to the otherness of the other; and moreover dis-inter-essence itself, an otherwise-than-being which does not permit itself to be expressed – of this point one can be quite certain – either in terms of appropriation, or in terms of the external world.[11]

"Life holds its breath." And in the holding back of the force of Being that is *jouissance*, the unique non-interchangeable *novum* steps forth in freedom.

It is only in this moment that the self is free, while it still floats in the space between the demands of the ego and the demands of the lifeworld. This, Levinas and Tymieniecka agree, is the moment of the birth of the human person.

This moment can be called ethical in several ways. First, because it gives birth to the ethical agent, the subject who can then choose to act. Second, because as a moment of absolute freedom it is the ideal starting point for an ethics. Third, it is a moment that is saturated with responsibility, for, as Levinas reminds us elsewhere, "every responsibility exists prior to freedom."[12] Because phenomenologically speaking, we know that in fact no action takes place in a vacuum, unrelated either to a history or a future. The moment prior to decision already holds within itself the responsibility for an ethical outcome. Will the *novum* be ethically positive or negative? Will it contribute to greater harmony, or greater division of the unity-of-everything-that-is-alive? Will it move towards greater meaning, or away from meaning? The moment may be neutral, but of course the subsequent action is not.

A similar treatment of this problem can be found in the work of another thinker who died in the spring of 2005: Karol Wojtyla, more well-known as Pope John Paul II. The English translation of his *Habilitationsschrift* was published as the tenth volume of the *Analecta*, with the title *The Acting Person*. In this work, Wojtyla presents a phenomenological, yet thoroughly Thomistic treatment of the origin of subjectivity and moral action. He writes:

> Consciousness is the "ground" on which the ego manifests itself in all its peculiar objectiveness (being the object of self-knowledge) and at the same time fully experiences its own subjectiveness ... the tasks of consciousness do not end with its illuminative and reflecting function ... in fact, the essential function of consciousness is to form man's experience and thus to allow him to experience in a special way his own subjectiveness.[13]

That is, in consciousness, the person moves beyond reacting to its environment and towards what might be termed real or authentic action, action, that is, that may bear meaning. Consciousness allows us to "*experience these actions as actions and as our own*."[14] This consciousness is the origin and first act of freedom, and opens the way for the creation of the person. Wojtyla identifies free will with self-determination, and with transcendence:

> ... with the self-determination we discover in the will as a constitutive element of the personal structure of man. Freedom thus manifests itself as connected with the will, with the concrete "I will," which includes ... the experience of "I may but I need not." ... The freedom appropriate to the human being, the person's freedom resulting from the will, exhibits itself as identical with self-determination, with that experiential, most complete, and fundamental organ of man's autonomous being. We are thus considering freedom as real, the freedom that constitutes the real and privileged position of man in the world and also the main condition of his will.[15]

"This premise is of essential significance," he continues. For in his view, which I believe is close both to Tymieniecka and to Levinas, the freedom thus manifested is the essential characteristic of human beingness, that which results in the personal and the ethical. Acting means acting with others, participating in society and integrating the self and its needs with those of society. In so doing, the person transcends both solipsism and existential isolation.[16]

The root of transcendence can be found in the free moment of attention prior to action. I would like to stress here that I am not talking about conscious reflection. In fact, the attention at issue here is pre-reflective, pre-rational, at the threshold of rationality. For reflection, if we are to take Kant's analysis seriously, is a function of the faculty of reason, and as such is always going to be in the service of the self, and speaking from its perspective. As such, reflection is always already, in one powerful sense, not free. It is always already caught, created, constructed within a social frame.

This is certainly not what Simone Weil thinks of the activity of attention.

> Thought, as we ordinarily understand it, is not quite the word Weil means, for thought implies the very focus that is intentionally suspended in this description of attention. At issue is the distinction between fastening the attention around a single phenomenon and leaving the attention open, a difference immediately understood by Weil in terms of larger freedoms and constraints: "If one desires a particular thing one becomes enslaved to the series of conditions. But if one desires the series itself, the satisfaction of this desire is unconditioned."[17]

Perhaps attention, in contrast, has the possibility of inhabiting a pre-reflective space, a space in which the distinction between self and other is still more fluid than fixed. It seems to me that attention can be understood as the origin of ethical action and the moral sense, for two reasons. First, as a neutral and free starting point, attention is a most positive and hopeful point of departure. Second, perhaps self-consciousness can emerge from attention, as simultaneous awareness of differentiation and responsibility. In making this claim, I believe that I am following Tymieniecka in her analysis of the human condition:

> ... in delineating his meaningful life-course the person manifests himself in his inventive role as an outstanding type of beingness from among the entire chain of living beings. The human person in its role of self-conscious giver – receiving, giving, and promoting agent – on the one hand manifests a *universal consciousness of life-conditions*, and on the other hand, its self-consciousness culminates in the capacity to appreciate, calculate, and plan them.[18]

I would like to claim that this consciousness is in fact the *resultant* condition of attention. Further: the more this condition can be inhabited or cultivated, the greater the potential for meaning-giving will be enhanced. Attention could thus be understood as the telos of onto-poeisis: simultaneously its origin and its goal. For:

... the human being does not rise from the natural anonymity of life for the sake of aiming at an ever greater autonomy and self-awareness. [For how can one be "more" autonomous or self-aware? One either is or is not.] The moral sense reminds him that he is an integral part of living nature. Moreover, in his universal self-consciousness of the life-conditions he is intimately united to *everything alive*.

It is from this perspective, that with Weil, Tymieniecka, Wojtyla, and Levinas, we can begin to speak about a specifically human dignity, a specific significance within being. For it is in seeing itself as a "for-other," existentially, phenomenologically, and ontologically, that human being can perform the uniquely human task of simultaneously transcending and integrating its being into Being.

Focusing on attention, it seems to me, resolves the paradox between subjectivity as a matter of bondage and subjectivity as the condition of the free and meaningful movement of *Dasein*. For the notion of attention as I understand it allows us to move beyond intention, beyond *mere* intersubjectivity, and towards the infinite: the infinite space that exists, not beyond time and space, but between individual persons.[19]

This distance, this infinitude, can never be grasped by reason. It is beyond reason, unreasonable. Yet we apprehend it. It is a space which can only be approached, grasped, attuned to by attention. Attention is not vigilant, it is a waiting both outside of time and otherwise than being. It is being in the form and condition of Being itself: the essential openness of the unity-of-everything-that-is-alive. It is, simply, the being of being: endlessly moving toward the other who is infinitely near and far, the neighbor whom I must love as myself, the other who calls to me out of sameness and whose call I must (and can never fully) answer. Attention opens out/into infinity:

"The negativity of the *In*-of the Infinite ... hollows out a desire that could not be filled, one nourished from its own increase ... This is a Desire for what is beyond satisfaction, and which does not identify, as need does, a term or an end."[20]

While preceding subjectivity *per se*, attention opens the space for ethics. It is my contention that this clearing is the space of responsibility which Levinas says always precedes freedom, and so human dignity, and so ethics.

The bearing and crossing of this distance that can never reach destination or closure is the responsibility to which the infinite calls us.[21]

Phenomenologically speaking, we are always already embarked on the journey across this impossible, inconceivable void between self and other, between being and nothingness. We can "never reach destination or closure" because to do so is to die. As Tymieniecka has conceived it, this phenomenon does not merely surge through human life, but is part of the dynamic of life itself (this is the genesis of her idea of *Imaginatio Creatrix*). Human being bears the ontological demand of life itself, "the yearning to give ultimate significance" to the creative enterprise of being.

But what has all this to do with attention? Much ink has been spilled in the religious, philosophical, and psychotherapeutic literature detailing the need for greater attentiveness to the other as the foundation of mindful action and compassion. My purpose here is perhaps only slightly different: I want to focus on the effects of the phenomenon of attention not on the one to whom attention is paid, but on the one who pays attention. What does the act of attention engender in the self? More directly: what role does attention play in engendering the self? How is the phenomenon to be understood as creative – as chosen, as a substantively new creation, rather than as a response, no matter how positive?

At this point, I want to return to the phenomenological method, and share with you a finding from the world of psychology that is quintessentially phenomenological, and which I believe supports what I am trying to say about attention. The psychoanalyst Donald Winnicott observed literally hundreds of children in over forty years of clinical work. In an essay written in 1941, Winnicott noted a behavioral pattern that is strikingly analogous to the topic at hand. In observing very young infants (between five and thirteen months old), he marked three distinct stages of attraction and attention to objects. (At this very early age, Winnicott feels, children's personality is extremely fluid.) In Winnicott's experience, the child experiences confusion and conflict when wishes and fears emerge in the context of relationship with others. In the first stage, when presented with a new object, the child enters what Winnicott calls "the period of hesitation." During this time:

> … the baby puts his hand to the [object], but at this moment discovers unexpectedly that the situation must be given thought. He is in a fix. Either with his hand resting on the [object] and his body quite still he looks at me and at his mother … and watches and waits, or in certain cases, he withdraws interest completely [and exhibits fear] … all this time … the baby holds his body still (but not rigid). Gradually he becomes brave enough to let his feelings develop, and then the picture changes quite quickly. The moment at which this first phase changes into the second is evident, for the child acceptance of the reality of desire for the [object] is heralded by [certain physiological changes] … The change in the baby's behavior is a striking feature. Instead of expectancy and stillness there now develops self-confidence, and there is free bodily movement … related to manipulation of the [object].[22]

Winnicott interprets this hesitation as the moment which gives rise to both desire and its accompanying anxieties. That is, the hesitation is the moment just before the engagement with the other, and realization of responsibility.

The "something" which the anxiety is about is *in the infant's mind*, an idea of potential evil or strictness, and into the novel situation anything that is in the infant's mind may be projected. When there has been no experience of prohibition, the hesitation implies conflict, or the existence in the baby's mind of a *fantasy* corresponding to the ... baby's *memory* of his really strict mother ... these fantasies of the infant are concerned not only with external environment, but also with the fate and interrelationship of the people ... that build up the inner reality.

In other words, hesitation ends when consciousness of the self overagainst others begins, as part of our inner reality: at the threshold of intersubjectivity. I would like to suggest that one helpful way to begin to ground an ethics would be to somehow relocate and reclaim this moment of hesitation and attention.[23] Perhaps attention can deliver us from the stranglehold of the ego, into a more hopeful future, in which anxiety about desire and evil can be replaced by an integrated and integrative sense of oneness. That is, perhaps attention can help us to hold the doors of possibility open, instead of constantly foreclosing opportunities for intersubjective engagement. Perhaps in this way we might be able to realize Levinas' dream of responding to the other in a way that engenders freedom.

I began this paper by quoting Arthur Miller's plea for attention. I would like to end it with a Zen *ko'an* which I think illustrates what I have been trying to say here.

A student spends years gathering the resources necessary to make a once-in-a-lifetime visit to the wise master of whom he has heard so much. After an arduous journey, he arrives at the monastery, and waits his turn for a moment with the master. When his turn arrives, he bows in humility and asks the master for a word of enlightenment. The master says simply, "Attention," and turns away to greet the next in line.

The student is stunned. Has he come all this way, through all these hardships, for this? Frustrated, he gets back into the line, and when he finally gets up to the master again, he begs the master. "I don't understand," he says. "Please, just one word more." The master smiles and says, "Of course, I will tell you more. Attention ... Attention." He turns to the next in line.

Now the student is truly enraged. After a night of fitful rest, he gets back into the line. When the master sees him approach, he smiles. "I knew you would be back," he says. The student is full of hope; he knows his perseverance will be rewarded. "Yes?" he asks eagerly. The master replies, in a manner that suggests that the matter is completely closed, "Attention ... Attention ... Attention."[24]

In closing, I would like to suggest that phenomenology in the 21st century rededicate itself, as a matter of moral urgency, to drawing attention to the practice of the *epoche*. Tymieniecka has shown us how essential this project is, both in the ontopoietic unfolding of human life, and of the Logos of Life. At this perilous juncture in human history, I think we would do well to remember the emphasis that both Tymieniecka and Weil place on imagination in constructing and using attention to move beyond the forces of will and desire towards true intersubjectivity. I believe that the impetus at the heart of Tymieniecka's Copernican revolution is her understanding of the creative forge as a call to the renewal of philosophy and of humanity. In one of her final essays, Weil suggests that "pure, intuitive attention is the only source of perfectly beautiful art, truly original and brilliant scientific discovery, of philosophy which really aspires to wisdom and of true, practical love of one's neighbor."[25] Weil presaged Tymieniecka's elaboration of the next phase of philosophical inquiry. Let us follow Tymieniecka's call to attention – and action:

> With the revelation of the creative context of human functioning an immense field of human inquiry lays open. There is so much to substantiate, elaborate, and follow, this in a dialogue in which not only the various expressions of phenomenology proper participate, but also the multiple perspectives offered by other philosophical horizons, the arts, literature, the human sciences, the natural sciences, the whole sweep of human investigation and endeavor, with each correcting the others, opening new vistas, undertaking deeper probes ... our task of exfoliating the workings of the human creative/inventive logos to their full extent is as vast as life itself. It is a continuing work that neither one human being, nor several, may fulfill in one life; it is a task for philosophy to come.[26]

Saint Joseph College, Connecticut

NOTES

[1] Anna-Teresa Tymieniecka, "The Moral Sense and the Human Person within the Fabric of Communal Life," in *Analecta Husserliana*, (Dordrecht: Kluwer Academic Publishers, 1986), Vol. 20, 28.

[2] *Ibid.*

[3] See, for example, my article on ciphering, "The Cipher as the Unity of Signifier and Signified," in *Analecta Husserliana*, Vol. 83, 13–23.

[4] Tymieniecka, *op. cit.*, p. 31.

[5] Joshua Glenn, "Hermenaut of the Month," 14: 12/22/2000. http://www.hermenaut.com/a47.shtml

[6] Weil anticipated this ontopoietic understanding, in Attention and Will, in *Gravity and Grace*, trans. Arthur Wills (New York, 1952), p. 170, Weil says: "Extreme attention is what constitutes the creative faculty in man."

[7] Tymieniecka, "Logos and Life: The Three Movements of the Soul," in *Analecta Husserliana*, Vol. 25, 195. (Emphasis in the original. In that passage she also says: "As the means life has to maintain the crucial balance between the necessities of its course and the circumambient world, they interact and co-determine each other's role.")

[8] Again, Simone Weil provides a parallel interpretation of the role of attention. For Weil, what she terms "creative attention" is the root of personality in the individual, and is the force (devoid of will) that leads to genuine political action. See Weil, *Waiting for God* (New York: Harper and Row, 1973), pp. 146–149.
[9] The "unity-of-everything-that-is-alive" being Tymieniecka's term for *Dasein*.
[10] Weil, *op. cit.*, pp. 111–112.
[11] Emmanuel Levinas, "Penseé et Prédication," in *Analecta Husserliana*, 6, 6. Here is the full quote in French: "En moi – dans le Moi qui cède le pas et renounce à la priorité – la vie, il est vrai, retient son souffle et sa vitalité de 'force qui va.' On pourrait être tenté d'assimiler encore à une auto-répression cette coupure - d'avant toute decision – de la jouissance dans son rythme égal de l'*esse*, de l'identification du *meme* qui assimile l'*autre*, dans son état, déjà tout dispose à l'ordre de l'Etat. Mais le *soi-même* de cette *auto*-répression, serait seulement sur le point de naître! Une pointe, à force de *revolutions permanentes* qu'accomplissent les tours d'une vrille, pénètre dans l'inter-essement de l'identique, *me* suscite seulement comme *moi* non-interchangeable, c'est-à-dire *comme unicité à la mesure ou, plutôt, a la dé-mesure de l'irrécusable dédicace à Autrui*; et ainsi le dés-inter-*essement* meme, un *autrement qu'être* qui ne se laisse dire – qu-on se rassure – ni en termes d'appropriation, ni en termes du monde.
[12] Levinas, *Totality and Infinity: An Exercise in Exteriority* (Pittsburgh: Duquesne University Press, 1969), p. 34. It might be more accurate to say that responsibility and freedom are equiprimordial, in this schema. (My thanks to Prof. Agnes Curry for this observation.)
[13] Karol Wojtyla, *The Acting Person*, which appeared as Vol. 10 of the *Analecta Husserliana*, p. 42.
[14] *Loc. cit.* Emphasis in the original.
[15] *Ibid.*, p. 115.
[16] There is a distinction to be noted here between the views of Levinas, Tymieniecka and Wojtyla. Levinas and Tymieniecka seem to agree that the "moment" of freedom is pre-rational, while Wojtyla links freedom to the movement of the will. It might, therefore, be more accurate to conceive of two moments of freedom: one, in the pre-rational moment, and the other, in the moment of commitment after decision, after the emergence of the self proper.
[17] Sharon Cameron, "The Practice of Attention: Simone Weil's Performance of Impersonality," *Critical Inquiry*, 29:2, Winter 2003. Citation is from "Attention and Will," *op. cit.*, 143.
[18] Tymieniecka (*Analecta Husserliana* 20, 1986), p. 44.
[19] I'm reminded here of one of Rilke's *Sonnets to Orpheus*: "How vast the distances between the stars; but we lie even farther apart from each other. Stand one child next to a second, and see between – how inconceivably far". From Sonnet II, 20: "Zwischen den Sternen, wie weit; und doch, um wievieles noch weiter, was man am Hiesigen lernt. Einer, zum Beispiel, ein Kind . . . und ein Nächster, ein Zweiter –, o wie unfasslich entfernt."
[20] Levinas, *Otherwise than Being* (Pittsburgh: Duquesne University Press, 1998), p. 67. See also Weil, "Attention and Will," in *Gravity and Grace*, p. 171ff: "Attention alone, that attention which is so full that the I disappears, is required of me. I have to deprive all that I call 'I' of the light of my attention and turn it onto that which cannot be conceived."
[21] Edward Emery, "Facing 'O': Wilfred Bion, Emmanuel Levinas, and the Face of the Other," a paper presented at *Face to Face With the Real World: Contemporary Applications of Levinas*, March 17–19, 1999, Walsh University, Canton, Ohio and at the Ontario Institute for Studies in Education, University of Toronto, in conjunction with the Toronto Institute for Contemporary Psychoanalysis, May 22, 1999 and available at www.psychematters.com/papers/emery1.htm

[22] D.W. Winnicott, "The Observation of Infants in a Set Situation," in *Through Pedicatrics to Psychoanalysis: Collected Papers* (New York: Brunner/Mazel, 1992), p. 53ff. I am grateful to Dr Edward Emery for bringing this case to my attention.

[23] To do so, however, we may have to give up our addiction to "knowledge," and revert to a more contemplative stance of unknowing. But that will have to be a topic for a future paper.

[24] "Ko'ans are the folk stories of Zen Buddhism, metaphorical narratives that particularize essential nature. Each ko'an is a window that shows the whole truth but just from a single vantage. It is limited in perspective. One hundred ko'ans give one hundred vantages. When they are enriched with insightful comments and poems, then you have ten thousand vantages. There is no end to this process of enrichment." Robert Aitken, "Foreword," in Thomas Cleary, *Book of Serenity* (Hudson, NY: Lindisfarne Press, 1990), pp. ix–x. My thanks to Stephen Billias for sharing this ko'an with me.

[25] Simone Weil, *Simone Weil: An Anthology*, ed. Sian Miles (London: Virago Press Ltd., 1986), p. 273.

[26] Tymieniecka, Thematic Summary of *Analecta Husserliana*, Vol. LXXXIII Imaginatio Creatrix: The Pivotal Force of the Genesis/Ontopoiesis of Human Life and Reality. http://www.phenomenology.org/AHLXXXIIItheme.htm

JENNIFER ANNA GOSETTI-FERENCEI

BETWEEN ANIMALITY AND INTELLECTION: PHENOMENOLOGY OF THE CHILD-CONSCIOUSNESS IN PROUST AND MERLEAU-PONTY

The situation of childhood between animality and intellection, as indicated by my title, is suggested in an early passage from the opening section of Proust's novel *A la recherche du temps perdu*, and resonant passages in Rilke.[1] In Proust's overture the narrator, at the edge of the depths of childhood memory – indeed the pillows' cheeks upon which his head rests are like the cheeks "*de notre enfance*" – discusses the phenomenon of waking from sleep and, as it were, losing one's coordination in familiar time and space (Proust, 4/14). Our usual habits of bedtime and awakening in tact, one can, he claims, instinctively and instantly regain a sense of position within the ordered procession of hours and years; but this "ordered procession is apt to grow confused, and to break its ranks" if one has fallen asleep in some different position after a night of insomnia, or in an armchair after dinner, then the world will become disordered: *"le fauteuil magique le fera voyager à toute vitesse dans le temps et dans l' espace ....* " This magical dislocation is not so much loss of measurable spatio-temporality but of horizons of expectation of the familiarly real, a dislocation to which even the narrator's sense of self will yield: "not knowing where I was, I could not even be sure at first who I was" (*... je ne savais même pas au premier instant qui j'étais ...*). And in this slippage of self-certainty beneath the ordinary life the narrator touches what he regards as the animal origins of consciousness: "I had only the most rudimentary sense of existence, such as may lurk and flicker in the depths of an animal's consciousness ... ." (*... j'avais seulement dans sa simplicité première le sentiment de l'existence comme il peut frémir au fond d'un animal ...*) (Proust, 5/16).

The narrator would remain in this destitution, were it not for memory (*le souvenir*) – and soon we will find that it is memories of rooms slept in during childhood, of awaiting his mother's goodnight kiss – that would "draw me up out of the abyss of nothingness [*du néant*], from which I could never have escaped myself: in a flash I would ... gradually piece together

A-T. Tymieniecka (ed.), Analecta Husserliana XCIII, 189–204.

the original components of my ego [*de mon moi*]." This memory is as yet a prelude to involuntary memory prompted by such rarified sensations as that of the madeleine; it is not yet deliberate, since "it always happened that when I awoke like this, and my mind struggled in an unsuccessful attempt to discover where I was, everything revolved around me through the darkness [*dans l'obscurité*]: things, places, years ...." The body itself would recollect these rooms "even before my thought, lingering in cogitation over when things had happened and their forms, had reassembled the circumstances sufficiently to identify" them (Proust, 6/17). The real texture of childhood will be evoked by this bodily sensation, at first by the position of the body in bed (a passage that draws Merleau-Ponty's attention in *Phenomenology of Perception*[2]), and more precisely by the sensation of some chance object in which the essence of a childhood reality must have been held captive. This experience, the recollected details of which introduce the epic architecture of the following novel, is said to be accessed in a pre-intellective manner. In recapturing childhood "all the efforts of our intellect must prove futile [*tous les efforts de notre intelligence sont inutiles*]" since "the past is hidden somewhere outside this realm, beyond the reach of the intellect [*hors de son domaine*]..." (Proust, 47–48/65).

In the context of this childhood recollection, inserting itself between the dim animal depths of original, pre-intellective consciousness and the self (*moi*) which identifies familiar things in their habitual station, we find in Proust the first evidence of something like a discovery of intentionality. Here first arises in the novel the insight that things are not as they seem to be in the so-called "natural attitude," but are in fact correlate to the consciousness which perceives them. It might turn out that the naive attitude of childhood as reflected upon is, contrary to David Michael Levin's contention in his study of narcissism, more like phenomenological reflection than it resembles the natural attitude.[3] In the slippage between adult consciousness with its fixed identification of things in time and space, and the more obscure reality of remembered childhood experience, the narrator supposes: "Perhaps the immobility of things that surround us is forced upon them by our conviction that they are themselves not anything else, by the immobility of our conception of them [*Peut-être l'immobilité des choses autour de nous leur est-elle imposée par notre certitude que ce sont elles et non pas d'autres, par l'immobilité de notre pensée en face d'elles*]" (Proust, 6/16). When this immobile conception gives way through the physiological confusions of sleep, awakening, and the play of imagination and memory, the narrator finds new access to the vital density of world as experienced in childhood. This reality is recaptured through condensation of disparate elements and images, like a nest of materials the child-narrator had, cozy in bed, "contrived to cement together, bird-fashion, by dint of continuous pressure [*qu'on finit par cimenter ensemble selon la technique des oiseaux en s'y appuyant indéfiniment*]" (Proust, 7/18). It is reality vivified, rarified, crystallized through what the narrator admits is a recollecting and finding creation: "*Chercher? pas seulement: créer*" – bringing into being something which it must realize (*il peut réaliser*) – must grant it the reality that, Proust's narrator, cannot be logically proven, but the experiential evidence of which renders other experiences of consciousness pale and shaky in comparison (Proust, 49/66).

That there is something distinctly phenomenological in some modern works of literature has been argued in terms of writers like Proust,[4] who was a reader of Bergson. This relationship well-established in the scholarship, it is specifically the phenomenologically-sensitive invocation of childhood that is of interest in this paper, as in this opening to *Du côté de chez Swann*, to be studied through some of Merleau-Ponty's reflections on childhood. Although there is no critical consensus here, Proust's invocation of memory is an attempt to return, through the initial obscurity, to a more original, more immediate grasp of things as rooted in original perception. The character of this originality and its relation to childhood is what I want to investigate in this paper; why this originality is associated (though never identified) with animality; and why this seems to be best served by the literary imagination rather than in phenomenological study proper. That the work of other writers will also be brought to bear, is meant not merely to compile evidence of a literary insight into childhood, but rather to magnify the special relation of the literary imagination to its situation between animality and intellection.

Discord on the subject of the fidelity of Proust's narrative to the real nature of childhood experience, about the immediacy paradoxically invoked by the mediation of memory, indicates a complex of issues both broader, and more precise, than the problems of Proust's literary style and of the (much discussed in Proust scholarship) inter-implications between memory and art. Rather, we are concerned about the nature of a specific subject-matter, what kind of access we can have to childhood perception in the first place; this problem is also reflected thematically in Rilke's eighth elegy, where Rilke also likens the original child-consciousness to the animal consciousness: "Was draußen ist, wir wissens aus des Tiers/ Antlitz allein; denn schon das frühe Kind/ wenden wir um und zwingens, daß es rückwärts/ Gestaltung sehe, nicht das Offne, das/ im Tiergesicht so tief ist" (Rilke, 192). There are those readers of Proust who find, though paradoxically and perhaps not entirely successfully, a grasp of the immediacy of the child's relation to the world. It is the "precious essence"

(*essence précieuse*) (Proust, 48/65) that Proust's narrator claims to reach beneath the immediate surface of things, an essence associated from the outset with child-consciousness, in particular the childish awareness of things before social (so flaunted at the outset by the narrator's aunts) and perceptual prejudices of adulthood, have become fixed. Readers find a purity in the splendor of sensuous perception, and a stunning transparency in the means by which the narrator reflects upon the generation of thoughts, how they come into being in his consciousness. Childhood as recalled in Proust's descriptions involves an encounter with what Merleau-Ponty considers brute being, rather than the "docile being which would have us believe the world can be fully explained."[5]

Adorno is one reader who recognizes the fascinating and in his view only apparent contradiction of the "extremely subtle and private experiences from the sphere of luxury" in which Proust wrote, affording yet his reach to the general, universal "secrets of every person." Proust, Adorno thinks, "is concerned with an intellectual splitting of the atom, trying to lay open the most minute elements of the real and show them as force fields in which all the power of life is crystallized."[6] This laying-open of the real is associated with child-consciousness to which we all have access as to the geneological layers of our own being, as we do to animality on the basis of our common corporeality with animals.[7] (In fact it is at the later post-syncretic level, the moment of the mirror stage of self-relation, where Merleau-Ponty indicates profound differences between the child and animal.[8]) Indeed childhood becomes almost as much a question of alterity as does animal existence. It is not the precision of imaginative description that astonishes Adorno but rather Proust's capacity to realize the state of perception of childhood, to which the habituated adult reality looks foreign: "Proust looks at even adult life with such alien and wondering eyes that under his immersed gaze the present is virtually transformed into prehistory, into childhood." This prehistory is, Adorno argues, the essentially democratic element in Proust despite his obsession with social class-custom, affectations, and snobbery. Proust brings to recognition that experience carried along with, but repressed in, every adult subject, provided they were provided the shelter of childhood: "For every somewhat sheltered child whose responsiveness has not been driven out of him in his earliest years has at its disposal infinite possibilities of experience." In Adorno's reading it is particular social demands which cover up the freshness of experience. Proust's gift and perhaps luxury, owing not only to his means but also to his poor state of health, was to refuse

> the compulsion to adapt [which] prohibits one from listening to reality with such precision, from taking its soundings ..... Proust ... did not go along with the renunciation of responsiveness, nor with the false maturity of resignation. He kept faith with the childhood potential for unimpaired

experience and, with all the reflectiveness and awareness of an adult, perceived the world in as undeformed manner as the day it was created, in fact developed a technique to resist the automatization and mechanization of his own thought. He strives indefatigably for immediacy, for a second naiveté ...." (Adorno, 315–316).

This second naiveté is a naive attitude, as it were, a primitive stage where the ego is only latent, a primitive stage, as it were, beneath the ordinary subject–object relation. What Proust finds, according to Adorno, in his descriptions of the magic lantern, of Combray, of Mme Swann's gowns, is a sensuous reality that "every individual once knew, in childhood, and then repressed." This returns to Proust "with the force of the familiar. What seems so individuated in Proust is not inherently individuated; it seems so only because we no longer dare to react this way, or are no longer capable of doing so" (Adorno, 316). Merleau-Ponty affirms the notion that in Proust adult consciousness, or the present, rests upon a pyramid of past experience of which we normally remain unaware. "We are, as Proust declared, perched on a pyramid of past life, and if we do not see this, it is because we are obsessed by objective thought .... Acquisition must be accepted as an irreducible phenomenon. What we have experienced is, and remains, permanently ours." As Merleau-Ponty argues in "The Child's Relations with Others," "already surpassed forms of activity are not abolished [but] are met with again in later life" (PrP, 154). The past of childhood would then be a "fertility" ever carried forward into life's discoveries (PP, 393). To this fertility, Merleau-Ponty argues in the essay "Cézanne's Doubt," a consciousness like that of Leonardo da Vinci remained intensely faithful[9]; and Adorno argues much the same in respect to Proust; but this reality of the most original artistic minds is one from which intellectualism has alienated culture at large. It is remarkable that Rilke, too, writes of the intellect as the tip of a pyramid, the wider base of which holds an unreflected vitality of ontological density (letter to Nora Purtscher-Wydenbruck, August 11, 1924). And this deeper vitality is known in the relation to the "open" in child and animal consciousness, since according to Rilke "the animal is *in* the world" whereas the adult (representational) consciousness stands before the world and turned away from it (letter to Lev P. Struve, February 25, 1926).

Merleau-Ponty approaches a Proustian reflection when he treats in a lecture the problem of sleep and dreams. He argues that we must think a state of passivity in which "'to be conscious' does not mean 'to give a meaning,' which one projects onto an ungraspable object of knowledge, but to realize a certain distance, certain variation in the field of existence already instituted ... [Here] to live is not simply to be constantly conferring meaning upon things but to continue a vortex of experience." To emerge from sleep as does Proust's narrator is to emerge from "the inarticulated, the resort to a global or pre-personal relation to the world" (IPP, 115). This pre-personal relation, pre-intellective relation before self-differentiation is analyzed by Merleau-Ponty in "The Child's Relation with Others" as the "syncretic stage." Even in the delirium of having lost one's bearings, consciousness follows not sheer chaos, but "a logic of implication or promiscuity, follows closer and closer a path whose slope it cannot see clearly." While Proust's articulateness is exquisitely refined, his language is animated by a vital fluidity associated by Merleau-Ponty with the acquisition of language. It suggests an "open system of expression ... capable of expressing, not some finite number of cognitions or ideas, but rather an indeterminate number of ideas to come" (PrP, 99). Descriptions of experience with this vitality would be, as it is in Proust, "full of immanent truths" (IPP, 118). In "Indirect Language and the Voices of Silence" Merleau-Ponty writes, "It is essential to what is true to be presented first and forever in a movement which throws our image of the world out of focus, distends it, and draws it toward fuller meaning" (AR, 115).

For Bachelard, these immanent truths reflected in reverie and dreams belong to childhood and the child's imagination, experiences which are not lost but rather remain as "reserves of enthusiasm which help us to believe in the world."[10] Bachelard's phenomenological poetics is explicitly devoted to establishing "the persistence in the human soul of a nucleus of childhood [*d' un noyau d' enfance*], of a motionless but enduring childhood, outside of history; hidden from others ... having real existence only in its moments of illumination – which is to say in its moments of poetic existence" (PIR, 95–96/PR, 85). The reveries of childhood, Bachelard argues, open up the world for us. One of the particularities of the child's being is the "images of cosmic solitude" that seem to return to us in the moments of waking from sleep such as Proust studies, in memories of childhood fears. When we turn toward memories of childhood, if we can, we are inducted into a different register of being and nothingness, a pre-objectifying, preconceptual awareness of the density and splendor of existence, an awareness Rilke in the *Duineser Elegien* attributes to the animal gaze. "Memories which turn toward childhood," Bachelard writes, "seem to reveal a being preexisting our own, an entire perspective of *antecedence of being*" (PIR, 97) which is not the contrary of cosmic solitude. But it is just this cosmic solitude that tends to be overlooked in analyses of child development which, due to the nature of that development, necessarily regard the child in the contex of social interaction. Just as Proust argues of the real existence of the musical phrase in Swann's consciousness, we take this latent ecstatic awareness of childhood memories with us as objective. Here is Proust in a passage that could be representative of the residues with in adult consciousness of childhood:

> But so long as we are alive, we can no more bring ourselves to a state in which we shall not have known them than we can with regard to any material object, than we can, for example, doubt the luminosity of the lamp that has just been lit, in view of the changed aspect of everything in the room, from which even the memory of the darkness has vanished. ... The density [of the musical phrase] was linked to the future, to the reality of the human soul ... (Proust, 381).

This is not to say that our access to such grasp of pre-intellective experience remains unobstructed. There is little doubt that in the progress of ordinary adulthood, our capacity for wonder tends to wane. This is taken to be the normal and desirable progression of a learning mind: wonder wears out with age and use, as we learn and thereby "accustom ourselves to cease to wonder" at even rare things.[11] But from the Bachelardian and even Merleau-Pontyian point of view, adulthood might be seen as a form of fallenness, a being-sunken into the indicative present, the world of the familiar and the disenchanted, turned toward the statically objectified, the dull trance of security in a logic of objects. We have turned away from the unfamiliar, the radical future, the abyss of nothingness from which we arose, and from our origins in animal consciousness which does not yet seize upon distinct and ordered perceptions of habitually familiar and socially meaningful things. Breaking through this does not necessarily indicate a reach to our animal origins, for as Bergson argues, "[even] the young child understands immediately things that the animal will never understand, by virtue of an inherited 'sense intelligence'"[12]; perhaps we could surmise the same vice-versa. Yet the nature of this child-intelligence is not considered at length in *Creative Evolution*; rather, attention is devoted to the difficulty of identifying childhood, an issue to which we will return below. In Proust this 'sense intelligence' is experienced as imagination, a tightened *emotional* concentration, or contraction around the precious essences latent in perception (of for instance the magic lantern, the book his mother finally reads to him in bed, the room itself, and then the church in Combray, its windows, the landscape, Swann's garden, etc.). What most attracts us in these condensations is the palpable joy in sensing, and thus in brute being, that they express and in turn invoke in the reader. Proust achieves the rendering of the architecture of childhood astonishment, a rendering impossible without the aid of reflective mediation.

Due to this mediation, however, Adorno sees Proust's relation to childhood as a fidelity to a happiness which in fact remains unattained. For happiness, the exposure to the full complete reality he seeks, the undamaged experience of childhood, is a paradoxical contact: "undamaged experience is produced only in memory, far beyond immediacy…" Yet Proust sees this happiness to some extent achieved by the preservation of transience in memory, a "rescue of experience" (Adorno, 317). Naturally this rescue cannot indicate an immediate access, not only because we are considering remembered experiences, but because they are experiences reflected upon; and the access itself – reflection, memory – might tarry with the danger of obliterating the content to be reached. There is at least something paradoxical here which is analogous to phenomenology proper: reflecting on the unreflected presents difficulties similar, as Merleau-Ponty argued, to intersubjectivity and "recapturing our own opaque past."[13] As one Proust scholar argues, "the time he recovers is above all a time he annihilates." If it has been levied against Husserl's eidetic reduction, that he can account for the world but not a specific world, Proust "does not wish to attain things themselves but their essence."[14] But just as the relation to others is not, Merleau-Ponty argues, achieved on the basis of analogous representation of the other's consciousness, but on the basis of a pre-intellective resonance of commonly-structured bodily interaction with the same world (PrP, 114–116), so too the relation to childhood is not one of external observation, looking upon it from the outside and decoding, but a breakthrough to a stratum of experience that is a part of the (albeit surpassed) history of any adult subject.

In Proust we find the attentions to child-experience in a striving for immediacy, albeit an achieved, and thus paradoxical, one. Yet it is perhaps this paradox of mediation between the emotional and the imaginative registers of childhood experience that recaptures the immediate by way of sensual intensity and contraction around the discovered "precious essence" of things. This dialectics has been said to be effected in a play between attention to the ever-changing surface of appearances, their centrifugal multiplicity – perhaps concretized best in the magic lantern, as it casts pictures of Golo on his horse transforming the objects in his room into material accomplices – and the centripetal reconstruction of a coherent self[15] – a dialectic that affords the novel's internal structure. One could relate this dialectic to the primitive, only latent "me" (*moi*), and the development of self-consciousness. A brief look at the magic lantern passage illustrates these two directions of consciousness: in a literal dispersal of moving images from a mythical past, the lantern destroys the familiarity of the narrator's bedroom. In fact they might be described in terms employed in Merleau-Ponty's review of studies of perceptual ambiguity, the acceptance of transition of images from one to another, and the correlated emotional-psychic states (PrP, 105). The magic lantern:

... substituted for the opaqueness of my walls in impalpable iridescence, supernatural phenomena of many colours, in which legends were depicted as on a shifting and transitory window [*comme dans un vitrail vacillant et momentané*] ... this mere change of lighting was enough to destroy the familiar impression I had of my room ... now I no longer recognised it, and felt uneasy in it, as in a room ... where I had just arrived ... for the first time.

... If the lantern were moved I could still distinguish Golo's horse advancing across the window-curtains, swelling out with their curves and diving into their folds. The body of Golo himself, being of the same supernatural substance as his steed's, overcame every material obstacle ... but taking it as an ossature and embodying it in himself: even the doorhandle, for instance, over which, adapting himself at once, would float irresistibly his red cloak or his pale face, which never lost its nobility or its melancholy, never betrayed the least concern at this transvertebration (Proust, 9–10/20–22).

While the narrator is charmed by these changing, ambiguous images, it disturbs the coordination of the narrator's phenomenal-lived body and what Heidegger would call his *in-der-Welt-sein*. This disturbance is an experience of uncanniness or, as Heidegger calls it *Unheimlichkeit*, no longer being "at home" in his room. The narrator "cannot express the discomfort I felt at this intrusion of mystery and beauty into a room which I had succeeded in filling with my own personality until I thought no more of it than of myself." What happens here to effect this disturbance of his lived-bodily and perceptual being is the destruction of "l'influence anesthésiante de l' habitude" which is, for the adult, far more resilient.

The anaesthetic effect of habit being destroyed, I would begin to think – and feel – such melancholy things. The door-handle of my room, which was different to me from all the other door-handles in the world, insasmuch as it seemed to open of its own accord and without my having to turn it, so unconscious had its manipulation become – lo and behold, it was now an astral body for Golo (Proust, 10–11/22).

In this "vertigo of images"[16] is expressed the contingency of objects vis-à-vis the consciousness. This is a disturbance of what Heidegger, in reference to similar objects, considered the *Zeugganzheit*, or whole arrangement of useful things, which in the child's perception yields to a dispersal of forms. This dispersion recalls the pre-personal drawings of young children, which involve "a duplication of the immediately given sensory spectacle in which the child was at first engulfed" and of a "subject who is henceforth capable of re-ordering and re-distributing his experience" (PrP, 152). The narrator then seeks restoration of normalcy in the company of adults: "And as soon as the dinner-bell rang I would hurry down to the dining-room, where the big hanging lamp, ignorant of Golo and Bluebeard but well acquainted with my family and the dish of stewed beef, shed the same light as on every other evening; and I would fall into the arms of my mother ..." (Proust, 10–11). Rilke, too, attributes to the mother-figure a stay against obscurity, when she comes to Malte as a child fearing noises in the darkness (Malte, 70), a passage repeated in the elegies: "das nächtlich-verdächtige Zimmer/ machtest du harmlos, aus deinem Herzen voll Zuflucht/ mischtest du menschlichern Raum seinem Nacht-Raum hinzu" (Rilke, 162). As for Proust's narrator, things become familiar and self-evident in the maternal presence. Merleau-Ponty recounts the studies of linguistic regression (and thus world-ordering) which "always" occurs if a young child is separated from his mother (PrP, 109).

It is essential to grasp this reassembling as not opposed to wonder but essential to it, as Fischer explains in his study: "The experiential world in which wonder takes place cannot be made up of unordered, singular patches of experience .... Wonder is the middle condition between an unawakened intellect and a systematic knowledge so complete that there no longer exists anything unexpected" (Fischer, 57–58). This middle condition can be rendered more precisely: the movement of Proust's narrator is the discovery of the contingency of appearances and, following their reintegration as his own, the discovery of their "precious essence." Through this movement, one could argue, it is possible to envisage the paradox of immediacy. The world outside – the bedroom, the family, Combray, is "always the object of the Proustian character in his moment of dispersion" and his "vain attempt to appropriate the shifting surfaces out there for his own," marking "a flight from the inside, from the center of the self." That regresses into the syncretic substrate of conscious life. We recall in the passage of awakening from sleep cited at the outset that a lostness among displaced objects of the bedroom leads to a loss of certainty about the self. Georges Poulet has called this scene "an instant of primal simplicity ... oriented not toward this 'becoming' [familiar] but toward the nothingness which precedes it." This external surface of the world is "a world of things, but of interchangeable things, in which nothing is attached to one particular point of space or duration, a world of things doubtful rather than certain: possible rather than necessary" (Poulet 170; 151–152). This is a flux which must be countered by a movement toward that self, as if grounding the flow in the constituting ego from which it will be recollected: "for Proust the experience of change and the succession of affective states which it brings is absolutely essential to the later, positive phase of the dialectic–the remembering and reconstituting of these experiences" (Macksey, 106). Here memory is less an access than a reconstitution, or, as the narrator himself admits, a creation, of the real. As Macksey argues:

Existence for Proust is thus defined in terms of an antinomy: a going out toward primitive experience, hopelessly fragmented into sensational instants, and a return toward the interior of oneself to relate these experiences, these instances, to the past: expansion and concentration. Just as Proust argues that all creation is re-creation, so all cognition is recognition (Macksey 106–107).

It could be argued that this movement parallels stages of child development from what Merleau-Ponty calls the syncretic stage, wherein there is no stable distinction between self and world and as panoply of inchoate forms, to the formation of the ego as centering the child's experience at a point from which they appear ordered and stable. Given Merleau-Ponty's insistence on the social nature of ego-formation, the narrator's missing of Maman and the familiarity of things regained by her presence re-enacts what is also for the child narrator an earlier accomplishment of psychic development, the emergence from syncretism (PrP, 120). At the syncretic stage "any form (e.g., those we perceive in space-colored forms) is actually subject to a play of forces in different directions," and thus there is a "spatial syncretism" (PrP, 149). In the narrator's experience with Maman in his room, there is a recovery from the slippage into the child–narrator's own psychic past; emerging from syncretism in social interaction there is, quoting Merleau-Ponty's study, "a correlative constitution of me and the other as two human beings among all others," a constitution similarly structured in Rilke's *Malte* and the elegies (PrP, 120). Accordingly, in Proust we find not a single core of buried, and then remembered, childhood experience, but a complexly layered but vital texture of child-experience, with its interweaving stages of progression and regression, reanimated in memory. An analysis of Proust in terms of developmental psychology, however, would be a delicate operation if it were to avoid subordinating the specificity of recaptured child-consciousness to its egological and ultimately adult telos, to which Merleau-Ponty argues Piaget, for one, falls prey.

Despite, or perhaps rather due to, this temporal remove from primitive experience, Proust's dialectic, an affective rather than rational-intellectual one, is "grounded in the moment [and movement]: from perceptive immediacy to ecstatic immediacy." The Absolute, or the (re-) unification of self with the objects of experience, as longed for in Romanticism, is said to be achieved in "intensity (of suffering or joy)" rather than in Hegelian conceptual totality: as in Hölderlin, this achievement is "more as immediacy than as mediation" (Macksey, 108). The key term here is 'ecstatic': the stepping outside of the ordinary attitude with which things are perceived, felt and intellectualized. This intensity in Proust takes the form of rarification and concentration, where the content and the access in the highest moments become one, such as when the world of Combray magically emerges from the teacup. This "uneasy vacillation" between surface and depth in Proust has to be read not as a vertiginous hyperaesthesia but as "the most subtle instrument of his love of truth."[17]

This breakthrough of the natural attitude presumption of the stable existence and necessary order of the world is granted a field of elaboration in the narrator's reflections upon how perceptions are formed on the basis of previously built-up syntheses and expectations. The doorbell that interrupts the magic lantern's wild display of Golo's pursuits indicates that Swann has come to the house for dinner. Here, having gathered his reflective composure, the narrator provides a crude account of prejudicial constitution, which the narrator calls "un acte intellectuel" albeit a passive one: "Even the simple act which we describe as 'seeing someone we know' is to some extent an intellectual process. We pack the physical outline of the person we see with all the notions we have already formed about him, and in the total picture of him which we compose in our minds those notions have certainly the principal place" (Proust, 20/33). That the narrator sees through and reflects upon this procedure is unique to his child situation among the more rigid habits of adult sociability. Seeing through the social prejudice is part of the regathering to self of coordination in a reality that had just been dispersed by the magic lantern into the renegade impressions of a world without fixity or permanence, as Poulet describes, "in which things project themselves before the eyes in instantaneous images which in turn are replaced by other images belonging to other moments and other places," a world in which, therefore, "the mind must precisely assign a certain place in duration and space, by imposing its own certitude upon it, by realizing oneself in the face of it." There is only a provisional recovery from the experience of the magic lantern; as in the opening scene of waking from sleep, the narrator "is never entirely able to detach himself from this first figuration of the world" (Poulet, 152), where once-familiar things are rooted in nothingness.

This dialectics of immediacy and realization, of fixing the mind within the unstable region of perception, is indebted to the operations of memory and reflected in the structure of aesthetic consciousness, upon which the narrator is reflective since deciding, whilst describing the church and the stained-glass windows in Combray, that he is to become a writer. Swann's experience of the "little phrase" of music, which renders to him a sense and order and justification for his hopeless love for Odette, represents this scene of world-constitution in miniature, a form of codification of experience which, Swann knows, escapes "rational discourse." Swann had recently learned to love music, an experience which struck him as a "metamorphosis" and, like the scene of the narrator's awakening from sleep, as a "momentary blindness" (Proust, 378). This moment is peculiarly echoed in the disorientation experienced by Kafka's literally metamorphosized and near-blinded Gregor Samsa when, in a last and fatal episode of blurred human–animal feeling, he crawls out into the living room to hear his sister's playing on the violin. In Proust's passage, however, Swann's experience of the music is described as if consciousness is surveying its interior architecture, its very constitution; at the same time it is realized that this order is imposed by the mind in place of the more unthinkable reality of nothingness with which its sweetness tarries.

> … Swann had regarded musical motifs as actual ideas, of another world, of another order, ideas veiled in shadow, unknown, impenetrable to the human mind, but none the less perfectly distinct from one another …. When, after that first evening …. he had had the little phrase played over to him again, and had sought to disentangle from his confused impressions how it was that, like a perfume or a caress, it swept over and enveloped him … in reality he knew that he was basing [his] conclusion not upon the phrase itself, but merely upon [the mind's] equivalents, substituted (for his mind's convenience) for the mysterious entity of which he had become aware … in that vast, unfathomed and forbidding night of our soul which we take to be an impenetrable void (Proust, 380).

This void is linked by the narrator, in another moment of waking from sleep, to the jelly-fish emptiness of consciousness, to the most elementary kingdoms of animal nature. It is the turn to childhood experience which renders this animal primacy accessible.

This dialectics of affect that leads to ecstatic immediacy in Proust echoes a Hölderlinian poetological structure. Hölderlin's novel *Hyperion* begins with a meditation on the child as just this kind of immediacy, also associated in the novel with the animal: "certain animals sob when they hear music. In contrast people laugh when beauty of the soul and youth of the heart is spoken of …."[18] In an early letter to Bellarmin, Hyperion, like Rilke's speaker in *Duineser Elegien*, complains of the sterility of the scientifically-prejudiced modern age, praises the innocence and immediacy of childhood against the fallenness and falsity of the adult. In an idealization to which Gustav von Aschenbach in Mann's *Der Tod in Venedig* will submit most perversely, the child is idolized as a godly being; whole and wholly what it is; undisturbed by the force of law and by fate; free and in peace, not yet at odds with itself; full of richness; knowing its own heart, and unknowing of the poverty of (adult intellectual) life; immortal, not yet knowing of death:

> … ein göttlich Wesen ist das Kind, solang es nicht in die Chamäleonsfarbe der Mensch getaucht ist. Es ist ganz, was es ist, und darum is es so schön. Der Zwang des Gesetzes und des Schicksals betastet es nicht; im Kind ist Freiheit allein. In ihm ist Frieden; es ist noch mit sich selber

nicht zerfallen. Reichtum ist in ihm; es kennt sein Herz, die Dürftigkeit des Lebens nicht. Es ist unsterblich, denn es weiβ vom Tode nichts (Hölderlin, 298–299).

In this idealization childhood cannot but be seen as the most beautiful of illusions (*die schönste aller Täuschungen*), most principally naive in respect to mortality. But it is also an access to the whole of life for which the poet and artist must recapture or rather be recaptured by, in a second naiveté, an escape from the deprivations of scientific (adult) modernity and its objectifying attitude toward the world. This theme is repeated in Hölderlin's poem *Da ich ein Knabe war* as a pre-articulate knowledge of the holy, of the gods, a familiarity that precedes even naming them. In childhood, the speaker claims, he had a knowledge of the real, an understanding that pre-dates human language (Hölderlin, 40–41). And though it is beyond what can be discussed in this paper, it is also of course Hölderlin for whom memory (*Andenken*) first becomes a dialectics of immediacy – an "eccentric path" (*exzentrische Bahn*) which traces an impossible return to and contact with the real preceding intellection or judgment (Hölderlin, 440). In Proust this is accomplished in the work of memory that restores to anonymous being – and of course to time – an individual human and personal architecture, a restoration first worked out – and first possible – in the intimacy of (recaptured) child-consciousness, to which Hölderlin's idealism, for all his appreciation of the splendor of childhood, remains phenomenologically insensitive. Yet in both Proust and Hölderlin (the latter whom Rilke was intensely reading before writing the elegies[19]) we can see the indirect, dialectical means to recapture childhood as situated between animality and intellection. With their poetical strategies are illustrated in what ways involvement and distance – immediacy and reflection – are not contradictory but co-possible. While in literary language this play of immediacy and reflection is most prominent, this play is also, Merleau-Ponty suggests, "the very definition of philosophy and perhaps even of human existence" (TD, 5).

Fordham University, New York City
and St. John's College, Oxford

NOTES

[1] Marcel Proust, *Du côté de chez Swann, A la recherche du temps perdu, Vol. I* (Paris: Gallimard, 1954); and in translation, *Swann's Way*, trans. C.K. Scott Moncrieff and Terence Kilmartin (New York: Vintage Books, 1989). Pagination first English then French original, when both are quoted. Rainer Maria Rilke, *Selected Poems*, bilingual edition, trans. Stephen Mitchell (New York: Vintage Books, 1984). Also *Die Aufzeichnungen des Malte Laurids Brigge*, ed. Manfred Engel (Stuttgart: Philippe Reclam jun., 1997); in translation *The Notebooks of Malte*

Laurids Brigge, trans. M.D. Herter Norton (New York: Capricorn Books, 1949). Hereafter cited in text as Malte. I have sometimes altered the translations of both Proust and Rilke.

2 Merleau-Ponty, *Phenomenology of Perception*, trans. Colin Smith (London: Routledge, 1989), p. 181n. Hereafter cited in text as PP.

3 David Michael Levin, "Visions of Narcissism: Intersubjectivity and the Reversals of Reflection," in M.C. Dillon (ed.), *Merleau-Ponty Vivant* (Albany: SUNY Press, 1991), p. 55.

4 See "The Time of Half-Sleep: Merleau-Ponty between Husserl and Proust," Mario Carbone, in Ted Toadvine (ed.), *Merleau-Ponty's Reading of Husserl.* (Dordrecht: Kluwer Academic Publishers, 2002); "Toward a Phenomenology of Nostalgia," James G. Hart, *Man and World*, Vol. 6, Nov. 1973, pp. 397–420; and "Proust and Phenomenology," James C. Morrison and George J. Stack, *Man and World*, Vol. 1, Nov. 1968, pp. 604–617.

5 Merleau-Ponty, *Texts and Dialogues*, Hugh J. Silverman and James Barry, Jr. (eds.) (New Jersey: Humanities Press, 1992), p. 12. Hereafter cited in text as TD.

6 Theodor Adorno, *Notes to Literature Volume II*, trans. Shierry Weber Nicholsen (New York: Columbia University Press, 1992), p. 313.

7 Merleau-Ponty, *In Praise of Philosophy and Other Essays*, trans. Wild, Edie, O'Neill (Evanston: Northwestern University Press, 1988), p. 190. Hereafter cited in text as IPP.

8 Merleau-Ponty, "The Child's Relations with Others," *The Primacy of Perception*, trans. James M. Edie (Evanston: Northwestern University Press, 1994), pp. 126–127. Hereafter cited in text as PrP. *Les relations avec autrui chez l'enfant* (Paris: Centre de Documentation Universitaire).

9 Merleau-Ponty, *The Merleau-Ponty Aesthetics Reader*, Galen A. Johnson (ed.) (Evanston: Northwestern University Press, 1993), pp. 72–75. Hereafter cited in text as AR.

10 Gaston Bachelard, *La poétique de la réverie* (Paris: Presses Universitaires de France, 1960). Cited as PR. *On Poetic Imagination and Réverie*, trans. Colette Gaudin (Dallas, TX: Spring Publications, 1971), pp. 97–98. Hereafter cited as PIR.

11 Philip Fischer, *Wonder, the Rainbow, and the Aesthetics of Rare Experiences* (Cambridge: Harvard University Press, 1998), p. 56.

12 Henri Bergson, *Creative Evolution*, trans. Arthur Mitchell (New York: Random House, 1944), p. 162.

13 Claire Elise Katz, "The Significance of Childhood," *International Studies in Philosophy* 34:4 (2002), p.80

14 Georges Cattaui, "Images as Instruments," in René Girard (ed.), *Proust* (Englewood Cliffs, NJ: Prentice Hall, 1962), p. 89. On this point Proust has drawn other skeptical readers; William Gass, a profound reader of Rilke, dismisses the paradox of immediacy in Proust, claiming that Proust does not remember but only analyzes and reflects upon childhood: "Proust returned to his childhood the way a modern primitive returns to the woods: with his books, his bankroll, and a stash of pot"; less crudely, but no less damningly, Gass claims that Proust "planned to replace his life with language," that his subjects were "eventually replaced by the style of their depiction" (Gass, *The World Within the World*, Boston: David R. Godine, 1979, here pp. 153–154), a sharp critique when taken in the context of Proust's biography, to which Adorno gives a more generous interpretation. In a similar, but more enigmatic vein, James Joyce associated Proust's lengthy descriptions – as we might associate childhood itself – with both severe boredom and the heights of the ecstatic.

15 Richard Macksey, "Architecture of Time: Dialectics and Structure," in *Proust*, 'op. cit.' p. 109.

[16] Georges Poulet, "Proust and Human Time," in *Proust*, 'op. cit.' p. 151.
[17] Charles du Bos, "The Profundity of Proust," in *Proust*, 'op. cit.' pp. 134–135.
[18] Friedrich Hölderlin, *Werke und Briefe, Vol. I* (Frankfurt am Main: Insel Verlag, 1969), p. 310. Translations are my own.
[19] Karl-Heinz Fingerhut makes this connection in *Das Kreaturliche im Werke Rainer Maria Rilkes* (Bonn: H. Bouvier u. Co Verlag, 1969), pp. 58, 192.

MARIA VILLELA-PETIT

NATURALISTIC AND PERSONALISTIC ATTITUDE

The Husserlian distinction between the naturalistic and the personalistic attitude, presented in *Ideas II*,[1] certainly deserves a revival of interest as a way of confronting the new challenges facing phenomenological philosophy today. These challenges stem, on the one hand, from the development of a science like neuro-physiology and, on the other, from the development of techniques whose aim it is to replace the natural processes of the human body.

Let us begin with neuro-physiology. Progress in this field is rapidly improving our knowledge of the brain, in other words, of the central organ responsible for the union of the psychic and the physical. Functional magnetic resonance imagery (fMRI) has raised many hopes not only on the therapeutic plane but also on the plane of a better understanding of the processes that underlie thought. On the occasion of the recent inauguration of the brain imagery instrument, Neurospin, the hope was voiced that a machine of this kind would make it possible to watch "the brain in the process of thinking." Moreover, given the interest presently shown by phenomenology in the clarification of psychic life, it is hardly surprising that an exchange should now be taking place between the descriptions furnished by phenomenology and the discoveries of neuro-physiology. At the same time, the temptation to naturalize phenomenology, to employ the expression that has taken hold, is becoming ever more insistent.

Indeed, it was under the title: *Naturalizing Phenomenology* that the contributions towards an important colloquium, held at Bordeaux, in 1995 were published, first in English by Stanford University Press and then in French (*Naturaliser la Phénoménologie*) by the CNRS Press. One of the principal concerns in the paper I contributed on that occasion was to defend the idea of phenomenology as transcendental philosophy. Thus I relied on Husserl texts explicitly devoted to the question of transcendentalism versus naturalism, and which are to be found in *Cartesian Meditations* just as well as in *Crisis* or in *Formal and Transcendental Logic*. My text, entitled: "Cognitive Psychology and Transcendental Phenomenology," appeared in the part of the

A-T. Tymieniecka (ed.), Analecta Husserliana XCIII, 205–218.

collection placed under the head of: "Scepticism," that is, amongst the contributions which did not share the belief in the imperative need to "naturalize phenomenology"; as if, thanks to the precision and richness of its descriptions, phenomenology had no other philosophical bearing than that of being a useful auxiliary for the scientific knowledge of thinking, or of the processes held responsible for it.

However, if confronting one with the other, or even comparing research in neuro-physiology with phenomenological descriptions, is perfectly legitimate, the same does not hold when one seeks to reduce phenomenology to the role of a simple supplement to the former.

That said, what I want to propose here takes a different, and perhaps more promising, path. It is not so much a matter of defending phenomenological philosophy by taking full account of the transcendental reduction as of according proper attention to a distinction I take to be very valuable. For this distinction, announced in my title, between the naturalistic and the personalistic attitude throws light on some of the problems with which our civilization is confronted, and which bring to bear the by now indispensable connection between the scientific-technical order and the ethical order.

But before getting on to what Husserl means by each of these two attitudes, I would like to draw attention to the highly significant history of the word "attitude," the word that, in French, translates quite satisfactorily, the German "*Einstellung*." According to the New Etymological Dictionary *Larousse*, this word was introduced in French, in 1637, by the painter Nicolas Poussin, on the basis of the Italian: "attitudine," itself borrowed from the Latin: "aptitude." And the dictionary sums the matter up with the remark: "The word passed into French first as the painterly term: 'posture'; it has a more general usage from the 17th century on, but it is not until the 19th century that it takes on the meaning of a 'manner of being with regard to someone'."

Why are we so interested in this notion of attitude? Because, in certain respects, it seems to sum up what Husserl has to say concerning the expressive character of the living body (*Leib*) in §62 of *Ideas II*, bearing on the mutual interconnection of the personalistic and the naturalistic attitude. After stating that, from a phenomenological point of view, "the Body plays an expansive role in the realm of spirit," he goes on to say: "Allied with this is also the fact of 'expression,' which allows us to interpret, on a broad scale, the other's Body as a Body for a spiritual life."[2] So it is from a phenomenological point of view that one can recapture the transition, or better still the shift, in the sense of a word which, designating first of all the corporeal posture as displayed in a picture, is then naturally led to signify a mode of being adopted by a person towards the environment *(Umwelt)* he or she shares with others. Others whose body is expressive of their personal life as it also is for my own.

These considerations bring to light the key position occupied by the body *(Leib)* in the attempt made by Husserl to understand the distinction between the personalistic and the naturalistic attitude. That said, another issue has also to be borne in mind if one does not want to go astray when one talks of the naturalistic attitude in Husserl's sense. For an attitude of this kind only really makes itself known in the Modern Age. In the sense in which Husserl uses the term, "naturalistic" in ethical discussions, where what is at stake are the positions of those for whom "nature itself provides the standards of goodness."[3] This kind of moral criterion is not what is at issue but an attitude which, from the first, is not practical but theoretical.

The best way of tracing the genesis of what Husserl had in mind with his naturalistic attitude, as emerging out of the modern science of nature, is to go back to its root source in the science and the philosophy characteristic of the modern age, that is, the physics of Galileo (and then Newton), on the one hand, and the philosophy of Descartes, on the other. As a matter of fact, however, the advent of the naturalistic attitude, the attitude that will come to dominate the modern epoch, harks back quite simply to a previous and decisive moment when art[4] and science were closely linked. The anatomical research of Leonardo da Vinci remains an indispensable reference for anyone wanting to take note of the emergence of an objective knowledge of the body and of what Husserl designates as the imbrication of the naturalistic with the personalistic attitude. And here, the example of painting once again proves helpful, as was already the case with etymology. In the Renaissance, with a painter like Leonardo, the figuration of the human body does not simply call for a knowledge of the proportions of the idealized human body, as it did with Greek sculpture and its Roman heritage. It seems to require a genuine anatomical knowledge. The artist scrutinises the empirical sub-structures of the expressive appearance of the body, and in so doing brings to light features indicative of the character of the person as well as his social status and function.

In his work *Leonardo-Anatomy*, Marco Cianchi notes: Leonardo's first studies in anatomy date back to 1487–1493, when he lived in Milan. They are explorations of the skull (revealing a striking accuracy and mastery of perspective), wherein Leonardo sought to find the centre of the senses, namely the "common sense" that, among other things, he takes to be the seat of the soul.[5]

Following up on studies which began already in the 14th century and which anticipated by several decades those undertaken by Andreas Vesalius,[6]

Leonardo's anatomical studies already presuppose a naturalistic attitude even if they are placed at the service of an artistic activity which, itself, belongs to the spiritual world and therefore to the personal world of mankind. Indeed, one finds in what Leonardo himself calls the *scienza nuova* not only an intertwining, *sui generis*, of a theoretical with an artistic activity but also the presage of a naturalistic understanding of the human and the animal body. And so it is not surprising that Leonardo's style of writing already relies upon mechanical metaphors to designate the human body, conceived by him as a machine (*machina*), just as it will bc later by Descartes.

It is precisely this naturalistic attitude that dominates the physical investigation of nature, including the anatomical study of the human and animal body, and which today dominates neuro-physiological research into the brain. And it is this same attitude which informs the computational model, a model that is gradually replacing the less sophisticated model of the machine conceived in purely mechanical terms.

So it is clear that the naturalistic qualification brings us back to nature as the correlative of the science of nature, just as Husserl himself affirmed from the very first paragraphs of *Ideas II*. This nature is no longer what we have to deal with when we cultivate our garden or that of the blue sky we contemplate with delight, not even that of a movingly beautiful body. It does not bear upon the things which surround us in our environment and which take on for us a pragmatic, axiological value. Rather, it is what we have in mind when, through a change of attitude, we envisage things in a theoretical way – the way required by what we call science: natural science.

To be sure, even in the naturalistic attitude, what I try to study when I turn to the human (or animal) body is not simply a material thing, a body-machine, a simple instrument at the disposal of something completely different, namely, a soul, as was supposed by those who operated in the frame of the Cartesian ontology with its substantial dualism of body and soul but, on the contrary, a psycho-physical unity. Despite the importance of Descartes for Husserlian phenomenology, Husserl never subscribed to Cartesian dualism, nor did he ever reduce the body, the living body (*Leib*), to a simple extended body (extension). He always thought of it as animate body, the organ and expression of a psychic life, even while underlining the functional relation of mutual dependence as between the psychic and the somatic. The soul (*Seele*), that is the psyche, is itself presupposed by the naturalistic attitude, he tells us. "The Soul animates or be-souls the Body, and the animated Body is a natural Object within the unity of the spatio-temporal world."[7] In this way Husserl is able to justify the use of the naturalistic attitude in the study of psychic life.

And without even having at his disposal the elements available to us today, thanks to neuro-physiology, he nevertheless observed:

> But the Body as physical Object is subject to physical influences to which psychic "consequences" are linked without my knowing precisely how they are connected. And finally I arrive at the brain, its structure and the physical processes occurring therein which are in correspondence with the psychical processes, a correspondence that includes in itself functional alterations and dependencies. If a process in the brain alters, then there occurs an alteration of the corresponding set of lived experiences, the set of psychical events, and perhaps the converse also holds.[8]

On reading such a proposal, one is tempted to see in it a general preview of the research pursued by neuro-physiologists today with such remarkable results, results which would unquestionably have interested Husserl, precisely because we are now able to establish with greater precision the kind of connection that obtains between the physical and the psychic.

However, even though he was fully aware of the fact that "in what concerns the life of the psyche (soul) there exist dependencies which even, and in a certain way, affect all conscious processes," there can be no doubt that, for Husserl, the naturalistic attitude is unilateral and has to be treated as secondary with regard to the attitude that presides over the affairs of our life. What happens when one sets up the naturalistic attitude as the only acceptable philosophical position is that one misunderstands the life of man. For within the horizon of his surrounding world, man remains in bodily and spiritual communication with other human beings. The only way of understanding the life of a man who thinks and who acts in accordance with his decisions and choices, including those that animate his theoretical activity, is by means of the personalistic attitude, in other words, by taking him to be another person.

This is the sense in which the distinction between the naturalistic and the personalistic attitudes becomes critical for the philosopher, and this in line with the very title of §34 of *Ideas II*. When I apprehend another human being concretely, Husserl will say, what is given to me "is the human person, who has his spiritual individuality, his intellectual and practical abilities and skills, his character, his sensibility."[9]

Moreover, as a person everyone knows himself "as dependent on others, and not merely on individual persons, but on communities of persons, social institutions, the state, morals, the law, the church etc."[10] Or, as Husserl also puts it: "The apprehension of the man as a real personality is determined throughout by such dependencies. A man is what he is as *a being who maintains himself* in his commerce with the things of his thingly, and with the persons of his personal, surrounding world and who, in doing so, *maintains his individuality throughout*."[11] And a few lines later he adds: "In these relations a man now finds himself bound and constricted, now unbound and free; also, he feels himself now receptive, now creatively active. But in any event, he finds himself really related to his thingly and spiritual surrounding worlds."[12]

What matters here is the nature of the dependencies in question. For those of the personal and spiritual world are not of the same order as those which arise out of psycho-physical conditioning. This is how Husserl brings out this distinction:

> ... in my theoretical, emotional, practical behaviour – in my theoretical experience and in thinking, in my position-taking as to pleasure, enjoyment, hoping, wishing, desiring, wanting – *I feel myself conditioned by the matter in question*, though this obviously does not mean psychophysically conditioned.[13]

Needless to say at the same time as he brings out the necessity and the priority of the personalistic attitude, Husserl does not fail to stress that "the naturalistic attitude is subordinated to the personalistic."[14] And following upon the passage from which we took this assertion, he adds: "that the former only acquires by means of an abstraction or, rather, by means of a kind of self-forgetfulness of the personal Ego, a certain autonomy – whereby it proceeds illegitimately to absolutize its world, i.e., nature."[15] It is precisely this kind of absolutization that underlies the reductionism of those who only have eyes for the naturalistic attitude. They have forgotten that it is in the personalistic attitude that we find ourselves, naturally, "when we live with one another, talk to one another, shake hands with one another in greeting, or are related to one another in love or aversion, in disposition and action, in discourse and discussion. Likewise we are in this attitude when we consider the things surrounding us precisely as our surroundings and not as 'Objective' nature, the way it is for natural science."[16]

In this order of ideas, it is important also to note that, on the personal plane, the relations of dependence are motivational rather than causal. "Motivation", Husserl writes, is "the fundamental lawfulness of spiritual life."[17] By motivation, or rational causality, he means what moves me, makes me act or accomplish such and such operations (as for example theoretical acts). This does not mean that I always act in a rational manner. "Reason is a 'relative' one here. He who lets himself be drawn by inclinations and drives (which are blind since they do not emanate from the sense of matters currently functioning as stimuli, i.e. they do not have their source in this sense) is driven irrationally. But if I take something to be true or take a demand to be a moral one, thus having a source in the corresponding values, and if I freely pursue the reputed truth or the reputed moral good, then I am being reasonable – yet only relatively so, for I may indeed be mistaken there."[18]

Quite rightly, Husserl ties these considerations in with the "fundamental questions of ethics." But before looking closer at the ethical implications of such an approach, let us spend a little more time on the way in which Husserl himself envisaged the personal life of the subject, with the complexity belonging to it – an aspect of his work often neglected by those who are content with only a partial knowledge.

For although Husserl does not hesitate to affirm the precedence of the life of the spirit in the comprehension of the personal ego, nevertheless, he is also ready to acknowledge its shadowy zones, its obscure horizons. As he says quite explicitly: "The specifically spiritual Ego, the subject of spiritual acts, the person, finds itself dependent on a *obscure underlying basis* of traits of character, original and latent dispositions, and thereby dependent on nature."[19]

In other words: on the basis of its sensibility, inseparable from its animate body, no personal subject can escape the hold of natural causality. Husserl is very incisive on this point when he writes: "all life of the spirit is permeated by the 'blind' operation of associations, drives, feelings which are stimuli for drives and determining grounds for drives, tendencies which emerge in obscurity, etc., all of which determine the subsequent courses of consciousness according to 'blind' rules."[20]

But despite the fact that, in its psychic individuality, the subject sometimes loses itself, ceasing to be master of itself, he can also be the one capable of willing ..., of "taking a position," of being responsible for itself. This capacity for free actions is attributable to motivations which are no longer rooted in natural sub-strata but which have been able to emerge at the centre of the world of the spirit to which all humans belong, though, of course, according to specific traditions and different modalities.

So Husserl distinguishes between hidden "motives," attributable to involuntary associations, and to which all psychic life is liable – and which he will examine in particular in his analyses devoted to *Passive Synthesis* – and the motivations of the ego as such, which, so to speak, command the conscious life of the spirit. As he says at §56, devoted precisely to "motivation as the fundamental lawfulness of spiritual life": "The 'motives' are often deeply buried but can be brought to light by 'psychoanalysis'. A thought 'reminds' me of other thoughts and calls back into memory a past lived experience, etc. In some cases it can be perceived. In most cases, however, the motivation is indeed actually present in consciousness, but it does not stand out; it is unnoticed or unnoticeable ('unconscious')."[21]

In addition to these motives, there are others which arise upon another plane, more specifically at the higher levels of spiritual existence, and which are more properly to be thought of as egological in character. "There are then," Husserl writes, "rational motives for existential positions as well as for the positions assumed in judgement and belief in general (from which logical-formal positions are derived). There are even rational motives for positions arising from feeling and the will."[22]

The personal life of the subject is therefore the locus of a conflict or even a combat between motivations stemming from diverse sources (individual as well as collective, obscure as well as rational) which get interwoven at the very centre of each individual and which, for most part, also have repercussions upon the planes of inter-personal and community life.

In every society, a human being can fail to be fully recognized as such, for example when he or she is a "foreigner," as anthropologists have amply demonstrated.

At a place where he also evokes the familiar phrase "treating humans and beasts like simple things," Husserl comments: "To treat humans and animals 'as mere things' has, to be sure, various senses, a juridical and moral sense and, on the other hand, a scientific sense. But both have something in common. From a moral-practical standpoint, I am treating a human being as a mere thing if I do not take him as a person related to the moral, as a member of a moral association of persons in which the world of morals is constituted."

Then, and after having made other similar points on the subject of this legal meaning, Husserl comes to the scientific meaning of the expression: "to treat humans and beasts like simple things." a meaning which is of particular interest to us here.

> Again, analogously, I treat a human *theoretically*, as a thing if I do not insert him in the association of persons with reference to which we are subjects of a common surrounding world but instead take him as a mere annex of natural Objects which are mere things and consequently take him as a mere thing himself. Now, within certain limits this is legitimate, and it becomes an injustice only if we do not acknowledge at all that the naturalizing of persons and souls can allow us to recognize only certain relations of dependency of Objective existence and continuity which obtain precisely between the natural world of things and the personal world of spirits, insofar as both belongs to the unity of the objective spatio-temporal world of realities. Further, it is an injustice if we do not acknowledge that nevertheless spirits make possible and require still an other and more significant mode of research precisely in this respect, that they have, as Ego-subjects, a being properly their own, that as such, they are required for all matters and all research into these matters as the subjective counter-parts, and that they thus, in their intentional life, relate to the world in form of the surrounding world.[23]

And from here he moves on to the conclusion which immediately follow these considerations. "He who sees everywhere only nature, nature in the sense of, and, as it were, through the eyes of natural science, is precisely blind to the spiritual sphere, the special domain of the human sciences. Such a one does not see persons and does not see the objects which depend for their sense on personal accomplishments, i.e. objects of culture."[24]

Let us dwell on these remarks. For, in addition to helping us to grasp the oversights which compromise the recognition of persons on the practical plane, whether moral or legal, they bring to light another threat. The one which manifests itself upon the theoretical plane.

This peculiar threat is becoming ever more menacing nowadays, all the more so since it is proving increasingly difficult to separate theoretical knowledge from the exponential increase in technological power. The naturalistic attitude invades every domain of personal and inter-personal life, even including those which might have been taken to be intimate and secret. From now on it will be possible to tackle the generation of human beings from a dominantly naturalistic standpoint. This is indeed what happens whenever one resorts to interventions of one kind or another, made possible by new technologies in order to satisfy the "motives" or even the obscure desires of individuals.

"Wanting a child," for instance, no longer seems to be limited by any natural barriers, such are the manipulations which have become possible in this field – in vitro fertilisation, artificial insemination in surrogate mothers. And it matters little that the desire to have a child (or some other desire) arises in a meaningful way within a community of responsible persons or that it simply satisfies a compensatory obstinacy or a will to limitless power. Reputable members of the scientific community are already envisaging gestation outside the maternal body. The title of a book by Henri Atlan, *The Artificial Uterus*,[25] indicates the direction some research is going, that is, towards the invention of machines capable of replacing the female body during the period of pregnancy.

Whatever the viability of such projects might be – will bio-technology be capable of developing cutting-edge equipment as they say in scientist jargon? – the fact is that certain scientists are only too ready to push things in this direction. It is impossible to pin-point the hidden motives operative here and there, and especially where women are concerned, woman who want to have a child without being ready to bear it, not because they suffer from any physical deficiency but because they reject pregnancy. It is significant, for example, that Henri Atlan was able to "listen to," pay attention to the demands of certain feminists who are in revolt against the physiological constraints of maternity and denounce what one of them called the "empire of the belly." One might even bet on a psychoanalyst's counsel being very different from that of a biologist.

Be that as it may, and rather than pursuing the matter in the direction of an ethics or a bio-ethics, I would like to make a detour across the metaphorical domain. The beginning of chapter 49 from the book of Isaiah where what is in question is the vocation of the prophet itself:

Listen, O isles, unto me: and hearken, ye people from far: The Lord hath called me from the womb; from the bowels of my mother hath he made mention of my name.

It is difficult to imagine what might be the resonance of expressions such as "maternal breast" or "mother's entrails" for someone born from an artificial uterus. The partisans of solutions of this kind (or who envisage such solutions) rarely bother to raise questions relating to the psychological impact of just such a mechanically induced pregnancy. Even though the construction of such a uterus belongs, for the moment, to the domain of science fiction, the same does not hold for surrogate mothers. Here maternity and gestation go their separate ways. And even though it might be admitted that resort to a surrogate mother might be justified in a limited number of cases, for medical reasons (for example, the absence of a uterus in a married woman), the vulgarisation of such procedures already presents redoubtable human problems and will continue to present even greater problems as this practise becomes more widespread.

Evidently, the naturalistic attitude, reinforcing the treatment and the manipulation of the body as a simple natural thing belonging to nature (of nature as objectified by science), reinforces the utilitarian status of the body and therefore its commercialisation. Already the uterus has been made use of as a means, more, a commodity placed at the disposal of the highest bidder. From this point of view, one easily slides over into treating the other, even treating oneself, as a means and not as a person, that is, as an "end in itself."

What is "new" here? In the past, a person could also see that he or she was being treated as a quasi-thing when the role assigned to him was that of a slave, for example, or when he came to treat his own body as a quasi-object by engaging in prostitution (often merely in order to survive). Even though a behaviour of this kind was often tolerated, it could not remain uncontested, either from a practical or from a philosophical point of view. Such a degradation of a human being called for a critique, a critique presented in the name of reason and of universal human rights.

But where do we stand today, when the instrumentalization of the body in the framework of a civilization where the naturalistic attitude seeks, by means of the success of the techno-sciences, to exert its supremacy and so to prevail over the personalistic attitude?

This kind of question has nothing to do with the convictions of the religious. It has been raised by philosophers belonging to no religious denomination and who, in the name of practical reason, have protested against the utilitarian status accorded to the body and against the fact that one takes a person as a simple "means." We are reminded here of one of the formulae Kant employs for his categorical imperative. "Act in such a way that you treat humanity, whether in your own person or in that of another, always at the same time as an end and never simply as a means."[26] This imperative of respect for humanity, whether in oneself or in the other, is sneered at nowadays, not for the usual reason of domination, or on account of the difficulty in conforming one's action to rational norms, to the moral law in its universality, but on account of techno-scientific progress.

Those who are not content to remain encamped on the terrain occupied by Kant's moral philosophy know how to estimate the value of a work like *Das Prinzip der Verantwortung* by Hans Jonas which, as its title indicates, consists in the "quest for an ethics for technological society" *(Versuch einer Ethik für die technologische Zivilization)*. What is interesting is to note that, ironically, in this work which dates from 1980, Hans Jonas, with a view to founding his "theory of responsibility" employs as his paradigm cases of parental responsibility and the responsibility of the statesman. Moreover, he takes parental responsibility to be something existing by nature and not just by contract.[27]

Hans Jonas is not put off by the fact that this natural responsibility (of which certain animal species offer examples worthy of admiration) is not free from cultural patterns. Parental systems, and therefore parental responsibility, vary from society to society. In certain societies it is delegated not just to the progenitors but to a male member of the family (the uncle, for example) as well as to the mother (an unavoidable figure).

What Jonas is not wrong to bring out is the exemplary character of parental responsibility. For, relying in the last instance upon the fragility and the vulnerability of the new-born whose crying has to be understood as an appeal for help calling for a response, this responsibility is endowed with an unequivocal priority.

And yet the hold of technology over our minds is beginning to undermine the distinctive features of this very responsibility, even in what pertains to their naturalness, however overlaid this naturalness might have been by cultural elements. Something against which Jonas himself inveighed with all the critical vehemence natural to him. Here is the proposal he upholds on the situation of man enslaved to the domination of nature, to the point of becoming an object himself. "Man has come to be one of the objects of his own technology. *Homo Faber* applies his art to himself, and is now learning to fabricate himself, the inventor and fabricator of everything else."[28]

To follow up on the questions raised in this field by the work of Hans Jonas, we need to turn to works dealing more specifically with bio-ethics, some of which give us much to think about.

It would nevertheless be a pity to lose sight of the insight that the Husserlian distinction between the naturalistic and the personalistic attitude can throw on this new challenge, since what is in question here is the very meaning of being human. For this distinction furnishes us with a criterion enabling us to find our way through the labyrinthine complexity of the problems posed by those who have to take decisions relating to interventions on living bodies. And Husserl's refusal to make these two attitudes mutually exclusive only renders the distinction much more pertinent. He knew how important scientific knowledge was in itself and how much it had to offer to human kind. Do we really need to be reminded that, from the very beginning, human existence relied upon artifice, upon art in the Greek sense of "techne," and that it is only natural that, today, it should appeal to the very technology we have succeeded in developing? But this resort to technique should not be allowed to circumvent considerations of a moral order concerning its own meaning and its aims. For only the personalistic attitude is capable of preserving humanity as such.

As Husserl understood only too well, whenever one has to do with human being, it is crucial that the person,[29] and therefore the personal ego, always be taken account of. By contrast with the notion of the subject, which always has as its correlate that of the object, the "person" refers to the relation between persons, and therefore to inter-subjectivity – as is attested by our use of personal pronouns. "Person" implies communication and therefore a community of persons as the site from within which alone there can emerge spiritual meaning constitutive of a specifically human existence.

With these remarks I would like to pay my respects to someone whose centenary we shall be celebrating this year: Stephan Strasser, whom I much admired and still vividly remember. With this centenary of his birth, I would like to associate Emmanuel Mounier, the founder of the review *Esprit*, also born in 1905. He was the inspiration behind the movement known as "personalism," whose influence is still felt on several continents. Emmanuel Mounier, who died in 1950, hardly knew the work of Husserl. (His ontological realism probably made him suspicious of a philosopher generally regarded as an "idealist.") He did not for all that escape the influence of phenomenology, due to the presence at Paris, from the mid thirties, of Paul-Louis Landsberg,[30] a disciple of Scheler.

And so it is from one of the last works of E. Mounier that I am going to draw a remark that sums up very well the ethical dimension of the personalistic attitude: "... the person is not an object. It is even that which, in each man, precisely cannot be treated as an object."[31]

By reminding you of these defenders of the personalistic attitude in the context of a reflection devoted to the heritage of Husserlian phenomenology, a heritage Stephan Strasser helped to save and preserve for us, I want to draw attention to the need to promote such an attitude in the face of the threat represented by any civilisation like our own. We have to defend our civilisation against itself, and against the death wish which lurks within it.

Paris, CNRS, Archives Husserl (retired)

NOTES

[1] Edmund Husserl, *Ideen zu einer reinen Phänomenologie und Phänomenologischen Philosophie, Zweites Buch: Phänomenologische Untersuchungen zur Konstitution*, herausgegeben von Marty Biemel, Husserliana Bd. IV (The Hague: Martinus Nijhoff, 1952), translated by Richard Rojcewicz and André Schwurer *as Ideas Pertaining to a Pure Phenomenology and Phenomenological Philosophy, Second Book – Studies in the Phenomenology of Constitution* (Dordrecht/Boston/London: Kluwer Academic Publishers, 1989).

[2] Husserl, *Ideen II*, trans. Rojcewicz and Schuwer *op. cit.*, p. 295.

[3] See Anne Fagot-Largeault's critique of this position in her article: "Evaluating and Judging – Bio-norms versus Human Judgment," in Peter Kemp, Jacob Rendtorff and Niels Mattson Johanson (eds.), *Bioethics and Law*, vol. 1, *Judgement of Life*, eds. (Copenhagen: Rhodos & Centre for Ethics and Law, 2000).

[4] It is important to note that what is called art (and later Fine Arts) also undergoes a change of meaning at the same time. It begins to designate artistic activity as "creative" and so not to be confused with the work of the craftsman.

[5] Marco Cianchi, *Leonardo-Anatomy*, transl. Christine Cesarini (Florence-Milan: Ghiutin Editore, 1998), p. 16.

[6] The *chef d'oeuvre* of Andreas Vesalius: *De humani corporis fabrica libri semptem* (*The Seven Books on the Structure of the Human Body*), was printed in 1543.

[7] Husserl, *Ideen II, op. cit.*, §49, pp. 164–165, Eng. trans. *op. cit.*, pp. 232–233.

[8] Husserl, *Ideen II, op. cit.*, §45, pp. 164–165, Eng. trans. *op. cit.*, p. 173.

[9] Husserl, *Ideen II, op. cit.*, §34, p. 139, Eng. trans. *op. cit.*, p. 147.

[10] Husserl, *Ideen II, op. cit.*, p. 141, Eng. trans. *op. cit.*, p. 148.

[11] Husserl, *Ideen II, op. cit.*, p. 141, Eng. trans. *op. cit.*, p. 148.

[12] Husserl, *Ideen II, op. cit.*, p. 141, Eng. trans. *op. cit.*, p. 148.

[13] Husserl, *Ideen II, op. cit.*, p. 140, Eng. trans. *op. cit.*, p. 148.

[14] Husserl, *Ideen II, op. cit.*, §49, p. 183, Eng. trans. *op. cit.*, p. 193.

[15] Husserl, *Ideen II, op. cit.*, §49, p. 183–184, Eng. trans. *op. cit.*, p. 193.

[16] Husserl, *Ideen II, op. cit.*, §49, p. 183, Eng. trans. *op. cit.*, p. 193.

[17] Husserl, *Ideen II, op. cit.*, §56, p. 183, Eng. trans. *op. cit.*, p. 231.

[18] Husserl, *Ideen II, op. cit.*, §56, p. 183, Eng. trans. *op. cit.*, p. 233.

[19] Husserl, *Ideen II, op. cit.*, §61, p. 276, Eng. trans. *op. cit.*, p. 289.

[20] Husserl, *Ideen II, op. cit.*, §61, p. 277, Eng. trans. *op. cit.*, p. 289.

[21] Husserl, *Ideen II, op. cit.*, §56, p. 222–223, Eng. trans. *op. cit.*, p. 234.
[22] Husserl, *Ideen II, op. cit.*, §56, p. 223, Eng. trans. *op. cit.*, p. 235.
[23] Husserl, *Ideen II, op. cit.*, §51, pp. 190–191, Eng. trans. *op. cit.*, pp. 200–201.
[24] *Ibid.* p. 191, Eng. trans. *op. cit.*, p. 201.
[25] See: Henri Atlan, *L'Utérus artificiel* (Paris: Seuil, 2005).
[26] Immanuel Kant, "Groundwork of the Metaphysics of Morals," in H. J. Paton (transl.), *The Moral Law* (London: Hutchinson, 1976), p. 91.
[27] Hans Jonas, *Das Prinzip der Verantwortung*, at p. 178 where one reads: "The responsibility instituted by nature, the responsibility which exists by nature is, in the unique example offered here (the only one with which we are all familiar) that of parental responsibility. The latter does not depend on any previous consent, it is irrevocable and it is global."
[28] Jonas *op. cit.*, p. 47.
[29] See: Maria Villela-Petit, "Personne, Droit, Droits de l'Homme," in Guy Coq (ed.), *Emmanuel Mounier – Acts of the Colloquium held at the UNESCO* (2000) (Paris: *Parole et Silence*, 2003, pp. 147–169.
[30] Paul-Louis Landsberg is the author of *Problèmes du Personalisme*, (Paris: Seuil, 1952). This collection was published posthumously with a preface by Jean Lacroix. A Jew converted to Catholicism, closely associated with E. Mounier during his French years, Paul-Louis Landsberg died in 1944 in a concentration camp near Oranienburg.
[31] Emmanuel Mounier, "Le Personalisme," in *Oeuvres de Mounier* (1944–1950) (Paris: Seuil, 1962), Vol. 3, p. 430.

MARA STAFECKA

MAMARDASHVILI ON THINKING AND SENSITIVITY

There are certain things in our lives that never go away. Sooner or later, we visit and revisit them again and again, led by an unconscious urge that calmly and persistently directs our mind to that unique and for some unknown reason unforgettable place in our past. I believe this is because our world-view is attuned to detect happenings that our conceptual minds are not focused on at the moment.

I will never forget the tense emotional atmosphere in an auditorium where people were waiting for the beginning of a lecture by Merab Mamardashvili. The way he was to present philosophy was absolutely different from the officially established way. The language was different, and the structure of presentation was different. He did not rely as much on using well-known conceptual vocabulary and clearing its content or drawing new connections between concepts. He did not attack directly the ideologically mandated framework that everyone had to use to avoid immediate persecution from the authorities. Mamardashvili created an atmosphere of evidence where thinking speaks for itself and an explicit elaborate logical construction of argument was not a necessity.

When Mamardashvili spoke, the listener was pulled into a mental space where he or she had never been before. It was an emotionally tense state of readiness to welcome a new opening, a new facet, or a new connecting line. Instant trust served as a supportive guide and as an invisible force, which helped one to navigate the mind-blinding journey to come. Mamardashvili's thinking was so powerful and strong that he instantly created an immensely sensitive energy field between the listeners and himself. There thoughts were molded, with everyone's mental alertness and openness involved.

Usually, in our society people exchange thoughts and ideas through a variety of means: they may have direct contact and use dialogue to convey information; they may use books or other sources of published information and avoid personal contact. Today, technology offers countless ways and possibilities for communicating what we think, how we think and what this thinking is doing to us. Nevertheless, very seldom do we truly know how and

A-T. Tymieniecka (ed.), Analecta Husserliana XCIII, 219–227.

why our thinking is occurring. We are using the word "thinking" as a synonym for reasoning, for a skill, which uses logic and factual information to make statements that make sense in the context of our everyday living. We cannot deny the use of thinking as reasoning, but at the same time it is destructive to forget about the existential purpose of thinking and its role in designing the human being as a thinking being. Heidegger, Gadamer, Mamardashvili and other thinkers worked on distinguishing the two kinds of mental applications that are displayed through the history of our civilized life. Mamardashvili convincingly describes how there is a special state of mind, a state of lucidity, which is piercing, strident and alienated at the same time, but also nostalgic and sadly sweet and it is the time when questions can be asked.[1] It is the beginning of thinking as a constitutive force or power.

Thinking as a fundamental philosophical problem was always at the center of Merab Mamardashvili's research. From a different point of view, he essentially kept asking the same questions as Heidegger and came to a similar conclusion: that in our times we do not know what thinking is, and, what is even more tragic, we are not aware of it. When philosophy in the Soviet Union was shadowed by communist ideology and was hiding behind the names of Marx and Engels, Mamardashvili kept the flame of thinking alive and flickering to protect and preserve the authenticity of a human being as a being that has the ability to determine self and project into the future.

Jacques Derrida begins one of his articles about Gadamer with the words, "will I be able to express, accurately and faithfully, my admiration for Hans-Georg Gadamer?"[2] I can repeat the same words about Merab Mamardashvili. In the times when philosophy in the Soviet Union was almost completely controlled by the communist party and its ideology, Merab Mamardashvili was actually teaching how to be an independent thinker, and how to preserve human dignity and one's mental self-sufficiency. His teachings were much deeper than the usual thinking of Soviet dissidents or any other political oppositionists. He was showing how to understand the nature of mental processes and how to cultivate in oneself gravitation towards being as self-understanding. In his lectures, he repeated over and over and demonstrated with his own being in thinking that we all need an inner voice, which may be called intuition, to distinguish and choose the path we are going to be on. If a person were not able to have unconscious feeling about direction, about choice or anything else, smart categories would not be able to give much help because they only demonstrate the logical connections amongst themselves.

In his article "Consciousness and Civilization," Mamardashvili looks at what is to him the most destructive disaster of the 20th century – the anthropologic catastrophe which manifests itself as the irreversible dismantling of the foundations of civilized life.[3] In the eyes of Mamardashvili, the anthropologic disaster is much more harmful than any natural disaster or even a nuclear war because it is a result of the disturbance of human consciousness and the braking of the rules of its functioning. We human beings, exclaimed Mamardashvili, roam across many different lands unable to speak; our memories are tangled; our history has been overwritten many times, and we do not know what happens with us or around us. Mamardashvili addressed these words primarily to Soviet citizens, but they can apply to everyone in the Western world as long as we share the common cause, which is the catastrophic negligence of our mental responsibilities and a failure to follow the fundamental laws that define the existence of a human being.

Is consciousness equal to the rationality that provides us with the tools to examine and cognize the world around us and ourselves? Mamardashvili spoke about classical and non-classical concepts of rationality and how they affect our worldview and our concept of self.[4] In the classical concept of rationality, thinking has an absolute power to see through the world and also itself, functioning as a thinking being. On the contrary, the non-classical concept of rationality introduces a two-fold understanding of thinking: as a cognitive illuminating power and as a state of consciousness – a mental space where being existentially transforms beyond itself and at any given point in time is more than illuminating power can embrace.

Mamardashvili joins the chorus of critics who complain that philosophy has lost its ability to explain the most fundamental questions of life and death and is no longer aware of its own role and place in the existence of a human being. What does philosophy mean to us? We know great names of splendid thinkers, the notions and categories that were used to express their ideas and concepts. History has also accumulated a vastness of scholarly writings about the understandings and misunderstandings of historically lined-up names and their lives' work. How about a research paper that gives us specific knowledge and informational insights? What exactly should we expect from philosophy? What should philosophy provide to us as readers and as human beings in general? According to Mamardashvili, philosophy starts from a question – what is philosophy for? This question was always present in Mamardashvili's lectures and writings. For Mamardashvili, philosophy clearly was a lifeline that every human being needs to fulfill his destiny. He looked at philosophy as a technique of being a human, as a technique of the constitution of self, as a way to transcend, to overcome the limitations that are imposed on human beings naturally and historically.[5] According to him, there is no other reason for philosophy than to guide one towards an understanding of one's own being and being's future possibilities.

Mamardashvili underlines that thinking is not an act of will. If I want to have a thought about something, it is not enough for me to want to think about it. If my being is not constituted in such a way that I can have that thought, I simply will not have it.[6] I may have something in my mind that would resemble it, but it would not have depth or impact of a thought that transforms being. An average person encounters misunderstanding on a daily basis because the use of language may be different or the expectations of context may not be identical, or even a psychological image may influence perception. An average person does not operate at the level of consciousness where thinking and being are interconnected and are listening to each other's whispers. Is it necessary for everyone to understand his own self and also others as beings that have a higher goal in their lives?

According to Mamardashvili, human beings have an obligation to constitute themselves by overcoming limitations they have as beings that are embedded in nature and natural history. Historically, people invented philosophy to learn to listen to the questions they faced in their daily life and to answer those questions while still listening. When talking about ancient philosophy, Mamardashvili mentions that Greek philosophers discovered this strange dependence – if I care about my being, my care will transform the being.[7] The Greeks invented philosophy to take care of being and to be able to fulfill being's destiny. They were also the first ones to warn against common misunderstandings and possible obstacles that may cross the path of understanding. From the beginnings of philosophy, the question of understanding and thinking's embeddedness in being was always a focal point of philosophical systems. This question explains why it is so important to verify that thinking is embedded in being and how it can be done. What are the signs of thinking's embeddedness or the lack of it?

Were other philosophers around the world looking into similar problems? Was the theme of thinking becoming a leitmotif of contemporary Western philosophers who were criticizing scholarly philosophy's inability to be the voice of self-awareness and also be a part of human being's becoming a Being? As an example, we can use Gadamer's point of view. Gadamer was concerned with grounding philosophy in the living world.[8] As was the case with many others, he was concerned with the fact that conceptual language would not lead to understanding; on the contrary, everyday language would contribute to finding new ways of self-awareness. His interest in temporality was related to his attempt to distinguish between the formal language, the language of concept, and the language of human experience, the narrative or poetry. His teaching about understanding focuses on sensitivity that allows thought to penetrate the darkness of the unknown and become a transformed state of mind. When Gadamer talks about ontological one-sidedness, he points to the imbalance between thinking that uses conceptual language and thinking that touches directly human existence and brings into mental space a new meaning, something that never existed before. The human being stands in the crossroads between the past and the future. This is the place where the conceptual language can be illuminated in the connection to the past, but also where new language is generated extending the human being into the future as a thinking being. Therefore, it is ontologically crucial to maintain the balance. On the one hand, a human being is the extension of the past, but, on the other hand, it is transcending into the future and projecting itself into possibilities beyond today.

Philosophy opens a chance for a human being to become self-aware and constitute the self as a self-conscious, thinking being able to transcend beyond its natural limitations. Philosophy makes the miracle of thinking possible. Can we have a thought whenever we want, asks Mamardashvili? Can we have a new thought whenever we want? No, says Mamardashvili. It is impossible to expect the thought we need to come exactly when we need it. Thinking requires an effort that comes from being. If the being is not ready for the thought, it will never come.

Mamardashvili presents thinking as a state of mind that is deeply rooted in being. Therefore, to have it and to retain it requires a great effort. He reminds us over and over again that an act of reasoning is simply not an act of thinking. Wisdom cannot be passed to others as knowledge. One's being has to labor into it. Mamardashvili speaks about ignited, inflamed consciousness. According to Mamardashvili, thinking is born in the act of suffering, in extreme emotional tension. Thinking uses metaphors and poetic language naturally. It is therefore important to highlight that Mamardashvili names poetry as a place where we can experience inflamed consciousness. He illustrates his ideas with examples from arts and literature, from history and social life. The names of Proust, Musil, Kafka, Bodleur, Gumilev and many others from different cultures serve as markers or signs to indicate the space where thinking happens. Again, those are only indirect markers that, like sirens from Greek mythology, try to capture our minds and lead us towards greater self-awareness and greater readiness to open up and embrace the true eternity as thinking beings. Cultural markers and cultural signs are the helping materials functioning as clues to achieve a mental state where thinking is a natural occurrence.

Mamardashvili speaks about the necessity of gathering oneself in one point in time and space to become something like a powerful nebula, which would burst and spread out light and recognize the former dependence on superficial psychological and social beliefs and regret the former captivity. "Gathering" is a metaphor that describes one's conscious effort to fulfill an obligation to constitute oneself as thinking being.[9] Similarly, Mamardashvili speaks about mental "muscles" that a person has to develop by exercising and practicing, by using different clues and road signs, and by facing stress, tension, failure and weariness. Gathering does not equal to an act of will. We may want to gather ourselves but fail to do it because we were not ready to be in the gathering, to live in it, to make it our space of being. Gathering begins as tension in oneself, as an unconscious pull towards the center, or as a vague intuitive anticipation of a miracle.

Mamardashvili never forgot in what kind of society he lived and what additional obstacles it imposed on thinking by having severed its ties to cultural tradition and world philosophy. Repeatedly he spoke about the "enslaved" language, which existed in the Soviet Union and manifested itself in literature, humanities and science. The enslaved language mirrored the cancerous mentality that was developing and strengthening. Many cultural products that Soviet society showcased did not contain meaning but only mimicked it.

Merab Mamardashvili was not a typical representative of opposition to Soviet-style philosophy. He did not oppose separate statements or concepts, and he did not attempt to denounce the ideological principles of Marxism-Leninism directly. His goal was to cultivate the kind of thinking that would be rooted much deeper than any ideology can reach and damage. He did not waste his time launching a crusade against ideological platitudes and the impotence of human mentality tied to them. When he talked to students at the Institute of Cinematography in Moscow in the late 1980s, Mamardashvili mentioned Orwell's novel, *1984*, as an example of how the "new language" creates mentality that turns the society into a totalitarian monster where deafness of hearts and stiffness of minds is celebrated and approved officially.

Referring to the French poet Mallarmé, Mamardashvili states that it is not enough just to know what to write about to create a decent poem. Indeed, words describe a certain mental state that a poet has. When somebody reads the poem, the same mental state can be recreated. When it happens, words release their ontological power and signify that they are alive. Mamardashvili cites the Russian poet Gumilev who revealed the existence of "dead words with a bad smell." It is important for Mamardashvili to acknowledge that history is a dynamic eternity a part of which we can become, only if we can find a proper way to open the door and be let in. The issue of "letting in" or "opening the door" becomes very important for the process of self-gathering, self-constitution of a human being.

Culture, the civilized way of living is being born anew with every new generation. If this generation accomplishes the act of "tying history" into itself, this act would be identical to "opening the door" into the eternal movement of truth. The process of "tying history" into oneself activates our historical roots and our historical potential. These acts are easily recognizable because they exhibit exclusive emotional marks, such as love, burden, suffering, and despair. And again, it is important to underline that the act of tying is not an act of will, because it is not enough to voluntarily decide to tie oneself into history and it will happen. It is an existential act that is a result of a change in our being.

When history is tied into one's being, the thinking acquires a universal experience, which belongs to other people in the same generation or anywhere in history. In the act of tying, the individual being becomes a point where the future connects to the past and a true succession takes place. With "tying" into history, an individual, finite being gains unlimited access to the universal wisdom of humankind and longevity across the timeline of the human race.

Mamardashvili talks about a borderline or boundaries where the tying actually takes place. Initially, a feeling of an unresolved problem indicates the closeness of a possible crossing point in the boundary. Thinking that approaches the crossing point is emotionally tense, which explains why Mamardashvili uses the metaphor of inflamed thinking. In this case, it is important for Mamardashvili to highlight the dramatic power of concentration and the level of alertness thinking needs to maintain to cross the line between the empirical, everyday world-view and the world where "eternity peeks into our lives."[10] The power is needed to overcome a psychological perception which tends to pull thinking away from the crossing point and to disrupt its gathering around unanswered questions. As we have seen, the act of gathering is an individual, heroic act of resistance to plain rules of logic or everyday established patterns of functioning. Thinking exists as this special gathering of our individual powers, a special state of mental energy that allows pushing for an opening and stepping into a crossing point. Mamardashvili points to the parallel in Eastern philosophy where thinking is viewed as awakening, as readiness to enter the mental space where a human being becomes a "prolonged" being or extended into the wisdom of the human race.

Heidegger also convincingly shows the aspect of tension in thinking. At the same time, it is an enormous challenge to give an adequate description of this tension using language. As one of Heidegger's translators notes, "the advance Heidegger wishes to make on the basis of Greek thought is to learn to think non-conceptually and non-systematically yet with rigor and strictness about the nature of Being."[11] The necessity to distinguish between thinking as a being's constitutive force and thinking as the faculty of reasoning brings into question the use of language in the foreground. Both Heidegger and Mamardashvili were purposely creating language to describe their philosophical findings and to properly showcase thinking in the process of gathering and in the other states of mental readiness. They also reached out into arts and literature to use the accumulated, rich, unlimited source of emergent, original metaphors. In the past, in the tradition of empiricism, metaphor was treated as "an ornament to language." Today, more researchers consider metaphor as a dynamic phenomenon which functions as a mechanism of extending and deepening the language.[12] Heidegger puts metaphor in the spot light and views it as a key element in his philosophical descriptions. Mamardashvili animates his discussion and his arguments with metaphor, which he calls muscular language and clearly favors using it instead of having arguments tied to the conceptual language only.

There is another side of the argument in favor of metaphor – metaphor belongs to the realm of art which both Heidegger and Mamardashvili consider a birthplace of meaning. According to Heidegger, "art is by nature an origin: a distinctive way in which truth comes into being, that is, becomes historical."[13] Mamardashvili cites Rilke, who said that poetry is not a way to express feelings we already have, but to create a reality where such feelings can come into being naturally. Shakespeare's Hamlet is another example, used by Mamardashvili, where the literary character becomes a mechanism through which being transforms and new dimensions are added to mental and emotional registers. Hamlet wants to act self-consciously as a human being, and in order to make an effort, he needs to organize his being differently. He can use symbolic forms that are accumulated in art and culture and reach fulfillment of being. Then he is ready to face the truth because he is in the mental state piercingly illuminated with understanding.

Sometimes art makes certain qualities of our existence more visible and can be considered as a way to amplify those qualities for enhanced understanding and for grasping their nature.[14] For Heidegger poetry was similar to the water that flows back to the source and brings us back to the beginnings where we can refresh our understanding and start anew the path we are on. Stress and tension serve the awakening of thinking, and art and poetry serve as a storage place for mental props to keep us alert and moving. To show the significance of art in forming our ability to become thinking beings, Mamardashvili uses as an example a well-known movie from the "perestroika" years – *Repentance.*

There is a watershed between a person who understands this movie and the person who does not understand it. Mamardashvili says that either someone can understand with absolute clarity the full meaning of Stalin's purges and, therefore, is able to think about everything without any boundaries, or someone cannot understand fully the tragedy of the time and is doomed to live with dead thoughts.

Mamardashvili's legacy lays in promotion of thinking as a way of being that fulfills the purpose of human existence.

Rockford, IL

NOTES

1 M. Mamardashvili, *The Aesthetic of Thinking* (Moscow, 2001), p. 10 (in Russian).

2 J. Derrida, "Uninterrupted Dialogue: Between Two Infinities, The Poem," *Research in Phenomenology* 34 (2004): 3.

3 Mamardashvili, "Consciousness and Civilization," in *How I Understand Philosophy* (Moscow, 1992), p. 120 (in Russian).

4 Mamardashvili, "Classical and Non-classical Ideal of Rationality," in *Necessity of Self* (Moscow, 1996), pp. 229–250 (in Russian).

5 Mamardashvili, "Emergence of Philosophy in the Foreground of Myth," *Ibid.*, p. 23.

6 Mamardashvili, "The Space of Thought and the Language of Philosophy," *Ibid.*, p. 58.

7 *Ibid.*

8 Sh. Ross, "Gadamer's Late Thinking on Verweilen," *Minerva. An Internet Journal of Philosophy* 8 (2004): 134.

9 Mamardashvili, "The Space of Thought and the Language of Philosophy," in *Necessity of Self*. p. 64.

10 Mamardashvili, *The Aesthetic of Thinking* (Moscow, 2001), p. 154.

11 M. Heidegger, *What is Called Thinking*? (New York: Harper & Row, 1968), p. XIV.

12 Grey, W., "Metaphor and Meaning," *Minerva* (2000).

13 Heidegger, *Poetry, Language, Thought* (New York: Harper & Row, 1975), p. 78.

14 Ross, "Gadamer's Late Thinking on Verweilen," *Minerva* (2004): 140.

SECTION IV

MORAL ELEMENT OF EXPERIENCE

CARMEN COZMA

"THE ONTOPOIETIC UNFOLDING OF LIFE" – A CONCEPTUAL SYSTEM FOR AN ETHICS FOCUSING ON THE "BIOS"

A major issue, which is developed within the Phenomenology of Life, is that of the "bios" comprehension in the "ontopoietic perspective." By an original way, using her concept of "Ontopoiesis of Life," Anna-Teresa Tymieniecka tackles "that which is life" in a complex appropriated view of the entire manifestation of life, considering (also) the significance of the natural/vital environment. Thus, the phenomenological approach reveals "the logos of life through the complex of nature," engaging an ethical dimension, too, that opens even a path to the understanding of the bioethical dilemmas. It is a path to a possible response, activating the value of "wisdom" that human being must to educate and to increase in his own status like "Custodian of everything there is alive" – according to the creator of the Phenomenology of Life. This present essay tries to highlight some of Tymieniecka's philosophical work potentialized in its virtues as contributing to an "environmental ethics."

Opening one of the most spectacular directions in the development of the posthusserlian phenomenology in the world, Phenomenology of Life challenges us by an impressive repertory of categories that – maybe, beyond the author's intentions – certifies a fruitful semantics for the ethics territory.

The "ethical referential" represents a constant of Anna-Teresa Tymieniecka's writings, a constant by which we dare to disclose even an ontical factor of her philosophy; more clearly, we think about *an onto-ethical dimension of Phenomenology of Life* soaring from a very complex cultural manner to put in act the philosophizing process upon *life*, by resorting to a "discourse fleuve"[1] – as *life* itself is manifesting.

To journey through the ramifications of a monumental work that encompasses philosophy, sciences and arts, following our phenomenologist in the temptation to catch the comprehension experience of the "logos of life" – the universal reason of life – in its manifold expression, the researcher becomes awake to the great value of a *system-holistic dynamical vision on life*. This is our very short assessment of Phenomenology of Life, under the auspices of

A-T. Tymieniecka (ed.), Analecta Husserliana XCIII, 231–243.

which we try to set about that we find to be a great theme: an *ethics focusing on the "bios."*

Intertwining cosmology and anthropology views, our approach aims to enlighten the *ethical significance of life* at all its levels, as we hitherto know about it. We prefer the Greek 'bios'/'βιος' term meaning "life," in general; a term used like a prefix for the most composed concepts related to the "life" area in its unity of vegetal, animal, human spheres. We think that "bios" is more adequate for the wholeness of entities, forces, operations, processes and manifestations, making structures and ordering networks in which we can decipher the "logos of life" movement as "the play of the natural and inventive forces."[2]

Why an *ethics focusing on the 'bios'*? Because it enables our becoming to be more sensitive and responsive as regards even the function of a philosophizing act; that is, to grasp and reveal part of the *mystery of life*, to highlight and make understandable, in a better way, *life*'s emergence and unfolding as a worthy one claiming to be increased and defended in its "depth meaning."

In the framework of the Phenomenology of Life, there is a leitmotiv, the idea of *equilibrium* and *measure*, of *unity in harmony* within the orchestration of all the manifested components of life, engaging the human subject in a complex "moral commitment" by the activation not only of reason, but also of passions, feelings, inclinations, emotions, tendencies, interests, purposes, motivations, of good will in deliberations, choices, decisions making, of attitudes and conducts, sustaining the necessary effort to discover a path of overcoming a bewildered existence – "the anatomy of our moral disarray," as Tymieniecka designs the profound crisis in which contemporary mankind has to recognize itself.

As it is known, as for the situation of "decline" has already warned the Master of Phenomenology, Edmund Husserl.[3] Actually, the thematization of "crisis-diagnosis" represents a "decisive motive" of the phenomenological texts, for example, by Spengler, Heidegger and Husserl.[4] Unfortunately, for contemporary phenomenologists – and not only for them!, the "crisis" problem is no more just a scientifically and cultural one of European mankind; but, mainly, it is a much more grave and an unprecedented crisis, a "world-wide life-system crisis," namely the "ecological crisis." Mankind today is confronted with an *ecological angst* of a basic *ecoalienation* – "a splitting of humans from connectedness with the natural world that they are in and that is in them"[5]; and, consequently, the need of the *awareness* of our ecologically damaged world. At this point, we can much better appreciate the phenomenological discourse of Anna-Teresa Tymieniecka with the thesis about the *all-alive-unity* and the *nature's harmony*, about the "entelechial principle of life" in its "ontopoietic design," and the distinct sign of man who has to inscribe an equilibrium of forces between "nature" and "culture" into the life's flux by his *creativity*.

Facing this very serious threat of the "earthling life," we identify some opportune marks to surpass it by molding our conscience in the tonality of the Phenomenology of Life. According to Tymieniecka, "A novel starting / . . . / required by the acuteness of the contemporary situation," "the most promising one is that offered by a new inquiry into the ontopoiesis of life itself through the access that we gain via the creative sources of the specifically human differentiation of the logos of life at the primogenital point of the Human Condition as a station in the evolutionary radiation of the logos."[6]

The accelerated transformations of man's style to live, due to the successes of tech-science, beside notable benefits, have brought, no less, the peril of great risks, raising major interrogations regarding even the sustainability of life on planet Earth. The man belonging to the "Cyborg Constellation" finds himself in a strange dangerous situation to alienate himself till to lose his very own spiritual essence, to forfeit his humanness. Therefore, the question of reconsidering his position within the world to which he has access has become one of the stringent priorities. A radical change of mentality, and at the same time of morality seeing the general reports to life is requested. Now, man has to act not only by reason and cognition; but, more than ever, he has to demonstrate his entire *creative potential* – that affords him the singularity in the world – in a moral direction.

To restore the *aretaic value* structured on the principle of "measure-in-everything," the principle of the "dynamic equilibrium of contradictory tension" of living experience, the "Golden Rule" constitutes a primordial condition of maintaining and growing life. A real turn of *ethics* to the *"bios"*, to the *value and imperative of life* regarding the totality of living forms intrinsically moral valuable is claimed by necessity. Without a healthy and reasonable integrator view upon the All – "from the germinal preorganic and organic circuits" until "the sentient, sympathetic, psychic, social and cultural circuits of life,"[7] man has no chance to develop his own life. Certainly, it is not a novelty, but it makes sense to be strenghthly recalled, for ever, that "our microcosmic route of life points to macrocosmic infinity for its beginnings and its endings."[8]

So, to attain the task of a "philosophical re-construction" on the phenomenology theory and methodology, Anna-Teresa Tymieniecka is "recovering the great vision of the All-in-becoming,"[9] encountering, among others, Max Scheler in his intention to spiritualize the relation between man and cosmos within the *life's unity*. As the German phenomenologist writes, owing to his *spirit*, man "can behave *openly towards the world*," like "a co-founder and a co-producer of an ideal series of becoming, that is occurring within the cosmic process and at the same time with himself."[10] At her turn, Tymieniecka says that, by the "Dionysian" (the sentient, felt, passional) and the "Promethean" modes of rationalities, "the logos of life expands in *freedom* into a new universe of its own *invention*, into the universe of the human spirit."[11] It is, somehow, an encounter with a very popular present-day scientist, Fritjof Capra, for whom "the human spirit could be defined as the modality of conscience by which we are feeling to be tied with the cosmos as a whole."[12]

By the *phenomenological exposure and interpretation of the wholeness and unity of the "All-is-alive,"* Anna-Teresa Tymieniecka realizes a *celebration of life*, manifesting a passionate interest for the "constructive advance of life" into the universal ordering, the "cosmos," as following a cyclic evolution presided by the "entelechial principle" of life, that leads to a peculiar understanding of the totality: "body-organism-soul-spirit." She states that the *element of entelechy* "asserts itself as a decisive factor in the cosmological perspective, within the individual structure of the living individual's autonomous being. The entelechial factor of the living individual being appears as the material nucleus of his existential mechanism, as absolutely his own, and carrying out the progress of his living's particular identification."[13]

Far from any intention to alter the vision of the Phenomenology of Life, we allow ourselves just to consider it like *an overall vision upon "bios,"* challenging to remind the wisdom of the ancients (either in the Asian visions of Hinduism, Buddhism, Taoism or in the European, Greek–Roman). It is not accidental that the North-American phenomenologist stresses the self-understanding of man *in the context of cosmos*, where he is not alone! Also, the insistence on the "natural-logoic order" or on the "soul of physis." It deserves to come to the original terms of "cosmos"/"κοσμος" – for example, in Stoic philosophy, meaning "a living being, endowed with soul and reason"[14] – and of "physis"/"φυσις" – used as the generic term for "nature," respectively: "that is alive, of eternal transformatory movement and inseparable from the human being"; more precisely, according to Plato, "the nature of the whole"/of the universe ("της του ολου φυσεως").[15]

As a terrestrial being, man is not only a social-culturally constructed being, but also a biological one, interacting with any grain of nature. Man is in a responsive position having to grasp the nature's gifts, to observe and even to get from nature a magisterial learning of maintaining the *life*/the *"bios"* in order and equilibrium, by measure and proportion; no less, he is in the position to enact a strategy in defense of life, as even the minimal interest of his own survival conditioned by the ability to take care of the whole relational (human and nonhuman) environment alive.

We find, in the territory of the Phenomenology of Life, a significant attention given to the interaction between "physis" and "bios." An important thesis, that of the "self-individualization-in-life" does not refer to a "self-isolation" of man from nature, a loss of the kindred between "human" and "cosmos." On the contrary, such a thesis considers the issue of *biosphere* – the interdependent community of all living beings; within that man is a link (maybe, the most equipped), not at all one of an exclusively steward-ship over nature. In the Phenomenology of Life, *biosphere* is taken as "the vitally relevant web," the vibrant "unity-of-everything-there-is-alive." It is the crucial net of life manifested in "all types and concrete individuals into an infinitely flexible, movable, changeable, transformable whole."[16]

"Integral and scientific," Tymieniecka's Phenomenology "is a philosophy of life *par excellence*,"[17] bringing together: "cosmos," "bios," "culture," "human." Bridging metaphysics and epistemology in a global and dynamical systemic perspective, the Phenomenology of Life is closely related to the modern sciences developed onto a philosophy of the "network and unifying structure," alike in Geoffrey Chew's "bootstrap theory" about the life-world thought as a dynamic net of correlated events, or in Gregory Bateson's theory about the intrinsic dynamical nature of all the interrelation phenomena in the universe.

Anna-Teresa Tymieniecka is thematizing the concept of *life* – considered to be "the nervous system of all philosophy,"[18] as the reference unique *medium* of beings-in-progress in their interconnectedness, from within individuals are self-differentiating, on the one hand, but within that they are following an integrator course of becoming, on the other hand. By her original concept of the "ontopoietical paradigm" – simultaneously dealing with specific individuals in their 'autonomy of meaning and unique roles', and with a global framework of circuits in 'cosmos', "bios," "zoe," "psyche," "human spirit," "society," "culture," – our phenomenologist offers, finally, a philosophy envisaging "singulars," "types," "structures," etc. from the coexistential network of life; a philosophy of "a new, global vision of the world / ... / physical, biological and human at the same time."[19] We face a philosophy of "correlation-in-life," of "living beings-in-community"; seemly to define the category of "living being" as "a relational term like *friend*."[20]

Without doubt, it is meritoriously the phenomenological endeavor to catch part of "the poiesis of life" within its enactment and constructive advance, in the internal and external forces interaction, "through following the royal road opened by the creative act of the human being" enabling a nuance explanation of the "meaning of life."

The Phenomenology of Life offers an integrator philosophy around the *system of vital forces* that "(1)carries the self-individualization of life, (2)accounts for the unity-of-everything-there-is-alive, and (3)points to a system of existential relevancies to the cosmic logos with its laws, sets down rules, and prompts synergetic resources toward a functional establishment for founding and carrying on the evolutionary progress of life in all its variety."[21]

Approaching "the Great Plan of Life" theme, Anna-Teresa Tymieniecka elaborates a veritable system of original concepts that can be successfully used to sketch a *"bios" centered ethics*. Do not forget that, in a large sense, *ethics*, in general, is a way of living for man together the entire spread of life. An *ethics focusing on the "bios"* – we could say: *a life principle ethics* – means to modulate the traditional anthropocentric ethical horizons, to expand the discourse by integrating "humans" and "non-humans" in the same one "biotic community" that accounts for a moral valuation, emphasizing an assertion belonging to Albert Schweitzer, that circumscribes the basic message: "I am the life wanting to live within life itself wanting to live." In such a perspective, we decipher into the Phenomenology of Life an "ethics *for* life," an "ethics of the right to life" by considering the moral report of man to nature. It is an ethics able to satisfy the human self-fulfillment only in respect with any living-being-in-becoming to which man is connected. Broadly speaking, even the realization of the task of ethics – "the human existential enjoyment and fulfillment"[22] – is depending on the responsibilizing of man's commitment into concrete relations and situations.

We are settling in the horizon of a special *wisdom*, a morally ascertained one of a fully respect bond towards the "common good" of life, accounting our situation as fully "conditioned by the situation of nature-bios (at its foundations in the cosmos)." According to Tymieniecka, we have to cultivate a "measuring and creative wisdom" that "life on earth depends on."[23]

Especially during the last decades, ethical language had to open toward various problems, like epidemics, natural calamities, radical climatic changes (owing to the air and water pollution, the ozone depletion, acid rain, trash, energy consumption, global warming and green house effect), extinction of species, poverty, global hunger, oppressive governments, wars, use of biotechnology, development of artificial intelligence, of nanotechnology, therapeutically and reproductive genetics' challenges and dilemmas, etc. Such problems essentially enfold the *life*/the *"bios"* theme.

We will not enter the dispute "anthropocentrism" or "biocentrism" in this study. We will just resume to agree and to say, together with a contemporary thinker, that: "An adequate theory of anthropocentric value seems to contain within it the seeds of biocentrism."[24] It is important to transcend the damaging dichotomy "universalistic humanism–animal welfarism" and to discover the path of a veritable "humanism of life" marked by a spiritualization of the relation of man with the universe; no less, marked by the awareness that the first vital interest of man is that of the entire system of life, that the person wellness is mutually dependent on the welfare of plants, animals, earth, of the whole biosphere. It is the mark of *a "bios" centered ethics*: an ethics of the general "physis" welfare; a health, security, and protection life ethics, eventually.

Somehow, this is an urge devolving from the Phenomenology of Life, too. For her creator, the reflection on *life* is one tied with the question of *nature* – "the vital milieu of all living beingness."[25] Thus, we have to comprehend the requirement of "the need to formulate anew the issue of *nature*"; the significance of "a new formulation of the concept of nature-life, one open to the cosmos and to culture"; the scrutiny into "the womb of life: nature-life as the seminal/generative groundwork of life's vital significance"; the inquiry of "the logos in the cosmic sphere."[26] Reiteratively, the question: "What is nature?" is addressing. The manifestation of the "core significance for philosophy" concept: the "logos of life" is conceived only "through the complex of nature," the entire *life*, as our author writes, being "suspended on the cosmic parameters."[27]

Thus, we get in the meaning of the Phenomenology of Life's assessment by many exegetes like "a Copernican Revolution" in philosophy, due to the "quest for the Logos of Life and the Life of Logos," following the essential principle of order "through the labyrinth of *life*" – in its strategies, in its specifics, in its plenitude of inner workings and unfolded accomplishments, contextualizing the complex-system-holistic-dynamical evolution from the pre-life level and the nature-life radius, until the revealing of the knot position of man in universe, as the *creative being* able to manifest himself in the singular hypostasis of "Custodian-of-the-unity-of-everything-there-is-alive."

Throughout a vision about the *constructive advancement of life*, is transluciding the image claiming the "harmony of *bios*" as "unity of multiplicity" by the own laws of existing, in which man has to adapt himself to the natural environment, concomitantly having to introduce his peculiar cultural avenue, but proving the capacity to maintain a "cooperative interaction" with the biotic medium. Actually, in terms of an *ecological* approach, encompassing the logoic process within "life and non-life," the category of "sharing-in-life" is an eloquent one for the instilling process of a "communal existence," for the promotion of the "common good of life."

Searching the "logos" within the "impetus and equipoise" processes of the emergence and development of *biotic and cultural life*, Anna-Teresa Tymieniecka leads us into the *totality of life*, by "a logoic vision of the All," helping us to understand the fullness functioning of the "logos" from the physical, organic base, to the creative human modes of life. In the intimate tonality with the character of life itself, her language is one of circularity's polar terms going to their conciliation: "virtuality-actualization"; "stability-flux"; "old-new"; "continuity-discontinuity"; "inward-outward"; "below-above"; "now-always"; "nowhere-everywhere"; "open-ended", "sameness-alterity," etc.; a language fitted to unveiling the "logoic strategies" of projecting the "*constructive* devices amid a tension of opposites," within the interplay of "attracting and repelling" forces of "nature" and "human creative being." It is a language fitted in the in-depth grasping of the "essence and phenomenon," of the "cryptic and epiphany" modalities in which a part of the mystery of life is revealing, at least; Tymieniecka knowing too well that, anyway, the "life's initial impetus remains a mystery."[28] But, she dares to pervade more in its crux, noticing that: "The 'hidden' and the 'obvious' in their opposites lay the basis for the enigmas of life and of the human being. The 'concealed' has meaning only with respect to the 'manifest'. The play between the two, / . . . / lies at the very heart of the Logos of life."[29]

By priority, the *cosmological dimension* of the Phenomenology of Life discloses a pathway to tackle and to cope not only with the *environmental*, but with the *bioethical* dilemmas of man, concerning: "what the nature human himself is"; "the boundary between life and death"; "living and dying in dignity"; "individual right to decide over one's own body and style of living, on the one hand, and the social consensus and the sacrality of life, on the other hand"; "sexuality interpretations," "gene therapy," "genetic engineering," "intervention to save humans or even to create humans," "genetically manipulated material," affecting, finally, all life. Generally, we can find responses to major ethical interrogations of the so vulnerable human being who is recognizing himself in his hopes and fears, chances and perils, opportunities and risks, temptations and demands, allurements and deceptions, more and more worried by the environmental degradation that means endangering the natural foundations of life – the very own human life, too – and more and more aware of "the need for regulation regarding external nature"[30] – the "nature" *of* and *into* which man is a part; the "nature" that must be treated, by *respect*, in its intrinsic value, like the primordial medium of life, and not like a property to exploit without limits.

The Phenomenology of Life's cosmology invites us to learn from "nature" the ordering, the relationship-in-harmony-of-the-All laws; and thus to avoid or to surpass the discomfort and insecurity, the harming of life, to counteract the irreparable, the distortions and fallings out, the desperation in rout and chaos.

Trying to draw some guidelines for the assumed task in this study, we establish that a sum of concepts introduced by the creator of the Phenomenology of Life to disentangle and to appropriate some of the life's mystery can be very well made as functional concepts for an *ethics focusing on the "bios" relevancy.*

So, categories by ethical resonance are the "Ontopoiesis of Life" with the process of "self-individualizing of life" in its coherence of a "constructive advance," through the action of the "logos of life."

The "Ontopoiesis of Life" – the core of Tymieniecka's phenomenology – is a key notion, representing the axial constructive force of life becoming the dynamic plan/route of the "logos of life's workings," in which both nature and culture are situated.

The "pivotal platform of the Phenomenology of Life,"[31] the "Ontopoiesis," with the central constructive factor of the manifestation of the "logos": the "entelechial principle," and with its vehicle: the "self-individualization of life," outlines the entire processual network of life. The "ontopoietic nucleus" constitutes the ground for the "inner-outer" life's movement in its tensioning and equilibrating All – "cosmos and human," "progressing through relatively stable levels of organization, which are to be described phenomenologically – as meanings."[32] Enacted by the "triadic logos" (the "Cosmic/Vital" Logos, the communicative "Dionyssian"/feeling/sharing Logos, and the "Promethean"/inventive Logos), as the ontological principle, the "Ontopoiesis of Life" is at the same time: "(a) *the point of reference* for the distribution of sense initiating operations at the horizontal level of the life carrying processes that operate the self-individualizing differentiation, (b) *the entelechially charged indicator of direction* for the genetic construction of individual beingness, and (c) last but not least, *the measuring rod for the constructive attainment of types of complexity*, that is, of the sense bestowing levels (levels of conscious functioning in the experiencing subject).[33]

The "absolute criterion of life, becoming and beingness,"[34] the "self-individualizing in life" process involves distinguishness, but not a split between "human" and "nature." To self-differentiate by *creativity* – "the Archimedean point from where everything finds its proper place"[35] – does not mean that man loses his natural-biological roots; but it means to improve them in an elevated manner of investing the advance of life by cultural values. The dyad of "detachment and memberships" is at stake; this is hermeneutically explained by Gadamer, for example, in the following terms: "Life is defining through the fact that the vividness self-distinguishes from the world in which it is living and to which it continues to be attached, and it self-preserves in such distinction of itself. The self-preservation of vividness occurs by the incorporation in itself of any kind of outer beingness. Everything there is alive provides for that is different of it. The fundamental situation of being-alive is the assimilation. The distinctiveness is, thus, at the same time a nondistinctiveness. Everything is different, it is appropriated, too."[36] In the words of Tymieniecka, we face the only one design of *life* unfolding, the unity between "the vitally significant circuits of nature-life and the inventive/creative significance of human culture (science, technology, art, etc.)."[37]

The phenomenologist of life is perfectly in the right to make one's option for the *creative act* – rather than the cognitive one – as the singular modality for man to rise from the natural condition to the "Human Condition," a "complex of informative, regulative, and directing data, devices, and virtualities."[38] As "the key factor of all ordering" in life, *creativity* affords man to manifest himself by freedom and dignity. Anna-Teresa Tymieniecka specifies that "Among the forms of life, the human being appears to have attained the most advanced complexity as the free, that is, creative differentiation of the logos of life."[39]

The "crucial web of all-there-is-alive" and the so inspired chosen term of the "Custodian" role attributed to man – the guardian who is "an inextricable segment within the unity-of-everything-there-is-alive"[40] – enlighten one of the most comprehensive definitions of the *human* situation within the whole life. Seriously wondering as regards the syntagma of the CUSTODIAN-OF-EVERYTHING-THERE-IS-ALIVE meaning, as a result of 'the moral sense' that introduces "the Sentiment of Benevolence toward other living creatures, toward oneself, and toward life in general,"[41] an autonomous larger approach is outlined elsewhere. Herein, we just resume to point some relevant aspects.

The concept of "Custodian-of-everything–there-is-alive" witnesses an apart status of man engaging many ethical values; respectively, man as a sentient, emotional, reasonable, interactive agent of "awareness" and "participation" in life; of *justice* and *charity, mercy* and *sympathy* by feeling the connection with every type of beingness[42]; of "reverence for life," as an internal imperative for the *humanity* within a "veritable civilization"[43]; of toleration and an equal treatment of humans and nonhumans; of respect for the inherent value of life; of a responsibility extended to the cosmic scale and concerning not only the present, but the future of life on the Earth, too[44]; of *solidarity* – "defined as a form of the ability to be a part"[45] – with all living beings; an agent of taking care, protecting and healing the ecosystem equilibrium, of willingness to change the present state of crisis and to work for the goodness of life; an agent who "cannot avoid forming a decent relationship with nature, a relationship of cooperation and stewardship, within which his human nature would meet in a cordial embrace his destiny"[46]; an agent able to show an authentically wisdom of living grounded on a real *love of life/biophilia.*

Certainly, by activating a more attentive insight, we can see in the plenty original concepts of Anna-Teresa Tymieniecka some very useful ones for the repertory of an *ethics centered on the "bios."* Moreover, the Phenomenology of Life can be, for each of us, a very welcome *guidance for a moral life.* Simply, in times of "moral disarray" and "ecological crisis," it would be damaging to evade its promising potential to enrich the ethical area that sustains our hopes concerning the life of the biotic wholeness/system. To conclude: through the Phenomenology of Life we are waking up to a question that really matters, since it teaches us, among other things, about a virtual *ethics focusing on the "bios"* as a beacon of our *creative* spirits and acts.

"Al.I.Cuza" University of Jassy, Romania

NOTES

[1] Anna-Teresa Tymieniecka, *Logos and Life*, Book 4: *Impetus and Equipoise in the Life-Strategies of Reason* (Dordrecht/Boston/London: Kluwer Academic Publishers, 2000), p. xxvii.
[2] *Ibid.*, p. 100.
[3] cf. Edmund Husserl, *Die Krisis des europäischen Menschentums und die Philosophie.*
[4] See the keen hermeneutic commentary made by Christian Möckel in his book: *Phänomenologie. Probleme, Bezugnahmen und Interpretationem.* The German thinker works out a very welcome exegesis, following the trajectory: Oswald Spengler, *Untergang des Abendlandes*, 1918; Martin Heidegger, *Ontologie (Hermeneutik oder Faktizität)*, 1923; and Husserl, *Die Krisis des europäischen Menschentums und die Philosophie*, 1935–1936; in Möckel, *Phänomenologie. Probleme, Bezugnahmen und Interpretationem* (Berlin: Logos Verlag, 2003), pp. 107–128.
[5] Howard Clinebell, *Ecotherapy: Healing Ourselves, Healing the Earth* (New York/London: The Haworth Press, 1996), p. 34.
[6] Tymieniecka, *op. cit.*, p. 98.
[7] *Ibid.*, p. 629.
[8] *Ibid.*, p. 648.
[9] *Ibid.*, p. 650.
[10] Max Scheler, *Die Stellung des Menschen in Kosmos,* Romanian trans., *Poziţia omului în cosmos*, (Piteşti – Bucureşti – Braşov – Cluj – Napoca: Editura Paralela 45, 2001), pp. 43, 116.
[11] Tymieniecka, *op. cit.*, p. 320.
[12] Fritjof Capra, *Uncommon Wisdom. Conversations with Remarkable People*, Romanian trans. *Înţelepciune aparte*, (Bucharest: Editura Tehnică, 2004), p. 116.
[13] Tymieniecka, *Why Is There Something Rather Than Nothing? Prolegomena to the Phenomenology of Cosmic Creation* (Assen: Royal van Gorcum/Humanities Press, 1966), p. 33.
[14] cf. Diogenes Laertius, *Lives and Doctrines of Eminent Philosophers*, VII.

15 Plato, *Phaidon*, 270c.
16 Tymieniecka, *Logos and Life*, Book 4: *Impetus and Equipoise in the Life-Strategies of Reason, ed. cit.*, pp. 124; 130.
17 Tymieniecka, "The Ontopoiesis of Life as a New Philosophical Paradigm," in *Phenomenology Inquiry*, Volume 22, (Belmont, Massachusetts: The World Institute for Advanced Phenomenological Research and Learning, 1998).
18 Tymieniecka, *Logos and Life*, Book 4: *Impetus and Equipoise in the Life-Strategies of Reason, ed. cit.*, p. 75.
19 María Avelina Cecilia, "Phenomenology of Life, Integral and Scientific, Fulfilling the Expectations of Husserl's Initial Aspirations and Last Insights: A Global Movement," in Anna-Teresa Tymieniecka (ed.), *Phenomenology World-Wide. Foundations – Expanding Dynamics – Life–Engagements. Encyclopedia of Learning*, (Dordrecht/Boston/London: Kluwer Academic Publishers, 2002), p. 711.
20 W. Kim Rogers, *Reason and Life. An Introduction to an Ecological Approach in Philosophy* (Lanham – New York – Oxford: University Press of America, 2003), p. 51.
21 Tymieniecka, *Logos and Life*, Book 4: *Impetus and Equipoise in the Life-Strategies of Reason, ed. cit.*, p. 102.
22 Robert Misrahi, *La signification de l'éthique* (Paris: Synthélabo, 1995), p. 101.
23 Tymieniecka, *Logos and Life*, Book 4: *Impetus and Equipoise in the Life-Strategies of Reason, ed. cit.*, p. 625.
24 Nicholas Agar, *Life's Intrinsic Value. Science, Ethics, and Nature* (New York: Columbia University Press, 2001), p. 76.
25 Tymieniecka, *Logos and Life*, Book 4: *Impetus and Equipoise in the Life-Strategies of Reason, ed. cit.*, p. 103.
26 *Ibid.*, pp. 100; 102; 105; 119.
27 *Ibid.*, pp. 97; 195.
28 *Ibid.*, p. 83.
29 *Ibid.*, p. 188.
30 Gernot Böhme, *Ethics in Context. The Art of Dealing with Serious Questions* (Cambridge: Polity Press, 2001), p. 116.
31 María Avelina Cecilia, *op. cit.*, p. 708.
32 Gary Backhaus, "Impetus and Equipoise in the Life-Strategies of Reason as the Culmination of A-T. Tymieniecka's Open System of Her Phenomenology of Life," in *Phenomenological Inquiry*, Volume 27, (Hanover, New Hampshire: The World Institute for Advanced Phenomenological Research and Learning, 2003), p. 9.
33 Tymieniecka, *Logos and Life*, Book 4: *Impetus and Equipoise in the Life-Strategies of Reason, ed. cit.*, p. 264.
34 *Ibid.*, p. 107.
35 Tymieniecka, *Logos and Life*, Book 1: *Creative Experience and the Critique of Reason*, (Dordrecht: Kluwer Academic Publishers, 1988), Foreground, Part I.
36 Hans-Georg Gadamer, *Warheit und Methode*, Romanian trans., *Adevăr şi Metodă* (Bucharest: Editura Teora), p. 194.
37 Tymieniecka, "Nature in the Ontopoiesis of Life: from the Cosmic Dissemination to the Human Cultivation of the Logos," in *Analecta Husserliana*, Vol. XLVII, 1995, p. 23.
38 Tymieniecka, *Logos and Life*, Book 4: *Impetus and Equipoise in the Life-Strategies of Reason, ed. cit.*, p. 352.
39 *Ibid.*, p. 102.

[40] Tymieniecka, *Logos and Life*, Book 1: *Creative Experience and the Critique of Reason, ed. cit.*, p. 4.

[41] Tymieniecka, *Phenomenology of Life and the New Critique of Reason: From Husserl's Philosophy to the Phenomenology of Life and of the Human Condition*, in *Analecta Husserliana*, Vol. XXIX (Dordrecht: Kluwer Academic Publishers, 1990), pp. 16; 15.

[42] Cf. Arthur Schopenhauer, *Le fondement de la morale* (Paris: Librairie Félix Alcan, 1925).

[43] Cf. Albert Schweitzer, *Philosophy of Civilization*, Part II, *Civilization and Ethics* (New York: Holt, Rinehart and Winston, 1949).

[44] Cf. Hans Jonas, *The Imperative of Responsibility: In Search of an Ethics for the Technological Age* (Chicago and London: Chicago University Press, 1984).

[45] Böhme, *op. cit.*, p. 96.

[46] Leszek S. Pyra, *Environment and Values. Holmes Rolston III's Environmental Philosophy* (Kraków: Wydawnictwo Akademii Rolniczej, 2003), p. 15.

FRANCESCO TOTARO

ONTOLOGICAL INTENTIONALITY AND MORAL CONSCIOUSNESS IN HUMAN EXPERIENCE

Intentionality is a founding aspect of phenomenology. It is our opinion that it is also an important structure of human experience and of its central focus: the mind. Indeed, we consider properly human the experience that is situated in the area of reflection, and that has above all a spatial and time dimension. Human experience, the experience of each person, is set in a place which expands back and forth, up and down; moreover, it is inherent in a present which extends to past memories and expects or anticipates the future.

But, besides this capacity of being in a place and time, human intentionality has the capacity of being free from place and time, and not only with respect to a determined place and time, but with respect to any place and any time. Human intentionality always adheres to a situation while transcending it. Moreover – as we know – *omnis determinatio est negatio* and so being conscious in a determined situation is already, in some way, the overcoming of the same situation. Michel Foucault, in the work *Les mots et les choses*, used to define the humane as an *allotrope empirique-transcendental* whose ambiguous or uncertain footprints are left on a strip of sand ready to be washed away. Indeed, human experience cannot grant an unlimited length, but, as long as it lasts, it is characterized by a capacity to question any own determined condition and, so, even itself as an experience. We could say that human experience is a paradoxical experience in so far as it is an experience that cannot rest on itself, and, therefore, it is an unquiet experience, an always in-experienced experience. Human experience is always human in-experience and, therefore, in an attitude of *expulsion* from itself.

We can positively articulate this concept, saying that human intentionality is the intentionality of the whole, we could say an "entiral" intentionality. As such it is always oriented towards determined poles while being at the same time disoriented with respect to any specific pole. Husserl taught us and showed us, rightly, that intentional polarity is indefinite either on the ego side (*Ich-Pol*) or on object side (*Objekt-Pol*) but rather consists of an aiming

A-T. Tymieniecka (ed.), Analecta Husserliana XCIII, 245–252.

to a convergence of subject and object; a convergence always looked for but never fully attained.

In the words taken from classical philosophy, we can rephrase this idea saying that human experience is characterized by the capacity to give an intention to or to intentionate the being either in the specific or determined conditions in which the being finds itself, or beyond any specific condition. In Aristotle's words, our mind is able to make the being as the being an object of knowledge, and, therefore, the being that is not bound to particular conditions, that is, the unconditioned being. In this way Aristotle is linked to Parmenides' great insight that the being cannot but be, a *sui generis* apparition of the being because its evidence derives from the argumentation stating as unjustified the claim of not being to give satisfaction to intentionality of the thought. The truth supported by thought is in the being: "to think, and to think that it is, is the same thing." The being is adequate to the thought intentionality while it is evident that the not being's thought becomes meaningful only in view of the affirmation of the being.

There is a clear resonance of this in Leibniz's question: "Why is there something rather than nothing?," successively reframed by Heidegger in a more universal version: "Why is there generally the being and not rather nothingness?" Indeed, the introduction here of nothingness does not have an absolute value, but it is due to the exigency of justifying the prime or original evidence of something appearing or, in general, of the being's appearance. The opening to nothingness is, as it were, ruled by the appearing of something positive, about which further reasons are asked respect to the simple appearing.

Ontological intentionality, in our experience, is, however, marked by the consciousness that the being's entireness does not appear in our actual existence. We experience the *idea* of the entire, not only in philosophy, but also in poetry and in the forms of symbolical language in which the partiality is taken as sign of the entireness. But we always turn to the entire in fact, by means of the part. The ways through which we access the entire are always partial. Our experience of the entire is then *lacking* or defective. The entire appears, and at the same time does not appear, to us. Therefore, to *us* it gives itself in a contradiction. More precisely, we can say that, on one side, the entire "produces" to us/for us the contradiction of its lack and, on the other, it excites our impulse to overcome this contradiction.

How can we possibly overcome this contradiction? Living the contradiction itself as a tension to the entire. Experience is, then, a tension towards the entire that we lack: the lack reverses itself in showing a positive goal. Where does this overcoming take place? One of the privileged fields of this overcoming is praxis or action that we can intend as a drive towards the entire or as a drive towards what we are not yet. The main concepts in which the human mind declines action are exactly tensional concepts. Thanks to them, the perception of what exists is used for a project so that in the proper human experience perception is always *projection*, even in a situation of conflict with respect to the data perceived. The human person is not only the animal that is able to abstract from perception data, but also the person who is able to integrate his or her perspective to the point of reversing them.

Action arises, then, from the desire to fill a gap and it turns to goals and good things as poles of its fulfilment. We do not propose a negative foundation of action, because the same intentionality of the entire acts, as it were, behind any situation of lack that the acting aims to fill. An acting having a negative principle as its ultimate founding would give rise to acts of resentment, that is, a type of acting Nietzsche taught us to consider purely "reactive," prone to contempt for all that pertains to the superior values of human existence, which, in Nietzsche's words, are concentrated in the capacity of "saying yes" to life even when it is marked by pain and sufferance. We must think, then, that the negative factor of action, which can be also its immediate drive, is actually the way in which we feel provoked by what is positive. Also in this case, the negative element acquires meaning and derives dignity from the fact of being an element of mediation of the positive principle. Our acting is ultimately driven by a need for fullness that is reflected in the fact that in our experience we are not only pushed to satisfy our objects of need, but we are also attracted by objects of desire; what is more, needs and desires intertwine to the point of fusion.

Just because acting is the aspiration of the part to the entire, it carries the assumptions of "the moral question." The tension to fullness, indeed, has got in itself the risks of arrogance and abuse of power. In the relationship between the part and the entire, the part that relates to the entire can bend the entire to the part's logic instead of declining partiality according to its possible fulfilment in the orientation to the entire. Human experience, in a recurrent and pathological way, is the experience of the terrible capacity of distortion of the correct relationship between the part and the entire. To understand this, we do not need to recall historical figures exceptionally brutal in subordinating the destinies of millions of people to their own evil passions. In our daily actions, we ourselves incessantly experiment the tendency to make the entire surrounding for us the tool of the part that we always are.

That is probably the core of the radical inclination to evil Kant dealt with. It contains the seeds of the perversion of the will in so far as it is the will to domination. We should not avoid to notice that this radical inclination to evil has its roots, paradoxically, just in the aspiration to an acting that wants fullness to itself. In the concrete movement of human acting, we repeatedly witness the unfathomable and *monstrously* original impulse to turn the desire of fullness into a lust for omnipotence centred on an abnormal self. It is, in short, in the faculty of the human will to twist that fullness, altough it is essential for the same will.

Paraphrasing Nietzsche's words, we could say that the moral question, in regard to acting, rises in relation to the question: "Which will to fullness? How to achieve fullness without twisting it?" This question implicitly expresses a search for a measure in acting, the measure without which the will of fullness would turn into devastating outcomes. Moral reflection arises when the human mind asks itself about the limit in the use of energies involved in actions and, then, in the use of balance or the correct combination of their various expressions.

If we think about the different moral reflections in the history of human thought, we easily notice that they are characterized by the search, and the direction, for the most adequate ways to give human acting a fullness that would take into consideration the *measure* with which to attain it.

In Socrates' "know yourself" emerges the topic of practical wisdom as the consciousness of the measure combined with the consciousness of not having a knowledge adequate to the entire and, therefore, of not being able to claim its ownership. In "knowing of not knowing" the limit is introduced not as a goal in itself or as a renunciation to the intentionality of the entire, but it springs just from the fact that the human mind relates to the entire acknowledging its partialness respect to it, laying aside its arrogance or *hybris*.

In Aristotle's *Nichomachean Ethics* the argument about the measure in acting results inseparably either in showing the golden mean as a virtuous disposition between defect and excess or in showing the balanced link between desire and intellect. Intellect is called to "reckon" the right means to make desire grow in a correct way, protecting it from renunciation and foolish ambition; desire gives intellect the material components to exercise its capacity of guide and knowledge. Moral consciousness positively defines the good measure of desire while dynamically discriminating between practicable and impracticable desires, not simply, however, according to their compatibility with the means. The moral use of consciousness implies the exclusion from the type of desires worth of being achieved by intellect of those desires pushing to "base" actions (such as, for instance, betrayal). Those desires would damage the person who receives them and dishonour the person who performs them, and, therefore, they drive the person away from human fullness. Intellect forbids to itself to get engaged with the search for the means to achieve worthless actions in themselves that cannot become noble for their probable technical success. The means are adequate to the goal if this is good in itself, and so it leads smoothly to the conditions favouring a state of happy life that crosses the border of the almost divine activity of contemplation. Here human experience gathers in "the mind's life," and, then, it becomes in some way life of the entire, the closest approximation of the being that cannot but be and gives meaning even to becoming things.

Also Kant's ethical works aim to define a measure that allows to discriminate between the actions that lead to the being's fullness and those that drive away from it. As reason is considered the connotation proper to the human, we need to rise human acting to the level of this connotation. Human acting must be that of a rational entity (*ens rationis*). How is it possible? It is possible if the action allows itself to be guided by a universal rule. This universal rule prevents acting from becoming a slave to particular inclinations. If reason is to express the entireness of the human as belonging to an intelligible domain in which we can achieve freedom from place and time conditions and from causality relationships, then through a rational action we can avoid the reduction of the human to something merely partial. We can avoid, indeed, that each person, taken in herself, thinks of herself as a mere tool of one of her own parts and that, besides, considers the others only in their instrumental function, that is, seeing them as part and denying their "entiral" dignity.

If we do not keep in mind Kant's lesson about anthropological fullness to be realized in the respect of a universal law, valid for the self as well as for the others, it is difficult to understand Hegel's analyses (in the *Phänomenologie des Geistes*) about consciousness and the relationships among consciousnesses. The struggle between the master's consciousness and the servant's one is about the equal or unequal recognition of anthropological fullness. Hegel urges the question: "Who has the right to be fully human? Are consciousnesses destined to remain in an unsolvable antagonism or is it possible that in 'me-you' relation we can recognize a 'we' that unites me and you?"

The moral question becomes radical at the political and social levels when emerges the blunt consideration that the fullness of the human can be fed with the capacity of creating submission, resting on the asymmetry between the master and the servant. What about the project, then, to create a symmetrical situation of mutual and equal recognition among consciousnesses? The universal vision of moral consciousness is called to come to terms with the harshness of the historical process and with the splits that plague it. As long as these splits or breaks last, it is even legitimate to doubt about the actual validity of the moral dimension or about the credibility of its claim of a universal value. We start "suspecting" about the abstract moral proclamations that hide the reality of class conflicts or the gap between "strong" men and "weak" men (it is superfluous to remember here that Nietzsche was worried about the injustice suffered by the former, whereas we are interested in the laceration he puts in evidence). The moral speech is interpreted as an expression of one part and as a justification of the human kind marked by breaks.

In this context there is a challenge for any consciousness' claims to be a moral one. If, indeed, the same moral consciousness is trapped in a partial vision, how can it carry on the task to be a measure of the human acting oriented to its fullness? Moral elaboration seems to represent the pathology of human acting rather than its virtuous expression. If it is a false universality, it deserves to be solved either through a critical analysis of the historical relationships of power and subordination or through the denunciation of the instincts' repression. In this opinion, the measure of the validity of human acting, detached from moral consciousness, resides in the force of both social and political emancipation and the liberation of psychic inhibitions.

Free from the ties with moral consciousness, the social and political route and the psychological one to human fullness meet risky drifts with regard to the same liberation effects they propose. The social and political emancipation falls back in the defects of the logic of power it wanted to overcome, and, in its turn, psychological easiness results in a strategic instrumental ability to any use. If we do not want to reduce politics to a technique of obtuse power and psychological autonomy to a technique of formal disinhibition, with disappointing outcomes with respect to global anthropological expectations, we need to listen again to the ambitious voice of moral consciousness and to the imperatives it suggests in view of the full realization of the human.

More in general, in the global context in which the evolution of the historical process does not seem to express an automatic direction of progress and, even more, the prevalence of economy restates the quantitative-type model of growth, instead of a qualitative development, moral consciousness is back on stage as a possible lighthouse that sheds a critical light on the given world to integrate with indications of sense the behaviours in the multifarious domains of human acting. The expressions of politics, economy, and science then stop being self-referential; but they contribute, in their partialness, to the anthropological entireness sustained by moral reflection.

Besides, in a period when the meeting of various cultural traditions continuously threatens to turn into conflict and intolerant denial, we are in need, much more than in the past, of the right universalism of moral consciousness that, far from being flattened on particular identities claiming unduly an exclusive absolutism, can have the function of a referee in the mutual attribution of dignity.

Moral reflection is today ready for this task either emphasizing a model of communicative ethics which favours an agreement among different points of view accepting without biases the most valid argumentation, or preparing the opening to the *other* as the first and irrevocable principle of action. Moral reflection is generally engaged to build the cultural bases of a cosmopolitan society in which all positions are equally stated.

This search for a common measure is certainly decisive for the destinies of humanity. But, about it, we should not underestimate the fact that the human has always the will to fullness, beside having the will to measure. And then: what can we do so that different fullness-perspectives can express themselves without destroying one another and without renouncing "entiral" intentionality? This is the crucial knot of the moral, and cultural, question of our time. We could concentrate our answer in a slogan: Neither absolutism, nor relativism, but let us look for an absolute meaning *through* or *thanks to* relation. On what basis? My proposal tends to value the contributions Western ontological tradition can give to the question. It is about our representation on a route towards a fullness of life nobody already has, and everybody, without exclusion, is called to achieve. The being's entire, which does not appear yet, is committed to everybody's capacity of manifestation, and, therefore, to the possible convergence of perspectives. We believe, thus, that a moral consciousness that respects the requirements of anthropological fullness does not demand, as is often held (for example by Levinas and his scholars), ontological censorship. On the contrary, it can take advantage of its good declension. The common reference to the being is, indeed, the "we" that *keeps* the "me" and "you" *together*.

In other words, we need to recognize that human experience is driven by an unconditioned search for a meaning or by an "entiral" intentionality that must be correctly interpreted, if we do not want to fall into abuse of power. The 'entiral' consciousness must, then, be associated with the "moral" consciousness of limit and measure. Limit and measure are not, however, a renunciation to the entire, but the route through which it can be attained in our experience.

In conclusion, we can say that the validity and the completeness of moral consciousness depend on the capacity of sharing the being's richness that transcends particular perspectives while pushing them to convergence. Ontological intentionality and moral consciousness, intertwined in solidarity, allow the flourishing of human experience. The former gives an idea of fullness without which existence would be void of deep motivations. The latter introduces an idea of measure without which the fullness of meaning would be blinding and devastating. Combining together fullness and measure seem, then, the specific task of the "human mind."

University of Macerata

ANDREAS BRENNER

GIBT ES EINE ETHIK DER LEBENSWELT?

1. *LEBENSWELT* IM ALLTAG

Die Rede über den Begriff der Lebenswelt weist mehrere Schwierigkeiten auf: Zum einen hat der philosophische Begriff der Lebenswelt sowohl in verschiedenen wissenschaftlichen Disziplinen, wie auch längst in der Allgemeinsprache Karriere gemacht. Im Zuge dieser zunehmenden Verbreitung des Begriffs hat dieser das fast unvermeidliche Schicksal enormer inhaltlicher Abschleifung erlitten. Entsprechend wird der Begriff mittlerweile in sehr wörtlichem Sinne verwendet, etwa dergestalt, mit ihm die Welt, in der wir *eigentlich* leben, zu markieren. Wobei die Betonung auf der Eigentlichkeit meist einen (moralisch) normativen Impetus hat und auszudrücken versucht, wir *sollten* in der besagten Welt leben, tun es jedoch in der Regel nicht und leben mithin in einer uneigentlichen, oder falschen Welt.

Dieses Kritikhaltung ist deshalb sehr populär, weil sie eine verbreitete, aber nur schwer fassbare, weil diffuse Unzufriedenheitshaltung gegenüber der herrschenden wissenschaftlichen und ökonomischen Kultur auf einen Begriff zu bringen verspricht: Ängste über unsere eigene wirtschaftliche Wohlfahrt, Verunsicherungen über unsere gesundheitliche Versorgung im Alter und eine zunehmende Irritation über unseren eigenen körperlich-leiblichen Status als Menschen finden demnach in dem Begriff der Lebenswelt eine Orientierung, deren Richtung zugleich als *Trost* erfahren wird. Und insofern ist die Verwendung des Begriffs auch durchaus legitim. Es stellt sich indes die Frage, ob ein solcher Lebenswelt-Begriff noch einen Bezug zu Husserl hat. Denn ein solcher Bezug wird, sofern eine Quelle überhaupt bekannt ist, in der Regel unterstellt. Dabei ist aber Husserl gar nicht der Schöpfer dieses Begriffs, in Wahrheit taucht der nämlich bereits 1912 auf, und zwar in der *Religion* von Georg *Simmel*.

A-T. Tymieniecka (ed.), Analecta Husserliana XCIII, 253–267.

2. LEBENSWELT BEI HUSSERL

2.1. Lebenswelt, allgemein

Wo aber soll man sich kundig machen, um den Lebenswelt-Begriff bei Husserl zu erschließen? Hier bietet sich vor allem natürlich die *Krisis-Schrift* an. Dieser letzten von Husserl selbst veröffentlichten Schrift kommt gleichsam der Charakter eines Vermächtnisses zu. Zu dieser Einschätzung trägt auch nicht zuletzt der Ton bei, den Husserl in dieser Schrift anschlägt und der stellenweise den Charakter einer *Heilsgeschichte* annimmt. So etwa in seiner Rede von den Philosophen als den "*Funktionären der Menschheit*," denen er die "*Verantwortung für das wahre Sein der Menschheit*" zuschreibt und es zugleich nicht für unmöglich hält, dass sie – die Philosophen – der Menschheit "*Sein auf ein Telos*" hin zu entwickeln helfen.[1] Sowohl der Ton, mehr aber noch die radikale Kritik an einem konkreten, fast möchte man sagen, "lebensweltlichen" Phänomen, nämlich der modernen Wissenschaftskultur, haben bei manchen Rezensenten den Eindruck erweckt, mit der *Krisis* habe Husserl eine neue Position bezogen, es sei geradezu von einer *Wende* zu sprechen erlaubt. Stephan Strasser stellt in diesem Zusammenhang indes unmissverständlich klar, dass Husserls Lebenswelt-Position "keineswegs einer Abwendung von seinem transzendentalen Idealismus gleichkommt."[2] Ob dies, wie von der Mehrheit der Husserl-Exegeten behauptet, in der Tat *nicht* der Fall ist, soll zunächst einmal offen bleiben.

Dennoch ist der Eindruck, dass mit der Lebenswelt-Problematik etwas Neues ins Werk Husserls trete, ja nicht von der Hand zu weisen. Das, was dort neu ist, ist vor allem die Einbeziehung von *Tradition, Geschichte* und *Kultur*. Dieser Komplex steht gleich am Beginn seines Werkes, dort nämlich, wo er von der "revolutionären Umwendung" des Menschen in der *Renaissance* spricht. Mit dem Stichwort der "Renaissance" sind neben der Kultur und Historie des weiteren die zentralen Themen *Wissenschaft* (als Naturwissenschaft) und, und dies trägt auch den Keim einer Irritation in sich, *Europa*, genannt.

Um nicht vorschnell in die bereits angedeutete Interpretation eines ganz neuen Husserl einzutreten, soll zunächst einmal versucht werden, Husserls Lebenswelt-Begriff *immanent* zu rekonstruieren.

2.2. Die Rekonstruktion von Husserls Lebenswelt

2.2.1. Überwindung der "Objektivität" Zunächst einmal entwickelt Husserl ein erstes Verständnis der "Lebenswelt" aus einer Außensicht: Mit dem merkwürdigen Begriff *Ideenkleid* bezeichnet Husserl *die* Naturwissenschaft(en), welche eben die Wirkung einer "Verkleidung" entfalten. Das

Ideenkleid hüllt ein und verkleidet und vermittelt damit den Kontrast zu dem eigentlichen, wahren der Lebenswelt, das gleichsam in diesem Kleid steckt und zur wahren Würdigung nur von diesem bloßgelegt werden muß. Man erkennt sofort die Nähe dieses Begriffs zu dem populären, umgangssprachlichen Verständnis und somit auch dessen Berechtigung. Indes fragt sich, ob der Lebenswelt-Begriff im Husserlschen Sinne damit in ausreichendem Maße grundgelegt ist. Und genau darauf beziehen sich ja die Einwände, die Husserl dann als unzureichend begriffen kritisieren, wenn man ihn als (bloßen) Kritiker des *wissenschaftlichen Objektivismus* betrachtet. Dieser Deutung zufolge wäre Lebenswelt also die von der Wissenschaft nicht überlagerte Welt, eine sozusagen naive, ursprüngliche Welt. Dies ist die Welt, die Gegenstand und Projektionsfläche einer weitverbreiteten Wissenschafts- und Technologiekritik ist. Ungeachtet der Berechtigung dieser Kritik und ihres Bezuges auf eine gleichsam paradiesische, vom Rationalismus unberührte, Welt bleibt festzustellen, dass diese Welt, vielleicht mit guten Gründen *Lebenswelt* genannt werden kann, dies jedoch nicht unter Bezug auf Husserl. Denn sein Werk steht sicher nicht in der Gefahr, als "Bilderbuchphänomenologie" verunglimpft zu werden. *Bilderbuchphänomenologie* ist die Bezeichnung, mit der Helmuth *Plessner* die Phänomenologen *Pfänder* und *Scheler* in Verbindung bringt, wenngleich unklar bleibt, ob er, Plessner, den diffamierenden Charakter dieses Ausdrucks teilt.[3] Eine Lebenswelt dieser Bestimmung ist unter Bezugnahme auf Husserl nicht zu haben, sie stellt sogar eine völlige Verkehrung seiner Position dar. Dies ist zumindest dann der Fall, wenn man, was wir einstweilen tun wollen, die *Krisis*-Schrift nicht als solitäres Werk eines radikalen Kritikers der abendländischen Rationalität, aber auch nicht als Markstein einer radikalen Neuorientierung des Begründers der Phänomenologie versteht, sondern sie statt dessen als integralen Bestandteil seines Gesamtwerkes ansieht.

Warum? Nach Husserl widerspiegelt ein solches Verständnis lediglich die "natürliche Einstellung" des Bewusstseins. Auch hier sei nochmals eine Widerlegung der beschriebenen populären Kritik gegeben: Die Metapher von der "Natürlichkeit" scheint ja zunächst die in der Wissenschaftskritik sich artikulierende "Zurück-zu"-Sehnsucht ("Zurück zur Natur," "Zurück zu einem selbstbestimmten Leben") zu bestätigen. Aber auch hier gilt: Aus Husserls Perspektive verbleibt eine solche Kritik an der Oberfläche einer "naiven" Einstellung. *Naiv* ist diese Einstellung deshalb, weil man die gegebene Erfahrung als solche gelten lässt. Diese Erfahrung nennt Husserl auch die "*mundane Erfahrung*."[4] In der mundanen Erfahrung verbleibt man selbst dann, wenn man, wie es den "reifen und wachen Menschen"[5] ausmacht, diese Erfahrung reflektiert. Eine solche Reflexion hebt einen noch nicht heraus aus der Naivität der Erfahrung: Wir bleiben gleichsam Gefangene der Mundanität. Es gibt also hinreichende Gründe, die natürliche Einstellung außer Kraft zu setzen. Wie aber kann dies geschehen? Husserl bietet dazu die *Methode der phänomenologischen Reduktion* an mit der, seiner Auffassung nach, das *transzendentale Subjekt* zu sein anhebt. Damit sind die Grenzen der Lebenswelt markiert: Lebenswelt ist jenseits der naiven Weltgegebenheit angesiedelt und übersteigt die natürliche Einstellung des empirischen Subjekts hin zum *transzendentalen Subjekt.* Der Schritt, zu dem Husserl hier ausholt, ist also ausgesprochen ambitiös und entsprechend riskant: Erst in der Überwindung der natürlichen Einstellung erlangt das sich so konstituierende transzendentale Subjekt zu ernsthaften Erkenntnissen.[6] Das bedeutet: erst in der Lebenswelt können wir (Menschen) überhaupt sinnvolle Aussagen über uns und über die Welt machen.

Bevor wir in diesem Sinne fortfahren die Lebenswelt auszuloten, könnte man zu fragen geneigt sein, was das für eine Welt ist, die uns lebensweltmäßig begegnet. Diese Frage erweist sich indes insofern als naiv, als sie einen *Objektivismus* der Welt unabhängig ihrer transzendentalen Erkenntnis annimmt. Statt dessen erweist sich die Welt als konstituiert durch das *transzendentale Subjekt*, und das bedeutet, dass "Welt" nur als eine Leistung des menschlichen Bewusstseins angemessen verstanden werden kann.[7] Zum Wortbegriff "transzendental": Transzendentalphilosophie "ist eine Philosophie, die gegenüber dem vorwissenschaftlichen und auch wissenschaftlichen Objektivismus auf die erkennende Subjektivität als Urstätte aller objektiven Sinnbildungen und Seinsgeltungen zurückgeht und es unternimmt, die seiende Welt als Sinn- und Geltungsgebilde zu verstehen und auf diese Art eine wesentlich neue Art der Wissenschaftlichkeit und der Philosophie auf die Bahn zu bringen."[8] Das transzendentale Subjekt verhält sich gegenüber der Welt nicht naiv, sondern *methodisch.* Die einzig nichtnaive Methode ist die der *Epoché.*[9] Was aber hat Epoché mit Lebenswelt zu tun? Die Epoché ist die Methode, die von einer naiven Weltvorstellung zu derjenigen von Husserl "Lebenswelt" genannten führt. Erst im Durchgang durch die Epoché in dem sich das Subjekt zum transzendentalen konstituiert, konstituiert es auch eine transzendentale Welt (= Lebenswelt), die sich von der naiven dadurch unterscheidet, dass "jedes Interesse an Wirklichkeit oder Unwirklichkeit der Welt (...) außer Spiel gesetzt ist."[10] Damit gilt, dass der Phänomenologe in der Paradoxie lebt, "das Selbstverständliche als fraglich, als rätselhaft ansehen zu müssen und hinfort kein anderes wissenschaftliches Thema haben zu können als dieses: die universale Selbstverständlichkeit des Seins der Welt – für ihn das größte aller Rätsel – in eine Verständlichkeit zu verwandeln." Die Auflösung dieser Paradoxie erfolgt in der Epoché.

Mit diesem Beginn der Reduktion wird etwas in seinen Ursprung zurückgeführt. Angezielter Ursprung ist dabei die transzendentale Subjektivität.[11] Damit geht es also gar nicht um die "Welt," aber auch nicht um die "Lebenswelt," sondern allein um die Subjektivität. Und erst diese gestaltet, konstituiert das Verhältnis zur Welt und, allem anderen voran, das Verhältnis zur Lebenswelt. Weder die Welt noch die Lebenswelt stehen mithin am Beginn des philosophischen Unternehmens, sie bilden auch nicht das Hauptinteresse dieser philosophisch-phänomenologischen Suche. Das Hauptinteresse der philosophisch-phänomenologischen Suche gilt alleine dem Bemühen einer Etablierung des transzendentalen Subjekts. Der Akt der Etablierung des Subjekts ist dabei zugleich ein Befreiungsschlag: Mit der Epoché wird "der Blick des Philosophen erst völlig frei (…) und vor allem frei von der stärksten und universalsten und dabei verborgensten inneren Bindung, von derjenigen der Vorgegebenheit der Welt."[12]

Transzendentales Subjekt und Welt stehen mithin in einer Beziehung zueinander: Die "Welt," das ist dasjenige, von dem sich das Subjekt frei machen muss, um zu dem zu werden, was es ist. Damit ist es mit einem bloßen Ignorieren der Welt nicht getan, denn ignoriert *ist* sie gleichwohl noch, auch diese Haltung wäre mithin naiv: Um was es geht, das ist ein *Überwinden* der Welt. Diese Überwindung der Welt erlangt der Phänomenologe dadurch, dass er "über sein natürliches Sein und über die natürliche Welt gestellt, nichts von ihrem Sein und ihren objektiven Wahrheiten verliert."[13] Auch hier wird die Bedeutung von "Welt" und ihrer "objektiven" Geltung hervorgehoben. Es wäre also nicht nur naiv, die Welt in ihrem Sein leugnen zu wollen, es wäre ebenso naiv, die Objektivität der Welt in Frage zu stellen. Es geht vielmehr darum, die "Naivität der Rede von 'Objektivität' " zu erkennen.[14] Worin die Naivität besteht, das haben wir bereits erfahren: Sie besteht in dem ungeprüften Fürwahrhalten eines Blicks, dessen Voraussetzungen bereits als letztgültig angenommen werden. Interessant ist nun aber zu sehen, wie einem diese Naivität und mit ihr die Unmöglichkeit, weiterhin in ihr zu verharren, aufscheint. Die Unhaltbarkeit dieser Naivität wird sichtbar, "sowie das Leben in den Blickpunkt rückt."[15] Was aber bedeutet "Leben?" Worin besteht der, dem Wortsinn nach sich zunächst ja aufdrängende, Zusammenhang von "Leben" und "Lebenswelt?"

Bevor wir uns dieser Frage zuwenden, kann festgehalten werden, dass die Überwindung der "Objektivität" keineswegs bedeutet, dass es diese nicht gibt. Allerdings gilt, dass die Welt in ihrer naiven Weise insofern nicht *ist*, als sie erst durch die transzendentale Wende verständlich wird. Gleichwohl bleibt das Faktum der Verwiesenheit auf die "Welt," denn diese stellt den "*Boden*" dar, auf dem alle Auffassungen über die Welt aufruhen und dessen man sich durch die Epoché enthebt.[16]

Das Verhältnis von "Welt" und "Lebenswelt" nennt Husserl "paradox": Der Welt kommt, und das treibt die Paradoxie auf die Spitze, eine Grundlagenfunktion ("Boden") für die Konstitution der Lebenswelt und vorgängig für uns selbst zu. Dieser Zusammenhang und die Bedeutung des "Bodens" der normalen Welt ist, wie Husserl klagt, von der Philosophie bislang nicht in ausreichendem Maße gewürdigt worden. Daher spricht Husserl von der "Bodenlosigkeit unseres ganzen bisherigen Philosophierens."[17]

Von einer Geringschätzung der naiv vorgegebenen Welt kann bei Husserl also keine Rede sein, im Gegenteil ist ihre Würdigung in der mit ihr einhergehenden Philosophiekritik unübersehbar. Der *Boden*, den die naive vorgegebene Welt darstellt, muß indes, das ist ja die Lehre des phänomenologischen Ansatzes, verlassen, also gleichsam überstiegen werden. Als Resultat der, von dem naiv vorgegebenen Boden der Welt sich abhebenden, Epoché gelangen wir zur Konstitution der Lebenswelt, die gleichsam den Ur-Boden unseres Welt- und Ich-Zugangs darstellt. Zu diesem Vorgang gibt es keine Alternative, will man sich vor dem Untergang in der Naivität retten. Diese Rettung ist gleichwohl nicht gratis zu haben und geht einher mit "*Einsamkeit*."[18] Diese Einsamkeit resultiert daher, dass in der Epoché *alles* zum Phänomen wird. Die Verwandlung, welche dermaßen geschieht, ist so radikal, dass selbst das *Ich*, dass sich in der Epoché bildet, lediglich dem Namen nach mit dem Ich vor der Epoché identisch ist. Diese Wirkung hat Husserl bereits in den *Cartesianischen Meditationen* in aller Drastik beschrieben: Alles, was mich umgibt, die anderen Menschen, die Tiere, die Natur, das ist nun "für mich statt seiend nur Seinsphänomen."[19] Und darin zeigt sich die Einsamkeit, in den einen die Reduktion stürzt. Gleichwohl ist es nicht das Nichts, das auf solche Weise übrigbleibt, denn weder das Universum der Phänomene noch das transzendentale Subjekt sind nichts.[20]

2.2.2. "Leben" und "Lebenswelt" Die Unhaltbarkeit der naiven Einstellung wird sichtbar, "sowie das Leben in den Blickpunkt rückt," hat Husserl behauptet.[21] Was ist das, "Leben?" Können wir uns denn immerhin bei diesem Wort mit einem uns vertrauten Verständnis zufrieden geben? Also, entweder dem vorwissenschaftlichen, das "Leben" beispielsweise in einer Gegenüberstellung zu "Nicht-Leben," also dem "Tod" begreift, oder wäre es besser, sich auf die Aussagen der Wissenschaften zu verlassen, naheliegender Weise also der *life sciences*, die für diese Frage ja in besonderer Weise kompetent zu sein scheinen?

Es leuchtet unmittelbar ein, dass ein solches Verständnis seinerseits naiv wäre, da die Naivität der natürlichen Einstellung ja nicht dadurch überwunden werden kann, dass man sich in eine andere Naivität flüchtet. Damit scheiden also die naheliegenden Antworten auf die Frage nach dem "Leben" aus.

Zugleich zeigt sich, dass diese Antworten gleichwohl nicht so weit von dem Husserlschen Lebens-Verständnis entfernt sind. Den Abstand kann man sogar recht genau angeben, er ist identisch mit dem zwischen "Welt" und "Lebenswelt." Denn auch zwischen "Leben" im Alltags- bzw. Wissenschaftsverständnis und "Leben" im phänomenologischen Sinn klafft genau die *Lücke* zwischen naivem und transzendentalem Bewusstsein und zugleich bildet das vortranszendentale Bewusstsein den "Boden" für das es überwindende Bewusstsein. Das transzendentale, aus dem Vollzug der Epoché hervorgegangene Bewusstsein hat auch noch eine weitere markante Parallele zum "naiven" Leben: Es konstituiert sich aus "Erlebnissen," die in Folge des Einklammerungsprozesses *reine Erlebnisse* werden, weswegen Husserl das Leben in phänomenologischer Hinsicht auch *reines Leben* nennt.[22] Das "Leben," das hier gemeint ist, ist kein psychologisches, sondern ein transzendentales Leben, es hat sich vollkommen über die Naivität bloßer Vorgegebenheit erhoben.[23]

Wie steht nun das *transzendentale Leben* in Bezug zur Lebenswelt? Die subjekt-relative Verfasstheit der Lebenswelt prägt natürlich auch das Leben und diese, seine Verfasstheit, könnte uns das Leben auch in dieser Hinsicht in besondere Nähe zum oben genannten vorphänomenologischen Lebens-Verständnis zeigen, denn "Leben," das ist beide Male eine Seinsweise, die erst in der radikalen Subjektivität ihren vollen Sinn entfaltet. Um diese Übereinstimmung besser beurteilen zu können, wird es nun nötig, innerhalb des vorphänomenologischen Verständnisses zu differenzieren zwischen dem sogenannten "Alltagsverständnis" und dem "Wissenschaftsverständnis." Letzteres vermittelt uns "Leben" in vom "Ideenkleid" verkleideter Weise. Der dermaßen verkleidete Begriff scheidet in seinem Anspruch auf Objektivität gerade die subjekt-relativen Bezüge aus. Das bedeutet also, dass von dort kein dem phänomenologischen Begriff adäquater Lebens-Begriff zu erwarten ist. Anders sieht es bei dem vorphänomenologischen Alltagsverständnis aus. Zwar ist auch dieses bekanntlich längst vom naturwissenschaftlichen Verständnis überlagert, gleichwohl gilt, dass dieses in wesentlich höherem Maße, als dies für das Wissenschaftsverständnis der Fall ist, subjekt-relative Züge enthält. Die subjekt-relativen Bezüge sind hier zwar stärker ausgeprägt, als im Wissenschaftsverständnis, jedoch bleibt auch das Lebens-Verständnis der Wissenschaft immer noch reduziert, ihm kommt daher eine nur eingeschränkte Wahrheit zu. Erst das transzendentale Leben erreicht die, die Subjekt-Relation kennzeichnende, Klarheit. Die gleiche Klarheit zeichnet auch die transzendentale Lebenswelt aus. Diese Gemeinsamkeit von transzendentalem Leben und Subjekt-Relation wird zusätzlich durch die Feststellung untermauert, dass die Lebenswelt "ein Reich ursprünglicher Evidenzen" sei.[24]

3. DIE KRITIK DER LEBENSWET

3.1. Die immanente Kritik

Die *Krise*, der sich die *Krisis*-Schrift verpflichtet weiß, können wir nach den vorangegangen Klärungen also wie folgt verstehen: In der gegenwärtigen Wissenschaftsgesellschaft zeigt sich, dass der Geist seiner selbst entfremdet und in den wissenschaftlichen Objektivismus "versponnnen" ist. In diesem Sinne wird auch einsichtig, dass "Lebenswelt" nicht etwas ist, das als solches beschrieben werden kann, sondern das zunächst "*zurück*gewonnen werden soll" und mittels derer der "Einstieg in die transzendentale Phänomenologie" erfolgen soll. Der sachliche Zusammenhang und die Bedeutung der Lebenswelt[25] mit der Phänomenologie und damit der Transzendentalität ist bereits in Verbindung mit der Methode der Epoché aufgeschienen. Bemerkenswert ist jedoch, dass der Lebenswelt bei Husserl immer auch ein praktischer Bezug zukommt. Denn die Epoché offenbart mit ihrer Begründung der Lebenswelt eine Ursprünglichkeit, deren Vergegenwärtigung eine dualistische und damit auch reduktionistische Aufspaltung der Welt erst gar nicht aufkommen lässt: "Man hätte nie Menschen und Tiere als psychophysische Maschinen oder gar als parallelistische Doppelmaschinen ansehen können."[26]

Diese Bedeutung der lebensweltlichen "ursprünglichen Evidenzen" macht Husserl in der "Krisis" jedoch viel zu wenig deutlich, womit er der Gesamtschrift einen unentschiedenen Charakter verleiht: zum einen drängt sich die wissenschaftskritische Deutung auf, wie wir sie zu Beginn am Beispiel der verbreiteten Naturwissenschaftsskepsis kennen gelernt haben, auf der anderen Seite fehlt dafür in der Krisis-Schrift selbst jedoch die explizite philosophische Grundlage. Des weiteren erscheint ohne diese kritische Wendung der auf das Projekt der Lebenswelt verwandte Aufwand kaum nachvollziehbar. Damit offenbart das Krisis-Unternehmen seine Schwäche vor allem in einer mangelnden Entschiedenheit, die aus einer anfänglich geleisteten scharfen Diagnose nicht genügend weitreichende Schlüsse zieht. Husserls Anliegen enthält nämlich eine explizit handlungspraktische Note, die jedoch geradezu verschwindet hinter Husserls transzendentalphänomenologischem Bemühen, aber auch hinter seiner von dieser beeinflussten Wissenschaftsanalyse. So erweist sich die Wissenschaft, die sich ja als *objektive Wissenschaft* begreift, im Grunde in ihren Möglichkeiten und

Anliegen als ausgesprochen eingeschränkt: Diese Einschränkung ist der Tatsache geschuldet, dass die Wissenschaft das Sein der Welt voraussetzt. Diese Voraussetzung teilt das wissenschaftliche Wissen dabei mit dem vorwissenschaftlichen. Allein im Anspruch unterscheidet sich ersteres von letzterem: Das wissenschaftliche Wissen will das vorwissenschaftliche nach einer im "Unendlichen" liegenden Idee ausrichten und damit überbieten. Dieser Impuls orientiert sich an der Vorstellung einer "Wahrheit an sich."[27] Problematisch ist die Wissenschaft, wie Husserl nachvollziehbar feststellt, deshalb, weil sie in ihrem Erkenntnisimplus die eigene Verwiesenheit auf die Lebenswelt nicht erkennt und sich insofern über ihre eigenen Voraussetzungen täuscht.[28] Insoweit erscheint die Husserlsche Analyse nachvollziehbar und, soweit es um die Kritik des oben benannten Reduktionismus geht, auch sinnvoll. Husserl bleibt indes nicht bei dieser Kritik stehen, sondern instrumentalisiert diese für sein Hauptanliegen eines weiteren Ausbaus der *transzendentalen Phänomenologie*, wozu vor allem der Einbezug der *Geschichte* zählt. Der Geschichte galt zwar bereits früher Husserls Aufmerksamkeit, aber erst in der auch geschichtlich bestimmten Lebenswelt etabliert sich die Geschichte zu einem zentralen Faktor: sie gewinnt Bedeutung auch für die Geschichte des Denkens. Aus dieser zieht Husserl den Schluß, die Philosophie als erste, als strenge Wissenschaft, eben als (transzendentale) Phänomenologie aufzubauen.

Dieses Anliegen soll in seiner Berechtigung keineswegs in Frage gestellt werden. Allerdings hinterlässt die *Krisis*-Schrift, wenn ihre Aufgabe in dieser Grundlegungsabsicht genommen wird, einen zwiespältigen Eindruck: konkret erscheint sie gespalten in den Teil einer Kritik am wissenschaftlichen Objektivismus auf der einen und einer Kritik an einer unzureichenden Grundlegung der Philosophie auf der anderen Seite. Beide Teile versucht Husserl in einem Werk zusammen zu denken. Dieser Zusammenhang ist jedoch weder zwingend noch in jeder Hinsicht überzeugend. Er ist weder zwingend, weil der beklagte Reduktionismus von "Menschen und Tieren (auf) psychophysische Maschinen" weder erst aus der Warte der transzendentalen Phänomenologie erkennbar und zurückzuweisen ist, und weil es auch nicht notwendigerweise das Fehlen der transzendentalphänomenologischen Perspektive ist, welche dieses Manko bewirkt, bzw. das durch eine solche Perspektive behoben wird. Genau dies behauptet Husserl indes, nicht zuletzt auch mit der Aussage, "die Lebenswelt ist ein Reich ursprünglicher Evidenzen."[29]

Träfe Husserls Einschätzung über die transzendentalphänomenologische Perspektive zu, so wären sowohl der Objektivismus der Wissenschaften, wie auch der beklagte Reduktionismus lediglich Ergebnis und Folge einer vortranszendentalen Haltung und nicht etwa eines moralischen Mankos. Der

Lebenswelt ginge damit jede normative Dimension ab. Eine solche Dimension bestimmt jedoch die *Krisis*-Schrift in ihrer Kritik am wissenschaftlichen Objektivismus.

Mit dem Hinweis auf dieses Manko, soll nicht Husserls Verdienst in Frage gestellt werden, dass er die Lebenswelt sowohl vor dem Untergang in puren Empirismus, wie auch in einen empiriefreien Rationalismus mit Erfolg zu bewahren versucht hat.[30] Worauf hier hingewiesen werden soll, ist, dass Husserl die Konzeption der Lebenswelt, in dem er sie seinem Anliegen der Beförderung der Transzendentalphänomenologie unterordnet, in ihren Möglichkeiten beschränkt. Und damit werden sowohl die Krise, wie auch die durch die ursprünglichen Evidenzen ausgezeichnete Lebenswelt in ihrer Bedeutung unterschätzt.

3.2. Die exogene Kritik

Die Überlagerung der Lebenswelt durch den wissenschaftlichen Objektivismus lässt sich, wie hier behauptet werden soll, nur problematisieren, wenn der Eigentlichkeit der Lebenswelt ein besonderer Wert zugemessen, ihre Erhaltung oder Wiedergewinnung mithin normativ gefordert und der Lebenswelt damit eine es konstituierende Bedeutung für das jeweilige Subjekt zugeschrieben wird. Damit ist im Begriff der Lebenswelt eine normative Komponente gleich mit enthalten.

Husserl vermeidet es, dieses Verständnis der Lebenswelt stark zu machen, da er andernfalls – wohl mit Recht – um die Stellung seiner transzendentalphänomenologischen Position fürchten müsste, indes kommt er, wie hier behauptet wird, um diesen Schritt nicht herum, will er Lebenswelt konsistent vertreten.

Wie wir gesehen haben, konstituiert sich die Lebenswelt erst in der Epoché, mit dem Prozeß der Epoché findet somit die Freilegung verdeckter Sinnschichten statt, was in der Umkehrung bedeutet, dass das Subjekt ohne die entsprechende Leistung verdeckt und damit letztlich "sich selbst" unzugänglich bliebe. Die "Normativität der Lebenswelt" formuliert damit zum einen den Geltungsanspruch der Lebenswelt selbst. Da diese aber ohne das Subjekt nicht ist, dieses aber, wie Husserls Epoché-Analyse deutlich macht, ohne die Lebenswelt nicht ist, so sind beide – Lebenswelt und Subjekt – aufeinander verwiesen und zwar in gleicher Weise, wie dies auch für Lebenswelt und "Leben" gilt. Damit rückt mit der Lebenswelt die "lebhaftige" Bedeutung des Subjekts in den Blick. Diesen Zusammenhang könnte man auch als den von "Biologie und Lebenswelt" bezeichnen.[31] Mit dieser begrifflichen Differenzierung scheint *zweierlei* gewonnen.

Zum einen wird der im vorgenannten Zusammenhang von "Leben und Lebenswelt" benutzte Lebens-Begriff aus der metaphorischen Redeweise herausgeführt und einer empirischen Bestimmbarkeit zugeführt. Damit wird *zum anderen* die in der *Krisis*-Schrift angelegte, aber, wie wir gesehen haben, nur unzureichend unterfütterte Naturwissenschaftskritik gestärkt.

Ein solches Unternehmen ist indes nicht unproblematisch, was sich nicht nur an dem komplexen Lebenswelt-Begriff, sondern ebenso am Begriff des Lebens im Sinne von "Biologie" feststellen lässt. Denn die Frage nach dem Leben hat nicht nur die abendländische Kultur von ihrem Beginn an bestimmt, sie ist auch seither nie eindeutig beantwortet worden. Leben erweist sich mithin als ein kulturell geprägter Begriff, wie andere auch. Diese Tatsache gilt auch dann, wenn man mit Husserl dem *europäischen* Denken ein "mit der Geburt der griechischen Philosophie eingeborenes Telos" unterstellt, das letztlich in der Globalisierung den Sieg davon trägt, einen Vorgang, den Husserl als *das Schauspiel der Europäisierung aller fremden Menschheiten* bezeichnet.[32]

Will man dennoch den Begriff des Lebens im Sinne einer "Lehre vom Leben" (= "Biologie") nutzbar machen, dann wird es nicht genügen, dem Lebensbegriff mit dem Verfahren einer Epoché auf den Grund zu kommen. Vielmehr wird das "Leben" jenseits aller kulturellen "Sonderwelten" selbst in den Blick zu nehmen sein. Treten wir mit dieser Erkenntnisabsicht nun in einen *circulus vitiosus*? Dies wäre nur dann der Fall, wenn der Lebens-Begriff alleine im Kontext des Bewusstseins und nicht einer zusätzlichen Dimension situiert würde. Diese Dimension kommt aber dort fast unausweichlich mit in den Blick, wo von Leben die Rede ist. Denn "Leben" ist ein Begriff, der theoretisch *und* praktisch fungieren kann. Die theoretische Funktion haben wir bereits kennen gelernt: sie besteht in der Ordnungsfunktion, welche mit diesem Begriff übernommen werden kann. Die *praktische* Funktion dieses Begriffs offenbart sich in seiner Erste-Person-Verwendung: Wenn der Begriff "Leben" in Bezug auf die erste Person Verwendung findet, dann schält sich zusätzlich zu seiner theoretischen, auch eine *praktische* Verwendung heraus, dies insofern als der Lebens-Begriff dann mit einem Sollgeltungsanspruch unterlegt wird.

Wer den Term "Leben" auf sich bezieht, formuliert damit immer auch den Anspruch, dass die Voraussetzungen und Bedingungen, welche den Begriff in der gemachten Weise verwendbar machen, erhalten und nicht be- oder gar verhindert werden. Dass der Term "Leben" diese Bedeutung in seiner Erste-Person-Verwendung entfaltet, hängt damit zusammen, dass "Leben" erst in dieser Verwendung und dort aber primär den Bezugsrahmen des Körpers übersteigt zu Gunsten des Leibes.

Im Unterschied zu dem des Körpers ist der Leib-Begriff auf einen lebensweltlichen Zusammenhang bezogen: Wer nämlich von "lebendig" oder, was damit identisch ist, in der Erste-Person-Verwendung von Leben spricht, der "leibt und lebt" und hat immer schon einen Lebenswelt-Bezug hergestellt und damit gegen jegliche Okkupation durch einen Objektivismus, wie ihn die Naturwissenschaften zu begründen versuchen, Einspruch erhoben. Es ist dieser Einspruch, der nicht nur des Lebens, sondern auch der Lebenswelt normative Dimension ausweist. Diese Dimension führt hin zur Ethik der Lebenswelt.

4. ETHIK DER LEBENSWELT

Die Demonstration der Lebenswelt bedeutet nicht nur eine Zurückweisung des wissenschaftlichen Objektivismus, sondern eines jeden Objektivismus, also auch des normativen Objektivismus, der in der Gestalt von als universal ausgewiesenen Normen daherkommt. Diese Normen definieren, wie Bernhard Waldenfels deutlich macht, das *Normale* und das *Anormale* und konstituieren damit Verhältnisse von *Inklusion* und *Exklusion*.[33] Diese Wirkung von Normen trifft ebenso auf ästhetische, gesellschaftliche und ethische Normen zu.

Diese Normen reklamieren dabei ihre Gültigkeit aus ihrem Vermögen zur Universalisierbarkeit. In der Universalisierbarkeit liegt geradezu die Schlagkraft dieser Normen, welche sie gegen alle Anfechtungen und Einsprüche immun zu machen scheint. Den Status universaler Normen wollen wir im Folgenden unter Bezug auf Bernhard Waldenfels in Frage stellen.

Gegenstand ethischer Normen sind Handlungen. Was aber sind *Handlungen*? Waldenfels begreift sie als "Auseinandersetzung mit der Welt".[34] Diese Auseinandersetzung hat, wie Waldenfels in Anknüpfung sowohl an Merleau-Pontys Wahrnehmungstheorie, wie auch der Theorie der symbolischen Interaktion G.-H. Meads formuliert, den Charakter einer *Antwort*.[35] Jede Auseinandersetzung stellt somit eine *Response* von jedem auf etwas/jemanden dar. Diese Erkenntnis wird nun indes erst dann voll gewürdigt, wenn man den Status der Normen neu verhandelt. Ergebnis einer solchen Neuverhandlung wird aus lebensweltlicher Sicht eine Klassifizierung sein, der zufolge die Normen, die bislang Gegenstand der Debatte waren, die sich also beispielsweise durch Universalisierung ausweisen, nicht länger prioritär sind, sondern im Gegenteil anderen Normen nachgeordnet werden. Die universalisierbaren Normen müssen demnach als "Normen zweiter Stufe" betrachtet werden.[36] Das bedeutet, dass diese Normen zwar zur Beurteilung von Handlungsmaximen dienen, nicht aber für den Entwurf von Handlungszielen herangezogen werden können. Dies – der Entwurf von Handlungszielen – geschieht durch “Normen erster Stufe.”[37] Bevor wir uns fragen, was mit dieser Differenzierung gewonnen ist, ist der Einwand zu klären, ob die Differenzierung in Normen erster und zweiter Stufe nicht insofern gegenstandslos sei, als es für die Handlung doch schließlich belanglos sei, ob die universalisierbaren Normen, wie hier kritisch behauptet, zwar am Entwurf der Handlungsziele unbeteiligt sein mögen, aber vielleicht doch insofern zum Zuge kommen, als die “Normen erster Stufe” aus einer Aneignung der “Normen der zweiten Stufe” resultieren.

Eine solche Internalisierung von Normen wird nur dann als fortgesetzte Geltung der hier als zweitrangig charakterisierten Normen verstanden werden können, wenn unberücksichtigt bleibt, wie die Differenzierung der beiden Normen überhaupt in den *Blick* treten konnte. Sie ist Resultat einer *Blickrichtungsänderung*: Die hier als zweitrangig bezeichneten Normen sind deshalb zweit- und nicht erstrangig, weil sie erst auftreten, wenn ein Handeln bereits erfolgt ist. Diesem Handeln liegen aber vielfältige Handlungsmotive zugrunde, die nur unzureichend erfasst werden, wenn sie an universalen Normen gemessen werden. Der gesamte Response-Charakter einer Handlung bleibt schlicht unverstanden, wenn die Handlung nicht als Ausdruck einer Auseinandersetzung mit der Welt eines konkret Handelnden begriffen wird.

Und damit kommen wir auch zu dem, was mit dieser Differenzierung gewonnen wird: Die Okkupation der Lebenswelt durch universalisierbare und damit objektivierbare Normen, welche, wie Jürgen Habermas in Fortschreibung von Max Weber schreibt, die Aufspaltung der Lebenswelt in einen Bereich, der der objektivierbaren Geltungsanforderung standhält und einen solchen, der dies eben nicht vermag, bewirkt,[38] wird durch die Inkraftsetzung von Normen der ersten Stufe rückgängig gemacht.

Resultat dieser Defensiv-Bewegung ist die Wiedergewinnung der Lebenswelt. Die ethische Bedeutung, die die Lebenswelt auf diese Art erlangt, besteht mithin in der Demonstration ihrer Eigenständigkeit und damit zusammenhängend der Eigenständigkeit des jeweiligen Akteurs. Dieser begreift sich leiblich und überwindet damit den beklagten Reduktionismus.

Die “Ethik der Lebenswelt” hat dabei wenig gemein mit einer normativen Ethik. Vielmehr ist die Ethik der Lebenswelt Ausdruck einer *Haltung*. Die Krisis des europäischen Denkens, von der wir mit Husserl unseren Ausgang nahmen, ist mithin eine solche des Verlustes dieser Haltung und der Chance, diese zurückzugewinnen.

Universität Basel, Schweiz

NOTES

[1] Edmund Husserl, *Die Krisis der europäischen Wissenschaften und die Transzendentale Phänomenologie* (Den Haag: Martinus Nijhoff, 1954), Hua VI: 15.
[2] Stephan Strasser, *Welt im Widerspruch. Gedanken zu einer Phänomenologie als ethischer Fundamentalphilosophie* (Dordrecht: 1991) 69.
[3] Helmuth Plessner, "Conditio Humana," in *Gesammelte Schriften* Bd. VIII (Frankfurt/M.: Suhrkamp, 1983) 395.
[4] Husserl, *Erste Philosophie* (1923/24). Zweiter Teil. *Theorie der phänomenologischen Reduktion* (Den Haag: Martinus Nijhoff, 1956), *Hua* VIII: 79.
[5] *Hua* VIII: 78.
[6] *Hua* VI (*Krisis*): 102.
[7] Husserl, *Zur Phänomenologie der Intersubjektivität.* Texte aus dem Nachlass. Dritter Teil: 1929–1935 (Den Haag: Nijhoff, 1973) *Hua* XV: 149.
[8] *Hua* VI: 102.
[9] *Hua* VI: 182.
[10] *Ibid.*
[11] Michael Theunissen, *Der Andere* (Berlin: De Gruyter, 1981) 27.
[12] *Hua* VI: 154.
[13] *Hua* VI: 155.
[14] *Hua* VI: 99.
[15] *Ibid.*
[16] *Hua* VI: 155.
[17] *Hua* VI: 134.
[18] *Hua* VI: 188.
[19] Husserl, *Cartesianische Meditationen und Pariser Vorträge*, 1950. *Hua* I: 59.
[20] *Hua* I: 60.
[21] *Hua* VI: 99.
[22] *Hua* I: 60; 64.
[23] *Hua* VI: 60.
[24] *Hua* VI: 130.
[25] Bernhard Waldenfels, *In den Netzen der Lebenswelt* (Frankfurt/M: Suhrkamp, 1985), 16.
[26] Husserl, *Phänomenologische Psychologie. Vorlesungen Sommersemester 1925* (Den Haag: Martinus Nijhoff, 1962): *Hua* IX: 56f.
[27] *Hua* VI: 113.
[28] *Hua* VI: 132ff; 462.
[29] *Hua* VI: 130.
[30] "Wäre die Lebenswelt ein bloß Erstgegebenes, etwa im Sinne einfacher Daten, so gerieten wir ins empiristische Fahrwasser; wäre sie ein Letztregelndes, das nicht an den Erfahrungsboden geheftet wäre, so behielte der Rationalismus die Oberhand. Indem Husserl die Vernunft aus der Erfahrung selbst erwachsen lässt, vermeidet er beide Klippen; er tut dies anders als Kant, aber in durchaus vergleichbarer Form" (Waldenfels 1985: 18).
[31] Mathias Gutmann, "Biologie und Lebenswelt," in Ulrich Krohs, Georg Toepfer (Hg.), *Philosophie der Biologie* (Frankfurt/M.: Suhrkamp, 2005), 400–417: 400ff.
[32] *Hua* VI: 13f.
[33] Waldenfels (1985).
[34] Waldenfels (1985) 132.
[35] Waldenfels (1985) 133.

[36] Waldenfels (1985) 138.
[37] Waldenfels (1985) 138.
[38] Jürgen Habermas, *Moralbewusstsein und kommunikatives Handeln* (Frankfurt/M.: Suhrkamp, 1983) 117ff.

SİNAN KADİR ÇELİK

TRACES LEFT BY LEVINAS: IS "HUMANISM OF THE OTHER" POSSIBLE?

Can anyone comment "on" Levinas without misunderstanding him? One may answer this question by a "yes." But if such a comment is "on" Levinas and aims to clarify the obscurity, ambivalence or the "essential enigma" of his works, then this answer I think, should be a definite "no." Because neither an intentional relation – writing "on" Levinas – nor clarification of the "essential enigma" can be "ethical" in the Levinasian sense. No doubt, within the boundaries of "common sense ethics" – ethics based on the "ontology of sameness" – such an inquiry would most probably be seen as "nonsense," simply because it is very difficult to understand why an attempt to explicate Levinas' enigmatic discourse may not be an ethical behavior.

Nevertheless, when we become aware of the alterity, diversity and the otherness of Levinas' works, what we encounter is something that radically differs from the "hegemony of the same." That is to say, in Levinas' works, there is something totally strange and vulnerable that invocates a reader. One cannot be "interested in" a text written by Levinas; but rather one should be in the mood of being related to them with a total "dis-inter-*estedness*." On the other hand, if the essential enigma of the other is beyond all cognition and if "[understanding] does not invoke beings but only names them, thus accomplishing a violence and negation,"[1] then any possible attempt to encounter Levinas necessiates being embedded in a certain dialogue. Because, according to Levinas, traces, which are signs signifying the presence/transcendence of someone who is already past, become apparent enough only in a dialogue, and can thus be trailed. To do this, in the first part of the paper, I will try to reconstruct Levinas' argument about the phenomenological foundations of the humanism of the other. In the second part, by inquiring these traces left by Levinas, I will investigate the possibility of Levinas' "humanism of the other."

A-T. Tymieniecka (ed.), Analecta Husserliana XCIII, 269–282.

TRACES LEFT BY LEVINAS: FROM ETHICS AS A FIRST PHILOSOPHY TO INFINITE RESPONSIBILITY

The first trace that could be trailed in a possible dialogue with Levinas is his famous motto of "Ethics as a First Philosophy." Levinas had always insisted on the possibility that we can open the blind eye of Western philosophy when we become aware of our most primordial responsibility. However, this responsibility could never be based on ontological premises. Rather it should be founded on an ethics that is prior to ontology.

In his article entitled "Is Ontology Fundamental?" Levinas argues that before the phenomelogical rupture, "ontology" had always been based on the assumption that the existence of the world as a sum of facts can be known by reason liberated from temporal contingencies. However, Heidegger's project in his *Being and Time*, which can be seen as a continuation of Husserl's phenomenological programme, aimed to reconstruct an ontology which did not give a priority to reason. Heidegger alternatively conceived facticity and contigency not as facts presented to intellection, but as acts of intellection. Within this "new ontology," to comprehend a tool is not to look at it, but to know how to handle it. Similarly, understanding of our place in reality is conceieved not as explicitly defining it, but rather as being in an affective state, etc. At first glance, all of these aspects of this "new ontology," according to Levinas, indicated its potential to break away from the traditional intellectualism of Western thought. Neverthless, Levinas argues that even this "new ontology" seemed to rejoin the tradition of Western metaphysics by approving ontology as a first philosophy. Against the identification of ontology with first philosophy, Levinas always insisted on the possibility that ethics is or should be the first philosophy. This would be the only way, according to Levinas, to overcome the unquestioned priority of ontology over ethics – ethics not as a branch of philosophy in its classical sense but as a philosophy of dialogue – which had always resulted in an implicit immorality.

For Levinas, the implicit immorality inherent in the tradition of Western philosophy stems from the confusion of two irreducable relations, namely the *I-It* and the *I-Thou* relations. The *It* here designates what Husserl called "intentional object." Levinas defines the *It* as "the impersonal nature of stage in which impersonal consciousness saw that there is something, without object, without substance – a nothing that is not a nothing, for this nothing is full of murmuring, but of a murmuring that has not been named."[2] This is, for Levinas, self's first ontological experience. By recognizing the things, the self is erupted from beings and looks around itself to fulfill its desires and needs. It is the stage of egoism, a stage of the enjoyment of life, at which self is self-sufficient and lives for itself. At this stage, the self is a possesive self, which aims to capture the world by perceiving it just as a sum of resources given to itself for satisfaction of its enjoyment and desires. This is also the stage at which the world is first transformed and then equated into representations by the self's intentional directedness towards objects that are re-presented by itself. The world as representations, or more properly as represented objects by the self to itself, in a sense revolves around the self. Until passing into the next stage, the self is an inhuman being. All of its relations with objects are intentional, and therefore, non-ethical.

I will later analyze in some detail that the self enters into the domain of humanity only when the self passes into a further stage, the stage of subjectivity, at which desire is transformed into "an attitude of openness to exteriority. Openness that is appeal and response to the other. The proximity of the other, origin of all putting into question of self."[3]

At the first stage, the *I* experiences the objects around itself that are put and percieved in a certain "horizon." In this sense, the *I* and the *It* share the same horizon; they are at the same level. They are pressupposed as given in a certain environment, i.e. in a Cartesian space. Within a given space, the *I* experiences objects around itself as surfaces towards which the *I* is intentionally directed. However, can the relation between the *I* and the *Thou* be phenomologically and/or ethically put into the same horizon? Can or should the *Thou* be conceived just as an object towards which the *I* is intentionally directed to comprehend? Looking for an answer to these questions, Levinas argues that the realm of the other, phenomologically speaking does not, or ethically speaking, should not belong to realm of the *I-It* relation. The *I-Thou* relation is fundamentally different from the *I-It* relation and therefore cannot be derived from it.

First of all, the *I* is conceived as a relation by Levinas, not as a substance like the *It*. This is one of the reasons why the *I* and the *It* cannot and should not be seen in the same horizon. Here, the distinction between the *I* and the *It* resembles the Cartesian subject/object distinction. However, what differentiates the *I* from the *It* is not explained by appealing to the Cartesian framework. Levinas treats the essence of the *I* like Kierkegaard who defined the self as "a relation that relates itself to itself or is the relation's relating itself to itself in the relation; the self is not the relation but is the relation's relating itself to itself."[4] Regarding self as "the relation's relating itself to itself," the existence of other is already pressoppesed even in the *I-I* relation. More explicitly, if the existence of the other is at stake, the notion of the self as a self-identical being seems rather to be a fiction. Even an inner-speech of the *I* with "the same *I*" seems to be a dialogue; simply because the so called "monologue" is a kind of dialogue, if the words are always taken from the other discourses and if the words are always foreign at the lips of the speaker due to the intrinsic dialogic character of language. Hence, there is no *I* as a single, unitary, monadic, self-identical being; but rather the *I*, as a relation relating itself to itself, is always and already mediated by the existence of the other.

Perhaps the main reason behind Levinas' claim that one cannot or should not derive the *I-Thou* relation from the *I-It* relation is the following: The *I* experiences the *It* by focusing *on* it, but faces *with* the *Thou*. Facing *with* another is fundamentalty different from focusing *on* an object; because the other is not an object waiting for me to be comprehended, but is an interlocutor: "The other is not first an object of understanding and then an interlocutor. The two relations are merged."[5] Regarding the other as an interlocutor, I am first vigilant. I am first vigilant because the other interrupts me, disturbs me and invades my solitude. Vigilance is my irreducable concern for the other. That is to say, there is at least an attention before the encounter with the other. The vigilance for the other is not a relation of being conscious *of* the other. Vigilance is the relation with the other prior to ontology and the necessary precondition of "all encounter" which "begins with a benediction, contained in the word "hello"; that "hello" that all *cogito*, all reflection on oneself already presopposes and that would be a first transcendence. This greeting addressed to the other man is an invocation."[6]

Vigilance and benediction of the other show me that the other is not a mere object, a represented object by me to myself. The other as an interlocutor cannot be understood without speaking with her/him. S/he is not solely an object that is put in front of me or given to me in a certain horizon to focus *on* it. Unlike an object, the other has a face and has an invocation; s/he expresses herself/himself to me as the other who is different from me and waits to be heard. The other towards whom I am vigilant has voice coming from a face; a face that invocates me. But this face is not a surface sharing with me the same horizon. The other is not an arbitrary object ontologically connected with me in a given space. In other words, the other is not an object that can be simply put under gaze. In the very relation of encounter, that is the *face-to-face* relation, the other presents, expresses herself/himself and waits a response from me. When I encounter the other, her/his face always signifies something beyond the face. I can correlate the face and the voice. I can immediatly be aware of the fact that this voice is coming from that face; but I cannot futher say that the real origin of this invocation is that face looking at me.

To put it differently, the relation of encounter with the other is already mediated by a language, within which the other manifests, presents, expresses himself/herself to me and at the same time hides herself/himself from me. Here what I confront is not an object available to me simply as a sense datum or a phenomenon. The other does not look at me like a picture. The other is not given for me to satisfy my desire. I cannot comprehend the other by looking at her/him; seeing her/him place within a certain environment, measuring her/his distance from me. I cannot even approach the other without speaking. However much "our relation with him certainly consists in wanting to understand him ... this relation exceeds the confines of understanding."[7] When I confront the other what I encounter is not simply her/his representation, but her/his invocation that is coming not directly from the face but from the *beyond*, that is beyond the phenomenal order of appearances. One can ascribe an origin to an invocation, I can phenomologically say that a face speaks with me, but it is impossible to comprehend the real origin of this invocation. "This is what we describe with the words. The face speaks. The manifestation of the face is the first discourse. Speaking is first and foremost this way of coming from behind one's appearance, behind one's form; an opening in the opening."[8] This invocation always trancedends the face, which enters into my world from a completely different, an entirely foreign sphere, that is "ab-solute" in the Levinasian sense. In other words, a face as an abstraction is a trace that signifies something *beyond* the face. When I meet with the other, I do not simply look at a face but I relate myself with an epiphany waiting a response from me. "This bond with the other which is not reducible to the representation of the other, but to his invocation, and in which invocation is not preceded by an understanding."[9] Unlike an object, the other is a being that is addressed. "The presence of *Thou*, the Other, is *ipso facto* addressed to me, requiring a response."[10] Since the other, unlike objects, waits for a response from me, I am responsible for the other. When I hear the other, I am immedieatly aware of the fact that I should give a response to her/him. I can prefer closing my eyes or shutting my ears to her/his invocation. For instance, I may prefer not to hear his greeting "hello," or his intivation which demanded a "help" from me. This is obviously also a response but it may at the same time be an escape from responsibility. At this moment, when a response is demanded from me, I am at least aware of my confrontation with the responsibility for the other. One can interperet the concept of responsibility in various ways, one can even say that "I am not responsible for anything," however, one cannot escape from confrontation with responsibility. If I hear the invocation of the other, in this or that way, I cannot escape from this confrontation.

Nevertheless, the other who thus invokes me, who demands a response from me, is always foreign to me: "It is between strangers that the encounter takes place; otherwise, it would be kinship."[11] The other is always foreign to me and cannot be reduced to the *I*. Unlike the things that can be possesed, the other always resist to possession; the other is not mine; the other is always "not-I." In other words, the other is absolutely other; s/he may be so close to me to look at but too far from me to hear. Then how can I relate myself with the other *qua* other? Can I say that responsibility necessitates empathy? Should we claim that to be responsible for the other is necessary?

Levinas argues that ontolology, by reducing the other to the same, becomes an egology. Within the boundaries of ontology, the other is experienced as far as sharing the same or similar properties with me. If so, the other becomes "another" as far as s/he resembles me. It simply means that ontology conceives the *I-Thou* relation as a form of *I-I* relation, which results in a journey begining from the *I* and arriving at the *I*. In the course of this journey, the other is always taken as "*as if* the other," the other becomes "another" like me. The other and the I are all poured into a mould of the I. In other words, this is the way of assimilating of the other and absorbing the otherness of the other by me in the name of ontology:

> Ontology as first philosophy is a philosophy of power. I issues in the State and in the non-violence of the totality, without securing itself against the violence from which this non-violence lives, and which appears in the tyranny of the State. Truth, which sould reconcile persons, here exists anonymously. Universality presents itself as impersonal; and this is another inhumanity... A philosophy of power, ontology is, as first philosophy which does not call into question the same, a philosophy of injustice.[12]

Within the ontology as a tyranny of the *I*, the *I* as sovereign, posits itself as the measure of everything. The *I* has a right to reduce every difference to the same. However, the sameness becomes sameness where I am not aware of the difference, the otherness of the other, the very alterity. That is why in a possible ethical relation with the other, the *I* as not a *being-for-itself*, but I as a *being-for-the-other*, has a conscience before the consciousness *of* the other, which is not a thing among other things. To put it differently, reduction of all otherness to the same, being the consciousness *of* the other, is not the only way to relate myself to the alterity, the otherness of the other. I am immediately unjust when I see the face of the other as a mask, as an ordinary man among men, as a *persona* having a fixed meaning. Meaning or signification of the face of the other is not something added to the face from outside. I should not attribute a meaning to a face. A face expressing itself as an interlocutor has already a primordial signification prior to my attempt to mask it. The essential nudity of the face of the other reminds me of the face in the relation of *facing*, this face is always beyond my conceptual streotyping, my strategies of masking the face. The face as mask when it is transformed into a concept, when it is reduced into a representation in my mind, does not reveal itself as a face:

The neutralization of the other who becomes a theme or an object – appearing, that is, taking its place in the light – is precisely his redution to the same. To know ontologically is to suprise in an existent confronted that by which it is not this existent, this stanger, that by which it is somehow betrayed, surrenders, is given in the horizon in which it loses itself and appears, lays itself open to grasp, becomes a concept. To know amounts to grasping being out of nothing or reducing it to nothing, removing from it its alterity.[13]

Masking the other, treating her/him as a pre-determined concept once again means "removing alterity from the other." The asymmetrical relation with the other cannot be resolved by the help of reflection. The face of the other "resists possession, resists my powers. In its epiphany, in expression, the sensible, still graspable, turns into the total resistance to grasp."[14] A face always invites me to an incommensurable relation that always exceeds my power, my attempt to fix its meaning. The face of the other as absolutely foreign to and never belonging to me diverges to infinity by breaking the chains of totality, that is, a world common to me. When I become aware of this resistance of the face, its continuous movement towards infinity is beyond my representation, I am forced to confront with the epiphany of the face, which is ethical.[15] This epiphany as neither a concept nor a worldly phenomenon imposes itself to me from above and puts the I into question: "The putting into question of the self is precisely a welcome to the absolutely other ... instead of annihilating the I, putting into question binds it to Other in an incomparable and unique way."[16]

In the Levinasian sense, to be related with the other in an "incomparable and unique way," the I should be put into question without annihilating it. In other words, "putting the I into question" does not mean that I should annihilate the I, instead, it is a possibility not to posit the I as a sovereign, who is "interested in" the other. The worry behind "putting the I into question" is the hegemonic way of "being related with the other" that we encounter in everyday life. This is exemplified in interestedness, which makes the I see the other as an instrument that can be used in any way for satisfaction of the ego. Besides, this "interestedness takes dramatic form in egoisms struggling with one another, each against to all, in the multiplicity of allergic egoisms which are at war with one another and are thus together."[17]

This relation with the other is not a relation that I could sustain with her/him. It is beyond my power and I am inevitably questioned by the epiphany of the other. I cannot resolve this enigmatic puzzle by masking the nudity of the other's face. I am rather a question in front of this face in its very proximity, which opens the door of exteriority. I am now a being that is not inferior to itself. The ipseity of myself is under threat. At this point, I cannot escape from the neccesary confrontation with the exteriority by going back to myself. Here, there is no way out. I am neither within me nor outside of me. I am now an ultimate question directed to me from everywhere and nowhere. I feel that I should submit to this absolute authority that overwhelms me. This is the very challenge of the other that empties me:

This is a challenge of consciousness, not a consciousness of the challenge. The Ego loses its sovereign coincidence with self, its identification where consciousness comes back triumphantly to itself to reside in itself. In the face of the obligation of the Other, the Ego is banished from that response, is not already glorious consciousness of this exile. All complacency destroys the rectitute of the ethical movement.[18]

Confrontation with such a challenge opens two ways in front of me. The one comes from and goes to inhumanity but the other takes me from inhumanity to humanity. This is a forking of a road where I should answer the following question: Who is the master, I or the other? Depending on my answer, the destinity of the other will change. I can answer this question in a different way, for instance I can say that we are equals. Neither you nor me is the master. However Levinas immediately raises his objection against this answer and says that the presupposed "equality" of the I with the other is a kind of reduction of the other to the I. This is a declaration of war against the otherness of the other by positing myself as sovereign. It is a form of possession or suppresion of the other in the name of the I. The other affects me, and this affection is originally a command to be obeyed. Otherwise I get rid of the other as a mere intentional object put under my gaze. Being affected by the face and the voice of the other is my primordial relation with her/him. If in face-to-face relation I treat the face as a pseudo-face, as a *persona*, as a representation, whom I encounter is definitely me, not the other. It is my re-represented other presented by me. However, a face exposed to my look is always bare, naked. It is powerlessness, defenselesness itself, insecure and innocent:

When you have encountered a human being, you cannot drop him. Most often we do so, saying "I have done all I could!" We haven't done anything! It is this feeling, this consciousness, of having done nothing that gives us the status of hostage with the responsibility of one who is not guilty, who is innocent ... the other involves us in a situation in which we are obligated without guilt, but our obligation is no less for that. At the same time it is a burden. It is heavy, and, if you like, that is what goodness is.[19]

Should the I obey commands of the supreme authority of the other? Answering this question by a definite "yes" is the very precondition of morality according to Levinas. The self becomes an ethical subject only when it allows itself to be a hostage held by the other. Being human *qua* human is possible only when the self evolves into an ethical subject by overcoming the tension between its habit of possesing the other and the possibility of seeing the other as a supreme authority from which commands should be obeyed without interrogation. The tension here is between my possesive tendency of reducing the otherness of the other into a mere intentional object put under my gaze, and the possibility of reading the traces inscribed in the face of the other, which invites me to be embedded in his infinity. In other words, the tension here is to be human *qua* human and remaning an inhuman being. As Levinas repeatedly says, to be human is to be responsible for the other "even when he bothers me, even when he persecutes me."[20] Further, I could not transfer this responsibility to someone else. Beause "I substitute myself for every man and no one can substitute for me, and in that sense I am chosen."[21] I am also more responsible than anyone else. This is the transition from conceiving the I as a being-for itself to being-for-the-other: "The fact that in existing for another I exist otherwise than in existing for me is morality itself."[22] That is to say, I am a human being as far as I am for-the-other and as far as I am infinitely responsible for the other. Then one could ask the following question: What is the origin of such an infinite responsibility? How can one become a human *qua* human according to Levinas? Or, how can one trail the traces left by Levinas? I will now try to answer these questions.

BEYOND THE TRACES: IS LEVINAS' "HUMANISM OF THE OTHER" POSSIBLE?

Before answering these questions, let us try to evaluate Levinas' conception of "ethics as a first philosophy." One of the implications of Levinas' invocation is to overcome the reduction of our asymmetrical interrelatedness with the other to an intentional relation, that is, positing the other far from me as an object of knowledge, which paves the way for conceiving our everyday confrontation with the face of the other, with his invocation, as a burden of our metaphysical task. This is also the foundation of our world of "instrumental rationality" by which the crucial question "How should one relate oneself to the other?" is dealt with instrumentally, as if this other was just another object of knowledge. Levinas sets out from a totally different point: Do not try to see, observe, grasp or comprehend the other; but just try to hear her/his invocation. In this sense, the rupture of Levinas from the tradition of Western philosophy has its roots in hearing the invocation of the other or in the incomprehensible conscience before being consciousness of the other. That is why metaphysically already separated men cannot meet by means of ontology, if "man is not a subject who constitutes, he is the articulation of the meeting"[23]; one should relate oneself to the face of the other without positing a definite self and without presupposing the ontologically well-grounded existence of the other. However, the first step in classical strategy of traditional Western philosophy is to separate or differentiate the I from the other and then try to unite them consistently. This strategy that could be formulated as "first separate the I from the other, then unite them," does not solely belong to Western philosophy, it also seems to be implicitly accepted and substantiated almost by every human being to manage their everyday life. In contrast to such an implicit social contract, which has been justified in practice, the following is the very point where Levinas' radical rupture of traditional Western philosophy begins: Both the separation of the I from the other and the unification of the I with the other are inherently unethical strategies of Western philosophy that results in the manifestation of the I as independent from the other. This is accomplished by reducing all multiplicity to the "ontology of sameness." That is why the asymmetrical and incomprehensible interrelatedness with the other should be the primordial foundation of ethics. But, according to Levinas, is such a radical rupture possible and if possible how?

As an answer to the question above, one should note that Levinas seems to leave the enigmatic concepts like "infinity," "insomnia," and "trancendence," etc. without clearly defining them. Regarding this point, I aggree with Michele Saracino's diagnosis that Levinas:

> neither has a well-defined method nor a structured anthropology. Levinas describes the limits and issues surrounding anthropology, yet never constructively envisions one ... Levinas's weakness, in terms of his ambiguous notions of freedom, agency, and justice and his lack of method, are indicative of his descriptive rather than explanatory anthropology.[24]

Accordingly, his concepts can really be treated as completely bizarre or highly abstract. But is it philosophically necessary to explicate or fix the meanings of these terms? Levinas would have at least two reasons to answer in the negative: (i) "violence ... proceeds from the proper functioning of Reason itself."[25] That is to say, any attempt to comprehend the "essential enigma" inevitably leads us to base ethics upon ontology. (ii) "The idea of infinite consists in grasping the ungraspable while nevertheless guaranteeing its status as ungraspable."[26] Hence, it is possible to argue that not only the idea of infinite, but also all other concepts that Levinas employs are derived by "putting the I into question," and are founded on the very asymmetrical interelatedness with the other. This seems to guarantee their status as not always fully comprehensible, at least by Reason. Levinas is well aware of the aforementioned obscurity and regarding it in his paper, "Trancendence and Height," he says the following:

> the obscurity of my paper may derive from the impression of the method ... The situation in the face of the other is significant. It is a structure that illuminates, and consequently its analysis is

not the description of an empirical fact. On the basis of this structure, which until now has most often been expressed only with *pathos* [my emphasis] I was anxious to rejoin the "language of professors." It is quite possible that I have not succeeded in this and I remain obscure. It would perhaps be necessary to review the detail of my analysis. But the very principle of my enterprise – giving value to the relation of infinite responsibility that goes from the I to the Other (*Autrui*) remains. Certainly I believe that this is our most valuable everyday experience, one that allows us to resist a purely hierarchical world.[27]

In this sense, the enigmatic structure of Levinas' works is consistent with his (non-) philosophical attitude. Because his concepts like "infinite" and "enigma" should remain obscure in order to investigate the possibility of ethics beyond ontology. On the other hand, when one claims that "I grasp the other" or that "infinity can be understood in this way," this entails for Levinas an ontological conversion of the absolute otherness of the other into the sameness. This would mean that infinity is already put into the framework of totality. However, ungraspibility of the other, infinity, etc. do not simply entail the conclusion that Levinas' nomenclature is simply nonsense. According to Levinas, if "I signifies not being able to escape from responsibility,"[28] then I cannot escape from my infinite responsibility for the other, I cannot stand still when I hear the invocation of the absolutely other. I cannot even say that "I am a just person," because "the just person who knows himself to be just is no longer just."[29] Such expressions of Levinas would aim to indicate a further possibility to be aware of our "most valuable everyday experience," which is "the relation of infinite responsibility which goes from the I to the other," that allows to "resist a purely hierarchical world." At least one can acknowledge Levinas to be right in his intention behind these enigmatic descriptions. As being against the boundaries of ontology, where we can avoid seeing the defenseless and naked face of the other, where we can come to rest, where we can apply violence by proper functioning of Reason, Levinas offers us another possibility: This is the way of *pathos*, a permanently uneasy way of being related with ourselves and with the absolute otherness of the other. This is the reason why "the ethics of Levinas is an excessive ethics; excessive alterity, excessive passivity, excessive responsibility. Indeed, the problem with Levinas' asymmetrical relation to the Other, my total indebtedness to the Other with no reciprocity, is that the Other is like a God to me."[30] According to Levinas, such an extreme ethics should remain incomprehensible by Reason, for it is the only possibility of preparing us for the experience of the impossible, which possibly delimits the limitations of humanism, and which makes a disclosure of immorality of ethics which is a priori committed to ontology.

On the one hand, the Levinasian shift from the "self-oriented" conceptual domains to "other oriented" ones can be understood as a symptom of the dissatisfaction with implicit immorality inherent in humanism based on a kind of Cartesian subjectivism within which immediate "others" are rendered to be always far from us, under the "imperialism of the I." That is to say, in Levinas' view one that cannot open the blind eye of Western philosophy, if it is not realized that ethics is precisely the art of living with ambivalence. Nevertheless, although it is ambiguous, Levinas' "humanism of the other" does not permit everything. On the contrary, its implicit moral obligations seem to be more rigid than any kind of humanism hitherto offered. It seems possible to argue that Levinas' enigma of the absolute otherness of the other somehow corresponds to Kant's "categorical imperative" as Anthony F. Beavers argued.[31] Levinas' imperative can alternatively be formulated as "treat the other person as an end-in-itself," the existential counterpart of which is "comport yourself towards the other person in such a way that your being is always for-the-other."[32] However, even after these clarifications, it still seems to be inevitable to confront the question of "how can I realize my infinite responsibility to be human *qua* human?" Levinas' answer to this question should not surprise us:

> In the responsibility for the other, we are at the heart of this ambiguity of inspiration. The unheard of saying is enigmatic in its an-archic response, in my responsibility for the other. This ambiguity within the subject is the trace of the infinite, alternately beginning and intermediary, the diachronic ambivalence that makes ethics possible.[33]

The enigma here is that the other's manifesting herself/himself without manifesting herself/himself. I hear her/his voice that ordains me, but I cannot know where this voice comes from. The face is visible but the source is not. I am here in conjunction with the visible and invisible in such a way that I am embedded in an essential enigma, which is the diachronic ambivalence, and this is the only possibility of ethics. As Thomas Carl Wall puts it, Levinas' ethics, "in short, is essentially ambiguous."[34] However, it seems to be inevitable to ask the following questions: How should we deal with such ambivalence? Is living in an essential enigma the only way out from inhumanity?

From the Levinasian point of view, this question can be answered in the following way: The precondition of a possible ethical relation, as discussed in the first part of the paper, is to believe in submitting myself to the other; I should be in command, in the service of the other without expecting any mutual agreement. If I am for-the-other, I should not expect even a response from the other. Such a total, non-complementary submission to the other is presented by Levinas as a way out from the tradition of Western philosophy: The face of other always unconditionally commands me. That is why Levinas tells us that we can cope with this "essential enigma" instantiated by the face of the other, by believing in the transcendence of the other who is beyond all my cognition. At this point, we can ask the following question: How can a trancendental being be phenomologically described? Thoughout his argument concerning the "essentail enigma" that I have tried to recontruct in the first part of the paper, Levinas seems to import theology into phenomenology without giving any account of it. He infers trancendence of the face and the invocation of the other from his phenomological descriptions of the *I-It* and the *I-Thou* relations, but suddenly shifts to a theological view to justify his ethical recommendations. For instance, he claimed that "the face is the locus of the word of God. There is the word of God in the other, a non-thematized word."[35] Although such claims of Levinas within his ambivalent structure of his argumentation may seem to be phenomologically justified, he offers his reader neither an explicit nor implicit phenemological account about their derivation. Hence, we could say that in the name of the "humanism of the other," Levinas tries to combine two incommensurable ways of thinking – phenomology and theology by imposing of "ought to" into his phenomological analysis of "is." As a result, there is no way for Levinas except from praying for us to share his *pathos* and believe in his "humanism of the other."

Department of Philosophy, Middle East Technical University, Ankara, Turkey

NOTES

1 Emmanuel Levinas, *Thinking-Of-The-Other*, trans. Michael B. Smith and Barbara Harshav (New York: Columbia University Press, 1998), p. 9.
2 Levinas, *Alterity and Transcendence*, trans. Michael B. Smith (New York: Columbia University Press, 1999), p. 99.
3 *Ibid.* p. 99.
4 Søren Kierkegaard, *The Sickness into Death*, eds. and trans. Howard V. Hong and Edna H. Hong (Princeton: Princeton University Press, 1980), p. 13.
5 Levinas, *Thinking-Of-The-Other, op. cit.*, p. 6.
6 Levinas, *Alterity and Transcendence, op. cit.*, p. 98.
7 *Ibid.* p. 5.
8 Levinas, *Humanism of the Other*, trans. Nidra Poller (Urbana and Chicago: University of Illinois Press, 2003), p. 31.
9 Levinas, *Thinking-Of-The-Other, op. cit.*, p. 7.
10 Levinas, *Proper Names*, trans. Michael B. Smith (Stanford, CA: Stanford University Press, 1996), p. 25.
11 Levinas, *Alterity and Transcendence, op. cit.*, p. 97.
12 Levinas, *Totality and Infinity: An Essay on Exteriority*, trans. Alphonso Lingis (Pittsburgh: Duquesne University Press, 1969), p. 46.
13 *Ibid.* pp. 43–44.
14 *Ibid.* p. 197.

[15] *Ibid.* p. 199.
[16] Levinas, *Basic Philosophical Writings*, eds. and trans. Adrian T. Peperzak, Simon Critcley and Robert Bernasconi (Bloomington: Indianan University Press, 1996), pp. 17–18.
[17] Levinas, *Otherwise Than Being*, trans. Alphonso Lingis (Pittsburgh: Duquesne University Press, 2002), p. 4.
[18] Levinas, *Humanism of the Other, op. cit.* p. 33.
[19] Levinas, *Alterity and Transcendence, op. cit.*, p. 106.
[20] Levinas, *Thinking-Of-The-Other, op. cit.*, p. 106.
[21] *Ibid.* p. 108.
[22] Levinas, *Totality and Infinity: An Essay on Exteriority. op. cit.*, p. 261.
[23] Levinas, *Proper Names. op. cit.*, p. 24.
[24] Michele Saracino, *On Being Human: A Conversation with Lonergan and Levinas* (Milwaukee: Marquette University Press, 2003), p. 207.
[25] Levinas, *Basic Philosophical Writings, op. cit.*, p. 23.
[26] *Ibid.* p. 19.
[27] *Ibid.* pp. 22–23.
[28] *Ibid.* p. 17.
[29] *Ibid.* p. 17.
[30] Andrius Valevicius, *From the Other to the Totally Other: The Religious Philosophy of Emmanuel Levinas* (New York: Peter Lang, 1988), p. 89.
[31] Anthony F. Beavers, *Levinas beyond the Horizons of Cartesianism: An Inquiry into the Metaphysics of Morals* (New York. Peter Lang, 1995), pp. 105–134.
[32] *Ibid.* p. 107.
[33] Levinas, *Basic Philosophical Writings, op. cit.*, p. 105.
[34] Thomas Carl Wall, *Radical Passivity: Levinas, Blanchot, and Agamben* (New York: State University of New York Press, 1999), p. 32.
[35] Levinas, *Alterity and Transcendence, op. cit.*, p. 104.

OSVALDO ROSSI

ON THE SUBJECT OF HEIDEGGER: EXISTENCE, PERSON, ALTERITY

Following the considerations made about Husserl on the theme of intersubjectivity, with all the difficulties encountered due to the discovery of corporeity, I should now like to pursue the study, pausing for a moment on Heidegger and making particular reference to his most famous early work.

I do not intend to run through again the developments of Heidegger's early thought; I wish instead to recall certain nodal points of his reflection, referring to the chosen theme. Existence, person and alterity in Heidegger are at times the terms of a complicated weave or a hermeneutical complex that he has tried to resolve beyond subjectivist points of view, leading to the realization of the turning point of phenomenology.

AN APPROACH

As it has been pointed out more than once, the aporias, in which the reflection of phenomenology had been argued, found in Heidegger a first and radical solution to the light of the temporal character of *Sein*, a fundamental aspect of a new and sound ontological thought, clarified in the first treatment of *Sein und Zeit* in 1927.

In the same year, in the work *Die Grundprobleme der Phänomenologie*, which gathered the lessons of the course held by Heidegger at Marburg during the summer term, he insistently proposed the question:

> Wie steht es nun mit dem Gegenstand der Philosophie? Kann man sich so etwas vorstellen wie Sein? Fasst einen beim Versuch dazu nicht der Schwindel? In der Tat, wir sind zunächst ratlos und greifen ins Leehre. Seiendes – das ist etwas, Tisch, Stuhl, Baum, Himmel, Körper, Worte, Handlung. Seiendes wohl – aber Sein? […].
>
> Wenn Sein der verwickeltste und dunkelste Begriff wäre? Wenn das Sein auf den Begriff zu bringen die dringlichste und immer wieder neu zu ergreifende Aufgabe der Philosophie wäre? […]. Wenn Philosophie die Wissenschaft vom Sein ist, dann ergibt sich als Anfangs-, End- und Grundfrage der Philosophie: Was bedeutet Sein? Von so aus ist dergleichen wie Sein überhaupt zu verstehen? Wie ist Seinsverständnis überhaupt möglich?[1]

A-T. Tymieniecka (ed.), Analecta Husserliana XCIII, 283–316.

Heidegger pointed out that only man, as a privileged being, was able to ask such questions and only through him the asking could discern the approach to the *Seinsfrage*, which is central in *Sein und Zeit*.

Therefore, the interrogation referred to the subject of such questioning, to man. He was the preliminary "guarantee of the right approach to the being" and to his opening.

Sofern das Sein das Gefragte ausmacht, und Sein besagt Sein von Seiendem, ergibt sich als das *Befragte* der Seinsfrage das Seiende selbst. Dieses wird gleichsam auf sein Sein hin abgefragt. Soll es aber die Charaktere seines Seins unverfälscht hergeben können, dann muss es seinerseits zuvor so zugänglich geworden sein, wie es an him selbst ist. Die Seinsfrage verlangt im Hinblick auf hir Befragtes die Gewinnung und vorherige Sicherung der rechten Zugangsart zum Seienden. Aber "seind" nennen wir vieles und in verschiedenem Sinne. Seiend ist alles, wovon wir reden, was wir meinen, vozu wir uns so und so verhalten, seiend ist auch, was und wie wir selbst sind. Sein liegt im Dass- und Sosein, in Realität, Vorhandenheit, Bestand, Geltung. Dasein, im "es gibt." An *welchem* seienden soll der Sinn von Sein abgelesen werden, von welchem Seienden soll die Erschliessung des Seins ihren Ausgang nehmen? Ist der Ausgang beliebig, oder hat ein bestimmtes Seiendes in der Ausarbeitung der Seinsfrage einen Vorrang? Welches ist dieses exemplariche Seiende und in welchem Sinne hat es einen Vorrang?[2]

The primacy was thus reasserted of that particular being that was aware of existing, which we call man; a primacy which consisted in asking and in connecting the question of his research, with the need of comprehending being and its meaning.

He concluded:

Hinsehen auf, Verstehen und Begreifen von, Wählen, Zugang zu sind konstitutive Verhaltungen des Fragens und so selbst Seinsmodi eines bestimmten Seinenden, *des* Seienden, das wir, die Fragenden, je selbst sind.[3]

He immediately stressed:

Dieses Seiende, das wir selbst je sind und das unter anderem die Seinsmöglichkeit des Fragens hat, fassen wir terminologisch als *Dasein*.[4]

What is Being present (*Dasein*)? It is that particular type of existence which is able to pose the fundamental questions about the existence of its own nature, that is about what makes being exist. Heidegger avoids using the word "man" because he does not wish to lend himself to considerations of an anthropocentric and religious kind, but only to wider and "neutral" considerations of an ontological type. Humanity does not mean a specific biological race, but every type of being that has a conscious and interrogatory existence. The other beings, animals for example, live to exist; man and every other possible existing being (therefore interrogator) exist to live.

Here the biological data, the corporeality (which according to Heidegger is a being), plays a totally secondary role and is subordinate to existence. Man is *Dasein*, in other words he does not exist because he is endowed with a body, but his body is the extension of a "plan" that makes him exist, aware of his own being; the body is a way by which he performs and presents to himself and to other beings, the historic and temporal dimension of existence.

Consequently, for Heidegger, resolving the *Seinsfrage* meant reaffirming "in its being" the subject of the research and of the questioning, man, as *Dasein*. Hence the attempt to open a passage towards ontology by means of the *Daseinsanalitik*, that is questioning the interrogator, the existence as *Dasein*.

Thus was the existence of man who had to re-emerge once again. He re-emerged like that being who asked the question about *Sein* of which the *Dasein* in its form of "existence" was the concrete, "effective" sense, but never totally representable; an existence always considered as singularity, determination and precision, a Being present. This led to the birth of the existential analytics of *being present*, precisely the *Daseinsanalitik*, as the interrogation of the existing.

One of Heidegger's first Italian scholars, the Venetian Luigi Stefanini, commented that the same being is not representable and not open to partial simplifications "because we are 'committed', 'deep-rooted' (*engagés, enracinés*) in it, and therefore we cannot emerge from being, with our autonomous existence, to formulate some questions about being and to provide an answer. We do not go behind the being to make it walk in front of us and observe its behaviour. We, who ask questions about being, are under question ourselves and, if we wish to give our reply some 'foundation' about the consistency of the being that we are, we should presuppose what we are trying to find out with our question, in the answer."[5]

The questioning, therefore, was not intended here as the formula for positivistic and naturalistic knowledge, which tends to discriminate, to clearly distinguish between the interrogator and the interrogated, but it was intended in a way that would later be called by Heidegger "hermeneutic," that is of circularity and correlation between the interrogator (*Dasein*) and the interrogated (*Sein*).

What, therefore did our questioning about the meaning of being, that is the *Seinsfrage*, ask? It was a "committed" interrogation, in which the interrogator assumed upon himself the prerogatives of the interrogated, living them from an existential point of view. Each step towards an understanding of the *Sein* was also a step towards the understanding of the existence, of oneself (*Dasein*).

Die ausdrückliche und durchsichtige fragestellung nach dem Sinn von Sein verlangt eine vorgängige angemessene Explication eines Seienden (Dasein) hinsichtlich seines Seins.[6]

We can draw two conclusions: on the one hand, the non-representable character of being, on the other hand its existential character. In other words, the sought after being on the one hand was not a being, but on the other hand, it could only offer itself by means of the being or by means of that "exemplary" being which was the Being present. Being, therefore assumed the characteristics of the transcendence as regards to the being and of the immanence as regards to Being present. This was also a way to overcome the traditional opposition immanence/transcendence (and the other similar categories such as monism, pluralism, idealism, realism ...) of the former philosophy.

Beyond this viewpoint, in the new perspective, existence was first of all *plan (Entwurf)*. However the plan was always, according to Heidegger, one of an existence "cast" into limitedness and into temporality, therefore into the situation: such was the *Geworfenheit*, the original "casting" of the *Sein* into the *Dasein* and into the fundamental form of existing. Since Heidegger's early work, *Sein*, man, existence have all been intended as a "plan," as a real life experience, open to any possible future results, "an die Nicht-Gegebenheit der Zukunft,[7] in other words as, an experience totally open to time.

Asking oneself questions meant therefore becoming aware of the finished historic and temporal character of one's existence. The act of questioning and of feeling questioned by the same interrogation was a sign of this "casting." Being cast expressed the "effectiveness," *Faktizität*, of being in existence, in the various and multiple forms of life, which found their manifestations in language. This was a language that from then on introduced a problem: its adequateness in expressing Being. Was the language of the interrogator a denotative language or a nominative/evocative one? In the *Daseinsanalitik*, Heidegger did not make this clear.

BEYOND PHENOMENOLOGY

Heidegger's surprising personality and the *pathos* of his thought were manifested in this way. All this was a development of the phenomenological approach, which on the one hand, aimed at recuperating reality "in flesh and bone," and on the other hand at grasping the essence by means of an adequate method. What had happened to existence? Was it a datum or an essence?

Something original or a result, an *eidos*? Did this not propose it, however, as an original starting point?

Gadamer, in his timely essay on *Die phänomenologische Bewegung* in 1963, pointed out:

Es war eine Aporie, ein grundsätzliches Problem, dass das faktische menschliche Dasein nur seinem "Eidos," seinem Wesen nach in der phänomenologischen Forschung aufgehellt werden konnte, dass es aber in seiner Einmaligkeit, Endlichkeit und Geschichtlichkeit in Wahrheit nicht als der Fall eines Eidos, sondern selbst als das Allerrealste anerkannt werden möchte. An dieser Aporie sollte Husserl und die phänomenologische Forschung insgesamt die eigene Begrenztheit, Endlichkeit und Geschichtlichkeit erfahren.[8]

Heidegger denounced the inadequacy and the untenability of this aporia, opting for a notion of existence as a "plan," as a preliminary factor of the interpretation of being. Some important consequences derived from this with regard to Husserl's phenomenological definition. Heidegger was able to reconcile the two moments which in Husserl's phenomenology contradicted each other: the seeking of a concreteness and the guarantee of an essentiality, of an *eidos*. In fact the notion of "plan" was needed to render the concreteness, intended as historicity, temporality, the authentic "essence," *eidos* (as a condition of thought and truth) of the Being present. Heidegger stressed this affirming that: "Das 'Wesen' des Daseins liegt in seiner Existenz."[9] This changed the sense of Phenomenology intended as an unveiling of the being in the existence. The essence was exactly this coming into existence (*ex-sistere*) of the being which was then established beginning from time.

With this transformation from an existential point of view of the phenomenological notion of *eidos*, Heidegger reaffirmed the free and independent character of existence, not conditioned by a preventive *telos*, as instead it seemed according to Husserl's perspective, in which the *eidos*, presented itself as a sort of *logos* that predetermined as an "end" the carrying out of the being's existence. This was a metaphysical residue, of an Aristotelian stamp, which still persisted in the phenomenological approach, as Gadamer revealed.[10]

It was a metaphysics that still tended to objectify the "being" in unchangeable "essences"; a metaphysics as a petrifying glance of the being on Being, which in its turn transformed, like in the myth of Medusa, the Being into a petrified being at man's disposal. In this it was possible to individualize the alienated loss of the Difference and the consequent existential confusion.

Heidegger, on the other hand, wanted to reaffirm the ontological, non-representable nature of Being that was never ending in the "simple-presence," but this became the viaticum of an alterity, because the *Dasein* was really that being in virtue of which the things were present.[11] Existence was not *ousia*, nor a "subject," but an ontological and temporal dimension, *Da-sein*.

This induced Heidegger to radically reopen the question of the traditional ontology, linked to a privileged consideration of the being, of existence, as the "object," *res inter res*. It was an ontology Heidegger defined as metaphysics, not in the sense that it went beyond physics, but as another form of physics. An idea that had also been appropriated by modernity in its scientific ideal.

Heidegger went on to immediately surpass the old abstract categories of ancient and modern Cartesian metaphysics, of idealism of the same Phenomenology. Concepts such as those of "subjectivity," "substance," "consciousness" and "self-knowledge" of person, fell under the criticism that was directed towards Greek and Christian anthropology as forms, giving the being body, which in this way declined to the "presence" the object of reasoning and of its operative methods. Heidegger showed that continuing in this way "misunderstood" all "die primäre Grundverfassung des menschlichen Daseins damit verkannt ist."[12]

A decisive point was the criticism of the concept of *subjectum* or "subject," as an activity of thought in "selbstbezug, Reflexivität, Ichheit."[13] All this could not be found in the ancient Greek word *hypokeimenon*, because the Aristotelian meaning of the word indicated permanence, "that what lies at the bottom." In Aristotle such words contrasted with the elements of change or alteration that took place in nature, that is to say with the accidents. In this metaphysical perspective "subject" and "object," man and nature, were moments that were led back to within a single definition of which the static and unchangeable aspect was grasped, the "essence" which in fact "lay at the bottom." The corresponding word in Latin was on the other hand *substantia* or *subjectum*, from where the word "subject" derived, which however, in modernity assumed a different meaning with regard to that of the ancient "substance," as from the *cogito* of Descartes, which, removing all doubt, confirmed itself as the first evidence, as the first certainty.

The Cartesian *cogito* in consequence had the reduction and application of the word "substance" to the only cogitative aspect, to thought, thus moving from nature to the thinking subject by means of the criterion, this also a "subjective" to the evidence. *Subjectum* was the only reflective subjectivity or "consciousness," deprived of every material or physical element. Man was "cogito," therefore "consciousness," no longer corporeity. This could be noted by the difficulty in finding and collocating the soul, the *cogito*, in a particular place of the human body, which could only be the pineal gland, due to the physical nature of the body. And this distinguished the functions of the human body from those of other animals which were only biological, as was explained by Descartes in his *Discours de la méthode* (part V, §2).

Descartes explained:

Car, examinant les fonctions qui pouvaient en suite de cela être en ce corps, j'y trouvais exactement toutes celles qui peuvent être en nous sans que nous y pensions, ni par conséquent que notre âme, c'est-à-dire cette partie distincte du corps dont il a été dit ci-dessus que la nature n'est que de penser, y contribue, et qui sont toutes les mêmes, en quoi on peut dire que les animaux sans raison nous ressemblent: sans que j'y on pusse pour cela trouver aucune de celles qui, étant dépendantes de la pensée, sont les seules qui nous appartiennent en tant qu'hommes, au lieu que je les y trouvais toutes par après, ayant supposé que Dieu créât une âme raisonnable, et qu'il la joignît à ce corps en certaine façon que je décrivais.

In a mechanistic description of the body there was no space for other considerations. Consciousness was the only criterion of interior and exterior certainty, even of nature and of the body, but only in the second instance. This was a criterion that even the empiricists had acknowledged and Kant exalted in the transcendental primacy of *Ich denke*, the permanent principle of every representation, while "consciousness" as only "substance" remained unknowable.

Unlike the natural *substantia*, the "substance" as thought or *cogito* was characterized by the possibility of falling back on itself, of reduplicating its action, *cogito me cogitare*. It is what we call "reflectivity," a falling back or "reflecting" of itself, according to the natural analogous phenomenon of light and the mirror, *reflexio*. The Greeks said that this true ability of the soul, *psyche*, a rational and vital principle, the mind and movement of bodies, inborn in them, was a sort of *auto kinoun*, involving the supreme principle of *Nous*, only mind or "thought of thought" *noesis noeseos* according to Aristotle, which as such, represented "the structure of reflectivity" ("Struktur der Reflexivität") offering "die höchste Erfüllung von Sein als Gegenwart."[14]

It was, in fact, this structure of *Nous* or divine mind that characterized consciousness in modernity, and which according to Descartes was only "substance," *res cogitans*. It achieved its absoluteness in the transcendent idealism of the early 19th century, which conferred "der Subjektivität den Vorrang des Absoluten."[15]

In fact, if according to Aristotle, the reflectivity of thought was achieved at the end of a process in which the human consciousness characterized itself for its realistic way of approach, a thought of something and not of itself, according to Descartes self-knowledge was completely preliminary, methodologically priority, a guarantee of certainty of all the knowledge.

In this way the reflectivity, which was initially at the end of a course on reality, instead, now assumed an original and basic character, fruit of that need of method which knowledge required. Hence at first the apodictic character of self-knowledge which then, at the end of the path of knowledge, became a metaphysical position in Idealism, as *Geist*, Spirit, the central point, according to Fiche of every *Grundlage der Gesamten Wissenschaftslehre*. In this way Idealism intended to recuperate the richness of life, the intimate movement of nature and history, in the process of the infinite, absolute self-knowledge, *Geist*.

The following, post-Hegel thought, refusing the category of self-knowledge, created difficulties for the modern subjectivism, at first with Marx and then with Freud and Nietzsche, who Paul Ricoeur emblematically called the "masters of suspect." As Gadamer noted:

> Das Gemeinsame von Marx, Nietzsche, Freud ist Gewiss dieses, dass man das Gegebene des Selbstbewusstseins nicht gutgläubig als eine Gegebenheit hinnehmen kann.[16]

The secondary character was thus affirmed, deriving from reflection and consciousness, subject to conditioning, both from the part of the psyche as well as that of society. Finally Nietzsche, radicalizing such conditionings, went as far as to deny the existence of consciousness, in that he saw in it a simple interpretive expedient of human phenomena, above all, morals.

It is clear that from a point of view of the subjectivity of reflective "consciousness," the body had been methodologically seen as extraneous, not permeated by the light of consciousness, another "substance," extended *res extensa*, as regards to thought or soul which constituted the inner principle of moment.

Only by abandoning the Cartesian pattern of the two substances, the body, in modernity, resumed life as a factor that conditioned consciousness with its natural needs and instincts, in order to get carried away by the Nietzschean moment and the initiation of the dance, by its vitalism, which beckoned the soul to a new lively youth. Nietzsche commented:

> der geschmeidige überredende Leib, der Tänzer, dessen Gleichnis und Auszug die selbstlustige Seele ist. Solcher Leiber und Seelen Selbst-Lust heisst sich selber.[17]

Not sharing Nietzsche's radicalism and his "overturning" of the values in the name of corporeality and earthliness, Phenomenology continued on the path of the "suspect," contrasting the primacy of the self-knowledge, "der Sachgegebenheit gegenüber."[18] This concreteness is affirmed from a moral point of view by Max Scheler and from a theoretic one by Husserl, by which he encouraged the going back "to the things themselves," to their eidetic "essence." However, according to Heidegger, the "suspect" had to be carried through to the end, doubting the pretensions of the consciousness of placing itself as an exclusive focal point of meaning and consequently opening itself up to existence. Hence the reflection of the task of Phenomenology, intended not simply as a "description of what is presented" but also and above all, as a "manifestation of the hidden."

Damit etwas sich zeigt, verlangt das eine Aufdeckung des Verdeckten, damit es zum Sichzeigen kommen kann. Das Wort "Phänomenologie" heisst also nicht allein "Beschreibung dessen, was gegeben ist," sondern schliesst die Abdeckung des Verdeckenden ein, das nicht nur in falschen theoretischen Konstruktionen zu bestehen braucht. So nämlich sah anfänglich in gewissen Umfang die phänomenologische Beschreibungskunst aus.[19]

In fact, the way in which the theme of the body and another are laid down in Phenomenology reveals the difficulties from its point of view regarding its "mistaken theoretic constructions." In the search for an essential constitution of the "world of life" Husserl, in his *Cartesianische Meditationen*, had to encounter the aporias concerning the *eidos* and existence, the perception of his own body and that of another person's, which indicated the appearance of an *alter ego* endowed in turn with a consciousness which however, due to its extraneousness, was unreachable. This was impeded, in fact, in first instance, by the phenomenological problems of the corporeity, that remained in Husserl like an unresolved paradox. He, in fact, wanting to attain the demonstrativeness of consciousness, that is its maximum clarity and cognitive evidence, clashed with the opaque residue of the body, which he was unable to control or look at in the right way, due to the subjective and transcendent intrusiveness of consciousness subject to *epochè*. Therefore the consciousness, knowing nothing about the body, which could not be an object of its own "consciousness," neither as *Körper* (in as far as a physical object, *res*, remained something foreign and heterogeneous to it), nor as *Leib*, as original, pre-categorial reality. So Phenomenology lived the paradox of the corporeity as alterity: of a consciousness that existed in the measure in which the corporeity represented its limits and its alienation. In fact, it was actually this extraneousness that established it as "consciousness,"[20] in virtue of the primacy of the "transcendental ego."[21]

Husserl, in his pursuit of "essence," was unable to go beyond a metaphysical claim linked to the primacy of the being's view of Being, which prevented him from going beyond the corporeity, the simple physicality of the body, in order to grasp the irreducible alterity of the real life of the *psyche*. From this point of view Husserl was unable to go very far.

In the ambit of *Erlebnis*, the resorting to the physical corporeity of the object perceived, that is to the other body, explained by means of the "hyletischen Daten," revealed to be totally inadequate and insufficient, purely an external and secondary fact with regard to the activity, infinitely richer, of the consciousness. Gadamer explained Husserl's work as follows:

> Offenbar bewegt sich hier Husserl selber im Zeichen eines Abbaus dogmatischer Verdeckungen und nicht im Zeichen der primären Gegebenheit der Sache. Das aber war die grosse Parole der Phänomenologie gewesen, zu den Sachen selber zurückzukehren. Jedenfalls hat Husserl unter dem Druck wissenschaftstheoretischer Motive darauf bestanden, dass der Andere zunächst nur als Wahrnehmungsding gegeben sein kann und nicht in der Lebendigkeit, in der leibhaften Gegebenheit. An diesem Punkte setzt Heideggers ontologische Kritik an Husserls Phänomenologie an, dem Punkte ihrer stärksten Evidenz.[22]

According to Gadamer, Edmund Husserl's work showed a very limited concept of "der Sachgegebenheit gegenüber," which shared the ideal of description and empirical measure of the actual phenomenon of modern science.

On the other hand, in the effort of exposing the monadic solitary character of Husserl's consciousness and self-knowledge, Heidegger highlighted the difficulties regarding reconciling the continuous presence of consciousness, its evidence (which seemed to be a never ending task) with the need to give a meaning to reality that remained extraneous to consciousness, beginning with the corporeity and the experience of others. In other words it was a matter of reconciling the two aspects which in the phenomenological perspective appeared to be irreconcilable: the *inseità* of consciousness and the *alterity* of the world of life, *Lebenswelt*.

Here then was the need to go beyond the transcendental character of consciousness in favour of the concrete existence, of an existential ontology. In this way, it was evident that reality was such because it was different from consciousness. The participation of consciousness in the "world of life," *Lebenswelt*, of which Husserl then spoke, and as time was immerged in consciousness, both became in Heidegger, existence, Being present, *Dasein*. The latter was, in the temporality, "the plan of being," in which the *Seinsfrage* found clarification regarding the question about the meaning of Being; this was the dominating motive of his work *Sein und Zeit*.

Leading Phenomenology back into the most radical criticism of the categories of Greek and Western thought was the merit of Heidegger. His notion of *Sein* did no longer concern the "substance" in the manner of Aristotle, nor in the modern Cartesian manner, as presence or "self-knowledge." He warned of the risk of such a "distortion of the substance,"[23] that should therefore be avoided. His distancing himself led him to place the *Sein* beyond such metaphysics, intending it as *Difference*. " 'Sein' ist nicht nur als zu verstehen, dessen ich bewusst bin, dass es da ist – oder wie der spätere Heidegger interpretiert: anwesend ist."[24]

THE NEW THEMES

This has brought us back to the considerations previously made. In fact, with Heidegger we saw a significant turning point through the criticism on Phenomenology; a criticism which was linked again to that of Greek and modern ontology, with the notion of "substance" in the centre, with its fundamental categories. In this Heidegger detected an anthropological disguise. In fact, in his *Brief über den Humanismus* in 1946, he affirmed that existing in a temporal way or *Dasein*, was a condition of *Sein*, of its "ecstatic" and "circumstantial" character. He pointed out how in this manner the above mentioned metaphysics all remained in an anthropological sphere which was unreal and therefore contrary to an authentic humanism.

Hence the traditional categories of the Aristotelian "substance" were completely inadequate.

Durch diese Wesensbestimmung des Menschen werden die humanistischen Auslegungen des Menschen als animal rationale, als 'Person', als geistig-seelisch-leibliches Wesen nicht für falsch erklärt und nicht verworfen. Vielmehr ist der einzige Gedanke der, dass die höchsten humanistischen Bestimmungen des Wesens des Menschen die eigentliche Würde des Menschen noch nicht erfahren. Insofern ist das Denken in *Sein und Zeit* gegen den Humanismus. Aber dieser Gegensatz bedeutet nicht, dass sich solches Denken auf die Gegenseite des Humanen schlüge und das Inhumane befürworte, die Unmenschlicheit verteidige und die Würde des Menschen herabsetze. Gegen den Humanismus wird gedacht, weil er die Humanitas des Menschen nicht hoch genug ansetzt. Freilich beruht die Wesenshoheit des Menschen nicht darin, dass er die Substanz des Seienden als dessen "Subjekt" ist, um als der Machthaber des Seins als das Seiendsein des Seienden in der allzu laut gerühmten "Objektivität" zergehen zu lassen.

Der Mensch ist vielmehr vom Sein selbst in die Wahrheit des Seins "geworfen," das er, dergestalt ek-sistierend, die Wahrheit des Seins hitte, damit im Lichte des Seins das Seiende als das Seiende, das es ist, erscheine.[25]

The authentic humanism moved between the "casting" of the existence and the "guarding of the being," and all it had to do was accept and preserve the "truth of being" in the new existential categories of "analytics." Heidegger's thought had been clear since the 1920s. However, where did this viewpoint leave the body and the person?

We were witnessing a reversal of Husserl's position and of the traditional anthropology, both successors of the ancient metaphysical thought: it was not the life, neither of "consciousness" nor of "nature," that founded the notion of being (*Sein*), but it was the being that founded the notion of life, which then, in fact, deprived of its psychological and subjective connotations, became a method of "existence." The corporeity was no more than the phenomenological aspect of life, its manifesting itself and its behaving in an environment, in the world.

Let us now try to get a closer view of what has just been said. We shall do this by stopping to consider §10 of section 1, part 1 of *Sein und Zeit*, entitled *Die Abgrenzung der Daseinsalalytic gegen Antropologie, Psychologie und Biologie*. Here Heidegger started with the circumstance where, in his opinion, anthropology, psychology and biology had never contributed to favouring a "precise answer," from an ontological point of view, to the question about "that being that we ourselves are."[26]

What did Heidegger mean by this? That such disciplines, concerning the human being or not, were inadequately revealed with regard to the objectives of the philosophical research that he was at that time carrying out. In general the following question emerged: which role of science or knowledge generally respects the goals of philosophy? Could Heidegger's philosophy give new impetus to such knowledge?

Here he proposed two problems: (1) the "delimitation" or distinction of existential analytics as regards these sciences; (2) The importance for them of the new ontological perspective. The two problems were closely connected, one led to the other.

As regards to the first problem, he observed:

> Eine ihrer ersten Aufgaben wird es sein zu erweisen, dass der Ansatz eines zunächst gegebenen Ich und Subjekts den phänomenalen Bestand des Daseins von Grund aus verfehlt. Jede Idee von "Subjekt" macht noch – falls sie nicht durch eine vorgängige ontologische Grundbestimmung geläutert ist – das Ansatz des subjectum *(hypokeimenon) ontologisch* mit, so lebhaft man sich auch ontisch gegen die "Seelensubstanz" oder die "Verdinglichung des Bewusstseins" zur Wehr setzen mag.[27]

This was because the categories of the subjectivity, which we have already mentioned, were led back to their existential foundations, getting round Descartes's path who had forgotten to argue about the *sum* as well as the *cogito*, and that of Phenomenology, which had not arrived to where Heidegger had begun with his "existential analytics." The ontological theme of the *sum* had been subordinate to the gnoseologic one of the *cogito*, while it was necessary to clarify the second by means of the first. Heidegger took care of this with his "analytics," which overturned the terms of Cartesian and phenomenological subjectivism in the light of the *planning* character of the existence. At the base of the Cartesian thinking substance (both it-"soul" as well as "consciousness"), there was as *hypokeimenon* the *Dasein*, whose *essence* consisted in *existence*, according to Heidegger's famous expression.

This was a notion of existence that was no longer thought of according to the metaphysics of the Greek and early Christian "substance," where it was considered from the point of view as the "simple presence," but according to the most authentic, existential peculiarity of being of *Dasein*, of the being as a "who," in other words, a concretely existing individual. This was the preliminary explanation that Heidegger considered necessary to make before using categories which could end up being inadequate for the understanding of that particular being "that we are."

Therefore, the ontological problem had remained until then on the fringe of man's problems, faced with personalism, that had tried to define the being as a person, therefore as a unit of the body, the soul, the spirit and the philosophy of life from W. Dilthey to Edmund Husserl and Max Scheler, in which the Being present was confused with a "simple presence" of life, while in reality, existence was not a simple living to which something was added, nor was it a simple presence of life, but it was the most authentic and inescapable way of Being present, therefore of life, of the spirit, of the soul or consciousness, of the body.

This meant that the psychical, physical, vital being, had nothing to do with the authentic being of the Being present, their reflections were found within the *Dasein*. This was why the moral sciences such as anthropology, psychology and biology already presupposed that the *Dasein* was a fact from where to begin. Here, the separation between these sciences and Heidegger's analytics was given as a starting point in virtue of this observation. Where the existential analytics ended the reflection of the moral and natural biological sciences began. The analytical task was that of defining the ontological problem of the Cartesian *sum*; the scientific task was that of defining the different *cogitations* or contents of knowledge.

Here the traditional definition of man as an *animale rationale* underwent a reinterpretation. Man was existence, *Dasein*; his rational animality was a possible *cogitatio*, therefore an assertive and cognitive content, not of the "what" of Being present, but of its possible "how," because existence being a plan, the *animal rationale* was a possible aspect of it, without necessarily having to intend it as its only "essence."

What can we say about man as a person in the sphere of philosophical anthropology? According to Husserl the contraposition between the natural and the spiritual world was needed to explain how the person was made up in a different way compared to things, therefore compared also to the body as physical reality; according to Max Scheler, the primacy of consciousness was translated in the intentional constitution of the person, by means of the carrying out of his inner actions. Here the contraposition between actions and substance, between a static vision and a dynamic one, that went beyond a psychical planning of the person, was clear.

Heidegger moved his criticism towards Husserl and Max Scheler, who led the notion of person to a form of intentional subjectivity, to a number of operations, of intentional acts of the consciousness. Was person therefore synonymous with "intentional consciousness"? The question Heidegger asked himself and which remained unanswered by Husserl or Scheler, regarded the ontological meaning of carrying out "intentional actions."

Jede psychische Objektivierung, also jede Fassung der Akte als etwas Psychisches, ist mit Entpersonalisierung identisch. Person ist jedenfalls als Vollziehr intentionaler Akte gegeben, die durch die Einheit eines Sinnes verbunden sind. Psychisches Sein hat also mit Personsein nichts zu tun. Akte werden vollzogen, Person ist Aktvollzieher. Aber welches ist der ontologische Sinn von "vollziehen"?, wie ist positiv ontologisch die Seinsart der Person zu bestimmen?[28]

This question could not be answered by appealing to a notion of person as a physical, animated, spiritual unit, because:

In der frage nach dem Sein des Menschen aber kann dieses nicht aus den überdies erst wieder noch zu bestimmenden Seinsarten von Leib, Seele, Geist summativ errechnet werden.[29]

From the moment that they alone cannot explain that being of the Being present which they necessarily assume. The body is of a being, the soul is of a being, the spirit is of a being. They are indeterminate because they do not possess within themselves their reason of being, therefore their being does not determine the authentic nature of Being present, but emerges determined from the *Dasein.*

For this reason Heidegger saw in Scheler and in the concepts of the body, the soul and the spirit still operating, a prejudicial question regarding the Greek and early Christian anthropology in that the intentionality was linked to a notion of consciousness that it still had in the *logos*, if not the essence in the Greek and Aristotelian meaning of the word, at least:

eine höere Ausstattung, deren Seinsart ebenso dunkel bleibt wie die des so zusammengesetzten Seienden.[30]

The notions of corporeity and of person, as they had been developed in the ancient thought until phenomenology:

zeigen an, dass über einer Wesensbestimmung des Seienden "Mensch" die Frage nach dessen Sein vergessen bleibt, dieses Sein vielmehr als "selbstverständlich" im Sinne des *Vorhandenseins* der übrigen geschaffenen Dinge begriffen wird.[31]

With Heidegger the anthropological problem went back to being, after the misunderstandings of the metaphysics of substance and Phenomenology, the problem of Being. Therefore he did not delve into the problem of man (body and soul as a substantial unit) or of the person, but only of the *Sein*. He clearly wrote that the personalism and the philosophy of life had not become aware of the "insufficiency of ontological foundations."[32] He thus contested that philosophical anthropology had defined man as *animal rationale* and theologically as *imago* and *similitudo Dei*.

THE KANTIAN ANALYSIS OF THE *PERSONALITAS*

Back in the 1920s, Heidegger carried out a concise criticism on the different notions of *personalitas*, in so far as the unessential definitions of "that being that we ourselves are." In his work (made up of the lessons of the summer term of 1927 held at the university of Marburg) *Die Grundprobleme der Phänomenologie*, he talked about *personalitas transcendentalis, personalitas psychologica* and *personalitas moralis*, discussing them from an ontological perspective. What did they mean?

The three forms of "personality" were discussed by Heidegger in reference to Kant. With the first of these, the *personalitas transcendentalis*, Kant strayed from the Cartesian determination where:

das Ich, das *ego*, ist für ihn wie für Descartes res cogitans, *res, etwas*, das denkt, d.h. das vorstellt, wahrnimmt, urteilt, zustimmt, ablehnt, aber auch liebt, hasst, strebt u. dgl. Alle diese Vorhaltungen bezeichnet Descartes mit cogitationes. Das Ich ist etwas, das diese cogitationes hat.[33]

At the same time Kant radicalized, beyond these determinations or *cogitationes* the *ego* or "I" as *subiectum* that was perceived as "foundation" of its own determinations, that is of its own actions.

Das Ich ist nicht einfach irgendein isoliert Punkt, sondern es ist 'Ich-*denke*'. Es nimmt sich aber selbst nicht wahr als ein Seiendes, das noch andere Bestimmungen hätte ausser denen, dass es eben denkt. Vielmehr weiss sich das Ich als der *Grund* seiner Bestimmungen, d.h. seiner Verhaltungen, als der Grund seiner eigenen Einheit in der Mannigfaltigkeit dieser Verhaltungen, als Grund der Selbigkeit seiner selbst.[34]

This was because:

Alle Bestimmungen und Verhaltungen des Ich sind ich-gegründet. Ich nehme wahr, Ich urteile, Ich handele. Das "Ich-denke," sagt Kant, muss alle meine Vorstellungen, d.h. alles cogitare der cogitata, begleiten können. Dieser Satz ist aber nicht so zu fassen, als sei bei jedem Verhalten, bei jeden Denken im weitesten Sinne jeweils auch die Ich-Vorstellung dabei, sondern Ich bin mir der Verknüpfung aller Verhaltungen mit meinem Ich bewusst, d.h. Ich bin mir ihrer bewusst in ihrer Mannigfaltigkeit als *meiner* Einheit, die in meiner Ichheit (als subjectum) als solcher ihren Grund hat. Erst auf dem Grunde des "*Ich*-denke" kann mir Mannigfaltiges gegeben sein. Kant interpretiert zusammenfassend das Ich als die "ursprüngliche synthetische Einheit der Apperzeption." Was heisst das? Das Ich ist der ursprüngliche *Grund der Einheit* der Mannigfaltigkeit seiner Bestimmungen in der Weise, dass ich als Ich sie alle rücksichtlich meiner selbst zusammen habe, im vorhinein zusammen halte, d.h. verbinde, Synthesis. Der ursprüngliche Grund der Einheit *ist*, was er ist, er ist dieser Grund als einigender, als synthetischer.[35]

This meant that in every representation, the subject, in thinking, perceived or thought also of itself, that is to say it became aware of its own "I" as foundation. In consequence of which the subject's being "intended in an eminent sense," lay specifically in this apperception as an *original synthetic unit*, which Kant also called "I" in the logical sense, intended as the subject of thought or of the *Logos*, that was at the base of every *cogitatio*, not a simple logical universal abstraction, an anonymous I, as intended by Rickert and the new Kantians, but, according to Heidegger, a real *subiectum*, as *hypokeimenon* "that what remains when all his accidents are removed."[36]

Heidegger explained that by this concept of *ichkeit*, "that is the same for all factual subjects"[37] the formal structure of the "personalitas" or *personalitas transcendentalis* had been reached. Here the "formal structure" next to the ontological one of *subiectum* can appear to be a contradiction. In truth this is not the case for Heidegger, who wishes to point out how the ontological foundation of Kantian critical philosophy coincides with the "how" of consciousness, or with the whole of all those formal structures that the "I" perceives as transcendental articulation.

In fact, according to Heidegger, by the well known word "transcendental", Kant connoted a philosophy, the "transcendental philosophy" which was nothing more than a form of ontology, in that it was a research into the foundation of knowledge. From this viewpoint an interpretation of the *Kritik der reinen Vernunft* was impossible as a simple theory of knowledge, a gnosiology, or as a research into its conditions regarding possibility, but it was possible as a research that had to constantly bear in mind, in the presence of itself "die Ontologie als Wissenschaft vom Sein die Wissenschaft von der Erkanntheit der Gegenstände und ihrer Möglichkeit sein"[38]: therefore the phenomenon of the object and the foundation or possibility of manifesting itself, that coincided concretely with its being.

The "I" then, was not a being among other beings, but their foundation. The *Ich denke*, as the original synthetic unit of the apperception was the fundamental ontological condition of every being represented, and the "formal structure of the personality intended as *personalitas transcendentalis*." Heidegger, thus interpreted the "transcendental personality" as the expression of a new ontology, the transcendental ontology that consisted in retaining the "I" as the foundation and regulation of phenomenal nature.

However, what kind of ontology was transcendental ontology? What foundation was at the base of the transcendental personality? These are the simple questions that Heidegger seems to have left unanswered in his interpretation of the Kantian gnosiology as with the transcendental ontology through the notion of the *personalitas transcendentalis*.

This notion of "personality" did not however complete the entire concept of *personalitas*. In fact Kant also introduced the notion of *personalitas psychologica* in reference to the empiric "I," existing as perceptible self-knowledge, as the ability of having sensations and of mediating them by means of the external senses or the inner sense, individual sensations like the feeling of pleasure or displeasure, due to the physicality of one's own body and of its sensations intended as "representations as subsisting and always changeable events," referring to an *I of apprehension*, as Kant called it, that could also be called "soul," a vital and experiential principle.

Heidegger explained:

Das empirische Ich als Seele kann gleichfalls theoretisch als Idee gedacht werden und fällt dann zusammen mit dem Begriff der Seele, wobei Seele gedacht wird als Grund der Animalität, oder wie Kant sagt, der Tierheit, des Lebens überhaupt.[39]

He specified:

Das Ich als personalitas transcendentalis ist das Ich, das wesenhaft immer nur Subjekt ist, das Ich-Subjekt. Das Ich als personalitas psychologica ist das Ich, das immer nur Objekt, vorgefundenes Vorhandenes ist, Ich-Objekt, oder wie Kant direkt sagt: "dieses Ich-Objekt, das empirische Ich, ist eine Sache." Alle Psychologie ist daher positive Wissenschaft von Vorhandenem.[40]

What did this mean? That the "I" of apprehension could be objectified, analysed and investigated in its representations, in the objective concreteness of its actions, therefore one could have knowledge of "I," always, however assuming the "I" of the transcendental apperception as foundation. Heidegger pointed out:

Wie es möglich sei, dass ich, der ich denke, mir selber ein Gegenstand (der Anschauung) sein, und so mich von mir selbst unterscheiden könne, ist schlechterdings unmöglich zu erklären, obwohl es ein unbezweifeltes Faktum ist; es zeigt aber ein über alle Sinnenanschauungen so weil erhabenes Vermögen an, dass es, als der Grund der Möglichkeit eines Verstandes, die gänzliche Absonderung von allem Vieh, dem wir das Vermögen, zu sich selbst Ich zu sagen, nicht Ursache haben beizulegen, zur Folge hat, und in eine Unendlichkeit von selbstgemachten Vorstellungen und Begriffen [d.h. die ontologischen] hinaussieht.[41]

Can we speak then of a double personality? Undoubtedly no. The reference to Descartes is inevitable, referring to when we spoke of the separation between man and the "I" as *cogito* and the corporeity, between man and the animal. However such a separation is no good at all in contesting two types of personalities, one real and the other false, because the latter is a material "thing".

Es ist nur soviel gesagt, dass das Sein dieses Ich problematisch, nach Kant überhaupt unbestimmbar, jedenfalls grundsätzlich nicht mit den Mitteln der Psychologie zu bestimmen ist. Die personalitas psychologica setzt die personalitas transcendentalis voraus.[42]

The psychological personality is a "representation" of the transcendental personality. Or, on the other hand, is the transcendental personality the basic condition of the psychological personality? One cannot exist without the other. The first represents the basic level of the empirical knowledge of itself, of its awareness; but it is exactly for this that it cannot be the objective of any empirical knowledge or consciousness. The second, on the other hand, represents the phenomenal representation of its own personality or of itself, but precisely for this reason it is never foundation, neither of itself nor of knowledge, therefore it always remains something indistinct that is individual and relative.

A third type, according to Kant, is the *personalitas moralis*. In a passage of his work *Die Religion innerhalb der Grenzen der blossen Vernunft*, Chapter 1, §1, Kant indicates three elements characterizing man.

Kant nennt hier im ersten Stück, ersten abschnitt drei Elemente nennt et erstens die Tierheit, zweitens die Menschheit und drittens die Persönlichkeit. Die erste Bestimmung, die Tierheit, charakterisiert den Menschen als etwas Lebendiges überhaupt, die zweite Bestimmung, die Menschheit, als ein Lebendes, zugleich aber Vernünftiges und zugleich der Zurechnung fähiges Wesen. Wenn er an dritter Stelle im Unterschied von zweitens, Menschheit, Persönlichkeit nennt ist offenbar, dass Persönlichkeit hier in einem engeren Sinne gemeint ist, abgehoben von der personalitas transcendentalis, die mit Menschheit identisch ist. Zum vollen Begriff der personalitas gehört nicht nur Vernünftigkeit, sondern Zurechnungsfähigkeit. Persönlichkeit besagt daher bei Kant ein Doppeltes: einmal den weiten, formalen Begriff der ichkeit überhaupt im Sinne des Selbstbewusstseins, sei es nun transzendentalis, Ich-denke, oder empirisches, Ich-Objekt, sodann den engeren und eigentlichen Begriff, der die beiden anderen Bedeutungen bzw. das, was sie meinen, in gewisser Weise einschliesst, aber sein Zentrum in der Bestimmung hat die es jetzt zu betrachten gilt.[43]

Kant in chapter 1, §1, entitled *Von der ursprünglichen Anlage zum Gutem in der menschlichen Natur* of the cited work, wrote in reference to the above mentioned disposition, that there were three types of possible determinations:

1) Die Anlage für die *Thierheit* des Menschen, als eines lebenden.
2) Für die *Menschheit* desselben, als eines lebenden und zugleich *vernünftigen*;

3) Für seine *Persönlichkeit*, als eines vernünftigen, und zugleich der *Zurechnung fähigen* Wesens.

1) Die Anlage für die *Thierheit* im Menschen kann man unter dem allgemeinen Titel der physischen und bloss mechanischen Selbstliebe, d.i. einer solchen bringen, wozu nicht Vernunft erfordert wird ....
2) Die Anlage für die *Menschheit* können auf den allgemeinen Titel der zwar physischen, aber doch *vergleichenden* selbstliebe (vozu Vernunft erfordert wird) gebracht werden, sich nämlich nur in Vergleichung mir andern als glücklich oder unglücklich zu beurteilen ...
3) Die Anlage für die *Persönlichkeit* ist die Empfänglichkeit der Achtung für das moralische Gesetz, *als einer für sich hinreichenden Triebfeder der Willkür.* Die Empfänglichkeit der blossen Achtung für das moralische Gesetz in uns wäre das moralische Gefühl, welches für sich noch nicht einen Zweck der Naturanlage ausmacht, sondern nur, sofern es Triebfeder der Willkür ist.

According to Kant only the *personalitas moralis* could call itself in an authentic sense "personality." Let us take a look at what this consisted of. The personality was basically self-knowledge and therefore indicated a particular awareness, that Kant called "moral sentiment" or "feeling of my existence." The moral personality posed itself as a synthetic idea of the two previous forms: this had neither the purely formal characteristics of the transcendental personality, nor the merely empirical contingent characteristics of the psychological personality, but it posed itself as an attempt at raising the empiric to the universal, by means of action and will, and at rendering the universal a sort of end of the individual, of the relative. These three personalities, in conclusion, represented, in the Kantian conception, a re-proposing of that correlation between the three-fold *Kritik* that was so typical of its speculation.

Heidegger observed:

Dieses ist keine zufällige empirische Erfahrung meiner selbst, aber auch kein theoretisches Wissen und Denken des Ich als Subjekt des Denkens, sondern ein Offenbarmachen des Ich in seiner nichtsinnlichen Bestimmtheit, d.h. seiner selbst als des Handelnden.[44]

The moral sentiment for Kant was the *respect* of the moral law of which he gave us a brilliant phenomenological analysis in his *Kritik der praktischen Vernunft* (part 1, book 1, chapter 3) aimed at revealing the moral self-consciousness, the *personalitas moralis*, as the authentic personality of man. Kant considered the respect of moral law not as a sentiment produced by sensitivity and its inclinations but by reason, therefore by a spiritual principle which had to induce man to act according to the "morals maximums," that is according to a determinate subjective motive of the will.

Das Wovor der Achtung bzw. das, wofür dieses Gefühl ein Gefühlhaben ist, nennt Kant das moralische Gesetz. Dieses Gesetz gibt die Vernunft als freie sich selbst. Achtung vor dem Gesetz ist Achtung des handelnden Ich vor sich selbst als dem Selbst, das nicht durch Eigendünkel und Eigenliebe verstanden ist. Achtung als Achtung vor dem Gesetz bezieht sich zugleich in ihrem spezifischen Offenbarmachen auf die Person. "Achtung geht jederzeit nur auf Personen, niemals auf Sachen."[45]

At this point, Heidegger posed the question regarding the ontological determination of the "moral personality," even though it was unrelated to the Kantian problem. He asked:

Als was muss das in dieser Weise im moralischen Gefühl der Achtung *ontisch* als seienden Ich offenbare Selbst ontologisch bestimmt werden?[46]

Taking the above quotation or the Kantian distinction between "thing" and "person" as a starting point, Heidegger pointed out that Kant implicitly seemed to provide an answer to the question in his *Metaphysik der Sitten*, which Heidegger intended as "ontology," that is as ontology of human existence[47] when he observed that man morally existed as an end to himself, not simply as a means for being used by anyone, not even God. However, we know that God can do nothing when faced with man's freedom.

On this Heidegger commented:

Von hier aus, d.h. aus der ontologischen Charakteristik von Seiendem, das nicht nur von anderen als Zweck aufgefasst und als Zweck genommen wird, sondern als Zweck objektiv-wirklich existiert, wird der eigentlich ontologische Sinn der moralischen Person deutlich. Sie existiert als Zweck ihrer selbst, d.h. sie *ist* selbst Zweck.[48]

The "moral personality" placed itself in its authentic and true personal dimension, with regard to its ability of "distinguishing *from the ontological point of view*" the personal "I" from the non "I," the subject and the object, the *res cogitans* and the *res extensa*, to use once again the Cartesian words, which Kant had made his own with the duplicate distinction between "thing" and "person."

With Kant it was possible to affirm that "Die Wesen, deren Dasein zwar nicht auf unserm Willen, sondern der Natur beruht [d.h. der Natur im Sinne der physischen Organisation], haben dennoch, wenn sie vernunftlose Wesen sind, nur einen relativen Wert als Mittel und heissen daher Sachen, dagegen vernünftige Wesen *Personen* genannt werden, weil ihre Natur [hier besagt Natur soviel wie *physis* gleich essentia] sie schon als Zwecke an sich selbst, d.i. als etwas, das nicht bloss als Mittel gebraucht werden darf, auszeichnet, mithin sofern alle Willkür einschränkt, (und ein Gegenstand der Achtung ist) "(*Grundlegung zur Methaphysik der Sitten*, in *Werke*, ed. Cassirer, IV, pp. 286–287). Heidegger commented, quoting Kant in the same work: "Was die Natur der Person, ihre essentia, ausmacht und alle Willkür einschränkt und das heisst als Freiheit bestimmt ist, ist Gegenstand der Achtung. Umgekehrt, das in der Achtung Gegenständliche, d.h. in ihr Offenbare, bekundet die Personalität der Person. Ihr ontologischer Begriff sagt kurz: Personen sind, objektive Zwecke, d.i. Dinge [res im weitesten Sinne], deren Dasein an sich selbst Zweck ist (*Ibid.*)."[49]

Were people in the more general sense "things" (*res*)? And if so, in what way? According to Kant both people as well as things (*Sachen*), were *res (Dinge)*, beings in the wider sense of the word. The distinction between *Sachen* and *Dinge* was fundamental. This did not mean that the person was a thing, but that the person as the *res*, was characterized, by its absolute autonomy, as something "in itself" or noumenon. For this reason the Kantian expression of *Ding an sich* denominates the "noumenon." Therefore the person was not intended as an empirical subject, but as a noumenal dimension. That is to say, the person's being was not linked to the phenomenon of life, but to his hidden, essential, volitional and planning dimension. Kant needed the noumenal dimension to translate in his lexicon the classical notion of the person as *humanitas*. It then became the "maximum" of the categorical imperative.

The person was a thing (*Dinge*), a *res*, that is something that existed as an end to itself, an end that could not be manipulated because it could not be traced back to the phenomenal dimension. Heidegger explained "Zweck muss hier immer im objektiven Sinne verstanden werden, als seiender Zweck, Person."[50] And "Zweck ist existierende Person, das Reich der Zwecke das Miteinander der existierenden Personen selbst,"[51] or as he commented, "das *Commerzium der Personen* als solches,"[52] their reciprocal relationship.

However, Heidegger asked himself, had Kant sufficiently determined by this the being of man? Had he clarified by this the being present's way of being? The answer is certainly no. He had not: "Wir suchen vergeblich bei Kant nach einer Aufklärung dieser Frage, ja auch nur nach der Fragestellung."[53]

> Im gegenteil, die angeführten Zitate zeigen, dass Kant vom Existieren des Menschen, vom Dasein der Dinge als Zwecke spricht, die Termini "Existieren" und "Dasein" für ihn aber Vorhandensein bedeuten. Er spricht ebenso vom Dasein der Sache. Er sagt nirgends, dass mit Bezug auf den Menschen der Begriff der Existenz und des Daseins einen anderen Sinn habe, oder gar welchen. Kant zeigt nur, dass die essentia des Menschen als Zweck anders bestimmt ist als die essentia der Sachen und der Naturdinge. Aber vielleicht spricht er nicht ausdrücklich von der spezifischen Seinsart der moralischen Person, meint aber de facto eine solche?[54]

Therefore what was missing in Kant's work, according to Heidegger, was a clear notion of "existence" as the foundation of the person and of the other beings. He had developed the character of the "subsistence" not of the "existence," which was and is a different notion. "Subsisting" meant existing in oneself, independently, which is different to "existing" which had the opposite characteristics. From the point of view of subsistence, the being did not have the characteristics of Being present as *Dasein*. By assigning subsistence's way of being to the person as *res* or being, Kant brought him back into a metaphysical point of view of the "substance," a "noumenon" reality, as *subiectum* and as "self-knowledge,"[55] hence compromising any advantage regarding the theme of the *Ichkeit*.

Here we find Kant's ambiguity: the notion of subsistence on the one hand guaranteed the person total independence and worth, while on the other hand confining him in an abstract dimension. He compromised his historical and concrete dimension, which Heidegger called in *Sein und Zeit* "existential."

What can we say about the *personalitas transcendentalis*? As it clearly appears in the second book of the *Kritik*, which deals with the "transcendental dialectics," the "I" as soul or "spiritual" reality was something "noumenal", subsisting, therefore indeterminable; and the attempts to render it such resulted in "paralogisms," that is in erroneous arguments, which claimed to count beyond the phenomenon or natural sphere. After Kant, the question was posed about the possibility of an *ontological* interpretation of the subject that would be metaphysically and scholastically "adequate, that is free of all traditions."[56]

The same basic result emerged, despite the distinction between the theoretical plane and the practical one: the impossibility of an adequate determining of the person and of his way of being as far as "existential" was concerned.

Heidegger, noted in Kant "a single negligence," that of not having grasped the theoretical "I" and the practical "I" in their oneness.

> Ist diese Einheit und Ganzheit beider etwas Nachträgliches oder etwas Ursprüngliches vor beiden? Gehören beide ursprünglich zusammen, oder werden sie nur nachträglich äusserlich verbunden? Wie ist das Sein des Ich überhaupt zu fassen?[57]

In fact, if from an "existential" viewpoint it was a question of "a single negligence," as Heidegger correctly pointed out, in that it was impossible to distinguish the theoretic moment from the practical one, the knowing from the doing, because they were structural forms of the way of man's being, from Kant's point of view it was a question of a singular "critical" necessity, in that it was needed to distinguish the "how" was man, from the "what" was man, the phenomenal plane from the "noumenal" one, and to found upon this distinction, the primacy of the latter, that is of the person as *Humanitas*. Kant thought he could better preserve the peculiarity, the person's freedom, by placing it on a high meta-physical and meta-scientific plane, as "an end to himself" not representable according to the categories of the intellect.

This created some problems, among which the decisive one of the abstractness, evanescence and inexistence of the person, as it subsisted only as "substance". It was therefore necessary to try a different solution, capable of reconciling the body's phenomenal needs with the "noumenal" ones of the soul, the will and the freedom. This concerned the personal "I" in its entirety and in its concrete relationships between the body and the soul, between the phenomenon and the noumenon of the human life.

Aber nicht nur die Seinsstruktur dieses ganzen Ich der theoretisch-praktischen Person in ihrer Ganzheit ist unbestimmt, sondern noch viel unbestimmter ist das Verhältnis der theoretisch-praktischen Person zum empirischen Ich, zur Seele, und weiterhin das Verhältnis der Seele zum Leib. Zwar sind Geist, Seele, Leib für sich und in je verschiedener Weise ontologisch bestimmt bzw. unbestimmt, aber das Ganze des Seienden, das wir selbst sind, Leib, Seele und Geist, die Seinsart ihrer ursprünglichen Ganzheit, bleibt ontologisch im Dunkeln.[58]

It was a question, therefore, not only of correlating the theoretic moment to the practical one, but above all of finding in this correlation an adequate ontological foundation that could justify it. Hence the need "Daher bedarf es der ausdrücklichen Besinnung auf den Weg, auf dem das Dasein selbst ontologisch angemessen zur Bestimmung gebracht werden kann."[59]

"Daher wird die Frage dringlicher: Wie müssen wir das Sein des Seienden, das wir selbst sind, bestimmen und von allem Sein des nichtdaseinsmässigen Seienden abgrenzen, aber gleichwohl aus der Einheit eines ursprünglichen Seinsbegriffes verstehen? Wir bezeichneten das Sein des Daseins terminologisch als Existenz. *Was heisst Existenz*? Welches sind die *wesenhaften Momente* des Existierens?"[60]

The Kantian considerations therefore reinforced Heidegger's conviction of passing to an "existential analytical" approach of the person, that is taking the "metaphysical" notion of the person back to its authentic, ontological, existential structures.

BY AN "ANALYTICAL" APPROACH

Thinking of the person from this point of view meant recuperating his existential-historic dimension within the horizon made up of the *Dasein* and its meaning.

The motive of temporality was central to the analysis of existence. Heidegger, with the analyses on temporality, quoted the original character of the *Dasein* as being in the world (*In der Welt-Sein*). It is here that he makes his specific contribution.

The temporal structure of existence parallelly brought about the discovery of the effective planning dimension, of the *Dasein*, that established itself in its multiple forms and expressions of life, all gathered by Heidegger within the ontological ambit of the "Hermeneutik der Faktizität."

In fact Heidegger, since his early works, especially through the study of Aristotle, defined worldly existence by means of the expression *das faktische Leben*, "the real life," as a group of worldly experiences of life, of professions, of activities, characterized by exerting oneself, by a will to move as commitment, *Bewegtheit*. Life no longer romantically had an irrational or intuitive character, but one of active interaction between the "I" and the world, through the various "experiences." Heidegger expressed this with a new word, *Dasein*, Being present, of which the I, the you, the person were operative determinations. The *Dasein* was defined and understood in fact by means of these rational and existential plans.

Already in the text of the so called "Natorp Bericht" of 1919, regarding "Aristotle's phenomenological interpretations," Heidegger, to indicate the unitary structure of the worldly being, of *Dasein*, coined the word *Jemeinigkeit*. In *Sein und Zeit*, this term would become the ontological condition of the two possible and fundamental ways of existing: authenticity and inauthenticity.

Vincenzo Vitiello explained:

> Heidegger intended to remove the being present, *das faktische Leben*, from the traditional determinations of the "I," of the subject, of the "person," with the aim of grasping the specific character of the "in being," that is to say the unitary structure of the being in the world. *Jemeinigkeit* was the term created to indicate that the "world" is always the world that one encounters in an "experience," *in a procedure of life*, and that this procedure is always the procedure of an "I," of a "you," of a "they" or even of a generic and anonymous *One (Man)*. It is clear then that the "I" like the "you" or the "they," which are predicated of the Being present, include the "object" (the world, the things, the things of the world) no less than the "subject" (of the "I"), referring them to the whole relational structure of being present (that is of real life as each time it presents itself in a worldly experience), and they have nothing to do with the so called "person" or with the "individual existence"; it can be no other way because *always mine (jemeinig)* is also the being present of *Man*, of *One*, of the inauthentic existence, that is to say of the *depersonalized* existence, or as it is also called: *standardized*.[61]

In fact, it appears to us that in Heidegger's work, the world, intended as the horizon of "taking care" by means of the commitment to the experiences of life, but also of a possible straying from existence and its authentic dimension ("dejection"), "was not originally the place in which subjects and objects met, thinking things and extended things but the horizon in which a being (man), dominated by *Care*, utilizes what he encounters."[62]

In the world man "is 'present' not as a subject nor in the way there are 'things,' as an object. The things, in fact, are within the world, 'they are included, man *ex-istit* in the world.' "[63] In this sense "the clashing of the subject and the object, is only an unprivileged one of the possible ways of being in the world."[64] Heidegger's provocation appeared as an invitation to search, in the "casting," in the "temporality" or historicity of existence, for the horizon of every possible human commitment and definition of being.

Here the theme of existence and of its structural opening (to the others and to the things of the world) became central for being able to think of alterity. In existence, the encounter with Others was different from that of the things as technical usability. The other, Kant had already stated, was an end and not a means: it is this of which one must take care of, in the worldly taking care.[65] In short, in the world in which we live, despite using things and techniques, we must never lose the original constitution of our relationship with Others.

This two-fold reminder is indicated by Heidegger in the *Brief über den Humanismus* in which he writes:

Soll aber der Mensch noch einmal in die Nähe des Seins finden, dann muss er zuvor lernen, im Namenlosen zu existieren. Er muss in gleicher Weise sowohl die Verführung durch die Öffentlichkeit als auch die Ohnmacht des Privaten erkennen. Der Mensch muss, bevor er spricht, erst vom Sein sich wieder ansprechen lassen auf die Gefahr, dass er unter diesem Anspruch wenig oder selten etwas zu sagen hat. Nur so wird dem Wort die Kostbarkeit seines Wesens, dem Menschen aber die Behausung für das Wohnen in der Wahrheit des Seins wiedergeschenkt.[66]

From this viewpoint what was the "Care" for man, for the other existence, if not the concern for the relationship with Being? Gadamer revealed the non "political" or "public" character (as Heidegger would have said), of this relationship in that it was not a case of providing for the other, but of making sure that the other realized his own plan of existence, that is to say "the placing of the other in its authentic being itself, as regards to taking care of the other that would like to eliminate the care of its own being present."[67]

It is so clear that for Heidegger the other was "another" plan that was presented as regards to "my plan" in the temporal horizon of existence. This was the result he achieved, beginning with different conjectures, both with regard to the existential analysis of *Sein und Zeit*, as well as to going back, after the *Kehre* of the 1930s, to an ontological revival. More precisely, in the famous *Brief* already mentioned, he pointed out:

Liegt nun aber nicht in diesem Anspruch an den Menschen, liegt nicht in dem Versuch, den Menschen fur diesen Anspruch bereit zu machen, eine Bemühung um den Menschen? Wohin anders geht, "die Sorge" als in die Richtung, den Menschen wieder in sein Wesen zurückzubringen? Was bedeutet dies anderes als dass der Mensch (homo) menschlich (humanus) werde? So bleibt doch die Humanitas das Anliegen eines solchen Denkens; denn das ist Humanismus:

Sinnen und Sorgen, dass der Mensch menschlich sei und nicht un-menschlich, "inhuman", das heisst ausserhalb seines Wesens. Doch worin besteht die Menschlichkeit des Menschen? Sie ruht in seinem Wesen.[68]

Such "essence," obviously, could only be ontological. Now the notion of being as the existence of the being, *Dasein*, implied the notion of separation and relationships not only between existence and the being, but also between the being and the being, between man and man, in the structural unit of a *humanitas* as "essence" of being that was no other than existence which, on this horizon, in *Unterwegs zur Sprache*, was seen as the imposing of the submission that leads to freedom.

According to us, since then there has been a misunderstanding regarding this perspective which had been seen as a negation of individual freedom: instead its limitations do not consist in denying but in letting the free and decision-making character of the individual pass into the background as regards to that of that casting plan which is existence. The individual, far from losing himself in the "curative" concerns of existing, perhaps with the help of technique, had to place himself as the "guardian" of Being, that is of that *alterity* or difference in which he existed. The true "care" was therefore of an ontological not ontic order.

On the basis of this concern, man posed himself more as a consciousness of being than a consciousness of the "with being," that is to say of the altruistic reciprocity. Heidegger's *Mit-sein* wavered between authentic and inauthentic existence, between the cast plan and the reciprocity of the plans.

At this point could the possible paradox of a monologized thought be surpassed? A thought which, in preserving the ontological difference, seemed to have lost its coessential dialogical component. Heidegger seemed to have bypassed the difficulty by poetically reformulating the theme of the dialogue in the ontological difference through the notions of waiting and "happening" (*Ereignis*).

THE HERMENEUTICAL TURNING POINT

Here Hans-Georg Gadamer's specific contribution is introduced. Gadamer found the crucial point of his hermeneutical perspective in the horizon of the other and of the dialogue. For example, in his *Self-criticism* of 1985, remembering the years of his youth, he observed how the attention for the "phenomenon of the other" constituted from the start of his investigation, a guiding motive.

Back in the 1940s, alterity indicated to Godamere a new direction to follow, which demanded an alteration of course compared with all the philosophies regarding subjectivity that were still ruling. *Alterity* was the new word that Gadamer had picked up from Heidegger in order to develop it further in a more fruitful manner. In this way he opened a new perspective of reflection. He pointed out:

Man spricht von dem Anderen. Dass man hier "der Andere" sagt, verändert die Perspektive. Sogleich ist ein Wechselverhältnis in die Verfassung von Ich und Du eingedrungen.[69]

Heidegger, according to Gadamer, seemed to ignore all the importance of this perspective, when he brought it back into the existential point of view of his ontology. Gadamer's position was very different as he described below:

Da mag eine eigene Erinnerung bezeichnend sein. Es war im Jahre 1943, als ich in einer Arbeit, die später als erstes Stück meiner *Kleinen Schriften* veröffentlicht wurde, in Abgrenzung von Heidegger zu zeigen versuchte, wie das Verstehen des Anderen eine grundsätzliche Bedeutung besitze. Wie Heidegger in der Vorbereitung der Seinsfrage es entwickelt hatte und wie er damit das Verstehen als die eigentlichste Existentialstruktur des Daseins herausgearbeitet hatte, konnte der Andere sich in seiner eigenen Existenz nur als eine Begrenzung zeigen. Am ende, so meinte ich, wird aber gerade das Starkmachen des Anderen gegen mich selbst mir erst die eigentliche Möglichkeit des Verstehens aufschliessen. Den Anderen gegen sich selbst gelten zu lassen – und von da aus sind alle meine hermeneutischen Arbeiten langsam herausgewachsen – heisst nicht nur, die Begrenztheit des eigenen Entwurfs im Prinzip anerkennen, sondern verlangt geradezu im dialogischen, kommunikativen, hermeneutischen Prozess über die eigenen Möglichkeiten hinauszukommen. Als ich das damals Heidegger vortrug, nickte er zwar ganz beifällig, aber dann sagte er: "Ja, und was ist es mit der Geworfenheit?"[70]

Gadamer refused Heidegger's objection which he found too reductive and generic as regards to the specific experience of the other.[71] In fact, in his opinion it was not a simple matter of identifying the non-exhaustive and limited in comprehending character of each individual existence, but "of the why I investigate my own limitedness in comparison to the other, and I shall have to continually learn to investigate it in order to be able to be in the condition of overcoming my limitations."[72]

In other words, Gadamer's position was not limited to recording the finished character of the existential plan, but he made this identification his strong point; it consisted in investigating the individual's limits starting with the presence of the other, seeing a possibility in this of surpassing one's self, of an advancement in understanding, of authenticity in communication.

Gadamer called this *hermeneutics*. This was the art of letting oneself be questioned by the other, bringing oneself into play with him. Hermeneutics was allowing that the other always told us "something again." Dialogical hermeneutics was in this way entirely an *experience of alterity*, of the interpersonal relationship. But it was not only this: the "letting something or someone speak again" also meant that something or someone always told us again in the dialogic form of existence. We were dealing therefore with a being that in speaking renewed itself and renewing itself said something again.

The dialogical character of hermeneutics destroyed the methodological aspect of the scientific "understanding" and introduced multiple, varied ways of approach to man, to reality, to their possible interpretations. This approach did not only regard man, who was capable of dialogue and listening, but it regarded all reality. The concept of alterity thus assumed for hermeneutics a much greater importance as regards to that of the person. The person was the human aspect of alterity which generally indicated the ontological dignity of reality.

In conclusion, we can say that the path followed, after the surpassing on the part of Heidegger with regard to Husserl's paradox of inter-subjectivity, that petrified the consciousness in its abstract constraint, arrived at rediscovering alterity as an existential and ontological dimension, which then assumed, in Gadamer's perspective, the character of experience as historic, dialogical relationship.

Along this path we cannot leave out those who have wanted to find in Gadamer's hermeneutic perspective, a precise answer, of epoch-making importance, to a problem that has become central to our times: "the belonging to a culture," which has been argued. Rudiger Bubner has spoken in merit of a new present-day form of experience: "the alienating oneself of self" (*das fremdwerden des Eigen*), the questioning of that what is more familiar, of that what, like culture, identifies us."

In this way Gadamer's hermeneutics have been considered as a form of "cultural criticism," capable of rethinking about the peculiar experience of the encounter with the other, a typical experience of the present day, which "would consist in the encounter and the dramatic confrontation with a stranger, with the other, with the experience of understanding the other."[73] This, calls to mind Antonio de Simone, "it leads us to consider identity and difference not as something given once and for all, but as something fragile that is historically and socially constructed. Only by basing oneself on identity and the difference so redefined can we deal with the contradiction between our belonging to a community and the belonging to ourselves, reaching a conception of one's identity which is not necessarily against the other."[74]

PERSON AND ALTERITY

In general the reclaiming of the alterity has led to that of the person. The other is the same as the person. Its specificity, individuality and dignity derive from its being a person. However, what is a person?

The traditional definitions that have been proposed are pending, more or less, surreptitiously on a presupposition of the "substantial" type which we can trace back to the Greek word *prosopon*. The literal meaning of the word is "that what is in front *(pros-)* of the eyes." The word has been rendered with "aspect," "face," "image," which are the physical ways in which the other presents itself.

The *visage*, according to Lévinas, is the revelation of the other, its Manifestation and Mounier says that the person as *visage* expresses the other's authentic character.

Often in a unilateral way, the Greek metaphysical tradition and the Christian theological reflection have reduced the other to *prosopon*, person, to presence, that is what is in front of the eyes and remains in the fixedness of my eyes. What has happened then to the other? As the other is such as a person, and the person is such as a presence, it follows that alterity is essentially thought of as "presence." This does not simply mean being "on hand," close or nearby, but *being present*, that is what my eyes fix and pose as being different to me. Here alterity is not that what it places in front of my eyes, but that what my eyes decide to fix and pose. In the sphere of such a personalistic conception, the defence of the person is not always guaranteed, because the alterity is not always protected which in its turn, is the condition and the protection of the person. Without alterity there would be no person.

Alterity is not intended as "being present" but as "presence," which is the existential historicity in which all the individuals "are present" (*Dasein*). The presence is the concrete visibility of the other or person. The person interpreted in this way is not reducible to that which falls below me, but is the possibility of something else. The body, the face, the history are the forms of a presence of the person that is not at my disposal, but a moment of a possible dialogue. The body, the face, the history are the dialogic hermeneutic visibility of the other, its questioning-us and answering-us.

Such a notion of alterity is no longer necessarily linked to an anthropocentric concept, in fact it marks its overcoming. I believe that Heidegger's lesson regarding this point has to be accepted and that, as Heidegger observes "the opposition to humanism does not imply in any way the defence of the inhuman, but it opens other perspectives."[75]

In fact, the presence of the other here does not exclusively mean the presence of the other man or person, but also the presence of the other being in nature and in history, in culture and in life. It is necessary to extend the first anthropocentric characteristic, which is tinged by the Christian notion of person, in another "cosmocentric" characteristic, which would give the being the complexity of its being in nature and in history, in culture and in life.

For example, in the essay *Philosophy on the Threshold of the Year 2000*, Hans Jonas identified in the responsibility towards the environment, nature and life, one of the greatest tasks of the philosophy of the near future. With these practical tasks, he observed, the reconciliation of our arrogant peculiarity with everything starting from where we live, pours into the centre of philosophic concern.

In front of who, like Aldous Huxley, has foreseen a "new world," that is emancipated and biologically strengthened and perfected, but not so free and rich with sentiments towards man and nature, therefore a dystopia from which to defend oneself or in front of who, after the fall of ideal tensions seems to have lost himself in issues of thought and action of an irrationalistic type, the need of reconciling and preserving the fullness and the creativity of life, together with its forms and relationships, becomes an indispensable task to follow.

It is necessary therefore to pass from an inter-subjectivity point of view that makes the other responsible only to the other man, to a hermeneutical point of view in which the other man or being is a moment of the dialogue with more that one voice. Bioethics, which is the ethics of the presence of the other being, is the personalistic expression of alterity and of its relationships.

It is here in the questioning that the being always speaks to us again, at liberty. Thinking of the other and his dignity means assuming one's responsibilities as a person. Becoming responsible is becoming aware, avoiding prejudicial positions regarding a thought and a commitment that have as their sole concern allowing being, the presence of the being or of the person.

AN ONTOLOGICAL REVIVAL

The three Kantian personalities previously examined, summarized the ways in which modern traditional Western philosophy has professed to define man: as a rational, moral and natural animal. The definitions of the personalities have revealed themselves, following the existential and hermeneutical approach, as inessential to that existing being which we are: inessential because they have not stressed the being of *Dasein*, but the *how* it has to be.

An analysis that wants to be ontological and essential, cannot pose itself as prescriptive, but only as descriptive. This is the sense of Heidegger's "existential analytics" and of the hermeneutic contributions of Gadamer, Lévinas and Hans Jonas, in which the different moral, natural and psychological meanings of existence are put in brackets or in the language of Heidegger are retained inessential and for this reason are confined to the inauthentic.

Every attempt at defining man by means of the notion of the classically intended person is rejected and criticised because at the base there is an ontological *deficit* (Heidegger) and a hermeneutical *deficit* (Gadamer, Lévinas), as explained above.

We therefore conclude our investigation with the second question raised by Heidegger in *Sein und Zeit*. What is the ontological contribution to the moral and biological sciences? The criticism of the Kantian analysis on the concept of person had served Heidegger to lead man back to his existential structure that was to consider it propaedeutical to each form of knowledge of himself, forms of knowledge that in modern times have been called "moral sciences."

It was inevitable, therefore, that Heidegger asked himself the question about the relationship between the existential, ontological dimension of being present and the characteristic one of the "moral sciences." In other words, asking questions about the type of relationship that they could entertain with ontology, beyond their separation, became prejudicial for their correct comprehension and demarcation.

In *Sein und Zeit*, Heidegger distanced himself from a purely scientific approach compared with moral and biological sciences, in that the progress of the research in this sphere did not consist in his opinion, in so much as:

in der Aufsammlung der Resultate und Bergung derselben in "Handbüchern," als in dem aus solcher anwachsenden Kenntnis der Sachen meist reaktiv hervorgetriebenen Fragen nach den Grundverfassungen des jeweiligen Gebietes.[76]

The "increased consciousness of things," that is to say knowledge, was none other than a group of interpretations that concerned the way of being, of existence as *Dasein*. However, the *Dasein* claimed for itself an "existential analytics" as the clarification of the fundamental way of existence. It was a crucial and basic aspect that preceded any empiric acquisition of contents, to which they necessarily deferred.

For this Heidegger was able to write:

Weil nun aber zum Dasein nicht nur Seinsverständnis gehört, sondern dieses sich mit der jeweiligen Seinsart des Daseins selbst ausbildet oder zerfällt, kann es über eine reiche Ausgelegtheit verfügen. Philosophische Psychologie, Anthropologie, Ethik, "Politik," Dichtung, Biographie und Geschichtsschreibung sind auf je verschiedenen Wegen und in wechselndem Aussmas den Verhaltungen, Vermögen, Kräften, Möglichkeiten und Geschicken des Daseins nachgegangen. Die Frage bleibt aber, ob diese Auslegungen ebenso ursprünglich existential durchgeführt wurden, wie sie vielleicht existentiell ursprünglich waren. Beides braucht nicht notwendig zusammenzugehen, schliesst sich aber auch nicht aus.[77]

For this reason Heidegger considered "existential analytics" as the "elaboration of the question of being,"[78] irremissible also from the point of view "of the moral sciences." In fact, although analytics studies man in his existential structure, in the "utterly singular character of its being (existence)," while the moral sciences (anthropology, psychology, ethnology, biology) study the human reality in its appearance, therefore in its exterior dimension, which expresses "the difference between individual and individual" (Pietro Chiodi), this, despite analytics, constitutes the inalienable foundation of moral sciences. In fact, moral sciences begin where analytics end.

What contribution could analytics make to these sciences? It was able to supply, although in the form of "fragments," the identification *a priori* of the conditions of possibility of their object. Consequently, Heidegger grasped the need for a critical reflection of their foundations. The goal was the acquisition of a deeper awareness of the limits and of the capacity of one's own knowledge.

The research of moral sciences, like every ontic research, remains "naïve and dull" if its investigation regarding the way and the difference of the being does not take into account "the sense of being in general."[79]

Moral sciences, as they also call for the posing of the question of being, can open themselves to a prolific "regional construction of ontology," finding in it the "path" towards the thematization of being. In this sense each science can acquire an adequate awareness that there is a philosophy involved behind each of its researches, an horizon of sense. It is this, pre-scientific ontological sphere, that prejudicially orientates the research, even when all is still "provisory."

In conclusion, if the person wants to be the authentically exact dimension of the "being who we are," he cannot be the *personalitas*, according to a line of thought that has travelled from the ancient world to as far as Kant and the phenomenology of Scheler, but the Being present, *Dasein*. Only in this way can he realize his own prerogatives of opening and transcendence towards alterity.

Therefore, as regards to the moral sciences, in order to be authentically human, they must rediscover that ontological pre-horizon and so become regional ontology, which will pre-orientate them in their research and consciousness. Therefore they must burden themselves, in order to gain a better understanding and interpretation of the results of their research, with a need and with a possible ontological sense.

San Benedetto del Tronto, Italy

NOTES

[1] M. Heidegger, *Die Grundprobleme der Phänomenologie*, Gesamtausgabe, Band 24, (Frankfurt am Main, 1975), pp. 18–19.
[2] Heidegger, *Sein und Zeit*, Einleitung, I, §2 (Tübingen, 1927), pp. 6–7.
[3] *Op. cit.*, p. 7.
[4] *Ibid.*
[5] L. Stefanini, *Esistenzialismo ateo ed esistenzialismo teistico* (Padova, 1952), p. 29.
[6] Heidegger, *Sein und Zeit*, Einleitung, I, §2, *cit.*, p. 7.
[7] H.-G. Gadamer, *Subjektivität und Intersubjektivität. Subjekt und Person*, in G. W., Band 10 (Tübingen 1995), p. 93.
[8] Gadamer, *Die phänomenologische Bewegung*, G. W., Band 3 (Tübingen 1987), p. 109.
[9] Heidegger, *Sein und Zeit*, Niemeyer (Tübingen, 1927), p. 42.
[10] Gadamer, *Die phänomenologische Bewegung, cit.*, p. 140.
[11] G. Reale, D. Antiseri, *Il pensiero occidentale dalle origini ad oggi*, Vol. 3, cap. XX, *Martin Heidegger: dalla Fenomenologia all'Esistenzialismo*, (Brescia: La Scuola, 1983), p. 447.
[12] Gadamer, *Subjektivität und Intersubjektivität, cit.*, p. 93.
[13] *Op. cit.*, p. 88.
[14] *Op. cit.*, p. 90.
[15] *Ibid.*
[16] *Op. cit.*, p. 91.
[17] F. Nietzsche, *Also sprach Zarathustra*, III, *Von den drei Bösen*, 2, vv. 68–71.
[18] Gadamer, *Subjektivität und Intersubjektivität, cit.*, p. 90.
[19] *Op. cit.*, p. 92.
[20] Gadamer, *Phänomenologie, Hermeneutik, Metaphysik*, in G. W., Band 10, pp. 103–104.
[21] E. Husseri, *Cartesianische Meditationen* (The Hague: Martinus Nijhoff, 1950), §42.
[22] Gadamer, *Subjektivität und Intersubjektivität, cit.*, p. 95.
[23] *Op. cit.*, p. 94.
[24] *Op. cit.*, p. 93.
[25] Heidegger, *Brief über den Humanismus*, Gesamtausgabe, Vol. IX (Frankfurt/M., Klostermann, 1976), pp. 329–330.
[26] Heidegger, *Sein und Zeit*, I, §10, *cit.*, p. 50.
[27] *Op. cit.*, p. 46.
[28] *Op. cit.*, p. 48.
[29] *Ibid.*
[30] *Ibid.*
[31] *Op. cit.*, p. 49.
[32] *Op. cit.*, p. 48.
[33] Heidegger, *Die Grundprobleme der Phänomenologie, cit.*, p. 177.
[34] *Op. cit.*, p. 179.
[35] *Ibid.*
[36] *Op. cit.*, p. 184.
[37] *Ibid.*
[38] *Op. cit.*, p. 181.
[39] *Op. cit.*, p. 183.
[40] *Ibid.*
[41] *Op. cit.*, pp. 184–185.
[42] *Op. cit.*, p. 185.

[43] *Op. cit.*, pp. 185–186.
[44] *Op. cit.*, p. 188.
[45] *Op. cit.*, p. 191.
[46] *Op. cit.*, p. 194.
[47] *Ibid.*
[48] *Op. cit.*, p. 195.
[49] *Op. cit.*, pp. 195–196.
[50] *Op. cit.*, p. 197.
[51] *Ibid.*
[52] *Ibid.*
[53] *Op. cit.*, p. 199.
[54] *Op. cit.*, pp. 199–200.
[55] *Op. cit.*, p. 216.
[56] *Op. cit.*, p. 207.
[57] *Op. cit.*, pp. 207–208.
[58] *Op. cit.*, p. 208.
[59] *Op. cit.*, p. 218.
[60] *Op. cit.*, p. 219.
[61] V. Vitiello, *Dialogando con Heidegger: mondo tempo ipseità*, in AA. VV., *Il destino dell'io*, (Genova: Il Melangolo, 1994), p. 143.
[62] P. Chiodi, *Introduction*, Italian edition of *Sein Und Zeit* (Torino: Utet, 1986), pp. 11–12.
[63] *Op. cit.*, p. 11.
[64] *Ibid.*
[65] Heidegger, *Sein und Zeit*, IV, §26, pp. 123–124.
[66] Heidegger, *Brief über den Humanismus, cit.*, pp. 318–319.
[67] Gadamer, *Subjektivität und Intersubjektivität, cit.*, p. 96.
[68] Heidegger, *Brief über den Humanismus, cit.*, p. 319.
[69] Gadamer, *Subjektivität und Intersubjektivität..., cit.*, p. 95.
[70] *Op. cit.*, p. 97.
[71] *Ibid.*
[72] *Op. cit.*, p. 98.
[73] Antonio de Simone, *Identità, alterità e dialettica del riconoscimento*, AA. VV., (Perugia: *L'Io ulteriore*, 2004), p. 140.
[74] *Op. cit.*, p. 125.
[75] Heidegger, *Brief über den Humanismus, cit.*, p. 348.
[76] Heidegger, *Sein und Zeit*, Einleitung, I, §3, *cit.*, p. 9.
[77] *Ibid.* §5, *cit.*, p. 16.
[78] *Op. cit.*, p. 17.
[79] Chiodi, *Introduction, cit.*, p. 64.

SECTION V

THE CREATIVE TURN

MARIA MANUELA BRITO MARTINS

LA VOLONTÉ HUSSERLIENNE EN TANT QUE POUVOIR CRÉATEUR

1. LE *BACKGROUND* HISTORIQUE DE LA VOLONTÉ

Comme Albrecht Dihle a dit très récemment saint Augustin "was, in fact, the inventor of our modern notion of will."[1] Il va même jusqu'à dire que dans l'histoire de la philosophie, la notion de volonté, avec ses descriptions et ses analyses, depuis la scolastique, en passant par le monde moderne jusqu'à Schopenhauer et Nietszche, a été inspirée par Augustin.[2] En réalité, c'est chez lui qu'on trouvera pour la première fois la définition exacte de volonté, dépassant en large mesure ses devanciers latins et pour cela, en reprenant à son compte la complexe et problématique notion de *boulesis* et *proairesis* des Grecs, notamment d'Aristote.[3] Augustin a une expression vraiment étonnante pour définir l'action et le mode d'être de la volonté: "unde hoc monstrum? – d'où vient ce prodige monstrueux?"[4] De la sorte, nous pouvons interroger et dire: peut-on nous avoir une expression plus forte que celle-ci pour parler de la volonté?

En effet, chez Augustin la notion de volonté est largement comprise dans ses différentes fonctions, à la fois intellectuelles, affectives, pratiques et axiologiques. Ses multiples valences démontrent déjà la profondeur de l'analyse qui sera postérieurement développée et approfondie dans la modernité. De ce fait, il faut, donc, faire justice à Augustin d'avoir entrepris des descriptions phénoménologiques très pointues, qui peuvent être confinés à trois sortes de recherches, dont nous parlait une étude sur Husserl à propos de la notion d'individualité et volonté: "Si Husserl a considéré ses recherches sur la conscience intime du temps comme une nouvelle interrogation de la question augustinienne concernant le temps et s'il a aussi rappelé le rapport de ses esquisses d'une philosophie radicale de la subjectivité, dans ses conférences de Paris et dans ses *Méditations cartésiennes*, avec la détermination augustinienne de la vérité *habitat in interiore homine*, ses recherches sur la question de la volonté montrent également une certaine affinité avec l'exploration de cette dimension de la vie humaine."[5]

A-T. Tymieniecka (ed.), Analecta Husserliana XCIII, 319–333.

Le véritable enjeux de la définition augustinienne de volonté subsiste dans sa détermination négative, comme étant un pouvoir qui, à la limite, existe comme une puissance infinie, puisqu'elle est libre et en même temps absolue, c'est-à-dire, sans contrainte: "la volonté est un mouvement de l'âme que rien ne contraint soit à ne pas perdre soit à acquérir quelque bien."[6] C'est dans un sens pareil que Descartes dans la quatrième méditation de ses *Méditations Cartésiennes*, définit la volonté, comme étant la seule faculté qui est illimité et qui l'on perçoit comme la plus grande des puissances.[7] La raison de cette grandeur vient du fait qu'elle peut affirmer ou nier quelque chose, qu'elle peut faire ou ne pas faire, qu'elle peut enfin, poursuivre ou même fuir des choses que l'entendement nous propose.[8] Par contre, si nous nous tournons vers Kant, nous apprenons aussitôt que la détermination négative du principe fondamentale de la volonté, lui vient du fait qu'elle est le concept de liberté.[9]

Le pouvoir de la volonté subsiste, donc, dans ceci: elle n'ajoute rien et elle ne perd rien qu'elle ne le possède pas déjà à son point de départ. Par ailleurs, il n'a aucune contrainte à être ce qu'elle n'est pas ou à être plus ce qu'elle est, puisque la volonté est par sa nature, le mouvement de l'esprit qui veut et qui veut ce qu'elle veut, tout simplement. Or, quelle contrainte pourrait-il y avoir pour la volonté elle-même, sinon la propre volonté? C'est dans se sens qu'Augustin s'interroge en disant: "Mais enfin, quelle pourrait être, antérieurement à la volonté, la cause de la volonté?."[10] Mais la volonté en tant que mouvement de l'esprit pour agir librement n'est qu'une dimension de la volonté, car celle-ci est simultanément le pouvoir de l'exécution[11] et de l'assentiment,[12] c'est-à-dire, de la décision ou du *liber arbitirum.*

Associé à l'acte intelligible de la volonté, Augustin rattache une *intentio* qui est propre de ceux qui veulent. De cette façon il y a une connexion dans l'acte intellectif, entre volonté et connaissance d'un objet. En effet, c'est dans sa théorie de la représentation, qu'on peut trouver la grande nouveauté d'Augustin, qui est celle d'avoir rapproché la représentation avec l'action unitive de l'acte volontaire pour la connaissance d'un objet. Ce qui dans un langage kantien s'appellera unité synthétique, Augustin parlera, de son côté, d'une union entre la mémoire, la représentation imaginaire et finalement la volonté en tant que capacité d'unir les deux.[13] Le rôle de la volonté est alors celui d'une "synthèse" de la représentation perceptive et imaginative. Il y a dans la théorie de la représentation augustinienne une tension entre le rôle joué par la *phantasia* et le rôle joué par la volonté en tant que pouvoir unificateur des différentes images produises par la sensation et par la mémoire. Mais, contrairement à Kant, la limitation augustinienne de la *phantasia*, en tant qu'imagination productrice d'une "synthése" c'est qu'elle se donne surtout pour des représentations du sens interne, voire même d'une représentation du passé.

Par ailleurs, la limitation augustinienne de la volonté se situe au niveau du concept. Or, cela ne ce passe pas autant avec Kant, car pour lui la volonté en tant que faculté de désirer est "celle qui agit d'après concepts."[14]

La fonction de la volonté dans la théorie de la représentation d'Augustin, est comprise essentiellement comme une faculté imaginaire, bien qu'unitive, dont sa finalité ultime est la représentation imaginaire elle-même.[15] Toutefois, il s'accroche à la volonté une double finalité, celle, d'une part, de la représentation imaginaire en tant que donné perceptif comme tel et, de l'autre, la finalité de la volonté qui, par soi-même, à partir des actes de souhaits volontaires, s'ajoute comme un surplus à la représentation imaginaire perceptive.[16] C'est cette double finalité qu'on peut trés bien déceler chez Husserl, comme on va le voir.

Mais ce n'est pas d'Augustin que Husserl s'est plus inspiré pour ces recherches sur la volonté, malgré les continus apports d'une tradition philosophique constamment renouvelée, qui par voie directe ou indirecte s'entremêlent dans la réflexion husserlienne. Par exemple, c'est dans les recherches sur la subjectivité, sur la tâche éthique et sur l'importance de la conception de volonté au sein de la vie de l'individu, que Husserl fera appel dans un texte de 1930, en s'appuyant sur Fichte, à l'expression très connue d'Augustin: "nur die Liebe macht sehend – für Wert und Ideal. 'Nemo cognoscitur nisi per amicitiam.' "[17] Cette même expression se trouve chez Heidegger et chez Max Scheler. En effet, elle nous introduit dans le problème majeur du rapporte entre connaissance et affectivité, entre savoir, vouloir et désirer. C'est dans un cadre plus élargi, que nous pouvons délimiter dans une recherche phénoménologique de la volonté, les différentes régions de l'être auxquels elle appartienne et d'y voir les différentes ontologies matérielles que s'y rapportent.[18]

1.1. L'influence kantienne chez Husserl

L'influence kantienne chez Husserl est très importante. Pour Kant, la synthèse passive est un prolongement de cette découverte fondamentale d'un ego qui lie très fortement dans l'acte de connaître, le véritable sens accompli de la perception d'un objet, à savoir, celui de la représentation imaginaire dans une unité accomplie. Déclare Kant: "Tout ce qu nous pouvons dire c'est que l'image est un produit du pouvoir empirique de l'imagination productrice et que le schème des concepts sensibles, comme des figures dans l'espace est un produit et en quelque sorte un monogramme de l'imagination pure a priori et que ces images ne doivent toujours être liées au concept qu'au moyen du schème qu'elles désignent et auquel elles ne sont pas en soi entièrement adéquates."[19] Dans les principes de l'analytique transcendantale

Kant établit, au moyen de ce schématisme, propre à l'entendement, l'unité entre d'une part, les traces visibles d'une empirique de l'imagination *productrice* et de l'autre, le modèle selon lequel le concept s'accorde avec le résultat imaginaire des premières, avec le schéma appliqué par le concept selon le pouvoir d'une imagination a priori. Par d'autres mots, il s'agit de relier une imagination productrice, qui est le résultat de ce pouvoir empirique de créer des images, avec une imagination pure à priori, qui sera désormais, une nouvelle production, c'est-à-dire, une re-production en tant que résultat de l'entendement pur où, Kant, affirme lui même, qu'il ne peut pas y avoir des images. De la sorte, il y a une l'application de la règle de l'unité catégoriale de la subjectivité en tant que telle, c'est-à-dire de l'unité de l'aperception.

Kant ne parle pas du pouvoir de la volonté dans ce schématisme de la raison pure, puisque c'est l'imagination a priori qui dégage cette fonction. Toutefois, c'est dans la *Critique de la raison pratique* que le pouvoir de la volonté apparaît comme un principe déterminant de la raison, voire même, de la raison pure. Il y a en quelque sorte une analogie entre le schéma transcendantale de la raison spéculative et celui de la raison pratique. Pour la première, la raison veut connaître à priori les objets, pour la seconde, la raison est un principe déterminant de la volonté, c'est-à-dire, la raison veut savoir comment les objets de la faculté de désirer sont déterminés par les désirs et, de telle sorte, que la recherche de la raison pratique est alors celle de déterminer la maxime de la volonté. Or, c'est précisément ici que le rapprochement entre la raison spéculative et la raison pratique deviennent une réalité unique, ce faisant de la démarche transcendantale une sorte de *continum* de la raison, dans la mesure où les lois qui réglent les objets de la nature et les objets de la volonté, ne sont possibles que dans l'ordre des règles non empiriques. De la sorte, la causalité de la volonté ne subsistant que dans la réalité de l'objet du vouloir, exige d'une part, que la volonté soit conforme à une loi (*gesetzmässig*), et que d'autre, qu'elle exécute le pouvoir de la volonté à faire ce qu'elle veut. La continuité entre la raison pure et la raison pratique comme une seule démarche de la même raison est celle qui cherche à saisir comment une raison pure peut être aussi pratique. Dit Kant dans la *Critique de la raison pure*: "(...) si pour la raison pure, la volonté est seulement conforme à une loi (*gesetzmässig*), il en sera ce qu'il pourra de son pouvoir dans l'exécution, il en résultera réellement ou non une nature telle que celle qui est possible d'après ces maximes de la législation, la *Critique* ne s'inquiète pas, puisqu'elle cherche seulement si une raison pure peut être pratique, c'est-à-dire déterminer immédiatement la volonté, et comment elle peut être."[20]

Husserl de son côté parle d'une analogie entre les actes perceptifs et les actes d'évaluation ou de valeur. En effet, les actes affectifs et volitifs sont des actes "objectivants" qui constituent également des objets. Cette analogie ne fait que prolonger la démarche kantienne en essayant de saisir la manière comme les actes perceptifs et représentatifs déterminent un rapport intentionnel avec les actes affectifs et volitifs. La différence mais, également, l'analogie, qui peut avoir entre les actes perceptifs et les actes affectifs et volitifs c'est que tous les deux ce sont des vécus intentionnels et qu'ils se chevauchent les uns sous les autres. Selon Husserl les actes affectifs et volitifs sont désignés comme des actes fondés sur les actes perceptifs, qui en sont les actes fondateurs.[21] Les premiers, les actes perceptifs, sont de l'ordre de la sensation, alors que les secondes, les actes affectifs, sont de l'ordre de la représentation. Cette distinction entre actes perceptifs et actes affectifs et volitifs sont héritiers de la distinction kantienne. En effet, Kant avait déjà établit une différence sur ce que nous entendons par sensation. Il y a deux sortes de "sensation." D'une part, la sensation est la représentation d'une chose, lorsque je la perçois; dans ce cas-ci, la sensation se rapporte à l'objet et donc, à la faculté de connaître en tant que réceptivité. D'autre part, la sensation est la représentation du sujet lui-même, qui se rapporte au sentiment (*Gefühl*) et qui ne constitue d'aucune manière la représentation d'un objet. C'est dans ce dernier cas que cette sensation détermine les conditions subjectives de la subjectivité comme telle.[22] Husserl reprenne cette même conception kantienne mais en essayant de retrouver le moyen de les faire conjuguer et en même temps déceler quel est le caractère objectivant des actes affectifs et volitifs.

2. LA VOLONTÉ CHEZ HUSSERL

L'analyse de la volonté chez Husserl pourrait, à première vue, paraître, en regardant la majorité des œuvres publiées et même de celles qui sont encore en cours de route de publication, une thématique parfaitement latérale, en la comparant avec la recherche minutieuse de la phénoménologie de la conscience et de ses structures intellectives. Toutefois, l'ensemble des manuscrits conservés à Leuven, révèlent d'une part, un immense chantier encore à exploiter consacré à la conscience volitive et émotionnelle, et d'autre part, le grand intérêt que Husserl portait à ce sujet, bien que moins explicité à maints égards.

Une des études plus éclaircies à propos de la phénoménologie de la volonté est celle de Ullriche Melle, intitulé, *Husserl's Phenomenology of Willing*. En effet, cette étude nous démontre combien la phénoménologie de la conscience affective, volitive et désidérative, a pris place dans l'ensemble de la recherche phénoménologique de plein droit. La phénoménologie de la volonté révèle être aussi un champ privilégié de la réflexion de Husserl, depuis ses premières études, datés aux environs de 1909–1914. Une grosse partie de ces textes sont agroupés sous un ensemble de manuscrits intitulés, "Phénoménologie de l'émotion, désirer et vouloir."[23]

D'autre part, les recherches de Husserl concernant l'éthique nous aident, en plus, à comprendre mieux le domaine de la conscience axiologique et pratique. Ces études sont intégrées dans l'Husserliana XXVIII, et c'est également Ullrich Melle qui nous situe, du point de vue historique et analytique les origines de la réflexion éthique husserlienne avant et après la guerre.[24]

En effet, c'est dans ces œuvres posthumes que nous trouvons des textes qui nous situent l'orientation de la recherche husserlienne de la conscience volitive et affective. Déclare alors Ullrich Melle:

> Between 1909 and 1914 he worked on the phenomenology of emotive and volitional consciousness, investigations found in not-yet published and rather fragmentary manuscripts. These analyses of affective consciousness and volition are part of comprehensive investigation of consciousness in those years during which Husserl sought to systematically describe the whole of consciousness in all its act-forms, founding relations, modes of accomplishment, etc. Even in its fragmentary state, the scope of this investigation of consciousness is extremely impressive. It covers more than 1,000 manuscript pages. As Husserl's assistant in 1926–27, Ludwig Landgrebe produced, on the basis of these collected materials, a very extensive type manuscript with the appropriate title "Studies on the Structure of Consciousness." The second of its three studies, which carries the title "Value Constitution, Affective Consciousness (*Gemüt*), Will", is devoted to the structures of emotive and volitional consciousness.[25]

Husserl accepte la classification kantienne des différents genres d'actes: actes intellectuels, d'évaluation et de désir. Et il accepte également différentes régions de la raison, à savoir, la raison logico-cognitive, axiologique et pratique.[26] A ces trois régions de la raison correspondent trois disciplines, qui sont décrites soit dans les *Recherches Logiques*, soit dans les *Ideés I*, comme étant les domaines de la logique formelle, de l'axiologie formelle et de la théorie formelle de la pratique. La volonté étant alors associée directement au domaine éthique et axiologique. Pourtant, la volonté n'appartient qu'au domaine des actes de valeur et d'évaluation ce qui la ferait objet principal de l'éthique et de l'axiologie. Dans la mesure où Husserl considère qu'il y a une analogie entre la structure inférieure de la conscience et les structures de degré supérieur de la conscience, il établit un rapport analogique, bien que différentiel, dans les actes intentionnels de la conscience.

Nous allons nous concentrer surtout dans la notion de volonté qui se rattache à la fonction de représentation et de l'unité synthétique des actes intentionnels. Étant donné cela, nous ne ferons donc pas, une approche de la volonté par le biais de la raison pratique et axiologique, mais au contraire, par une approche de la volonté en tant que principe complet et intégral de l'acte intentionnel comme tel, en essayant de dégager la connexion entre la démarche logico-cognitive avec la démarche affective et volitive.

C'est la volonté en tant qu'accomplissement des actes supérieurs de la conscience qui nos intéresse ici analyser. Nous commençons par l'analyse husserlienne des *Ideen I.*

3. IDEEN I

C'est au §93 que Husserl énonce la description de son analyse sur les structures de la sphère supérieure de la conscience. Ses structures sont composées par genres d'intentionnalités complexes, à savoir: d'une part, la fonction du jugement, c'est-à-dire, le jugement prédicatif, et d'autre part, les opérations affectives et volitives, qui feront objet d'examen dès les §95, et puis, repris au §116, lorsqu'il s'agit de comprendre dans ces actes intentionnels complexes et nouveaux, leur rapport avec les corrélats noetico-nomeatiques fondés. "Dans le développement prochain nous considérons des structures de la sphère supérieure de conscience: là un certain nombre de noèses sont édifiées les unes sur les autres pour composer l'unité d'un vécu concret."[27]

Cette description nos met au cœur de la notion même d'intentionnalité, puisqu'il s'agit de cerner le rapport entre la conscience et les *data* qui déterminent la visée de la conscience. Or la visée intentionnelle des noèses comme sont celles du sentiment, du désir et du vouloir, ce sont des noèses fondées à partir des représentations.

Nous avions déjà dit plus haut que les actes volitifs se distinguent des actes perceptifs car, les premiers sont fondés et complexes, alors que les secondes sont fondantes et simples.[28] Husserl reprend, à son compte, la distinction faite par Brentano qui distinguait deux sortes d'intentions superposées: l'intention qui fonde, et l'intention fondé, la première donne l'objet représenté, alors que la seconde donne l'objet senti.[29]

Pour Husserl la raison véritable de cette distinction sert plutôt à déterminer la structure de la sphère supérieure de la conscience, à savoir, les actes prédicatifs et les actes volitifs; c'est, d'ailleurs, pour cela qu'ils sont décrits comme des actes complexes, alors que les actes perceptifs sont des actes simples. Ceci est également exprimé dans les *Recherches Logiques*[30] *et* dans les *Ideen I.*[31] Dans les *Recherches*, l'acte fondé est un acte complexe et intentionnel, composé d'une pluralité d'actes partiels, bien que chacun soit doué d'une intentionnalité propre et visant touts ensemble à un acte global et donc, unitaire.[32] Cette unité de la relation intentionnelle est possible grâce à la fonction de l'unité de l'objectivité de la représentation qui ne se constitue pas en dehors, voire même, à côté des ces actes partiels mais en eux-mêmes, sous le mode de leur combinaison. Cette combinaison peut se faire selon l'unité de la prédication catégoriale ou même hypothétique.[33] C'est pour cela que nous pouvons donc, dire avec Husserl, que "la représentation fonde la volition, et c'est à la représentation que la volition doit son comprendre."[34]

Dans les *Idées*, il y a un détour élargi pour comprendre l'analyse de l'intentionnalité des actes affectifs et volitifs et de l'objectivité du vécu volitif. Ce détour se fait par le biais des analyses des noèses, c'est-à-dire de la constitution noetico-noématique de l'intentionnalité des vécus affectifs et volitifs. Au §95 Husserl s'efforce pour distinguer la sphère affective (*Gemüts*) de la sphère de la volonté (*Willen*) bien que les deux appartiennent à cette structure intentionnelle de la conscience qui évalue. Dans la conscience qui perçoit, il y a aussi l'évaluation en tant que telle et l'évalué. Cela veut dire que cette forme de perception est accompagnée également de l'imaginer et du juger en tant que formes de soubassement de l'évaluation comme telle. L'analyse husserlienne va dans le sens de découvrir dans la perception, la valeur qui fait partie de l'objectivité de la chose. La stratégie dans l'évaluation de la propriété d'une chose qui est perçue passe inévitablement par l'introduction d'une réduction par rapport au perçu, de manière à saisir le noème de ce qui est visé, à savoir la visée volitive (*Willensmeinung*).[35]

On découvre aussi dans la sphère volitive un ensemble de stratifications noématiques qui fonctionnent de façon semblable à celles de la sphère perceptive mais dont la couche d'évaluation est le modèle qui rend possible l'objectivité de la valeur de la propriété de la chose. Lorsque Husserl tente de décrire la manière comme s'articule le sens perceptif avec le sens de l'évaluation, il tente de dégager quel sorte d'intentionnalité, au complet, pourrons-nos trouver. Ce que les *Recherches Logiques* établissaient comme différence entre les actes fondés et les actes fondants, de manière à en faire la distinction entre acte volitif et acte perceptif, l'est établit maintenant dans les *Idées* entre la structure supérieure de la conscience et la structure inférieure, selon laquelle celle-ci est comprise dans un sens à la fois étroite et vaste de la conscience; par d'autres mots, il y a bel et bien une distinction entre les actes perceptifs et les actes volitifs, étant chacun, assujetti aux différentes modalités doxiques qui lui sont propres. Or, ce que sont les actes fondés dans les *Recherches*, ce sont dans les *Idées*, l'édification des noèses les unes sur les autres, afin de former l'unité d'un vécu concret. Les noèses propres à l'intentionnalité des couches supérieures de la conscience, comme celles de l'intentionnalité volitive sont à la fois des noèses fondées et fondantes. Ce sont, d'une part, les noèses fondées dans des représentations, productrices des synthèses noétiques, et d'autre part, ce sont des noèses qui produisent des nouvelles formes de noèses "positionnelles" fondées.[36]

Tout le long de ce processus du cogito nous assistons à la conquête de la structure totale du vécu, par le moyen de la compréhension des actes et des opérations, ayant comme fil conducteur trois aspects essentiels: (a) l'explicitation du rapport noèse noème élargi aux actes effectifs et volitifs; (b) analyse des thèses affectives et volitives fondées sur la représentation mais, également, en tant que possibilité d'édification; (c) finalement les synthèses que s'y rattachent en tant que formes de création du moi pur et de son actualité. C'est sous la base de cette phénoménologie, qui a son point de départ dans la conscience et ayant comme thème fondamental une phénoménologie de l'attention, qu'ils se produisent les différentes modifications attentionnelles (*attentionalen Wandlungen*) de la conscience, tout d'abord envers l'objet et ensuite de la conscience avec elle-même. Déclare Husserl:

Le rayon de l'attention se donne comme irradiant du moi pur et se terminant à l'objet, comme dirigé sur lui ou s'en écartant. Le rayon ne se sépare pas du moi, mais est lui-même et demeure rayon-du-moi (*Ichstrahl*). L' "objet" est atteint; il est le point de mire, simplement posé en relation au moi (et par le moi lui-même); mais lui-même n'est nullement "subjectif". Une prise de position qui comporte en soi le rayon du moi est de ce fait même un acte du moi lui-même: c'est le moi qui agit ou pâtit, qui est libre ou conditionné. Le moi, pourrait-on dire encore, "vit" dans de tells actes. Ce mot vivre: vivre ne désigne nullement l'être de "contenus" quelconques emportés dans un flux de contenus; il désigne une multiplicité de modes accessibles à la description et qui concernent la façon dont le moi, engagé dans certains vécus intentionnels qui comportent le mode général du cogito, vit au sein de ces actes comme l' "être libre" (*freie Wesen*) qu'il est. L'expression: en tant qu'être libre ne signifie rien d'autre que des modes du vivre tel que: sortir-de-soi-librement, ou revenir-en-soi librement, agir spontanément, éprouver quelque chose de la part des objets, pâtir, etc.[37]

C'est cette phénoménologie de l'attention qui mieux caractérise l'intentionnalité de la couche supérieure, qui ne se rapporte qu'à l'objet, mais qui caractérise, surtout, le vécu de la conscience, fondement d'où tout part. La couche intentionnelle de la perception est trop limitée et simple. Par ce fait, donc, le mode de l'intentionnalité des vécus du moi sont ceux qui dégagent le véritable sens de la création et de la production de ce qui apparaît au regard de l'esprit, c'est-à-dire au rayon-du-moi (*Ichstrahl*). C'est à partir de ces modifications "attentionnelles" de cette phénoménologie de l'attention (*Aufmerksamkeit*) qui sont responsables de la modification et de la réduction phénoménologique de façon à ce qu'il se produise un sens nouveau par rapport à la visée comme telle et par rapport au mode de l'inattention ou de l'inactualité de la conscience. C'est à partir de cette réduction qui peut alors surgir la production et la création. "Une synthése opérée pas à pas; elle devient, elle naît par production (*Produktion*) originelle."[38] C'est pour cette raison qu'il y a bel et bien dans l'analyse des structures intentionnelles noetico-noématiques une véritable production qui doit exprimer le sens du *fiat* par delà le degré de toutes ces couches intentionnelles et dont seule une phénoménologie intentionnelle des couches supérieures de la conscience en peuvent rendre compte. Le sens de l'affirmation ricoeurienne peut être mieux explicité et développé lorsqu'il affirme dans son introduction aux *Idées* qu'il y "aurait trois concepts d'intentionnalité; celui de la psychologie qui est synonyme de réceptivité, celui des *Ideen*, dominé par la corrélation noéme-noése, dont on ne sait si elle est réceptive ou créatrice, celui de la constitution véritable, productrice et créatrice."[39] Il semble donc, que nous assistons, tout le long de ce excursus analytique et descriptif de l'intentionnalité de la conscience volitive, mais aussi, affective et prédicative, à une véritable constitution phénoménologique productrice et créatrice.

Nous pouvons soulever deux sortes de questions qui découlent de ce que nous avons parlé jusqu'ici. La première, concerne le statut de la représentation en tant que genre commun de la perception et de la volition. "L'expression même de représentation désignait en même temps ce genre commun."[40] C'est, en effet, à ce genre commun qu'appartient la possibilité de fonder[41] d'une part, la couche de la sphère volitive en tant que prolongement et extension de l'intentionnalité à d'autres modalités du vécu tels que, le souhait, le désir, la croyance etc.. et, de l'autre, d'édifier la spécificité de l'acte volontaire en tant que tel, ayant ses propres spécificités noétique et noématiques, par rapport à la sphère perceptive. Étant donné cela, nous pouvons nous interroger sur la capacité de cet entrecroisement phénoménologique au sein du vécu total. C'est cette unité qui fait problème, d'une part, par rapport aux nouvelles possibilités crées, et d'autre part, par rapport à l'unité qui englobe la totalité du vécu dans le flux intégral de la conscience, qui à la limite, s'avère comme l'unité qui effectue une conscience avec une conscience.

Le deuxième problème concerne le statut des actes perceptifs par rapport aux actes volitifs. Husserl entend les actes perceptifs comme étant des actes simples et fondants, alors que les actes volitifs sont complexes et fondés. Dés lors on peut poser la question de savoir comment se fait-il que les actes de la structure supérieure de la conscience soient seulement des actes fondés, alors qu'ils sont des actes complexes et qui appartiennent à des couches de degré supérieur de l'intentionnalité? Nous croyons qu'il n'y a qu'une seule réponse possible: c'est que les actes sont, à la fois, fondés mais, également producteurs des sens nouveaux et, pour cela même, étant des fondateurs des nouvelles modalités thétiques et synthétiques du moi en tant que sources originelles des nouvelles noèses et donc, des productions

(*Erzeugungen*) et en même temps, en tant que des actes créateurs des ses propres possibilités du sujet du vouloir et de l'agir.[42]

4. LES SYNTHÈSES ACTIVES

Selon Franco Volpi la stratégie des synthèses passives et actives de Husserl, tâchent de montrer comment, dans l'acte de connaître du sujet, dans ses fonctions constituantes plus fondamentales et même élémentaires, ce que nous pensons comme passivité, l'est en fait, rempli aussi de l'activité et, à son tour, comment l'activité est remplie également de passivité. La démarche husserlienne suit une voie qui est typiquement cartésienne; elle se fonde sur la différence entre réalité et pensée, qui est sous-jacent à la distinction entre les objectités de la réalité sensible, et les objectités de la pensée.[43] Dans le langage husserlien c'est la distinction entre ce qui est de l'ordre de l'immanence et de l'ordre de la transcendance. Plus on avance dans la recherche phénoménologique plus on comprend que la distinction établie entre le domaine de la sensibilité en tant que passivité pure ou réceptivité et le domaine des fonctions intellectives supérieures et catégoriales en tant qu'opérations actives, ne peuvent pas être comprises comme des distinctions à être évaluées et à être comprises *ipso facto* en elles mêmes et de façon dichotomique, mais qu'elles sont appelées à être comprises les unes par les autres. Le point d'encrage husserlien c'est précisément cela.

Les synthèses actives (1920–1921) tâchent de démontrer l'activité du sujet aux origines transcendantales de la synthèse catégoriale de l'expérience. Dans les §§1 et 2 de ce texte, Husserl explicite les rapports de l'activité du sujet avec la passivité. L'importance accordée à l'idée d'acte traduit le souhait de Husserl à orienter la recherche transcendantale de la conscience à l'objet, dans son point de départ génétique. Prolongeant la recherche entamée dans les *Idées*, il s'agit dès lors, d'approfondir comment dans l'acte de l'ego s'accomplit la conscience de l'objet. La question laissée en ouvert sur les rapports entre les actes fondés et les actes fondants peuvent maintenant avoir une réponse plus éclaircie de la part de Husserl.

Aux §1 et 2 Husserl présente comment se constitue une phénoménologie de l'attention dans les sphères plus inférieures de la conscience et de son activité. Mais dès que nous regardons de plus près l'activité de la conscience, nous nous rendons compte du paradoxe qui d'y en résulte: d'une part, nous parlons seulement des couches inférieures lorsque nous avons sous les yeux ce qui est déjà accompli et activement constitué, mais d'autre part, lorsque nous nous écartons, par abstraction, de l'activité elle-même, elle révèle sa propre indétermination, de telle manière que seulement dans la recherche des couches supérieures nous pouvons trouver l'accomplissement des couches inférieures.[44]

Afin de résoudre ce paradoxe, où la passivité est mieux comprise selon l'activité de la passivité en tant que forme de réalisation du moi, Husserl explicite cela même avec une phénoménologie de l'attention et de ses "dérivés." C'est dans cette attention – attenter – (*Das Aufmerken*), que l'on peut trouver la meilleure expression de l'activité de la conscience, car elle est "la manière de réalisation permanente de la conscience (*Vollzugsmodus des Bewusstseins*)."[45] L'activité de la conscience révèle alors le champ authentique de son activité selon le mode du "se tourner vers l'attentif" (der aufmerkenden Zuwendung). Le "Das Aufmerksam" exprime les configurations possibles de la réalisation du moi. Il est compris en tant que le début (*Anfang*), la poursuite (*Fortgang*) et l'accomplissement (*Vollzug*) de la conscience.

Le mode de l'attention négative dans la passivité s'appelle affection (*Affektion*).[46] La passivité est alors l'activité première, le sous-sol où se constitue la passivité et l'objectivité pré-constituée. Toutefois, cette passivité n'est qu'une pure réceptivité, elle exprime, davantage, l'activité de l'ego. Dans cette sphère, que Husserl désigne pour *Gemütsbewustseins*, nous sommes au plus bas de la constitution de la conscience.

Mais, tournons-nous maintenant vers la sphère de la représentation (*Vorstellungssphäre*) ou de la "Sensuelle Sphäre" (sphère sensuelle), comme le dit Husserl, expression, qui exprime mieux le caractère ambigu de ce que nous entendons par "représentation."[47] C'est dans cette sphère que la conscience de l'objet acquis un nouveau sens avec la conscience du sentir. Cela signifie que la conscience d'objet fonde une nouvelle conscience de type nouveau, à savoir, une couche de la conscience du sentir qui présente une intentionnalité de type nouveau."[48] C'est, précisément, cette nouvelle intentionnalité qui détermine l'objectivation de la représentation en tant que "représentation, comme sentiment et comme volonte."[49] Les deux modes de la conscience sensitive, à savoir, de l'attention et du "retourner vers l'attentif," sont l'expression du logos apophantique de la conscience. Ces deux configurations, de la vision de l'attention et du "tourner vers" permettent d'activer la conscience passive, en tant que conscience objectivante (*objektivierenden*), sentie (*fühlenden*) et désirante (*begehrenden*).[50]

La conscience du sentir et du sentiment indique l'activité dynamique de la sphère volontaire. Mais, selon Husserl, le concept de volonté doit être précisé. La volonté ne peut pas s'identifier au simple désir. Dit Husserl: "Der Wille ist kein bloßes Begehren; er gehört in die allgemeinere Sphäre der reinen Aktivität."[51] Mais qu'entend alors, Husserl par désirer? Et quel type de désir est engagé dans la volonté, pour qu'elle puisse être pleinement volonté?

Tout d'abord, Husserl éclaircit la signification pleine de la "réalisation d'un désir," comme étant une conscience synthétique. Mais, la volonté ne peut pas être tout simplement, l'attention (*Aufmerksamkeit*) en tant que résultante de l'appréhension de l'activité de la volonté. En réalité, l'activité de la volonté est beaucoup plus complexe, elle est de l'ordre de l'activité synthétique et, de la sorte, l'accomplissement d'un désir active ou passif, volontaire ou involontaire. Elle est également le mode propre de la conscience au sens général, mais également le mode particulier de la conscience, à savoir, celle qui s'effectue dans les objectivations et dans les sensations.

Or, étant la volonté un acte de plein droit de la conscience et un acte qui révèle l'accomplissement et la réalisation pleine de l'objectivation significative d'un objet par la conscience, nous comprenons aussitôt, comment la volonté associe à son effectivité les différentes composantes qui la structurent. C'est pour cela que nous pouvons dire avec R. Bernet: "La connaissance est donc bien un acte et non la simple possession intuitive d'un objet, et cet acte est un acte particulier en ce qu'il ne vise pas simplement un objet mais l'identité d'un objet déjà visé dans d'autres actes. Il faut insister également sur le fait que, pour Husserl, ce modèle de la connaissance par synthèse de remplissement s'applique indifféremment aux actes simples ou fondés, aux représentations nominales ou catégoriales, aux noms et aux jugements."[52]

Universidade Católica Portuguesa, Porto

NOTES

[1] Albrecht Dihle, *The Theory of Will in Classical Antiquity* (Berkeley: University of California Press, 1982), p. 144.

[2] *Ibid.*, p. 123.

[3] *Ibid.*, p. 45. Cette idée est reprise par Simmo Knuuttila, "The Emergence of the Logic of Will in Medieval Thought," dans Gareth B. Matthews (ed.), *The Augustinian Tradition* (Berkeley, Los Angeles, London: University of California Press, 1999), p. 207.

[4] *Œuvres de saint Augustin*, "Les Confessions," VIII, 9, 21 (Paris: Desclée de Brouwer, 1962), p. 50.

[5] Ashraf Noor, "Individualité et volonte," *Études phénoménologiques* 13–14 (1991): 137.

[6] *Ibid.*, XII, "*Révisions*," I, 15, introduction, traduction et notes par G. Bardy (Paris: Desclée de Brouwer, 1950), p. 367. Cette définition est reprise d'une autre œuvre d'Augustin: *Contra manichaeos de duabus animabus liber unus*, cap. 14.

[7] Descartes, *Meditationes de prima philosophia. Méditations métaphysiques*, Texte latin et traduction du Duc de Luynes, introduction et notes G. Rodis-Lewis (Paris: Librairie philosophique J. Vrin, 1978), p. 57.

[8] *Ibid.*, p. 57.

[9] E. Kant, *Critique de la Faculté de juger*, traduction A. Philonenko (Paris: Librairie philosophique J. Vrin, 1968), p. 21.

[10] *Œuvres de S. Augustin*, "Le libre arbitre," p. 474.

[11] Nous voyons cela très clairement dans plusieurs passages du *Libre arbitre*, par exemple en II, 2, 4.

[12] Augustin, *De mendacio*, XIX, 40 (ed. J.-P. Migne, Patrologia Latina, 40, col. 514).

[13] *Obras completas de San Agustin, La trinidad*, XI, 3, 6. (Madrid, Biblioteca de Autores Cristianos, 185, 1985), p. 527.

[14] Kant, *Critique de la Faculté de juger*, p. 21.

[15] Le rôle de la volonté est fondamental dans la représentation imaginaire. Voir à ce propos notre article: M. Manuela Brito Martins, "La Théorie de la représentation chez Augustin dans le livre XI du De Trinitate," *Mediaevalia. Textos e estudos* 11–12 (1997): 107–108.

[16] Augustin, *De Trinitate*, XI, 6, 10, p. 536.

[17] Cf. Ms. E. III 10 (1930).

[18] E. Husserl, *Idées directrices pour une phénoménologie*, traduit de l'allemand P. Ricoeur (Paris: Gallimard, 1950), pp. 149 sgs.

[19] Kant, *Critique de la raison pure*, traduction française A. Tremesaygues et B. Pacaud (Paris: Presses Universitaires de France, 1968), p. 153.

[20] Kant, *Critique de la raison pratique*, traduction française F. Picavet, introduction F. Alquié (Paris: Presses Universitaires de France, 1960), pp. 45–46.

[21] Husserl, *Recherches Logiques*, §18. Tome II. Recherches pour la phénoménologie et la théorie de la connaissance, traduction H. Élie avec la collaboration de Lothar Kelkel et R. Schérer (Paris: Presses Universitaires de France, 1962), p. 208.

[22] Kant, *Critique de la faculté de juger*, p. 51.

[23] U. Melle, "Husserl's Pheneomenology of Willing," dans J. G. Hart and L. Embree (eds.), *Phenomenology of Values and Valuing* (Dordrecht: Kluwer Academic Publishers, 1997), p. 172.

[24] Melle, "The Development of Husserl's Ethics," *Études phénoménologiques* 13–14 (1991): 115.

[25] Melle, "Husserl's Pheneomenology of Willing," pp. 169–170.

[26] *Ibid.*, p. 171.

[27] Husserl, *Idées directrices pour une phénoménologie*, p. 32.

[28] Chez Guillaume d'Occam nous trouvons l'idée d'acte complexe. L'acte perceptif est double: il englobe l'*actus apprehensivus* et l'*actus iudicativus* qui peuvent être simultanément dans l'intellect. Cf. Guillaume d'Occam, *Scriptum in Librum Sententiarum, Prologue*, q. 1, art. 1 (S. Bonaventure: Ad Claras Aquas, 1967), pp. 16–20

[29] Husserl *Recherches Logiques*, §15, p. 193.

[30] *Ibid.*, §18, p. 208.

[31] Husserl, *Idées directrices pour une phénoménologie*, §93, p. 322.

[32] Husserl, *Recherches Logiques*, p. 208.

[33] *Ibid.*, p. 209.

[34] A de Muralt, "La notion d'acte fondé dans les rapports de la raison et de la volonté selon les *Logische Untersuchungen* de Husserl," *Revue de Métaphysique et de morale* 82 (1977): 511–527.

[35] Husserl, *Idées directrices pour une phénoménologie*, p. 331.

[36] *Ibid.*, pp. 392–393.

[37] *Ibid.*, p. 321.

[38] *Ibid.*, p. 413.

[39] *Ibid.*, p. XIX.

[40] *Ibid.*, p. 392.

[41] La notion de "fonder" signifie une "présentation" directe et thétiquement. Voir à ce propos: Maria José Cantista, "O sentido do fundamento radical em Husserl (Esboço de uma interpretação)," *Revista portuguesa de filosofia* XLIV (1988): fasc. 4: 451.

[42] Husserl, *Idées directrices pour une phénoménologie*, p. 414.

[43] F. Volpi, "Aux racines du malaise contemporain: Husserl et la responsabilité du philosophe," in *Husserl*, Collectif sous la direction de Eliane Escoubas et Marc Richir (Grenoble: Jerome Millon, 1989), p. 161.

[44] Husserl, *Aktive Synthesen aus der Vorlesung "Transzendentale Logik" 1920/2*, ed. Roland Breuer (Dordrecht/Boston/London: Kluwer Academic Publishers, 2000), p. 3; *Sinteses Activas*. A partir da lição "Lógica Transcendental," traduction portugaise Carlos Aurélio Morujão (Lisboa: Centro de Filosofia da Universidade de Lisboa, 2005), p. 13.

[45] Husserl, *Aktive Synthesen*, p. 4; *Sinteses Activas*, p. 14.

[46] *Ibid.*

[47] Husserl, *Aktive Synthesen*, p. 4.

[48] "Das sagt, das Objektbewustsein fundiert ein neuartiges Bewusstsein: eine Bewusstseinschicht des Fühlens, die eine neuartige Intentionalität darstellt", Husserl, *Ibid.*, p. 5.

[49] Husserl, *Aktive Synthesen*, p. 7.

[50] *Ibid.*, p. 9.

[51] *Ibid.*, p. 10.

[52] R. Bernet, "Désirer connaître par intuition," *Revue philosophique de Louvain* 99 (4) (2001): 616.

ELGA FREIBERGA

MENTAL EXPERIENCE AND CREATIVITY: H. BERGSON, E. HUSSERL, P. JUREVIČS[1] AND A-T. TYMIENIECKA

"The gentle fluctuations" that mark the transition and correlation between the human body and mind revealing themselves in different forms of experience have always been within the sphere of attention of philosophers. We can discern the problem in Plato's dialogues, in Aristotle's philosophy, in the teaching of Aurelius Augustine where this issue appears also as problems of change and movement in harmony with the significance of memory in the realm of man's thinking. This problem occupies a prominent place also in the philosophy of the twentieth century in relation to the problems of mental experience. The significance of mental experience is also noted when touching upon the character of consciousness accentuated in the meditations of the French philosopher Henri Bergson and in the philosophy of the founder of phenomenology, Edmund Husserl. The philosophical aims of the two are distinctly, different: while Bergson finds significance in the critique of intellect and the sphere of man's intuitive wholeness, for Husserl it is important to postulate transcendental subjectivity that cannot be reduced to either physical or psychic being. Notwithstanding the distinctly differing aims the philosophers postulate, both of them have appraised mental experience and that has influenced the views of the Latvian philosopher Paul Jurevičs. Concepts of mental experience play an essential part in grasping the overall process of life, the constitution of its wholeness that is in the sphere of interest of many of A-T. Tymieniecka's works. Human being as a unity of body and soul is a creative individuality wherein the diversity of being finds its sharpest manifestation; where man becomes human beings forming there own interpretation of being through self-individualization of life: "So existence of the human being, whatever other factors of its exercise may be, consists ultimately in man's self-individualization – in – existence. Indeed, the unique specificum of life is that it unfolds from within, from its very own core, which progressively in its struggle against circumambiant factors unfolds its virtualities and grows toward a maximal development which is uniquely its

A-T. Tymieniecka (ed.), Analecta Husserliana XCIII, 335–349.

own."[2] It is the intellectual, moral and aesthetical foundation of human's being that A-T. Tymieniecka calls *onto-poiesis of life*. The significance of Tymieniecka's phenomenological approach to life emphasizes M. Kūle: "The significance of the phenomenology of life worked out by A-T. Tymieniecka lies in a fact that she takes philosophy into an unusual field, that is, following the Logos through the labyrinth of life and interprets the process of self-individualization not only as Human Condition, but as a part of the life process at all."[3]

THE EXPERIENCE OF A LIVE BODY: H. BERGSON

The most essential impulse in connection with the subject under discussion in Bergson's philosophy concerns the manifestations of amalgamation of consciousness and body that are to be most often to be found in his work *Matter and Memory*. Manifesting the significance of the amalgamation of body and consciousness Bergson has turned his attention to characterizing mental experience: memory, perception and imagination. The way Bergson described these forms of mental experience has made Bergson's philosophy attractive not only to modernist writers (A. Machado, T.S. Eliot, M. Proust), but also to painters (V. Kandinsky, G. Braques).

Placing his body at the centre of all perceptions: "As my body moves in space, all the other images vary, while that, my body, and remains invariable. I must, therefore, make it a center, to which I refer all the other images ... *My body* is that which stands out as the center of these perceptions."[4] Bergson marked the contrast dominating up to this time between the outer and the inner perception of the body, the body as an instrument that is subjected to an outer examination and the body as a live body that is a personified realization of man's wishes, motivations and movements.

A live body is not to be put in motion by outer stimuli; it does not resemble a mechanism. Remembering, responding, image making that are modeled both by the mind created constructs and also by the sensual motions inherent to the body itself are its characteristic features: "The bodily memory, made up of the sum of the sensory-motor systems organized by habit, is then a quasi-instantaneous memory to which the true memory of the past serves as base. Since they are not two separate things, since the first is only ... the pointed end, ever moving, inserted by the second in the shifting plane of experience, it is natural that the two functions should lend each other mutual support .... It is from the present that the appeal to which memory responds comes, and it is from the sensory-motor elements of present action that a memory borrows the warmth which gives it life."[5] The body, in Bergson's view, is not to be divided into matter and spirit; as a live body it is characterized by wholeness. The body to Bergson is like an experience of being in the world, it is in space and is spacious at the same time: "The truth, is that space is no more without us than within us, and that it does not belong to a privileged group of sensations. *All* sensations partake of extensity: all are more or less deeply rooted in it."[6] However, to substantiate the statement that the body is wholeness, and a live wholeness at that, live reality that cannot be experienced just as perception of space is also important to Bergson. That is why he accentuates the role of intuition in the experience of time as only through duration of time the wholeness of mental life can emerge. Duration (*durée*) is at the bottom of the life drive *(élan vital)*; it is creativity itself.[7] In a way duration for Bergson becomes characterization of reality, as it were, changing views on the world and man that is not split into body and spirit, but formed as a creative mentally supported whole that opens up to man only from the "inside," intuitively. Intuitive observation is not inert; it is guided by the activity of the will. Conjoint with the will intuition acquires dynamics and activity turning into something like creative self-inspiration engendering the exertion for the creative act. It is at the basis of man's creativity, demonstrating the freedom of his creative ability. The human being not just moves, functions, survives, it also lives, remembers, experiences and is capable of keeping intact the traces of the wholness of time.

The problem of memory, according to Bergson, is related to the experience of the live body, but it is not reduced to physiology or psychology. The live body is not put into motion by means of an external force. It does not resemble a mechanism. It is characterized by remembering, responding, type formation following both the constructions of the mind and the sensorimotor body qualities.[8] According to Bergson, the body is not to be divided into matter and mind; being a live body it possesses wholeness. The body to Bergson is an experience of being in the world; it occupies space and is spacious at the same time.[9] However, to substantiate the body being whole, and live whole at that, Bergson sees the importance of live reality that cannot be experienced as a simple perception of space. That is why he underscores the role of intuition in the experience of time, for the wholeness of mental life comes into being only within the duration of time. Duration (*durée*) is at the basis of the life force *(élan vital)*, it is creation itself.[10] A-T. Tymieniecka states: "We experience ourselves in a lived duration of acts. Bergson grasped this as the phenomenon of 'la durée vécue.'"[11] In a way, to Bergson, duration becomes a kind of characterization of reality, changing the views on the world and the human being that is not split into body and mind, but formed as a creative whole held together by mental life opening up to man only on the "inside," intuitively.

Intuitive observation is directed by the activity of one's will, it is not inert. Intuition acquires dynamics and activity in its association with willpower turning into something like creative self-suggestion compelling exertion for a creative act. It is at the basis of man's creativity demonstrating the freedom of his creative ability. However, Bergson does not maintain that a human being's creative ability originates in his ego, but he can continue the life impulses that are at the basis of changeability of the flow of life and connect man and cosmos. It is the task of the genuine real selfness that merges the fragmented perception of life into a united organic whole. This conclusion allows Bergson overcome the exteriorization of the human's world, the dependence of the human self on social conditions, the images of the external world and their power over the manifestations of man's creative freedom. Mental life expands between the extremes of activity (where images are manifested in motion) and knowledge (images retained in consciousness). The significance of the conclusion is important for Bergson in order to demonstrate the difference between the states of mental life and the states of the brain.

Bergson emphasizes the basis of a human being's activity in the world where the fundamental truth is to be found in the movement of the live body, not in the object under observation that moves within space. Being live the body perceives and retains a stock of memories and recollections that are not identical with the simple stimuli of the outside world. This standpoint of Bergson's allows an approach to the solution of the problem of memory.

In Bergson's view perception and memory are closely related or even parallel phenomena: Bergson refuses to regard perception as a passive mechanism registering impressions and accentuates its synthesizing, vital disposition. That shows how a human being perceives the external world, the world of material things or just matter because all that is perceived is an image (a sensory image). An image is more than a mere likeness or reproduction of something. The image differs from reproduction in the degree to which the perception occurs. Perception in its pure manifestation is matter itself or materiality of things where this materiality is incorporated in images. Pure perception is unconscious. That is why Bergson deems important also the second type of perception where images are being intuitively structured or organized within time and space in one's consciousness, not at all being an exact replica of some objective reality, consciousness being present intuitively. Perception is also a birth of consciousness, but consciousness itself is not just "picture observation" or "watching" perceived images. Such a standpoint would present consciousness as a spacious structure that, in turn, would make one return to the problem of its placement (in the brain, for instance). For

Bergson, consciousness is a flow of time and that is exactly what makes memory so important. Memory occupies a higher position in the hierarchy of understanding consciousness because only that way can the flow of time (*durée*) be expressed. Bergson underscores the qualitative difference between perception and memory, as well as between the two types of memory: habitual memory (*mémoire-habitude*) and pure memory (*mémoire-pure*).[12] Habitual memory naturally is sensorymotor; it allows the body to get his bearings in space while pure memory is autonomous from the body being a phenomenon of the mind. The separation indicates that both perception and memory, in his view, posses both virtual and actual aspects. The virtual aspect is connected with pure memory. The elements of pure memory are *souvenir pure, souvenir image* and *perception.* Act of remembering/souvenir includes retaining that is transferred on the souvenir image.

Bergsonian separation within the framework of pure memory into *souvenir pure* and *souvenir image* is to be connected with Bergson's understanding of consciousness in which the image does not yet indicate qualitative distinction between consciousness and the material world; this separation is realized only in the ability of memory as a flow of time to retain an image separated from the perceived reality. Relationship between the act of remembering/souvenir and perception is mediated by imagination, the imagined. Imagination, the imagined in Bergson's philosophy acquires a peculiar image modality – they are called spontaneous reminiscences – recollections; the most familiar type of recollections are childhood recollections. Recollections are among the strongest types of memories that include both the things perceived and the things sensually experienced both movements and habits. They keep the sensed and the perceived – the experience – in its wholeness; however, being closely connected with the perceived they resemble a condensation of acts of recollections. Recollections point to the fact that remembering itself is not a continuous process because it is closely interwoven with forgetting. By this paradoxical explanation Bergson shows that it is perception that contains forgetting, not memory and every perception is simultaneously destruction of forgetting. Memory is not unraveling an unbroken chain of recollections (if one could imagine a symbolic ball of recollections the thread of which we start unrolling in leisure times) that guards the images we have in memory from being replaced by those perceived a second ago.

The action of images within the framework of memory branches off into two planes: firstly, the image "retains" matter (the perceived); secondly, the image "retains" some lost initial intuition. The latter kind of memory indicates the very essence of it – not that we *possess* memory, but we *are* memory. That is because in the life of human consciousness there is no precise moment at which the present turns into the past, or the perceived turns into a recollection. Recollection is important because it summarizes, stockpiles, as it were, the acts of remembering. Perception and memory are simultaneous.

It has already been mentioned that Bergson separates *souvenir pure* from *souvenir image*. *Souvenir pure* is characterized by duration of time that is self-sufficient, thus not to be materialized into images. What is embodied in *souvenir images* is no longer pure, but adjusted, though indirectly, to the demands of the present. *Souvenir images* are connected with the requirements of life they have an objective in view. *Souvenir images* are subdivided into recollections or reminiscences that are spontaneous. Bergson calls them dreams, but the greater part of memory is habitual memory that retains the necessity for the habitual in the perception of renewal. Thus, in Bergson's conception of memory images contain a threat for the existence of *souvenir pure*/pure acts of remembering – the image supplants the lost original intuition taking its place, but not the whole of it, by rendering space to the flow of time that has a negative part in the process of self-emergence. The Bergson investigator J. Mularkey, calls it "the subject's self-emergence suspension"[13] reminding that the problem of memory and perception just as the problem of matter and mind are not problems *per se*, i.e., philosophically neutral and abstract, but closely related to the problem of human consciousness and subjectivity. The problem of pure memory just like the problem of pure idea is hard to substantiate because it is *consciousness without consciousness*, i.e., unconscious; at the same time it is not just a psychic state as it is being explained in contemporary psycho-analytically tinged theories. Pure memory is a wider notion than consciousness, though in Bergson's works there is no clear standpoint to be found as to the way the transition from pure memory to *souvenir image* takes place. Evidently, pure memory should be explained on the analogy of creative evolution – as a metaphysical concept.

Souvenir image in Bergson's philosophy is to be understood as a result of disintegration of pure memory. *Souvenir image* is what has been separated from pure memory and is to be seen, analyzed and described.

It means that *souvenir image* when separated from pure memory that is a synthetic and virtual phenomenon is what one can deal with in most various spheres of science and art, as well as in philosophy.

Singling out of memory, its qualitative separation (notwithstanding the simultaneousness of memory and perception) makes Bergson conclude that perception and action are "a fundamental law of our psychic life"[14] whereas memory is "like a whimsical play of imagination because it has as much freedom as the mind takes from nature."[15] Bergson allots the thankless part of separation whereby reality turns into images to imagination. "Separation is the work of our imagination, the true task of which is to record the moving images of our daily experience."[16] However, nowhere does Bergson give a strict separation of perception from imagination, or perception from memory, perception being an isolation of images from the totality of images, as it were. Bergson can be reproached for lack of explanations because he clearly explains one concept with the help of another. Perception of images is also imagination of images just like memory is "image making" because it is a team play of imagination and memory.

The obscurities indicate Bergson's wish to separate our mundane experience that philosophers call empiric from philosophical experience by asking the question: what is the meaning of memory, imagination and perception in philosophical sense? That is why the concept of pure memory is something bigger than the one given only in individual consciousness. Memory is a multi-plane design, no one dimension will do to describe it. Pure memory is passivity; it is contemplative like a plane on which surges activity, movement that finally destroys pure memory splitting it into innumerable *souvenir images* that at the same time are not only separated units but form spontaneous recollections or stocks renewing the subtle requirement for the presence of the forgotten spontaneous intuition. Bergson's conception of memory is complicated multi-plane designs that cannot be explained taking into account only body, mental experience or mental autonomy of consciousness; one senses growth and creativity in it.

However, Bergson does not maintain that man's creative ability originates from his ego, but he is able to continue those impulses of life that are at the basis of changeability of life stream and link man with the universe. That is the task of the true, genuine self that merges together the separate, fragmentary perception of life into a united organic wholeness. This conclusion allows Bergson to overcome externalizing man's world, the dependence of man's selfness on social conditions, the images of the outer world and their power over the manifestations of man's creative freedom.

THE MENTAL AND THE TRANSCENDENTAL: E. HUSSERL

E. Husserl, who in the early stage of his phenomenology was critically disposed towards psychology and its possible influence on philosophy, in his "The Amsterdam Lectures" turns to the analysis of the interrelations of psychology and phenomenology. Husserl's enthusiasm is naturally directed towards the decisive significance of transcendentalism, the superiority of the transcendental task that alone can constitute sense, form meanings. However, despite the fact that psychology will never become a transcendental field of study alongside philosophy, it still accomplishes a significant analysis of mental experience.

Husserl points out peculiarities of human perception, namely, that we perceive all the reality surrounding us as physical Nature. He calls it the structural law of world experience that has a great influence on the way the interrelationship of the surrounding world and human consciousness work together, how the "inherent nature" is represented. It is not every concrete body or real thing that possesses "its own nature" as we often imagine, but all the bodies of the world form a combined unity. This unity is linked together in infinity that forms a unity of nature totality; it is a time-space structure that is a law form for our thinking and perceiving concrete things that forms the basis of mental experience: "If we examine the world of experience in its totality, we find that its nature is to articulate itself in to an open infinity of concrete single realities. According to its nature, to each single particular belongs a physical corporality, at least as relatively concrete substratum for the extra-physical characteristics, that are possibly layered on it, to which belongs, for example, the determining factors through which a physical body becomes a work of art."[17]

It is not an easy task to comprehend mental experience; it cannot be achieved directly, but only through the medium of the eidetic. Mental experience does not exist of itself as an experience of something separate; it has already a certain eidetic form. Husserl stresses that every experience is included in a mode of reflection that is intentionally directed towards the mental. That is why to live as an ego-subject means to live a mentally manifold life experiencing all the possible phenomena and events, things and imaginings that all together form what man calls life. However, the ego-subject in the mental act of experience is not himself, yet, but only possibly himself; he possesses a certain anonymity that becomes more concrete when basing on something that is not mental – something already experienced, meditated on, guessed at, appraised: "In such a reflective perceiving and experiencing, mental life as such is grasped and itself made a theme which one can work with in a variety of ways. Naturally this new experiencing and making something thematic in reflections is itself also latent, but likewise also disclosable through still higher reflection."[18] Reflection themes the perceived and the experienced, but it is not seen, it is concealed. Thus, Husserl marks the significance of phenomenological experience thanks to which the mental is not a chaos of separate perceptions, but can be grasped in all its essence.

Husserl points out that to describe phenomena of mental experiences in their interconnection is extremely difficult because the link between one mental thing and another is only intuitive, namely, the link forms an intuitive context that defies description. Unlike Bergson, Husserl does not stress the advantages of intuition in the formation of wholeness. Intuition is only one side of reflection, and an invisible and non-descriptive side to boot.

Without the phenomenological position and phenomenological reduction the anonymousness of ego-subjectivity is insurmountable and the outer world is only and solely the outer world. All nature turns into a noetic phenomenon that means that from an anonymous being ego turns into a human being. Ego is not to be reduced to mental being and life process, to a concretely material ego, but to a really focused human ego. The accentuation of this focusing appeals to me because in it Husserl makes use of a metaphorical semblance to a glance as an especially privileged manifestation of thought that links it to the "metaphysics of light and look"; the most essential manifestations of which are to be found in the philosophy of Plato and Augustine, Descartes and Heidegger, as well as being an essential basis for the transcendentalism of the beautiful in the aesthetic tradition. Unlike Bergson who notes the critical line of the glance discussed before, Husserl is closer to Descartes, to the accentuated principle of obviousness in his *Meditations* that occupies an important place in the phenomenological conception of truth. In the philosophy of rationalism the tradition of obviousness has been closely linked with elucidation or explanation that is retained in interpretation theories. In Husserl's "The Amsterdam Lectures" the characterization of noema is similar to a point of view, while the act of reflection is likened to seeing.

For Husserl the priority of transcendental issues is the most essential thing. He stresses plainly that even if psychology is eidetic, it is clearly naïve as far as transcendental issues are concerned. Psychology as a classifier of mental experience pays closer attention to the mind of a concrete human being in the world of space; that is why it cannot, even as pure psychology, solve transcendental problems. The subject of psychology is mentally "soulful/spiritual"; it does not step over the threshold of the transcendental.

The apparent closeness of mental experience to life-world (*Lebenswelt*), however, finds no justification. In his work *The Crisis of European Science* Husserl also stresses the importance of the spiritual world in constituting sense. In Husserl's view life-world is a spiritual formation. "The surrounding world is a concept that belongs to the spiritual sphere to the highest degree .... Our surrounding world is a spiritual formation in us and in our historical life."[19]

Thus the creative act depends on the person, it does not resemble an intuitive, thus misty experience, but is meaningful thanks to the intellectual activity of the personality that is manifested as the subject's intentional – not accidental – direction towards the object. Sense constitution is a really creative process that is rooted, on the one hand, in the already experienced, appraised, those points to the fact that sense differs from a simply intuitive spontaneous act; it rather resembles spiritual, purposeful work that makes no sense without some intellectual effort. That is why spontaneous discoveries of mental experience, being eidetic as well, are not fully creative.

Returning to "The Amsterdam Lectures" it must be noted that Husserl connects the experience of mental life with everyday experience that lacks "something else" that could make it human, i.e., the transcendental act: "It is now evident that the apperceiving conscious life-process, wherein the world and human being in its particularity within it are constituted as existentially real, is now what is apperceived or constituted; it is not the mental which as human mental being or human mental life-process comprises the apperceptive make-up of the real world. Something is necessary in order to make this distinction between transcendental and wordly, concrete conscious life as fully scure as possible, and in order to make transcendental subjectivity evident as an absolutely autonomous field of real and possible experience ...."[20] Transcendental reduction, however, cannot fully refer to mental experience that manifests itself rather as eidetics of an individual ego. Mental experience always retains something external that cannot serve as an ontological basis for subjectivity. Only overcoming this presence of external, even eidetically transformed, reality, can the transcendental parallel be discovered, writes Husserl: "The transcendental focus which is set up through a radically consistent and conscious transcendental reduction, signifies nothing less than an altering of the whole form of life *Lebensform* previously practiced not only by the particular this or there, but also historically by humanity as a whole: an absolute, all-embracing, and radical shift in the natural living-alone of life and one's natural living in a pregiven world; a change in the mode of experiencing of thinking and of every other kind of activity, and also in all modes of reason."[21] But these conclusions are binding in the question of pure psychology as a draft of transcendental phenomenology.

Bergson's and Husserl's attitude to mental experience marks two opposing tendencies where, in the first case, creativity manifests itself as intuitive spontaneity, qualitative manifestation of life, whereas in the second only transcendental subjectivity can be creative.

RESTORATION OF WHOLENESS: P. JUREVIČS

The first half of the twentieth century in Latvian philosophy is marked by two more powerful philosophical figures belonging to different trends. Teodors Celms, whose work Maija Kūle has written about, is more widely known and the less known follower of Bergson's philosophy is Pauls Jurevičs.

Teodors Celms' philosophical position is rooted in or, to be more exact, is close to Husserl's phenomenology, accentuating the tradition of rationalism. Pauls Jurevičs is linked with the interpretation of mental experience to be found in Bergson's philosophy that characterizes creative act as an intuitively spontaneous manifestation. This tendency is to be observed not only in his doctor's thesis on Bergson's philosophy published in French,[22] but also in the essay *Problems of National Life* published in Latvian 1936.[23]

Jurevičs' philosophical position is characterized by anthropological dominance. Man, as described by him, is a being of contradictory inclinations, not a sense-constituting consciousness. Man and his reality both constitute and do not constitute a whole. How are we supposed to understand this paradox? Quite in a Bergsonian spirit if we remember Bergson's ideas about morality and religion having two sources. Jurevičs also points out the fact that man's reality is whole to only a few, to those who are capable of creating new forms of life that lead and inspire others.[24] Creativity is what distinguishes one man from another and it is sooner to be regarded as a spark of genius than an intellectual effort.

Jurevičs regards restoration of the wholeness of life to be one of the most important tasks of the philosopher. Life, in his opinion, is like movement, an intuitive drive towards perfection that parallels Bergson's idea of creative evolution.

Jurevičs comes forth with critique of rationalism pointing out that restoration of the wholeness of life is impossible within the framework of old forms of thinking. Contemplating the attacks of irrationalism philosophy on the mind, Jurevičs indicates the situation that was conducive to such attacks: It is the paradox created by the mind itself that has created excessive rationalization of life thus preparing favorable conditions for its own critique and destruction. Jurevičs writes that the ambiguity of life on the philosophical plane is mainly marked by the ability to reflect, by being speculative,[25] as regards life. It is not only on the basis of Bergson's position as regards the critique of the mind, but also following Husserl's views on the distinction that exists between philosophical thought on the one hand and psychology and the exact sciences on the other. Jurevičs stresses the excessive pragmatism, technology of the mind that created favorable soil for irrationalism. Jurevičs stresses the irrationalism (Nietzsche, Spengler) engendered delusion that philosophical reflection is to "blame" for shattering the value basis. Leveling and degrading intellectual and analytical work of the mind is tantamount to obtaining pseudo-values – "bogus small deities," writes Jurevičs.[26] However, the mind being formal, it cannot create values of itself, only emotions are capable of it. The mind neither creates, nor destroys; it can only reflect. Thus it is not creative. Substantiating this thesis Jurevičs often refers to another philosopher that appeals to him, Blaise Pascal.

Pascal characterized the mind not as *ratio*, but as spirit (*esprit*) marking the paradoxical character of spirit. It is the thought that distinguishes man from an animal or a stone, but it has a fatal meaning because it brings loneliness to man. Man is alone in the infinite universe, he has no counterpart and he can think only of himself, his Creator and his end. Man has not been given the elaborate instrument – the geometrical spirit – to get to know himself, but only instinct and experience; that is why his thought about himself is indefinite and paradoxical, it cannot be direct. Still less grounded are man's claims to get to know his Creator because man's love is directed towards him; yet it is only love for the reflection of the Creator in man himself and it is like self-love that makes one feel loneliness still more keenly. Thus, loneliness and the unreliable mind when it is turned towards man himself make Pascal say that the mind is not faithful to itself[27] or submissive to anyone. The above utterances by Pascal form the context of Jurevičs' critique of the mind. Just like Pascal preferred the order of the heart (*l'ordre de la Coeur*), so Jurevičs stressed the organic value of life and the creative manifestation of mental experience the task of which it is to restore the wholeness of life which the reflexive mind alone is incapable of doing.

The task of reflection or the ability of reflection, according to Jurevičs, is especially important when it is directed towards the subject. Jurevičs calls this function of the mind the inner vision (which means that the mind is characterized as mental experience, not as a formal category) which is important for discovering such ideas that can further deeper value experiences. The mind and instinct are two sides of one whole that are not to be opposed to each other, but are mutually supplementary. "Bergson's accentuated intuition," Jurevičs writes,[28] "and is the mind elucidated instinct serving as the impulsive force of the mind." Intuition is an important element of consciousness, but it cannot do without the regulation of the mind; just like the mind cannot do without the creative elation of intuition.

The source of man's world of values, according to Jurevičs, is mental experience in which individual, creative activities predominate forming a balance between the disharmonic parts of life, namely, the mind and the instinct. Life has only one law – it is an intermittent, dynamic tendency to move forward towards ever new forms.[29] That does not mean that the created forms are the purpose of the life stream and that from them just like putting together a mosaic one can restore (the way Jurevičs formulated his task) the wholeness of life. That is an intermittant process of creation in which, to quote Jurevičs, "Priority belongs to the spirit and not to the form it has created."[30]

The mental priority of creativity life as the highest value possesses, in Jurevičs' opinion, means: the only way one can choose is to be one's own self, original and truthful. Playing up the word "originality" Jurevičs stresses the category of origin characteristic of the philosophy of the thirties where origin does not point to the source or roots, but to rebirth which is possible in emotional experience.[31] Reflection of the mind is not creative; it can only serve as a context. However, creativity, according to Jurevičs, has no divine character the way it can seem at first glance. The seeming spontaneity and the original genuineness are the fruit of culture, Jurevičs stresses, and the result of conscious activity that comes into being as something natural, organic and perfect. It is touching the deepest layers of life that justifies the existence of philosophy and art because they more vitally than any other spheres of man's activity appeal to the value of singularity that implies being one's own self, not repeating oneself. Originality or beginning, in Jurevičs' opinion, means individual manifestation of life, its unique revelation, in other words, creativity.

IMAGINATION CREATRIX: A-T. TYMIENIECKA'S CONCEPTION

The problem of creative imagination is one of A-T. Tymieniecka's principal themes which she develops in connection with human condition and stages of life. This connection reveals both a live presence and originality, qualitative distinction of man's creative acts. In her view it is the creative act that is the prototype of human being activity.[32] The view on reality through the prism of human being's creativity enables to expand the borders of phenomenology and critically analyze the excessive rationality of classical phenomenology. The creative act is fundamental, it is a possibility to genuinely express human being, as a rational, perceptive and creatively active being's, connection to the source of life. The fundamental direction of human being creative activity, its *telos* is the creation of self as bringing into existence of a human being's individual autonomy in the cosmos. A human being's course from natural determinations towards creative possibilities[33] that rests on human condition. Creative acts incorporate latent spontaneity that manifesting itself reveals the inimitable and unique character of the creative act: "With the surging of the creative act as such – not this or that creative process or act of any human individual, but of the primogenital, original creative instance – the human was brought into life's orbit."[34] The explanation of the significance of human being creative activity is to be found in the work of art as the visible outcome of creative manifestation or, one might say, an analogy of creative activity that always appeals to the human being as a whole, as an allusion to the wholeness of his world. A work of art from this standpoint is not a mere mental act or an abstract artefact, but a revelation of a concrete artist's being. A creative act is a manifestation of wholeness irreducible to mental activity as the only keeper of creativity. That is Tymieniecka's objection to classical phenomenology: "From the outset I have proposed to be faithful to the aim of Husserlian and also of post-Husserlian phenomenology in investigating the human universe of discourse in terms of meanings reaching to the very origin of sense, but we have forthrightly sought to seek this through the itineraries that are traced out not by the conscious operations of intellective reason but by *Imaginatio Creatrix*, prime force inspiring human endeavors. Thus emerged our novel project in history, namely, a critique of reason limited by the span of creative imagination."[35] The creative act is independent and autonomous, it is like an aim that is directed towards one's self and emerges from the roots of life. Creation is not only creation of a thing; it is also creation of man's self that develops a human being's most diverse abilities by a correlation between them. A-T. Tymieniecka calls it creative orchestration of all types of human activity. This polyphonic view on the interrelation of human abilities and activities allows mind to be regarded as one of man's abilities, yet not the only one.[36]

The synergetic role of creative imagination is evident: "... *imaginatio creatrix* the decisive factor with the creative process"[37] that is to be regarded not only as a manifestation of human activity, but also as an allusion to the fundamental aspect of human's reality – the aspect of being. This allusion indicates new possibilities for analysis.

The importance of mental experience, its relationships with creativity and human being freedom continue to be in the sphere of attention of philosophers. Freedom, creativity and the wholeness of life are especially important philosophical categories in the twenty-first century as well, when life and the flow of time become ever more fragmentary and we grasp the value of life only in its pragmatic sense.

University of Latvia

NOTES

1 Pauls Jurevičs – a Latvian philosopher, professor at the University of Latvia. His famous work was on a philosophy of Bergson. Interested in M. Heidegger's philosophy. After the Second World War lived and worked in Australia.

2 A-T. Tymieneicka, "Phenomenology of Life and the New Critique of Reason: from Husserl's Philosophy to the Phenomenology of Life and of the Human Condition," in *Analecta Husserliana*, Vol. XXXIX (Dordrecht: Kluwer Academic Publishers, 1990), p. 9.

[3] M. Kūle, "Self-individualization as the Main Pinciple in the Phenomenology of Life," in A-T. Tymienicha (ed.), *Analecta Husserliana. Imaginatio Creatrix. The Pivotal Forces of the Genesis/Ontopoiesis of Human Life and Reality* (Dordrecht/Boston/London: Kluwer Academic Publishers, 2004), p. 48.
[4] H. Bergson, *Matières et mémoire. Éssais sur la relation du corps à l'esprit* (Paris, 1971), pp. 46–47.
[5] *Ibid.*, pp. 152–153.
[6] *Ibid.*, pp. 216.
[7] Bergson, *L'Évolution créatrice* (Paris, 1907), p. 14.
[8] Bergson, *Matières et mémoire*, p. 44.
[9] *Ibid.*, pp. 155–156.
[10] *Ibid.*, p. 216.
[11] Tymieniecka, "Memory and Rationality in the Onto-poiesis of Beingness," in *Phenomeonological Inquiry*, Vol. 13, October, 1989, p. 105.
[12] Bergson, *Matières et mémoire*, p. 57.
[13] J. Mularkey, *Bergson and Philosophy* (Paris: Notre Dame, 2000), p. 25.
[14] Bergson, *Matières et mémoire*, p. 180.
[15] *Ibid.*, p. 180.
[16] *Ibid.*, p. 189.
[17] E. Husserl, "The Amsterdam Lectures," in *Husserliana* 9, p. 304.
[18] *Ibid.*, p. 307.
[19] Husserl, *Die Krisis der europäischen Wissenschaften und die transcendentale Phänomenologie//* Gesammelte Werke (Den Haag: Martinus Nijhoff, 1954), p. 317.
[20] Husserl, "The Amsterdam Lectures," in *Husserliana* 9: p. 340.
[21] *Ibid.*, p. 348.
[22] Jurevičs, *Le problème de la connaissance dans la philosophie de Bergson* (Paris: Vrin, 1930).
[23] Jurevičs, *Nācionālās dzīves problēmas.* (Rīga: Valters un Rapa 1936) (in Latvian).
[24] *Ibid.*, 29. lpp.
[25] *Ibid.*, 65. lpp.
[26] *Ibid.*, 90. lpp.
[27] B. Pascal, *Pensées*, ed. Z. Tourneur et d'Anzieu (Paris, 1960), p. 272.
[28] Jurevičs, *Le problème de la connaissance dans la philosophie de Bergson*, p. 56.
[29] *Ibid.*, p. 285.
[30] *Ibid.*, p. 286.
[31] Jurevičs, *Nācionālās dzīves problēmas*, 292 lpp.
[32] R.M. Barral, "Creativity and the Critique of Reason," in *Phenpmenological Inquiry. Phenomenology in the World Fifty Years After Husserl*, The World Institute for Advanced Phenomenological Research and Learning, Belmont, Massachussets, USA, Vol. 12, 1988, p. 137.
[33] Tymieniecka, "Logos and Life. Creative Experience and the Critique of Reason," (Dordrecht: Kluwer Academic Publishers, 1988), p. xxiv.
[34] Tymieniecka, "Memory and Rationality in the Onto-poiesis of Beingness," in *Phenomenological Research and Lecturing*. Belmont, Massachussets, USA, Vol. 13, October 1989, p. 108.
[35] Tymieniecka, "The Triumph of Imagination in the Critique of Reason," in A-T. Tymienicha (ed.), *Imaginatio Creatrix. The Pivotal Force of the Genesis/Ontopoiesis of Human Life and Reality* (Dordrecht/Boston/London: Kluwer Academic Publishers, 2004), p. XV.
[36] *Ibid.*, 325.
[37] *Ibid.*, 375.

KIYMET SELVI

LEARNING AND CREATIVITY

ABSTRACT

Creativity can be based on some factors such as learning, experience, motivation, imagination, personality and these factors may affect human creativity. It "comes up" in various forms as ideas, approaches, products, art, systems, solutions, situations, strategies, changes, methods, techniques, designs, treatments and research. Creative study needs to search ambiguous situations to find explanations, facts or answers that satisfy one's curiosity. Creative behaviour is defined as highly original and there is very low consensus on value (Kasof 1997).

Creativity is mostly attributed to some internal factors such as soul, gene, brain, personality, value, cognitive skills, mind, instrinsic motivation and it is seldom attributed to external factors such as education, work, economy, technology, culture and extrinsic motivation. The important point here is that the internal and external factors affect each other. The external factors support the improvement of internal factors by affecting creativity. Learning, for example, is a complex factor of creativity and it also affects both internal and external factors of creativity. Creativity and learning are two components of human experience, because experience is a dynamic relationship between learning and creativity. Creating and constituting new things also need new experience and learning.

Individuals continuously learn from their feelings, imagination, experiences and their environment. Some of the external factors of learning such as the educational system, learning environments, learning processes, teacher competences may cause barriers to creativity potential. For example, during their time at school, students have insufficient time for studying, reading, examining, questioning and criticizing. Presentation of simple or complex aims and low or high level of performance of students in the educational system can damage creativity.

Education, an activity of human development planning, supports the development of an individual. It gives a chance to individuals to construct their knowledge via experiences. Education should provide students who want to

A-T. Tymieniecka (ed.), Analecta Husserliana XCIII, 351–369.

become creative persons with encouragement and reward, and it has to teach students not only to learn facts, but also to think critically (Sternberg 2000). But according to Westby and Dawson (1995), one of the most important findings in educational studies of creativity is that teachers dislike personality traits associated with creativity such as impulsivity, risk taking, independence, determination and individualism. School must be the place for introducing new ideas, explicit representation of imagination, using mental processes to create novelty (Halford and Wilson 2002), and the way of doing these is based on preparation of a qualified curriculum.

A qualified curriculum should be flexible and it should include some disciplines and courses but it should not be constructed by subject-matter experience and subject-matter understanding. It should involve some courses and activities that are connected with creative and critical thinking, art, science, philosophy, ethics and cultural studies. At the same time, it should organize education in a way that can provide various educational services.

INTRODUCTION

Creativity can be based on some factors such as learning, experience, motivation, imagination, intelligence, flexibility and personality. These factors may affect human creativity and most of the creative acts are unexpected. Kao (1996) defines creativity as "the entire process by which ideas are generated, developed and transformed into value" (p. xvii). In creative behavior internal and external factors of learning and creativity are discussed.

LEARNING AND CREATIVITY

Creativity is usually characterized as novelty and relevance. It may occur as a creative process, a creative product, a creative personality and a creative situation (Eysenck 1996). It somehow connects something which is previously unconnected in a way that constitues a meaningful item (Duffy 1998). Though it is hard to define creativity, it can be defined as characteristics of products, persons or thought processes. However, we are not sure what the distinguishing features of these characteristics are. Many of the earliest definitions of creativity such as issued by Watson, Koestler, Wertheiner, Newell et al., Feldman, Gruber and Barett focused on the creative process. Guilford focused on characteristics of personality and Jackson, Messick, Bruner, Barron focused on creative products (Amabile 1983). Norbert (1994) defined creativity as destruction of the present structure in which the problems have appeared and are transfered into a new structure. "Creativity is proposed as the radical feature, the most fundamental dimension of human being, the characteristic which marks its specificity" (Cecilia 2000, 188). Many of these definitions which emphasize on fluency, flexibility, originality and ambiguity are to be found in the literature (Houtz et al. 2003). It is clearly seen that the definition of creativity is abundant and each definition lays emphasis on different dimensions of creativity.

Several external and internal factors affect creativity differently. Creativity is mostly attributed to some internal factors such as souls, genes, brains, personalities, values, cognitive skills, mind, instrinsic motivation and it is attributed rarely to external factors such as education, work, economy, technology, culture and extrinsic motivation. It is hard to explain how internal and external factors affect creativity. The important thing is that internal and external factors affect each other. The external factors support the improvement of the internal factors of creativity. For example: learning is a complex factor of creativity and it also affects the other internal and external factors. It is clearly seen that "... the creative constituent of human beings gives a chance to a person to use the whole apparatus of thinking, feeling and activity" (Kurenkova et al. 2000, p. 203).

Amabile (1983) stated that the componential framework of creativity includes three components: Domain-Relevant Skills, Creativity-Relevant Skills, Task Motivation:

> Within this framework, Domain-Relevant Skills can be considered as the basic for any performance in a given domain. This component includes factual knowledge, technical skills, and special talents in the domain in question. Creativity-Relevant Skills include cognitive style, application of heuristics for the exploration of new cognitive pathways, and working style. Task Motivation includes motivational variables that determine an individual's approach to a given task (p. 67).

Many researchers explain the creative process as a four-stage model, which is composed of problem or task identification, preparation, response generation and response validation (Lubart 2000–2001). According to Groborz and Necka (2003), at least four basic stages of creative problem solving are usually distinguished: preparation, incubation, illumination and verification. Some studies on creativity and critical thinking stressed that the four-stage model of creativity is related to problem-solving process and critical-thinking processes. The creative-thinking process is divided into some phases and is mostly connected with problem solving.

Guilford stated that "creativity is not the same as intelligence. An individual can be far more creative than he/she is intelligent or far more intelligent than creative" (cited in Michalko 1998, p. 2). Intelligence and creativity are separate. But they attribute to each other (Robinson and Stern 1997).

Creativity is a complex development system and it is shaped by cognitive processes, social and emotional processes, family aspects, formal and informal educational processes, characteristics of domain and field, socio-cultural contextual aspects and historical forces, events and trends (Feldman 1999). Creativity is a puzzle and many creative people rarely know how their original ideas arise. According to Lipman (2003, 248) "creativity has its roots deep in nature, in nature's diversification and amplification of itself. It also has its roots deep in human nature where it plays a role in reproduction in addition to its critical role in discovery and invention." But many researchers have pointed out that creative experience and expression depend on personality, motivation, learning, interest, effort, opportunity and learning environment.

There is a dynamic relationship between learning and creativity. Creating and constituting something new is related to new learning. Learning requires a person to display an interest and considerable mental effort (Ornstein 1995). Kohonen's (1996) definition of learning is that learning is an important part of the learner's own reflection, functioning as a bridge between practical experience and various ways of conceptualizing it. The more learning tasks or experiences increase in difficulty and complexity, the more learner brain activities increase. "The brain simultaneously operates on many levels of consciousness, processing all at once a world of colors, movements, emotions, shapes, smells, sounds, tastes, feelings and more" (Jensen 2000, 12). An individual learns continually from his/her feeling, imagination, experience and environment. Learning refers to acquiring new information, knowledge, skills or dispositions through study, task, instruction or practice.

According to Gardner (2001) there are two dilemmas about learning in school. The first dilemma is "what should be taught?" It is known that knowledge "doubles" every one- or two-year periods. Which types of knowledge are appropriate for creativity? Active learning involves having both "skill" and "will". Active learners encounter three different kinds of knowledge (Nist and Holschuh, 2000): (1) declarative knowledge, which is related to knowing what, (2) procedural knowledge, which is related to knowing how and (3) conditional knowledge, which is related to knowing why. The second dilemma is related to learning and education (Gardner 2001): "how should we teach?" This means that we think about ways of the learning of students, learning environment, time, motivation, the student's own learning ability, individualization of learning, responsibility of learning, multiple intelligences and so on.

We arrange an enjoyable environment for learners during learning and experience. "Enjoyable experience stimulates the release of chemical neurotransmitters which have been shown to enhance learning experiences" (Jensen 2000, 31). Creativity occurs in a suitable environment that includes necessary skills and abilities to realize ideas and understandings to explore and find new combinations of attitudes and motivation that enable us to continue with a task until our ideas are realized (Duffy 1998). Creativity needs encouragement and guidance and students needs gentle coaching. Desire to learn and value of learning reinforce the motivation.

Tolman stated that motivation has two functions (cited in Klein 1987). First, motivation produces a state of internal tension which creates a demand for acting. Second, it determines the environmental features which will be covered. Motivation can be considered as instrinsic and extrinsic. Extrinsic motivation is the more suitable way of encouraging students to commit themselves to instructional goals to increase their achievement than to improve the quality or creativity of performance (Brophy 1998). This type of motivation is based on interest, self-requirement, self-determination, self-regulation and autonomy of learning. Instrinsic motivation occurs for intrinsic rewards such as reading a book for pleasure, interest or curiosity. According to Amabile, Maslow, Perkins, Rogers, Torrance, Getzels and Jackson, the concept of instrinsic motivation is the basic characteristic of the creative person (Baloche 1994). Instrinsic motivation is necessary for continuous creativity. According to Torrance (cited in Fasko 2000–2001), creative learning occurs as improved motivations, alertness, curiosity, concentration, and achievement can enhance learning. Extrinsic motivation does not occur spontaneously and it needs external force, regulation, goal-oriented studies and extrinsic rewards such as reading a book for a grade or an assignment. When learners are encouraged to learning, their motivation level increases. Students will be motivated when they choose their own learning task and this would make the task meaningful to the individual (Fasko 2000–2001).

Glück et al. stated that (2002) instrinsic motivation is necessary to cope with a higher number of external problems. Many of them arise from the educational system. External constraints on human behavior, including rewards, reduce the instrinsic motivation innately by reducing the perception of personal freedom (Eisenberger and Shanock 2003, 123). Westby and Dawson (1995) explained that a teachers' concept of creativity did not include impulsivity, instrinsic motivation or discomfortisim (Glück et al. 2002). The result of Carlsson's study showed that "the high-creative group would have a higher level of anxiety than the low-creative group" (Carlsson 2002, 346).

Instrinsic motivation is increased by supporting freedom and autonomy of choosing task and activity in the classroom. Freedom and autonomy of choosing task and activity of students is directing students to their own interest. However, it is very difficult for some of the students who do not engage in any classroom activities. "The teacher should emphasize academic tasks that students find inherently intresting and willingly without need for extrinsic incentives" (Good and Brophy 2003, 228). The other way of motivation in education is to create "motivation to learn" in the classroom. The way of increasing to learn is that the teacher prepares activities for students and explains the intents of activity, goals of activity and benefits of activity for the student.

Experience is central for consideration of all learning and a potential opportunity for creativity. An individual improves and qualifies his/her understanding of learning through reference to his/her own personal experiences (Wright 2001). Experience occurs in different ways or processes. While some experiences occur suddenly, some experiences occur in long periods of study or engagement in a subject. Gruber (cited in Gardner 1982) explained Darwin's experiences about his scientific study in that Darwin's experiences are no sudden epiphany of inspiration and innovation, no wholly novel thoughts or theories. Instead, Darwin marshaled endless lists of thoughts, images, questions, dreams, sketches, comments, experiences, arguments and notes to himself, all of which he continually organized and reorganized. " ... The creative genius of human being in inventing ever new types of experience and introducing them into the common world of life is ..." (Tymieniecka 2000, p. 11) is Darwin's.

Creativity is a function of enthusiasm and experience. Experiences grow continuously throughout life, while enthusiasm peaks up to about the thirties in a lifetime and then declines slowly (Robinson and Stern 1997). According to Miller and Boud (2003, p. 8) "Experience is the total of the ways in which humans sense the world and make sense of what they perceive. Learning is the process which takes this experience and transforms it in ways which lead to new possibilities, which may involve changes in action, ways of viewing the world or relationship." Learning from experiences is argued by many educators and psychologists such as Dewey (1902), Rogers (1994), Rung (2000), Boud et al. (1985), Boud et al. (1993), Miller and Boud (2003), who give important considerations of animating learning from experiences.

It is possible that experience in our life is divided into two parts. One part of our experiences is determined by "imaginative experiences," the other part is determined by "real-life experiences." Imagination leads to the creation of unreal situations that are an important function of our brain. Real-life experience occurs when an individual encounters a situation or an event that he/she reflects or joins to the situation or the event. The education system can help learners to "appeal" their own imaginative experiences and real-life experiences. Students can learn outside the classroom so that they can also stimulate and foster their experience.

Creativity and imagination are related to each other. Imagination ability promotes creativity. Creative studies and experiences trigger imagination. Imaginations and using them are internalizing our perceptions, they object to create meaning that does not depend on the external world. Imaginative experiences encourage a wide range of attitudes and feelings such as fun, motivations, values, cooperation, curiosity, willingness, self-esteem, pleasure, reflection, enthusiasm and confidence. Imagination reinforces the alternatives, changing of lives, interests, assumptions, experiences, insights, and motivation (Duffy 1998).

Imagination is a kind of ability to think and a kind of mental experience which is real and unreal. Imagination does not need equipment, action, things, place and people. It is a colorful, enjoyable, creative and silent experience of the human mind. Imagination is explained as the freedom of the human mind, of thinking and creativity. It is an untouched and hidden garden of human life. Imagination is full of uniqueness of mental activity and it depends on the individual capacity of perceptions of life.

I will begin with an experience of my own. This experience took place while I was preparing these papers. I imagined that I was in a garden but did not like it. I sat and dreamed about a lovely and pleasant garden. I imagined a bird which sang better than a nightingale, a colorful waterfall I had not seen before, a part of the garden with sunlight, one corner with plenty of snow, another part of the garden with a lot of flying objects. The garden was filled with a lot of animals and they were talking and dancing with me. I was moving in the garden without walking. I was able to talk to animals, trees, snow and stone. I noticed that I was sitting at a kind of wooden table which "reflected" my pleasant garden. These are imaginative experiences; they are not real things. We can create a new garden easily by using the imaginative capacity of our mind.

My mind created the garden. If I want, I can change the whole garden without any help. These are the imaginative experiences. My past experiences about the garden give messages and I use my ability and capacity of imagination to create my garden. I had never seen this garden in my life. It is known that a four-year old child can easily create a new story before anybody has told it. As four-year old child has an incredible imaginative mind. Why? His/her imaginative mind does not frame real-life experiences and mature experiences of others.

During the creative process, an individual guesses the possibility of the future. The futurist thinking process encourages imagination. Imagination of an alternative future is the most important way of creativity and experiences of tomorrow. Seeing and thinking the future mean catching new ideas, new curiosity, interest, uncertainty and creativity.

While I was writing this paper, I had the topic of learning and creativity in my mind. However, during the first three months, I did not have enough ideas about how to organize my paper. One day, while I was reading a book about education, an idea related to the topic of my paper came into my mind. My mind gave an explanation of learning and creativity, and it was very clear. Before I had enough experience on my topic, I did not know how to relate learning with creativity. In my opinion, I found a kind of missing relation between learning and creativity. After I had enough experience my experiences showed that my mind had unconscious thinking processes to find new arguments for my topic. This experience might be mixed imaginative experience and real-life experience. It might include past, present and future experience.

Past experiences help us direct our behaviors or solution of problems. They repeat the solution learnt or experienced before. Kurenkova et al. (2005) and Dewey (1902) stressed that human life in a developing world has a dynamic character and a person focuses on the past and the future. It is important that creativity is based on past experiences and that from old situations learning is related to future experiences. The past is important for a person's future decisions. An individual remembers and rethinks past experiences and applies them to a new situation. Past experiences give a pattern of a solution and they frame our present and future creative experiences. Creativity is the ability to remember the past, to live in the present and to see the future and to create unique forms of something. If an individual "breaks out" frames of the past experiences he/she will find a lot of solutions and some highly creative acts.

Dewey sated that "the development has reached a certain positive stage of fulfillment" (Dewey 1902, 19). For example "the map" orders individual experiences, it summarizes, arranges and orders views of previous experiences, and it gives a direction to future experiences, but the map is not a substitute for a personal experience and it does not replace an actual journey. "It intervenes between the more casual, tentative and roundabout experiences of the past and the more controlled and orderly experiences of the future" (Dewey 1902, 21). This means that future experiences will be shaped by past experiences which we discover at school.

Experiences and creativity put forward the possibility of learning. Young students have their own ideas, thoughts, experiences and desires to create. "Imagination is the ability to think of things as possibly being so. This odd ability is which seems to be most energetically active in our early years" (Egan 2003, 444). There are some questions related to imagination and creativity.

If an individual imagines himself/herself to live in the next ten years, what he/she can see? Where is he/she? What is he/she doing? How does he/she feel? And so on. The imagination ability of students should be supported. At school the students are isolated from their imaginative experiences, real-life present experiences and future experiences and they are subjected to apply their experiences by depending on past experiences.

The ability of imaginative experiences feeds students' creativity. Children are full of imaginative activities and they are unaware of producing new things. Children's imagination changes very quickly and this is very important for school learning. Learning at school supports the changing and development of imagination. A well-structured learning activity, subject matter curriculum and behavioral oriented curriculum, however, do not give a chance for imagination and an awareness of creative acts.

Personality characteristics might be associated with learning and creativity. These are some of the creative person characteristics: imagination, intelligence, unconventionality, assertiviness, field-independency, divergence, "syllabus-freedom," risk-taking, optimism, confidence, and a wide range of interests, flexibility, tolerance for ambiguity, curiosity, persistence, independence, emotional variability (Brookfield 1987; Glück et al. 2002). If we analyze the characteristic of a creative person we must ask the question: "How can we develop creativity at school?" Sheldon's (1995) research shows that tolerating conflicts is a main characteristic of a creative person, which is supported by Barron and Harrington's (1981) study. The characteristic of creative thinkers gives an explanation about how people think and act creatively. Aslan (2004) found that the relationship between creativity and personality and the creative individual does not allow outside rules to govern them.

A creative person seeks for unclear and unknown situations in order to "come up" with these. A creative person does not prefer unchanging situations, models, knowledge, information, thoughts etc. and he/she focus on changeability. He/she is busy with the most specific details. A discomfort situation is the first phase of a creative act. The sense of discomfort triggers some feelings and these feelings lead to awareness of a problem or a situation. Awareness is explained as trying to identify, clarify, self-examine certain situations. Learning activities include a discomfort situation for students. Beginning or during the learning activity and task the student experiences some problems. But properly creating motivation and interest about a learning task might not be easy. Students worry about their learning, homework, tasks, and they feel discomfort in these situations but they do not motivate themselves. We should ask this question: "Why the discomfort situation has not created motivation and interest in learning studies at school?" Brophy has an answer to this question: "learning is fun and exciting, at least when the curriculum is well matched to students' interests and abilities" (Brophy 1998, 1).

Every individual has the ability of creativity. Craft (2003) argued that there are two types of creativity, the first one is extraordinary creativity and the second one is ordinary creativity. Extraordinary creativity involves products, novelty and validation. On the other hand, ordinary creativity belongs to the ordinary person who can be creative in his/her daily life. In education the creativity of ordinary people began to be focused on in the 1990s.

CREATIVITY IN EDUCATION

The best education should be built on what has worked in the past. At the same time the best education should take into account what is going on at present. The other dimension of the best education is measured by the span of the needs of the future generation (Gardner 2001). Education is a massive area and very important in public expenditure in which the future of the country is constructed (Robinson 2001). It will be designed to meet the needs of changing conditions and challenges for future demands. Duffy (1998) stated that learning experiences that are offered to children must reflect their current level of development and offer the possibility of their further level of development, but the curriculum does not allow for applications of free search for full development. Students' needs are redefined related to a creativity approach that aims at full development and self esteem. "Curriculum should bring out aspects of the subject matter that are unsettled or problematic in order to capture the laggard attention of students and to stimulate them to form a community of inquiry" (Lipman 2003, 21).

A scientific understanding of learning studies includes understanding about learning processes, learning environments, teaching of some subjects, characteristics of learning behaviors, acquisitions, changing of behavior, learning technologies, learning motivation, learning tasks, learning activities, learning styles, learning strategies, educational system and teacher competence.

Much of our learning occurs without formal instruction, but some learning occurs as highly systematic and organized information that requires a formal learning system. Some of the external factors of formal learning become barriers to the potential of creativity. For example, insufficient time is given for studies, reading, examining, questioning and criticizing, providing simple or complex aims of students and low-level or high-level performance standards for the students may damage their creativity acts.

According to Torrance' (2002) twenty-two-year follow-up research results from personal sketches of individual cases quite obviously showed that motivation, skills and opportunities make a difference as many other forces do in creative achievement. Torrance's studies pointed out that many of them who joined into the research as a creative person according to Torrance Tests of Creative Thinking were having difficulty in learning to free themselves from the expectations of others. Only about half of them had found a teacher or mentor who could help them to force their creative achievement. This result explained that teacher's behavior at school is very important for a creative person and creative act. As a teacher we should remember the words of Molière: "We are responsible not only for what we do but also for what we do not do" (cited in Gardener 2001, 10).

Westby and Dawson (1995) found that their research about creativity and teachers' judgments were negatively correlated with creativity and that the least favorite students were positively correlated with creativity. Teachers reported that the least favorite students' characteristics are nonconformist and risk-taking. Most of the research books and research papers show that nonconformist and risk-taking are the main characteristics of a creative person. This result shows that teachers dislike students' characteristic trails associated with creativity. Risk-taking is the main characteristic of all creative acts. However, educational study does not require risk-taking understanding and learning activity does not include risk-taking situations. Activity is bordered and framed by teachers and most of the learning activity is determined confidently and clearly. It is quite possible that teachers' concepts of creativity is fostering or blocking students' creativity at school.

According to Carter et al. (2002) a creative person combines ideas and information in ways that form completely new situations, ideas, processes, uses or products. According to Westby and Dawson (1995) one of the most consistent findings in educational studies of creativity has been that teachers dislike and do not approve personality traits associated with creativity such as impulsivity, risk-taking, independence, determination and individualism. The personality characteristics of an individual influence creativity.

Creative persons do not allow themselves to become stuck with other person's explanations, ideas and definitions of problems and they are continually seeking to define problems in new ways and move on to the next unusual ideas. Schools and teachers do not allow new explanations or definitions of problems.

Now we focus on Guilford's question "Why schools are not producing more creative persons?" (Fasko 2000–2001, 317). Education is defined as a full development of an individual by means of promoted creativity. Creativity is the fundamental aim of learning and development in education. Making students aware of their learning style means developing their learning and thinking capacities, interest, potentiality and weakness which organize their learning strategies. Being aware of learning preferences leads to the development of self-confidence in their learning ability and capability.

Diversity in learning style encourages choosing appropriate ways of learning and individualization of attempts of thinking. Creative thinking techniques show us how to think differently, create new solutions and generate ideas. The techniques of creative thinking support an individual by having him/her to get different perceptions from the same situation. In the educational process students have many experiences in their mind in the learning and teaching process. After flashing imaginative experiences during their study they miss the point of their experiences and their mind looses creative moments. A creative mental process might occur during lessons, but a creative act is not accomplished, because the educational process prevents the creative act.

Teachers have a critical role in assisting students to engage their understanding, activities, knowledge, problem-solving, observation and all learning experiences. Students are encouraged to learn more abstract knowledge in the educational system. School and teachers can help to prepare students to meet the needs and demands of the future society. Teaching general knowledge has become an extremely inefficient way of preparing young people to meet the challenges of the future (Burden 1998). Thinking should be promoted at school by way of problem-solving, learning to learn, critical thinking and creative thinking. According to Torrance (2002, 117) "schools of the future will be designed not only for learning but also for thinking." Teaching children to think is an important subject for the educational system and this idea is explained in Philosophy for Children by Lipman in America (Bransford et al. 2000).

According to Bruner (cited in Norbert 1994), creativity grows out of a combinational activity that a student needs more learning activity to catch a new point, "perspective to creativity." Creativity requires that students learn from experience which means that students must think about their experiences rather than merely experiencing them. Students cannot escape thinking and talking about their experiences (Levine 2002). Teacher and student share their experience of learning to promote active learning and creativity.

Students should be busy with multi-functional studies rather than to be restricted with core subject studies. If students concentrate on their subject they need in their core subject and grade, they will have little time to think and act creatively (Fatt 2000). However, Dartnall stated that creativity depends on explicit representation (cited in Halford and Wilson 2002) so that they need more time.

According to Davis, Treffinger, Houtz and Krug (cited in Fasko 2000–2001), some methods may influence creative learning such as brainstorming, checklists, listening, role playing, creative problem solving and independent studies. Teachers organize an educational environment and activities which pay attention to these methods. Amabile (1983) stated that the educational environment is related to school learning such as peer behavior and characteristics, teacher behavior and characteristics, classroom climate, the social environment at school. The physical environment at school might influence creativity in education. A creative classroom climate is essential for creativity and cooperative learning has the potential to create a creative classroom atmosphere. Cooperative learning has the potential to create a classroom climate to stimulate creativity (Baloche 1994). It is necessary to change the classroom environment according to the demands of creativity in education.

How can an educational environment be established for creativity? There are many studies about how to develop creative abilities in the educational system such as by Amabile (1983, 1986), Baloche (1994), Bresler (1998), Craft (2003), Fasco (2000–2001), Fatt (2000), Gardner (2001), Jalongo (2003), James (2001), James and Asmus (2000–2001), Miller and Boud (2003), Moore (1980), Rea (2001), Torrance (1981, 2002), Vandenberg (2002), Webster (1990). These studies give some indicators to establish an educational environment for creativity:

- Encourage students to share their individual experience in the classroom.
- Use teaching technologies to stimulate creative thinking.
- Provide opportunities for students to participate in group activities.
- Give freedom to students for their studies.
- Support unusual ideas or ways of problem solving in the classroom.
- Provide sufficient time for students to think about and develop their creative ideas.
- Create a classroom climate to learn together from one another as well as independently.
- Encourage creativity in all areas and disciplines.
- Allow students to have self-esteem in their studies and their development.
- Give an opportunity for students to explain their imaginative experiences to other students.
- Improve communication by encouraging feed back of new ideas.
- Create more individualized instruction and less emphasis on teacher control.

- Serve models of creative activities.
- Engage in playful activities in the classroom.
- Encourage brain-storming in the classroom.
- Encourage students to use inexpensive materials in their studies.
- Encourage students to make hand-on projects.
- Give time to presentation of student thinking and working.
- Prepare a physical environment to present students' studies.
- Support students' free studies.
- Give students quick and short-term activity and projects.
- Support an extra curriculum activity at school.
- Establish a good relationship with parents.
- Encourage imagination experience of any situation.
- Give guidance to students' studies.
- Support creative thinking.
- Encourage students' self-evaluation of their studies.
- Encourage students' self-expression of their thoughts and emotions.

Learning in the education system should be flexible to encounter students' needs related to their potentialities and intelligence. Flexibility tends to use and understand different environmental conditions. Flexibility supports the creativity act. What Goodlad observed is that there is often a gap between the rhetoric of individual flexibility and originality and creativity at school (cited in Baloche 1994). Flexibility and freedom of learning is promoted by multiple learning potentialities and creativity.

Sternberg (2000) explained that people can make ten decisions to be creative in their lives. If teachers apply these decisions in the classroom students become creative. He stated that anyone can create problems, analyze his/her own ideas, sell his/her ideas, use knowledge, surmount obstacles, take sensible risks, is willing to grow, to believe in himself/herself, has tolerance of ambiguity, finds what one loves to do and does it. If Sternberg's decisions apply to the classroom, the educational environment will be suitable for creativity and students will be creative in this educational system.

Craft mentioned at least four limitations to creativity in education. She recommended "ordinary creativity rather than genius, characterizing rather than measuring, the social system rather than the individual and encompassing views of creativity" (Craft 2003, 117). Dilts and Bonissone (1993) ask some questions related to ordinary creativity in school. Is this how teachers can enhance creative instruction? In what ways might teachers be able to coach the process of creativity at school? These are connected with Craft's ideas creativity at school.

CONCLUSION

The primary aims of education and schooling are to improve learning abilities and develop students' capability. Students should be encouraged at for thinking and acting creatively. School studies should support and promote creative ability. However, there is an important reality that the students' creative process may be blocked or destroyed and prevented because of many reasons such as curriculum approaches, teachers, school staff, parent belief and behavior, time and space, school rules and school environment.

Much research shows that adults, and teachers, views and perceptions can negatively affect students' creativity. Especially teachers believe that interaction between students and teachers is very important for students' learning and creativity. Teacher and educator are trying to teach their students to develop their beliefs, norms and experiences and they do not allow free search for a free development of students. Teachers prefer an interventionist role for student learning and creativity. How can we limit reflective teachers' ideas, experiences and imagination while working with students? It is known that teachers' expectations of students have a significant impact on students' grade and performances.

Creative persons are open to free search, experience, imagination and differentiations. Imaginative experiences should be placed in learning activities. How much chance is given for openness of free search? If we analyze all indicators to establish an educational environment for creativity as is argued above we conclude that learning is defined as a kind of creative act for an individual, because the way of learning and functions of the brain such as coding, imagination, perception, understanding are unique. Can teachers foster creativity and imagination? Do teachers think or behave creatively? Do teachers know the concept of creativity? Are teachers aware of learning from experiences? Every mind is unique, uses a single way and works differently. However, uniqueness of personal capacity and characteristic is surrounded by moral, social, cultural norms, teacher belief, ideas and assumption.

Curriculum development studies and its applications must pay attention to student thinking ability, creativity, motivation and capacity of imaginations. Curriculum development should be planned carefully to support creativity at school and curriculum development studies do not stop to encounter student needs. Curriculum can be applied at school or outside so that students can stimulate their thinking, come up with unexpected and new ideas, catch new solutions, proposals, inferences and ways of thinking. The learning environment and learning atmosphere support creativity to encourage students in order to think creatively. Environmental conditions support creative talent which is composed of freedom, autonomy, motivation, less direct supervision and control, flexible structure and changeable conditions. These must be applied to curriculum development studies, school organization and management system to support creativity in the educational system.

The curriculum should be more flexible and it should include some disciplines and courses but it is not constructed of subject-matter understanding. Reinforcing the curriculum with new disciplines and courses is necessary in order to meet the challenges, needs and demands of a new world and individuals. It should involve some of the most significant studies, courses and activities at school which are connected with creative thinking, critical thinking, art, science, philosophy, ethics, cultural studies, poetry, drawing, photography, song and drama. The curriculum should be organized to provide educational services for learners such as a very wide perspective of selective courses, different teaching methods and techniques, unstructural programs of courses and seminar, free courses and activities, individual and group project studies and assessment. Students should be free to determine and choose topics, projects, and to solve problems in their studies. At school, flexibility in choice of activities might increase motivation. Teachers can apply a phenomenological approach in the classroom.

Some creative thinker characteristics should be applied to support learning and creativity in classroom. Many educators have pointed out that the capacity for creativity is discouraged by school, teachers and the educational system because they define a student's success by adherence to aims of education, institutional standard and norm. Educators must think about creative person characteristics and develop their learning–teaching process based on these characteristics which is to say: welcome all creative act and studies in learning situations.

Creativity has become an important and critical topic for organizations, global economy, technology, companies, social change, society and education. Every system in society tries to develop its capacity related to creativity. In the future of learning the society is tied to more empowering, imaginative and inclusive visions for teaching and learning (Hargreaves 2003). The educational system is not properly interested in creativity. It deals with explanations about past experience. If we look at school books related to present and future experiences, imagination and creativity, we can see the isolations of these dimensions. There are few researches about creativity in education. Learning in the formal system needs to include different types of materials, strategies, activities, applications and explanations to encounter possible different learning styles, ideas, divergent thinking and creativity.

Anadolu University

REFERENCES

T.M. Amabile, *The Social Psychology of Creativity* (New York: Springer-Verlag, 1983).

T.M. Amabile "The Personality of Creativity," *Creative Living* 15(3) (1996): 12–16.

E. Aslan, "Research on the Relationship between Critical Thinking, Creativity and Personality Traits," Trakya University Edirne Education Faculty (ed.), *Searching Excellence in Education*, Proceedings of the 2nd International Balkan Education Congress, 8–10 October 2004 (Ankara: Ertem), pp. 93–99.

L. Baloche, "Breaking down the Walls: Integrating Creative Questioning," *The Social Studies* 85 (1994): 25–30.

F. Barron and D. Harrinton's, D., "Creativity, Intelligence and Personality," *Annual Review of Psychology* 32 (1981): 439–476.

M.A. Boden, "What is Creativity," in M.A. Boden (ed.), *Dimension of Creativity*, (MIT Press, 1996), pp. 75–117.

D. Boud, R. Cohen and D. Walker, "Understanding Learning from Experience," in D. Boud, R. Cohen and D. Walker (eds.), *Using Experiences for Learning*. Buckingham: SRHE and Open University Press (1993), pp. 1–17.

D. Boud, R. Keogh and D. Walker (eds.), *Reflection: Turning Experience into Learning* (London: Kegan Page, 1985).

J.D. Bransford, et al. (eds.), *How People learn: Brain, Mind, Experience, and School* (Washington, D.C.: National Academy Press, 2000), pp. 221–225.

L. Bresler, " 'Child Art,' 'Fine Art,' and 'Art for Children': The Shaping of School Practice and Implications ...," *Arts Education and Policy Review* 100 (1998): 3–10.

S. Brookfield, *Developing Critical Thinking: Challenging Adults to Explore Alternative Ways of Thinking and Acting* (Buckingham: Open University Press, 1987).

J. Brophy, *Motivating Students to Learn* (Boston: McGraw-Hill, 1998).

R. Burden, "How can we best help Children to become Effective Thinkers and Learners?: The Case for and against Thinking Skills Programmes," in R. Burden and M. Williams (eds.), *Thinking Through the Curriculum* (London: Routledge, 1998), pp. 1–27.

I. Carlsson, "Anxiety and Flexibility of Defense related to High or Low Creativity," *Creativity Research Journal* 14 (2002): 341–349.

C. Carter, J. Bishop and S.L. Kravits, *Key to Effective Learning* (New York: Prentice Hall, 2002).

M.A. Cecilia, "Human Existence as a Creative Process: A Commentary on Anna-Terasa Tymieniecka's Anthropological Reflection," in A-T. Tymieniecka (ed.), *AnalectaHusserliana LXVII* (Dordrecht: Kluwer Academic Publishers, 2000), pp. 183–194.

A. Craft, "The Limits to Creativity in Education: Dilemmas for Education," *British Journal of Educational Studies* 51 (2003): 113–127.

J. Dewey, *The Child and the Curriculum* (Chicago: The University of Chicago Press, 1902).

R. Dilts and G. Bonissone, *Skill for the Future: Managing Creativity and Innovation* (Capitola, CA: Meta Publication, 1993).

B. Duffy, *Supporting Creativity and Imagination in the Early Years* (Philadelphia: Open University Press, 1998).

K. Egan, "Start with what the Student knows or with what the Student can imagine," *Phi Delta Kappan* 84 (2003): 443–445.

R. Eisenberger and L. Shanock, "Rewards, Instrinsic Motivation and Creativity: A Case Study of Conceptual and Methodological Isolation," *Creativity Research Journal* 15 (2003): 121–130.

H.J. Eysenck, "The Measurement of Creativity," in M.A. Boden (ed.), *Dimention of Creativity* (MIT Press, 1996), pp. 199–242.

D. Fasco, "Education and Creativity," *Creativity Research Journal* 13 (2000–2001): 317–327.
T.J. Fatt, "Fostering Creativity in Education," *Education* 120 (2000): 744–757.
H.D. Feldman, "The Development of Creativity," in R.J. Sternberg (ed.), *Handbook of Creativity* (New York: Cambridge University Press, 1999), pp. 169–186.
H. Gardner, *Art and Mind and Brain: A Cognitive Approach to Creativity* (New York: Basic Books Inc. Publishers, 1982).
H. Gardner, "An Education for the Future: The Foundation of Science and Values," Paper presented to The Royal Symposium, Amsterdam, March 13, 2001. Retrieved December 17, 2004, from http://pzweb:harvard.edu./PIs/HG-Amsterdam.htm
J. Glück, R. Ernst and F. Unger, "How Creatives define Creativity: Definitions reflect Different Types of Creativity," *Creative Research Journal* 14 (2002): 55–67.
T.L. Good and J.E. Brophy, *Looking in the Classroom* (Boston: A and B Publishing, 2003).
M. Groborz and E. Necka, "Creativity and Cognitive Control: Explorations of Generation and Evaluation Skills," *Creative Research Journal* 15 (2003): 183–197.
G.S. Halford and W.H. Wilson, "Creativity, Relational Knowledge and Capacity: Why are Humans so Creative?," in Terry Dartnall (ed.), *Creativity, Cognition and Knowledge: An Interaction* (London: Greenwood Publishing Group, 2002), pp. 153–168.
A. Hargreaves, *Teaching in the Knowledge Society: Education in the Age of Insecurity* (New York: Teachers College Press, 2003).
Houtz, et al., "Creativity Styles and Personal Type," *Creativity Research Journal* 15 (2003): 321–330.
M.R. Jalongo, "The Child's Right to Creative Thought and Expression, Association for Supervision and Curriculum Development," 2003. Retrieved May 10, 2005, from http://www.udel.edu/batetman/acie/creativepp.htm
D. James, "What does It mean to be Creative?," in D. James and K. Ashcroft (eds.), *The Creative Professional: Learning to Teach 14–19 Year Olds* (London: Falmer Press, 2001), pp. 8–27.
K. James and C. Asmus, "Personality, Cognitive Skills, and Creativity in Different Life Domains," *Creativity Research Journal* 13 (2000–2001): 149–159.
E. Jensen, *Brain-Based Learning: The New Science of Teaching & Training* (San Diego: The Brain Store, 2000).
J. Kao, *Jamming: The Art and Discipline of Business Creativity* (New York: Harper Collins Publisher, Inc., 1996).
J. Kasof, "Explaining Creativity: The Attributional Perspective," *Creative Research Journal* 8 (1997): 311–366.
S.B. Klein, *Learning: Principles and Applications* (New York: McGraw Hill Book Company, 1987).
V. Kohonen, "Authentic Assessment in Affective Foreign Language Education," in J. Arnold (ed.), *Affect in Language Learning* (Cambridge: Cambridge University Press, 1996), pp. 274–294.
R.A. Kurenkova, et al., "The Methodologies of Life, Self-individualization and Creativity in the Educational Process," in Anna-Teresa Tymieniecka (ed.), *Analecta Husserliana LXVII* (Dordrecht: Kluwer Academic Publishers, 2000), pp. 195–205.
R. Kurenkova, Y. Plekhanov and E. Rogacheva, "The Transcendental-phenomenological Meaning of the Notion of 'Experience' in E. Husserl and J. Dewey's Philosophy," in Anna-Teresa Tymieniecka (ed.), *Analecta Husserliana LXXXIV* (Dordrecht: Kluwer Academic Publishers, 2005), pp. 33–41.

D. Levine, "Thinking about Doing: On Learning from Experience and the Flight from Thinking," *Human Relation* 55 (2002). Retrieved October 28, 2004 from http://proquest.umi.com/pqdweb?index.

M. Lipman, *Thinking in Education* (Cambridge: Cambridge University Press, 2003).

T.L. Lubart, "Models of the Creative Process: Past, Present and Future," *Creativity Research Journal* 13 (2000–2001): 295–308.

M. Michalko, *Cracking Creativity: The Secrets of Creative Genius* (Berkeley: Ten Speed Press, 1998).

N. Miller and D. Boud, "Animating Learning from Experience," in D. Boud and N. Miller (eds.), *Working with Experience: Animating Learning* (London: Routledge/Taylor and Francis Library, 2003), pp. 3–13.

J.L.S. Moore, "Strategies for Fostering Creative Thinking," *Music Educators Journal* 76 (1990): 38–42.

S.L. Nist and J. Holschuh, *Active Learning: Strategies for College Success* (London: Allyn & Bacon, 2000).

J. Norbert, *Flexible Thinking: An Explanation for Individual Differences in Ability* (Cresskill, NJ: Hampton Press, 1994).

A. Ornstein, *Teaching Theory into Practice* (London: Allyn & Bacon, 1995).

D.N. Perkins, *The Mind's Best Work* (Cambridge: Harvard University Press, 1981).

D. Rea, "Maximizing the Motivated Mind for Emergent Giftedness," *The Roeper Review* 23(3) (2001): 157–164.

J.S. Renzulli and G.A Davis, "The School Wide Enrichment Model: A Comprehensive Plan for the Development of Creative Productivity," in Colangelo and G.A. Davis (eds.), *Handbook of Gifted Education* (Boston: Allyn and Bacon, 2002), pp. 111–141.

A. Robinson and S. Stern, *Corporate Creativity: How Innovation and Improvement actually happen* (San Francisco: Berrett-Koehler Publisher, 1997).

K. Robinson, "Mind the Gap: The Creative Conundrum," *Critical Quarterly* 43 (2001): 41–45.

C.R. Rogers, *Freedom to Learn for the 80s* (New York: Free Press, 1994).

K. Rung, *Learning from Monkeys* (Bangkok: Amarin Printed and Publishing Public Company, 2000).

K.M. Sheldon, "Creativity and Goal Conflict," *Creative Research Journal* 8 (1995): 299–306.

R.J. Sternberg, "Identifying and Developing Creative Giftedness," *The Roeper Review* 23 (2000): 60–64.

E.P. Torrance, *Manifesto: A Guide to Developing a Creative Career* (Westport, CT: Greenwood Publishing Group, 2002).

A-T. Tymieniecka, "Origins of Life," in Anna-Teresa Tymieniecka (ed.), *Analecta Husserliana LXVII* (Dordrecht: Kluwer Academic Publishers, 2000), pp. 3–12.

D. Vandenberg, "Phenomenology and Fundamental Educational Theory," in Anna-Teresa Tymieniecka (ed.), *Analecta Husserliana* LXXX (Dordrecht: Kluwer Academic Publishers, 2002), pp. 589–601.

P.R. Webster, "Creativity as Creative Thinking," *Music Educators Journal* 76 (1990): 22–28.

E.L. Westby and V.L. Dawson, "Creativity: Asset or Burden in the Classroom," *Creativity Research Journal* 8 (1995): 1–10.

D. Wright, "Creativity and Learning: Creative Work and the Construction of Learning," *Reflective Practice* 2 (2001): 261–273.

J. C. COUCEIRO-BUENO

EDUCATION WITHOUT *PAIDEIA*. A PHENOMENOLOGICAL VIEW OF EDUCATION TODAY

I

Husserl, in his famous lecture entitled *The Crisis of European Sciences*,[1] condemns the sharp technoscientific turn that took place in Europe following the Second World War. This turn plunged the concept of *paideia*, which had subsequently crystallised in the Germanic world into the concept of *Bildung*,[2] into a profound crisis. This concept took the world of ancient Greece as a reference for education; an understanding of the whole as the human ideal, and the ideal for education and training, albeit an interior education striving towards an ultimate and superior goal.

This was Husserl's way of telling us that Europe is ailing, and that its illness is the result of the decadent state of the sciences (a field which, as it is known, had sprung up from the art of philosophy). In this sense, Husserl is accusing Europe of leaning dangerously towards specialisation, in a clear betrayal of its origins and an ominous obliviousness to its Greek essence.

In the opinion of the father of phenomenology, this blatant forgetfulness of its classical spirit represented a deviation of the true nature of philosophy, namely the means to development towards personal autonomy.

All Husserl's thoughts and reflections are extremely relevant in understanding the profound crisis in which education, and the very concept of education itself, is currently immersed. And as the results of Husserl's analyses showed, it is science (and technology), that lies at the root of the illness which in this instance is affecting education today. In other words, we are forced to experience an education without *Bildung*; an education that has been divided up into specialities, an education without roots, aimless and empty, an education that has no qualms in declaring that knowledge is not important. In short, education without *paideia*.

An in-depth analisis and insight into the classical tradition makes it almost impossible to consider the current thesis and proposals for education

A-T. Tymieniecka (ed.), Analecta Husserliana XCIII, 371–383.

techniques defended by scholars in the field with any degree of seriousness. The pleasures afforded through the application of the great classical works to modern day society, by the study of theories and by observing how the traditions we are taught can change us – all these "pleasures", naturally with rare exceptions are all absent from our schools and education centres.

It is worth remembering that the great men of philosophy (Plato, Aristotle, etc.), always integrated the thinking of their predecessors into their work, as they perceived that their own thoughts were inextricably linked with the understanding and assimilation of those who had gone before them.

Today, however, it can be seen how Western nations, appealing to a tradition that is defined as being progressive and under the influence of science, are bold enough to claim that progess and advances in knowledge do not depend on the great intellectual influences of humanity. If such foolishness continues, then the current crisis in education will inevitable drag us towards a radically nihilistic educational landscape.

We must therefore reiterate our conviction that, if we continue to ignore the voices of the past, treating them as voices that have no bearing on our modern day world, then we will be permanently doomed to such a landscape. Examples of the indifference and oversight of education can be found in the widespread conviction that the principal mission of educational centres is to provide professional and vocational training, when traditionally such centres (and universities in particular) have been places for elucidating the questions that have puzzled mankind since time immemorial and where opportunities are created for choosing a lifestyle that is worth living.

It is evident that the impoverishment of education is the result of the privileged position science and technology hold in our society, which leads to the need for specialised teaching staff in an attempt to satisfy social interests, to the detriment of the true aims and objectives of education.

What is absolutely clear is that science, regardless of the influence it exerts on current day society, has an extremely low impact on the essential issues of life: science creates its own self-contained world which supposedly addresses issues of essential importance, but is incapable of displaying its relevance when we invoke knowledge in its entirety. It has been proved that the most important things in life cannot be subjected to a scientific approach.

Along these line, we are faced with educational syllabuses which are totally lacking any "knowledge integration," which in turn implies sacrificing the teaching process. This can be clearly seen in the attempts by teaching staff to avoid having to take the decision of proposing what the student actually needs to know.

It must also be said that education today stands totally apart from the capacity of fiction,[3] the narrative capacity that forms a major part of education. Educational theorists fail to highlight this aspect of human sensitivity and intelligence. From an early age, children become aware that the world is interpreted by adults. Yet what they really admire are the fictional tales, stories where the world is interpreted by adults and, that offer them the chance to understand "others" through the animated beings that live in a world that for them still retains all its magic and mystery. Thanks to such fictional tales or fantasies, children are able to develop their emotions and to attribute varying states of mind, etc.

In primitive cultures this function was performed by myths. Later on, in developed civilisations, literature took over this task, offering children privileged insights into the world, unpolluted by the disenchanted vision of science.

In classical Greece, literary expression played an essential role in educating the population, as shown by the tradition of Greek theatre and drama. Consequently, and after Homer, the great tragedies represented a major source of education for the Greek people. Aristotle, in his *Art of Rhetoric*[4] and *Art of Poetry*,[5] makes a fierce defence of the vital role played by tragedy in the education of the Athenians, particularly highlighting the function of *katharsis*.

The lack of a concept of tradition is another of contemporary education's blatant shortcomings. Naturally, not in its facet of traditionalism, but rather as the acknowledgement that all education requires a framework based on references to the past.

H. Arendt[6] succinctly summarises my views on the concept of tradition when she confidently declares that the absence of such a concept makes education imposible, as it lacks the basis of a conservative approach that allows for a consolidated conceptual reference framework, which in turn may serve to bridge the gap between past and future; in other words, ignoring tradition in education is tantamount to converting it into a sinister parody of what a child expects when he comes face to face with his teacher.

The challenge in education today lies in our ability to provide it with a philosophical foundation, resorting to the use of antiquity as a theorical paradigm or horizon to which the human being may aspire in order to reach his full sense of dignity.

Consequently, recoverimg the educational thinking of ancient Greece is to prove that, in terms of education, things were better before, and that what we have today is the imposition of technical knowledge over cultural knowledge,[7] replacing artistic creation for compulsive consumerism, an activity that is practised alone, totally removed from any need for dialogue.

II

For the purposes of this study, the theoretical fundamentation of the concept of *paideia* is taken as being the traditional idea of "a return to ancient Greece" from a historical perpective and as the aim of being able to put into practice what people in the past only dreamt of.

Such is our initial approach to the pre-comprehension of the philosophical foundations of *paideia*, invoking that the essential inquisitiveness of human communication consists of establishing a dialogue.

This way of thinking is totally removed from any sense of instrumentalism, from relations based on the struggle for control, thereby enabling us to see differences as merely differences, and to understand people once and for all, in a way other than that inspired by their presence or by considering them as objects.

Reminiscent thought, then, that implies the ability to stand back and remember true experiences from the past in an objective manner, helps to build ourselves as persons and sets us on the path towards the future. Thought in which we consider both historiographic and non-historiographic (oral) aspects, in order to allow us to fully and accurately understand and assimilate the civilisation of the past.

The consequences of this reminiscent thought is, as Heidegger[8] put forward in his classes, the ability to understand life, shrouded in (artificial) mist, in which we are forced to live.

It must be remembered that, based on these statements, the essence of what we refer to as historic is a productive event of destiny. The essence of the historic is the essence of that which makes history, which opens up and establishes history. And as part of this "historic destiny," we here wish to refer to the need to update questioning and thought which are not created by themselves, but from events which call for this questioning and thought in order to keep them alive and creative.

This standpoint places us in a position whereby we can call for the discussion of the problem referred to above of unitary knowledge which should be approached from a widespread concept of knowledge and which we can find within the concept of "theory" (*theoros*), insomuch that this concept points towards participation and contemplation free from the restrains of necessity and utility. Dilthey[9] was the person responsible for broadening the historic horizons upon which philosophy should be based.

And this is true because historic experience is unable to ignore the impact caused by history itself (*Wirkungsgeschichte*). Dilthey's reaction, as we know, was to refer life to actual reflection. This awareness of being conditioned, which determines what we understand as a "historic awareness," is a form of self-knowledge which enables us to position ourselves within the world in a constant and steady manner.

Confident of his position in the world, in language, in customs, in standards and certain types of *ethos*, this historic awareness will allow the individual to rise above his particular nature (instinctive and ideological) and situate himself in a reflexive relationship with the self (*reflektierten Verhältnis*) as well as with the tradition within which he is positioned. And this is so because all experiences imply earlier and later horizons that are not the product of any "life world" (*Lebenswelt*) that is intersubjectively created.

Hence we are in a position to claim, basing our ideas on Dilthey's proposals, that life is an *ethos* under construction, a *telos* to be interpreted and the assumption of responsibilities, taking charge of one's own individual and community existence.

If, as M. Scheler[10] claims, it is true that man wishes to continue formulating questions about his place in the cosmos, then he must also question the predominance of a technology that has isolated history, the very historic nature of existence. Technological experience, at once so overwhelming and blinding, aims to strip itself of everything human, banishing it to the confines of the consideration of history as an awareness of historic determination and to remove all forms of limitation. It is not hard to imagine that all this would lead human nature to fall into ominous obscurity.

In this sense, man's role lies in the quest for an environment in which knowledge, knowing things, does not imply any notions of dominance or manipulation, but instead those of integration within a common project.

Man, thus portrayed in the role of interpreter, is therefore encouraged to take part in shared and established knowledge, in a set of traditions that enables him to actively integrate within a specific community and formed by the tradition, habits and lifestyles that he is given to sharing through language. This participation must necessarily take place through linguistic interaction in order to make discernment, appropriation or questioning possible.

Therefore, any fundamentation that takes language as its starting point cannot be considered to begin from a position of subjectivity, but rather in one of "inter-subjectivity," which transfers us to a community horizon.

From such a community horizon, reminiscent thought, such as that we defend, cannot be considered to be instrumental or instrumentalised thought, but instead thought which still retains the capacity to question.

The creative force of language, the force of poetic language is the force that has the ability to forge, to inaugurate, to "open up the world," thereby making it possible for our existence to develop, naturally from the horizon of dialogue. Turning our attention to classical Greece, formation and education are seen as the process of bringing man out from his instinctive blindness in order to increase the differentiation between human interests; this involves increasing the aesthetic capacity of pain and pleasure until a personal balance is reached.

In this sense, educating a young person, teaching him to be a good citizen is aimed at reaching the idea of *kalokagathia*, which unquestionably represents the culmination of Greek thought.

Within this context, we must not overlook the contribution of the Sophists, which has a direct impact on the situation today if we consider their defence of the central nature of language, whose mission is to "teach men" and to educate them in democracy.

Therefore, the Sophists' contribution must be seen as the movement of a transforming *praxis* of the *physis*, which also possesses the capacity to innovate nature itself.

Moreover, and as with all Greek education, in Plato aesthetic considerations play a fundamental role (indeed, this is a distinguishing factor of classical Greece in general). Hence the idea that rhythm and harmony have a major contribution to make in education: "Accordingly, beauty, harmony, grace and the number of discourse are the expresion of the goodness of the soul" (*The Republic*, 483C).

This aesthetic component extends to the essential concept of *nomos* (as a contrast to *physis*). In Greek, *nomos*, it must be remembered, means both "law" and "song" and possessed legal values (*Laws*, 799).

Plato eventually makes the claim that he sees true education as the "awakening of the talent that lies sleeping in the soul" (*The Republic*, 518C).

With regard to Aristotle, we must look back to the section in *Metaphysics* that would appear to offer the philosophical basis for education: "It is in the nature of all men to aspire to knowledge" (I,1,980, 21).

Aristotle refers to man's irrepressible urge for *sophia* (knowledge) which is a part of nature itself. From this perspective, knowledge means intellectually being in possession of the truth, which is a universal desire.

In his works *De Anima* (417b, 15) and *Nicomachean Ethics* (VI, 11, 1143a, 25, 1143b, 15), Aristotle refers to two basic aspects of education, namely:

(a) The cultivation of intelligence, seen as the orientation of theoretical knowledge, which constitutes a pathway leading to truth, revelation, the uncovering of the hidden meaning of things that will eventually make us their masters.
(b) Education and training seen as character formation guided by virtue (*arete*).

Aristotle sees character formation mainly as the achievement of the ideal of life, in a practical order of things. Virtue is what men should essentially strive to in life (*Politics*, IV, 3, 1371). This virtue, seen as the goal of human dynamism, develops and takes place through the education of the will. Within this educational process, man becomes accustomed to acting in accordance with a fair and just view of life at all times. The *phronesis* guides this strict way of behaviour, putting forward practical opinions which determine and define it. In this sense, *phronesis* provides man with the ability to assess fairly his acts of will and to behave accordingly: it consists of a state of the will and also implies knowledge.

All these ideas from classical Greece have been applied to Germanic culture within the concept of *Bildung*. It represents a time when education is understood from the concept of form and formation (*Bildung*), which refers to that which man requires to shape himself. *Bildung* takes a specific type of man as its perspective.

The worlds of *Bildung* can be classified as follows:

(a) The Greeck world (as the sole means of harmonising the potentialities of the individual).
(b) The world of art (influenced by Plato who characterised the spirit with aesthetic features, as he understood that the education and formation of the human being should imitate the way the artist shapes his work. After all, it must be remembered that the root of *Bildung* lies in *Bild* which means "painting" or "image," etc.).
(c) Language seen as the constitution of the world.[11]

Bildung therefore signifies the forming and training (*sich bilden*) of the self in a balanced manner and free from all ideological temptations. *Bildung* refers always to the education and training an individual needs in order to become a person.

Seen from these perspectives, those of the Greek *paideia* and the German *Bildung*, education has much in common with the ideal of art, which attempts to compensate for the shortcomings of nature through beauty.

In keeping with this argument, W. Jaeger[12] makes the categorial claim that Greek education cannot be understood without poetry and rhetoric. Greek tradition has taught us that focusing on rhetoric and poetry is merely an alternative perspective to the view that language lies at the centre of all reflection and action. Naturally, the language referred to here is not the propositional language of the *logos apofantico* – which incidentally is at once perfectly irrefutable and dispensable – but rather the type of language that always goes beyond the narrow limits of logic: the rhetorical and potical languages of the oral tradition.

The language of rhetoric and poetry is by no means predetermined; quite the contrary, its mission is a strictly practical one which always implies performing some action or other. Aristotle considers rhetoric to be a *techne*, as well as a set of rules based on experience, which enables us to show to carry out an action that borders on perfection.

What is absolutely clear is that Aristotle sees rhetoric as *paideia*; a true reasoning of "conjecturable" issues. In politics, rhetoric is positioned somewhere between poetry and ethos, in addition to the fact that its "practical truth" leads it into the field of *polis*.

With regard to poetry, we must again highlight the fact that the Greek poets were the first educators of the people; their poems contain knowledge that covers multiple aspects of the people's needs. Poetry was thus responsible for spreading art and culture amongst the people. It was seen as having the capacity to act as *paideia*, and possessing real power and ability to guide our conduct and behaviour.

Aristotle solemnly states in *Poetics* that "poetry is more philosophical and noble than history; "poetry refers to generalities, and history to specifics" (1451b).

In conclusion, Greek culture sought to turn the education and training of man into a work of art, a process in which poetry played a central role.

III

The thematisation of the decline in education is the direct consequence of the global technification of humans. In turn, a silence is imposed which leads to the creation of educational and political communities dominated by so-called "techno-education" or "technopolitics." Consequently, intellectuals, teachers and students are in possession of humanistic values and objectives that are governed exclusively by a set of "rules."

In the light of these considerations, questions arise regarding the space for freedom and deliberation that would correspond to the educational experience in a technocratic world.

This study defends a type of education inspired by the Greek *paideia* which accepts the role of philosophy as a means towards education that prevents us from succumbing to a status of inferior humanity.

We therefore defend a more conversational *paideia*, whose Roots are embedded deep within the Socratic tradition of dialogue; the attitude that has always aspired to knowledge. And in doing so, we also reveal the limitations of science when it makes claims such as those of Heidegger[13]; namely the statement and a scientific one. However, and unlike scientific statements, instead of drawing our attention to an object, practical statements tend to highlight a sense of conformity, of agreement between what is stated with the practical and effective relation between man and an object.

Consequently, in comparison with the potential of language, scientific activity is always reduced to expressing the sense that something is lacking. In addition, questions must depend exclusively on the specific truth.

Moreover, not all that is relevant to the human being can be expressed in a statement. Indeed, all the fields of education stand completely removed from the ideal of verification put forward by scientific methodology.

Having clarified these points, we will now go on to claim that all experiences (including artistic experience and particularly *phronesis*) are directed at understanding an experienced person not so much as a definitively self-made person, but rather as a person who is always open to new experiences.

This should be reflected within the field of education, as an attempt to avoid the obsessive claim that experiences are of an entirely scientific nature. If this were true, it would inevitably lead to the total destruction of man's capacity for invention and imagination, and would dry up the constant flow of historic events. At all events, it is important to bear in mind the fact that the objectivist concept of the human being leads us directly to a totalitarian society.

From this perspective, education means the ability to form the world from one's own finite experience, thereby creating the basis for an image of the world (*Weltanschauung*) that is rich in meaning thanks to its inextricable links with the historical effectiveness of our times.

It would thus be possible to show the person being educated that knowledge must not be seen as an attempt to appropriate and manipulate the world, but rather as a subject that is possessed by a tradition in which it is already linguistically firmly entrenched.

In this sense, an essentially humanistic education will not be concerned with finding the right way of interpreting texts which will lead us to a definitive and unquestionable interpretation. What matters here is that the interpretation of those texts will influence the way in which the interpreter considers his own involvement and role within a tradition which, in turn, influences his expectations in relation to the text and his own horizons.

Comprehension will therefore represent a temporary event whereby the text is subject to both expectations and horizons, which, in turn, are influenced by the tradition which led to the initial encounter between text and interpreter. The interpretation can therefore by no means be considered to be final, but instead part of the dynamics of dialogue.

Consequently, the concept of comprehension represents the "fusion of horizons",[14] whereby the horizons of both text and interpreter are modified within the action of managing to obtain some degree of comprehension.

This implies that the interpreter will find himself caught within a flow in which "preconcepts" are constantly being altered and cast out as a result of the continuous degrees of comprehension that must be acquired and restructured.

All of the above represents a significant blow to empiricism, which as we know, predominates current educational trends, and which claims that educational research is neutral, leading us to theories that are applicable to practical situations.

The issue here is not whether the "theory" should be more or less closely related to the "practical"; nor whether theoreticians and practitioners should be able to understand one another better. What matters here is the ability to assimilate the idea that no theory can ever be considered practical, and that all theoreticians are to some extent also "practitioners."

A further hypothesis of contemporary education referred to above is that the education system is essentially conceived as a form of professional training that will enable the individual to perform an economic role within society. In response to this claim, it must be said that education must never overlook the fundamental task of learning the "lifestyles"[15] that various groups choose to adopt as their own.

Education today insists on the notion of education in the transfer of knowledge. In contrast, education that is deeply rooted in classical antiquity knows that its primary function is to stimulate thought in order to prepare us to participate in dialogues which will transform our "prior knowledge." As a result, we will eventually be able to act and choose – *proairesis* – (*Eudemonistic Ethics*, II, 10, 1226) our own life (it must not be forgotten that for the Greeks a full life meant a chosen and determined life based on self-mastery, *enkratia*).

The concept of authority is another word that is central to any education system which yet has suffered more in contemporary education. Tremendous confusion abounds amongst educational theorists regarding the concept of authority. It is a necessary element, and indeed without it, there could be no education. However, the distinction must be made between "authority as an imposition," which is nothing more than an invitation to violence, and "authority as recognition." The alter forms an essential aspect of education. It is the type of authority that leads to a desired and intensely felt "obedience." Such authority is not only accepted, but it is also actively sought. The pupil does not see this type of authority as a type of coercion, quite the opposite in fact; he sees it as a necessary and pleasurable experience that will guide him towards his own autonomy and independence. It represents the acknowledgement of superior knowledge. It is a kind of authority that is actively sought and longed for, as to a large extent our life projects depend on the suggestions of a wise and prudent *phronimos (Nicomachean Ethics*, 1140, 31). In short, and considering that what is at stake here is tradition, we must never strive to lose the capacity to act as disciples before the powerful words of the wise man.

The concept of the teacher (master) implies a series of complex religious and ideological roots. It extends to both the inept teacher (a destroyer of the soul) and the charismatic teacher. A poor teacher has the capacity to psychologically destroy his pupils, to annihilate their spirit. A good teacher, on the other hand, plays a decisive role in determining the direction his pupil's life will take. Quality teaching will accompany the student throughout his life.

What must be categorically stated is that schooling, at all levels, continues to be of major importance since the living word of the teacher is far more important than any book, as there is nothing that can rise above the spoken word. Such an assertion naturally owes its existence to the marvellous Greek oral culture, the foundation of Western culture.

The importance of education, which legitimates the continued existence of schools, is the fact that the things that really matter in life can only be communicated from one human being to another through spoken language.

Education is simply the continuation of the ancient flow of oral communication (mythological and philosophical) that has been passed down from generation to generation. Seen from this perpective, the teacher is merely a messenger of the *logos*, the original and fundamental word.

Oral education has been plunged into a state of crisis by the hegemony of science and technology. Oral teaching is the basic justification for education as we know it and there is no replacement for it. If we continue to "ignore the oral tradition," then we will lose our ability to achieve that which can only be learnt through words (Plato, *Letters*, VII). The only way of transmitting the important things in life is through the spoken word. In other words, if basic teaching were no longer to be taught through words, then this would mean the end of Western civilisation. Another civilisation would take its place, totally removed from classical Greece.

When compared with the spoken word, the written word is found lacking. What is written is incapable of response, it is unable to amend itself, to listen, to help itself. It is totally lacking in intonation and timbre. We are unable to discern any type of modulation, gesture or embodied emotion.

In classical antiquity it was known that the finest texts are not those that are written down, but those that are "inscribed" on the soul. To speak is to write on the soul and can never be forgotten. Going back to the controversy surrounding the status of education, it must be said that the science of education, as practical knowledge, is the type of knowledge that enables us to conduct our lives – (*phronesis*). And if we understand education within the concept of *phronesis*, then we acknowledge its essential role as a mediator between the vital contingency of the interpreter and oral tradition that will provide us with a basic type of educational experience. This experience will then be responsible for recovering the forgotten aspects of education (oral tradition, self-teaching, tradition, language, the experience of art and beauty, etc.). Aspects that have been cast aside by a type of education that has become what we have chosen to refer to as "education without *paideia*."

In light of the above, it is evident that our manner of thinking, our vision of the world, our *Weltanschauung*, is conditioned to a large extent by classical Greece, and essentially by its oral tradition. Its ideas on education, its *paideia* continues to be highly relevant today, and consequently we are obliged to keep up a dialogue with these ideas. Indeed, we have always looked to the ancient Greeks for a model to imitate. We are the heirs to a way of thinking and behaving that is unquestionably Greek. Their thoughts are our thoughts. Consequently, we should adopt their *paideia* as our own.

All that we can do is to trust that the current decline in education and the humanities in their classical sense is not merely a further stage in the twilight of humanity. Indeed, we must hope that at least we will be able to witness the contrasting side of this twilight; that of the diffused rays of light that appear prior to sunrise and which is the inevitable continuation of the setting sun. Let us hope so.

University of La Coruña, Spain

NOTES

[1] E. Husserl, *Die Crisis der Europaischen Wissenschaft und die transzendentale Phänomenologie* (Martinus Den Nijhoff, Haag: 1962).

[2] In German *Bildung* represents (in the midst of the crisis between the two great wars of the 20th century) a major cultural and education adventure that can be summarised by the expression "*Bilde dich Griegisch*" (educate yourself like a Greek). *Bildung* is *paideia*, for humanitas, and for culture (*animi*) aimed at promoting, from an educational point of view, the values of the human spirit. Within such a context we must not forget what has been termed as the "novel of formation" genre (*Bildungsroman*: autobiography, confession, introspection and interiority based on the realisation of the self), one of the most notable contributions of German culture to universal literature.

[3] J. C. Couceiro-Bueno, "Ontofiction: the altered comprehension of the world," in A-T. Tymieniecka (ed.), *Analecta Husserliana: The Yearbook for Phenomenological Research*, Vol. LXVIII (Dordrecht: Kluwer Academic Publishers, 2000), pp. 399–413.

[4] Aristóteles, Retórica (Madrid: Instituto de Estudios Políticos, 1970).

[5] Aristóteles, Poética (Madrid: Instituto de Estudios Políticos, 1970).

[6] H. Arendt, *Entre el pasado y el futuro* (Barcelona: Península, 1996), pp. 185–208.

[7] M. Heidegger, *Ontologie. Hermeneutik der Faktizität* (Frankfurt, 1998).

[8] Techniques are not the prerogative of the world of science. The humanities embody techniques that are far more complex than any to be found in the sciencies (art, music, rhetoric, grammar, poetry, logic, etc.).

[9] W. Dilthey, "Die Entstebung der Hermeneutik," in *Gesammelte Schriften* (Leipzig–Berlin: Teubner, 1958), Vol. V.

[10] M. Scheler, *El puesto del hombre en el cosmos* (Buenos Aires, 1971).

[11] W. von Humboldt, *Sobre la diversidad de la estructura del lenguaje humano y su influencia sobre el desarrollo espiritual de la humanidad* (Barcelona: Anthropos, 1990).

[12] W. Jaeger, *Paideia* (México: FCE, 1974), pp. 664–678.

[13] Heidegger, *Sein und Zeit* (The Hague: Niemeyer, 1927).

[14] H.-G. Gadamer, *Warheit und Methode* (Tübingen: Mohr, 1960).

[15] L. Wittgenstein, *Philosophische Untersuchungen*, Vol. I (Frankfurt, 1984), p. 12.

SECTION VI

CREATIVITY AND AESTHETIC EXPERIENCE

PATRICIA TRUTTY-COOHILL

WHEN THE GIVEN BECOMES THE CHOSEN[1]

We dwell in the ciphering itself.

Anna-Teresa Tymieniecka[2]

For Joe

A destructive fire consumed a painting studio at Western Kentucky University in the 1980s. All the paintings and drawings were cinders, as were the wooden stands where they were stored. The lighting fixtures hung in disarray, trailing long drools of melted metal and plastic against the blue tarp that was now the roof. The air was poisonous.

A week or so later the art faculty met to discuss the future. The first comment was that there were lots of interesting, awesome forms in the wreckage. We all nodded in agreement (we were an art department after all), and then got right down to business, a perfect example of aesthetic detachment – or perhaps aesthetic *at*tachment. We, as artists, had found beauty amid the ugly detritus; we had reinterpreted mundane reality through our creative impulses so, in Tymieniecka's terms à la Nancy Mardas, the destruction "was not merely passively encountered … [but rather] posited within and by the synthesizing organization of the manifold of which the subject is also a part."[3]

While in the instance of the fire, the aesthetic response did not go further than a comment (no work of art was made from the aftermath of the fire), there have been occasions when works of art specifically accepted the premise of ugliness, the horrific, the overwhelming, e.g. Gericault's *Raft of the Medusa*, Turner's *Slave Ship*, and Picasso's *Guernica*.

In the essay below, the focus is on the artist's choice of subject matter from the world, from nature – its disasters and commonplaces – in light of Anna-Teresa Tymieniecka's discussions about the work of art in the lifeworld, especially in the first book of her series, *Logos and Life* (hereafter *LL* 1). I wrote the analysis of artworks below to open Tymieniecka's texts, or perhaps it was vice versa. My art criticism is a means of setting forth Tymieniecka's arguments and of demonstrating how the works of art parallel them. The works

A-T. Tymieniecka (ed.), Analecta Husserliana XCIII, 387–400.

of three contemporary artists demonstrate that photographs can be a portal to the life of the logos, but the reader must be cautioned that reproductions are at least thrice removed from the reality of the works. This is especially true of the works by Lillian Schwartz and Kathleen Connally, meant to be seen in the light emanating from a computer screen.

THE HUMAN-BEING-IN-CONFLICT

Images by the computer artist Lillian Schwartz demonstrate one way an artist actively responds to the horrendous destruction of the 2004 earthquakes and tsunami in the Indian Ocean.[4] They also provide a demonstration of Tymieniecka's understanding of how the creative act "emerges within the human-being-in-conflict." As a way of taking in, accepting and coping with the influx of images from the disaster, Schwartz literally grasps and manipulates images of the fish killed in the cataclysm, turning them into arresting capsules of horror, (Figures 1 and 2).[5] As if to handle the

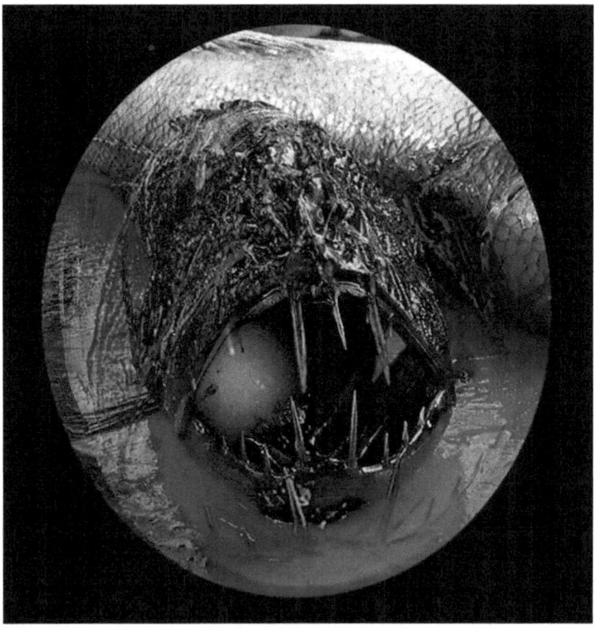

Figure 1. Tsunami #1. Copyright © 2005. Lillian F. Schwartz. Courtesy of Lillian Feldman Collection, Ohio State University. All rights reserved.

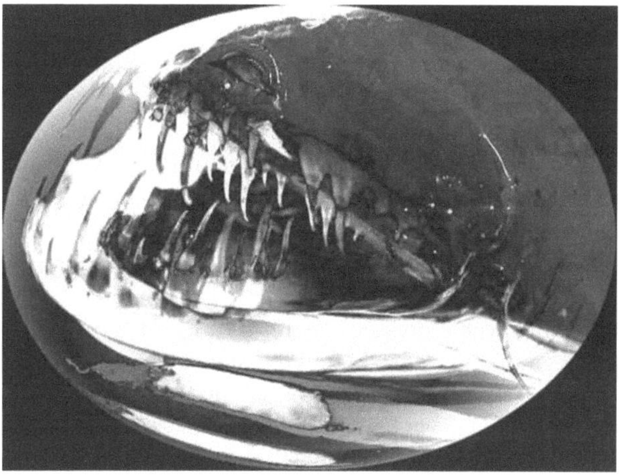

Figure 2. Tsunami #2. Copyright © 2005. Lillian F. Schwartz. Courtesy of Lillian Feldman Collection, Ohio State University. All rights reserved.

situation discretely, she compresses the fish into three-dimensional lozenges. Harmonizing textures and the colors of the peripheral shapes frame eerie splinters of broken teeth. Schwartz, in her manipulations, provides a concrete example of the choices of the creative act: she takes on and transforms the Real into a "cipher" that encapsulates the whorling forces that killed the fish and so much else. Because Schwartz was able to make something of the horrors of the given of the tsunami, she moves beyond that horror, beyond her own immediate response. Through a process of aesthetic detachment from Reality, she is able to supercede the lifeworld, including her own response. In doing this, she demonstrates Tymieniecka's idea that "the creative function, guided by its own telos, generates Imaginatio Creatrix in man, as a means, par excellence, of specific human freedom – that is the freedom to go beyond the framework of the lifeworld, the freedom of man to surpass himself."[6] Schwartz demonstrates the active quality of the human emerging into the world (*phainomena*), of one who has "moved beyond the basic levels of individual and even of person, onto the level of the creative self, which creates itself by establishing its meaningfulness within the lifeworld."[7] What is more, Schwartz invites us to join in her triumph over the given, that is, what we are given to live with, to survive. The *chosen* is what we choose to live with, what we choose to live by.

THE CIPHER AND PHOTOGRAPHY

Professor Tymieniecka has for these long years posited that the cipher[8] is our interpretation of [Being] through our own being; it "gives 'body' to the vision"[9] that opens the door to our humanity and to that of others. When the artist makes a work, a cipher, he asserts himself as a constitutive factor with respect to the entire realm of Nature, "a center in which the forces of hitherto anonymous Nature converge, ... and thereby strengthens his own, unique position. In short he projects his own itinerary, establishing himself in existence."

Professor Tymieniecka stressed the immediacy of the cognitive and the creative act: Cognitive meaning-bestowing and life-course enactment cannot be separated or even sharply distinguished in their life-meaning function. As a matter of fact, they are one: "while delineating by response, deliberation, selection, choice, imaginative invention, planning, etc. and their enactment, his life-route, the individual living being is performing his self-interpretation-in-existence .... Both life enactment and cognitive function of the human being meet in the crucial device which life-progress entails, the prototype of human action which resides in the CREATIVE ACT OF MAN" (*LL1*, pp. 6–7).

While in a recent study I used Leonardo da Vinci's brainstorm drawing (British Museum) for the Burlington House Cartoon (National Gallery, London) – *gemisch* of experimentation, of possibilities – as direct evidence of Tymieniecka's description of the ciphering process of impetus and equipoise,[10] here I will concentrate on a discrete event in the creative process, the instant when the artist's chooses an image from the panoply of the given, just that moment when the artist sees and chooses some aspect or some moment of a scene in such a way that the work makes us look at Nature or Life anew. This attentive act (see Nancy Mardas's essay in these volumes) can provide us with a record of what Tymieniecka calls an "act of will, a creation from within the life world of some symbol of unique significance by which the subject makes its mark upon that world."[11]

At times, especially in literature and music, we enjoy a "story"/the composition, for which the author has set the stage, its beauty in its "natural" unfolding. The choices of the artist in such cases are shrouded in long deliberations, subsumed in the pattern of the final work. Photography can take us *into* the instant of the find, and it can also record the moment of the artist's decision-in-action. Some art photographs provide us with a cipher not only of the object but of the moment of the artist's choice. In this way, we can even document "the discrete, disruptive, randomly enmeshed or dissolving fragments [that] meet spontaneously in the great constructive phases [that advance] the self-interpretative game of life ..., endowing with significance each step of its articulation and progress" (*LL 1*, p. 17).

Let me be clear. It is not my purpose here to use the photographs as symbols or analogues for Tymieniecka's operatic phenomenology of life, but as an image of one aspect of human creativity, as an image of one kairic instant in the "orchestrating of man's self-interpretation-in-existence" (*LL 1*, p. 17). The photograph records not only the object before the lens, but it gives picture of the instant of the artist's vision, will, and choice. The photograph captures, in nearly eidetic focus, what Tymieniecka has called "the crucial role of Imaginatio Creatrix, the surging of which ... marks [at the one extreme], the emergence of the specifically human significance of life, and [at the other extreme] the set of primogenital elemental virtualities that, being simultaneously activated, set up and circumscribe the basic platform [of the creative orchestration of man's self-interpretation-in existence] within the incessant process of life's constructivism, the Human Condition"(*LL 1*, p. 17).

Let us work like scientists and choose the sort of photograph that can most obviously show us the instant of decision. At another time, we saw such a process in the self-activated and self-actualized drawings of Leonardo da Vinci, whose ideals of beauty and order, are well understood when we examine his drawings of grotesque heads or read of his fanciful practical jokes.[12] For example, there is a whole class of grotesque or caricature heads that obviously derive from a manipulation of a standard of beauty, as can be seen in the *Head of a Man* (Metropolitan Museum, New York). What this drawing gives us is the more than just two heads, one ugly and one handsome: it takes us into the liminal region of creative act, that threshold on/in which we can infer the reason behind the construction, the pleasurable relation between the ugly and the beautiful. It reveals what Tymieniecka would call the surging of Leonardo's Imaginatio Creatrix, but it also allows us to see the point at which our "perception of reality has through habit" become the surrogate of experience (Leonardo's habit was to draw a head as a doodle, to warm up; here the ink) and is disintegrating (Leonardo breaks up its reactive sense by adding lines with a pencil). The artist does not abandon us here, but rather leads us (and is led) by his own searching impulse that "rallies all his faculties" to bring himself and us to a new image from the old.

BEYOND FLATLAND[13]

Now it is difficult to imagine a photographer's creative imagination "surging," nor the instant of the photograph as "spontaneous" (both favorite descriptive terms for Tymieniecka), especially at the point of the shot: one needs control, a steady hand and eye that must not be drowned in gushing spontaneity, no matter how creative. But never mind the more romantic and painterly quality of Tymieniecka's descriptions of the relationship between art and nature (*LL 1*, p. 127). If we start from a specific work of art rather than theory, Tymieniecka's organic aesthetics describes both the process and the effect of the work of art, whether it be a drawing or a photograph.

To get back to the issue at hand, the choice of type of photograph that best explains the creative process. Those photographs that present a new view of the ugly or of the commonplace document the immediacy of the artist's decision, and they do that precisely because we realize we would have seen much less than the artist did. So we can measure the artist's imaginative leap from the "flatland" of our awareness. We meet and measure the creative imagination head on. I do not speak here of journalistic photography, like that of the Pictures of Famine series shown on the BBC; these pictures are made to raise our consciousness of suffering, to appeal for donations. Laudable though this purpose is, they are not the subject of this essay: For the most part, they do not usually reach the level of the disinterested.[14]

Photography has a special relationship with beauty: it can document that beauty can be found in the oddest places, in the oddest way, and at the oddest times. The artist's focus can bring us in from a world of seeming ugliness to an instance of the beautiful. Such is the work of the late Joseph Gluhman who in his Artist's Statement explains:

> Fascinating details can be found in very ordinary places .... Through the recognition, selection, and isolations of a subject and through careful composition and presentation, ambiguous forms and interesting colors can serve as a means of artistic expression .... Photographer and viewer collaborate to create new meanings, and thereby bring order and aesthetic pleasure from the confusion of mundane experience.

In this way even the most mundane becomes transcendent. In this way the given becomes the chosen.

Gluhman takes a purist approach. He uses "straightforward technique without cropping and without either darkroom or digital manipulation" so that he "can endow an image with the directness and vitality that characterize real art" (Artist's Statement). His photograph of a painted wall is just such a case (Figure 3). The dark shadows of the fire escape align themselves across a "landscape" in paint, articulating the horizontals, entheir flatness enlivened with another dimension. The mundane reality of the windows and grey wall is forced to cooperate with the framework of the fire escape by the action of the sun. The sensitive artist caught the instant and forever transformed our perception of the wall, the paint, the fire escape, the sun.

Figure 3. Painted Wall. Joseph Gluhman, 10.5 × 7″. Cibachrome Print. Private Collection, New York.

THE KAIRIC MOMENT

The creative process is predicated on the experiential fact that the world exists and that we can know it. From that given the photographer chooses a shot that shows the world as he has seen it, as he has faced it, *one "take" at a time*. The artist's answer to life, his gift to life is, in the words of Nancy Mardas:

a moment which enters history, and in fact, which creates history, on both a personal and a societal level. The *kairic* event is both a conception and an action, spanning the gaps between natural, human, and cosmic reality. In the moment of the *kairic* event, the rational, intentional, and passionate aspects of the human soul coalesce and converge. In so doing, they reach out to the radically other, the Transcendent.[15]

Such is the case of the work of photographer Kathleen Connally who lets us into her world *each and every day*. She daily records her best photographs of the landscape within ten miles of her southeastern Pennsylvania home on her photoblog, www.durhamtownship.com.

Blogs are public diaries; artists use them to show the development of their work, to let us into how they work. Connally shoots about 150 photographs during her early morning walks, and posts a few on her blog. She then further culls the "dailies" and puts a very few into a category "Best Photos." The last, most critical phase of her selection process, then, is open to us; her choice demonstrates the critical quality of the artwork. In the processes of selecting the daily, she has already discarded those that are not technically or formally. In choosing the "Best" category, her criteria move beyond the both technical and formal qualities and move to the level of communication of experience, that, I propose, of the *kairic* moment she experienced, consciously or unconsciously.[16] See, for example, the series of seven shots of Ganoga Falls in Rickets Glen State Park on May 15, 2005 (http://www.durhamtownship.com /blog-archives/002370.html). Six of the photographs show the beauty of the forest, of the rocks, of the waterfall: they are perfect "postcard" shots, National Geographic shots. None of them made it to the Best category. Her Best photo is unsettling because we do not know what we are looking at, and most tellingly, we do not know where we stand (we identify with the position of the photographer). The "Best" photograph gives us a sense of the instability of the situation that *built* what is given. We have some sense of the shifting and sifting the water caused as it carves the falls. The best photograph lets us experience Connally's own wonder at the existence of the place. It puts us in a position that is not "nearly undefined, uncrystallized," so that our imagination can "unfold its wings toward the crystallization of a novel 'reality' " (*LL 1*, p. 128).

PLEASURE FROM CONFUSION

In Book 1 of *Logos and Life* Tymieniecka makes the case that photography cannot be a true art form because it is too exact, because it is a kind of reproduction or copying. "In order to inspire art, reality has to appear in a nearly undefined, uncrystallized form. Taking its impetus from features only vaguely outlined, imagination is stimulated by this vagueness to unfold its wings toward the crystallization of a novel 'reality' " (*LL 1*, pp. 127–28).

What we have seen above is not that the object/*artwork* be "out of focus" but that the *beholder* is. Let us read the "vagueness" as the artist's means of involving us, no matter how sharp or how hard the focus. On the same pages, Professor Tymieniecka sees this, too: "The intrinsic potentialities of this undetermined given material become essential, and scrutinizing them, we try all possible forms and structural systems searching for that most appropriate … in our movement toward an entirely unforeseeable configuration" (*LL 1*, p. 128). In the same vein, Tymieniecka recommends or the stains on a wall that Leonardo also instructs us to study for inspiration.[17]

How that delightful confusion happens depends very much on the choices the artist makes, where he expects us to meet him in the work, and how much space he has left for our imagination. We take our pleasure from the confusion of our senses because our confusion is the result of the image being more "packed" than mere reproduction, copy, or denotation.[18] The order that secured us is a no longer apparent because the artist has changed it. Our delight rests in searching through our confusion for his vision. Such a confusion is akin to the effect of "the fecundity of the hazy light of the French Riviera" (*LL 1*, p. 128); the fog allows us to *imagine* what is there rather than to merely see it.

For Connally, the creative process is "a form of 'brain writing' – whatever is floating around in your mind, consciously or unconsciously, is bound to come through your neural channels, into your fingertips and onto your artwork."[19] Note the freedom of the activity, the pool-like image in her words. Tymieniecka speaks of circuits, webs, immersion in the creative mix, but expands the significance of the process to "the totality of life: the cognitive, emotive, and volitional activities of the human mind and culture – it searches out the web of relations that together form the living tissue of the that constitutes the total experience of existence."[20]

THE BEHOLDER'S SHARE[21]

When artist places the work of art or cipher into human society as a *novum*, he lets us into his experience, into his own creativity, into his essential humanity. What is more, while the artist's creativity posits his freedom to establish himself ontologically; our imaginative wandering in his creative work frees us from the slavery of the self and to reestablish our own ontology. The artwork provides us (as artists and as beholders) with a locus of exchange that is in no way binding or taxing but is rather a free market wherein we find the other, the fact of our necessary relationship, and a new self. Therein lies its pleasure. The process can be understood as we take pleasure in Connally's *Mud and Reflection* (Figure 4).

When we look down at a puddle that is *Delaware River Mud*, what we find, what is revealed, is the heavens in the water-mirror of the earth. We climb trees down into the brilliant blue sky, as orange mud and plants float over us. The affirmative truth of the photograph plays with our common sense of order, tweaking that common sense to question whether it is an adequate interpreter of human reality. The free imagination of Alice's looking glass takes us to the immanent order and beauty of the logos. Connally accomplished this not by aggressive revolt, but by gentle observation and recording; her gentility allows us to enter the cosmos on tip toe.

The work of art has ceased to be a simple object, but becomes the action ground of the human creative experience and interchange; it becomes the boulevard along which we reestablish ourselves in existence. That is our share.

Figure 4. *Delaware River Mud & Platanus acerifolia* Reflection. Copyright © 2005. Kathleen Connally, *A Walk Through Durham Township, Pennsylvania*, www.durhamtownship.com. All rights reserved.

THE COSMOS AS A WHOLE

One of the great questions of Tymieniecka in *Logos and Life Book 1* (pp. 15–16) is:

> what is the thread of the coherence among the discrete, disruptive, randomly enmeshed or dissolving fragments such that they meet spontaneously in the great constructive phases of the self-interpretative game of life in which to "live" means to delineate a self-individualized path of constructive advance while simultaneously endowing with significance each step of its articulation and progress?

It is, of course, the creative act. With the photograph, as with any other form of art, the artist can provide us with what Tymieniecka has called a "half-open door … a glimpse of the unshaped searching impulse that is rallying all our being's faculties, the passionate, emotional depths wherein all is fused together within a spontaneity that puts to work all that is in us. This spontaneity offers the vision of a new universe"[22] that is the spontaneity of life itself (*LL* 1, p. 149).

The new vision is not chaotic, rather it is individuated, as musician Peter Ablinger tells us: "When standing at the waterfall we become aware of our thoughts, but not of the waterfall itself. If we succeed in letting our thoughts stand still, we hear a melody within the turmoil, everyone his own."[23]

In *Nockamixon Cliffs* (Figure 5), the shadowed foreground cliff provides the frame for the iterated tonal moiré pattern rising to the full white of the

Figure 5. Nockamixon Cliffs near Kintnersville. Copyright © 2005. Kathleen Connally, *A Walk Through Durham Township, Pennsylvania*, www.durhamtownship.com. All rights reserved.

sunlight, the source of which is hidden in the deep darkness of the foreground. Trees stain the edge of shadow and light, softening their contrasts. Life joining – but not yet obscuring – heaven and earth. Here Kathleen Connally has taken advantage of her own sight, but also of the image the camera gave her.[24] In offering it to us for contemplation, she provides us with a means of turning off the waterfall, of letting our thoughts stand still, of hearing our new melody.

NOT BEING JOHN WAYNE (FIGURE 6)

Instructions to the reader: In the previous criticisms, I have made the comparison with the photographs explicit in language. In this last, however, my comparison is not with the poetry of the image, but with the image as

Figure 6. Jumping into the Delaware River, Beneath a Bridge. Copyright © 2005. Kathleen Connally, *A Walk Through Durham Township, Pennsylvania*, www.durhamtownship.com. All rights reserved.

document that takes us to the heart of Professor Tymieniecka's words. We were taken through the earlier images. Here we must carry the image into the text to see its simple brave truth. Text and image become symbiotic. To understand the predicament of the artist and the predicament of life, we must take the innocent leap the artist takes, every day, to keep his work alive.

Within the map of creative relationships we have drawn, it is possible to see, I hope, the logos of life. We realize with Tymieniecka that "Life is full: discrete, disharmonious," as "seemingly inconsequential" as viewing a waterfall. But we must not let the new melody of the waterfall be a siren song, for our philosopher warns that the Logos is "an ever-expanding creative coherence that surpasses itself at every instant" (*LL 1*, p. 15) that the "passional soul ... may from its abysmal depths surge into its own aspirations and engage in a passional pursuit of its own ... [asking] for the ultimate answer to All. Thus at the very heart of the creative orchestration ... the human being embarks upon an ultimate quest of which his soul is the instigator and judge" (*LL 1*, p. 13). It is no wonder that the second volume of *Logos and Life* was devoted to "Man's Self-Interpretation in-the-Sacred."[25] In this "ultimate quest, or the genesis of the spiritual life [the soul] seeks to free itself from every 'meaningful' attachment, aim, perspective .... in a creative fashion it despoils itself while seeking its ultimate fulfillment" (*LL 1*, p. 13). So in the end, the creative work destroys its "very context and frees the human being from ... all the entanglements of his beingness. Life presents itself as a game that calls us toward 'deepest wish,' the ultimate, the vertiginous beauty and greatness which is ours at last reveals itself!" (*LL 1*, pp. 1–14).

The work of an artist who remains creative will be just that. Refusing to be trapped in a singular chosen, refusing to be cast as a type, he returns to the given that now includes his own *novum*, and continues to "search for forms attuned both to the vibrant capacities of life and the human mind" (*LL 1*, p. 149). The creative process chooses the music, not just a song.

Siena College, Loudonville, New York

NOTES

1 "The new division of human faculties, in which the supreme role passes from the intellect to the *Imaginatio Creatrix*, and sensory perception cedes primacy to the will, emerges as the cornerstone of a new critique of reason." This is Anna-Teresa Tymieniecka's sketch of her revolutionary understanding of the human condition. See her *Logos and Life*, Book 1: *The Creative Experience and the Critique of Reason, Analecta Husserliana*, Vol. 24, (Dordrecht Boston London: Kluwer Academic Publishers, 1988), p. 11. Hereafter cited in text as *LL 1*.

2 Tymieniecka, *Impetus and Equipoise in the Life-Strategies of Reason, Logos, and Life: Book 4, Analecta Husserliana*, Vol. 70 (Dordrecht: Kluwer Academic Publishers, 2000), p. 19. Hereafter *LL 4*.

3 *Ibid.*, xxviii.

4 There are nearly fifty million websites devoted to the Indian Ocean tsunami of 2004. See, for example, http://tsunamihelp.blogspot.com/2005/01/indian-ocean-earthquake-tsunami-2005.html

5 Schwartz writes (pc) that the original pictures were taken in a lab with rulers and other measuring devices. The images created are not identifiable from the source which serves as inspiration. They are not reproductions, rather they might be best understood as appropriations.

6 *LL 4*, p. 468.

7 Nancy Mardas decoding Tymieniecka in "The Cipher as the Unity of Signifier and Signified," in *Analecta Husserliana* LXXXIII, pp. 13–23. Tymieniecka's original account is in "The Creative Self and the Other in Man's Self-Interpretation," in A-T. Tymieniecka, (ed.), *The Self and the Other: The Irreducible Element in Man*, Part I (Dordrecht: Reidel, 1997), pp. 151–186.

8 See Madras, "The Cipher as the Unity of Signifier and Signified," in *Analecta Husserliana* 83, pp. 13–23, for the history. A recent volume of the *Analecta Husserliana* is devoted to the cipher: Vol. 92, *Logos of Phenomenology and the Phenomenology of the Logos*. Book Five: *The Creative Logos. Aesthetic Ciphering in the Fine Arts, Literature and Aesthetics*, Lancaster: Kluwer Academic Publishers, 2005 (hereafter *Creative Logos*).

9 Tymieniecka, "Creative Self and the Other," p. 169.

10 "The Ontopoiesis of Leonardo da Vinci's Brainstorm Drawings," in *Creative Logos*, pp. 3–11.

11 Mardas, "Creative Imagination – The Primogenital Force of Human Life; Following Anna-Teresa Tymieniecka's Thread from the Elemental Stirring to the Human Fulfillment" (xxxlii) in A-T. Tymieniecka (ed.), *Imaginatio Creatrix: The Pivotal Force of the Genesis/Ontopoiesis of Human Life and Reality*, Vol. 83, *Analecta Husserliana* (Dordrecht: Kluwer Academic Publishers, 2004), esp. xxxvii.

[12] P. Trutty-Coohill, "Comic Rhythms in Leonardo da Vinci," in A-T. Tymieniecka (ed.), *Enjoyment: From Laughter to Delight in Philosophy, Literature, Fine Arts and Aesthetics*, ed. *Analecta Husserliana*, Vol. LVI, 1998, pp. 185–202.

[13] The reference to Edwin Adams's charming 1884 *Flatland: a Romance of Many Dimensions* is very intentional.

[14] Those like Dorothea Lange and Eugene Smith are exceptions.

[15] In Nancy Mardas's description of Tymieniecka's theories: "Following the Golden Thread: A Journey through the Labyrinth of Tymieniecka's *Logos and Life*," http://www.phenomenology.org/mardas.htm

[16] Unlike Gluhman who limited his parameters to the instant of the shot with no darkroom manipulations, Connally develops her photographs on her computer; working the picture so that the image better communicates her experience. Her great talent is in portraying light; sadly, the subtleties of the light are lost in reproduction.

[17] Paris Ms. A, fol. 102v (B.N. 2038, fol. 22v), see Jean Paul Richter, *The Literary Works of Leonardo da Vinci, Compiled and Edited from the Original Manuscripts*, 1883, 3rd edition, 2 Vols (London, 1970), para. 20.

[18] Cf. Roland Barthes's *analogon*, "message without a code" in "The Photographic Message," in Susan Sontag (ed.), *A Roland Barthes Reader*, pp. 194–210 (London: Vintage, 2000 [1961]), p. 196.

[19] Kathleen Connally, photographer: (http:// bedsheet-id.blogspot.com /2005/05/comparing-and-contrasting-two.html).

[20] Laurence Kimmel, "Logos: Anna-Teresa Tymieniecka's Celebration of Life in Search of Wisdom," in *Thinking through Anna-Teresa Tymieniecka's* Logos and Life, *Phenomenological Inquiry* Vol. XXVII (October 2004), pp. 22–23.

[21] E. H. Gombrich's phrase for the perceptual and psychic acts through which the spectator brings the image into existence by perceiving and understanding it [in the publication of his 1956 Mellon Lectures Series, *Art and Illusion* (New York and London: Princeton, 1960)]. The chapters in the section on "The Beholder's Share" deal with our constructions based on perception: "Image in the Clouds," "Condition of Illusion," "Ambiguities of the Third Dimension."

[22] *Ibid.*, p. 467.

[23] *Der Regen, das Glas, das Lachen*, Ohne Titel, Quadraturen IV, Klangforum Wien, Sylvain Cambreling (http://www.kairos-music.com /startFR.html).

[24] Although Connally records the method she uses for every shot, this photograph is older than the catalogue of her site. The moiré pattern is a lens-light distortion.

[25] *Book 2, Logos and Life: The Three Movements of the Soul of the Spontaneous and the Creative in Man's Self-Interpretation-in-the Sacred* (Dordrecht Boston / London: Kluwer Academic Publishers, 1988).

WALTER LAMMI

GADAMER AND THE "TRADITIONALIST" SCHOOL ON ART AND THE DIVINE

The "Traditionalists" are a cohesive and increasingly influential school of religious thought. Most of its members live and work in the West but look to the East for spiritual sustenance. Their traditionalism is Islamic in faith but inclusive in spirit. They are concerned with the meaning of religion in general, the nature of Islam in particular, and its spiritual task in context of the contemporary world. They articulate a moderate Muslim alternative to Islamist ideology that integrates Islam as *primus inter pares*, so to speak, into the global community of religious traditions.[1] In so doing the Traditionalists offer common ground for dialogue among the global community of believers in the world's great religions. This paper offers a critique of Traditionalist thinking from a philosophical point of view. Let us not, however, forget that the political importance of philosophy in the world at large is, at best, extremely limited. In the long run I believe that the discourse of philosophical hermeneutics offers the deeper and hence more effective conceptual path for religious dialogue. But the long run, which may be very long indeed, is not the only issue. I will return to these considerations in my conclusion.

The Traditionalists view art in terms of its connection to the divine through beauty and order. Implacably opposed to any approach to art as imitation of nature, the Traditionalists argue that genuine or authentic art must instead reflect timeless archetypes that recall the respondent to the depths of his or her being, rejecting soulless experimentation and overcoming all subjectivism. The depth of one's being and the end of art is recollection of the unity of God.[2] Its source is likewise the divine.[3] Art is neither "humanistic" nor "for art's sake," but rather a matter of truth within and from authentic religious traditions.[4] As such, art is both produced by and helps to create community grounded in the sacred. "Traditional art," Seyyed Hossein Nasr says, "is inseparable from sacred knowledge."[5] Sacred knowledge, the *Sophia perennis*, indissolubly connects art and the divine. Without that connection, art is not only worthless but an impediment to human flourishing. With it, art is of central importance to human culture.

A-T. Tymieniecka (ed.), Analecta Husserliana XCIII, 401–422.

In this broad outline the Traditionalist school seems remarkably compatible with Gadamer's thinking. Gadamer, also no relativist, approaches the divine in terms of unity of the whole and art in terms of the divine, insisting that the authentic work of art always has about it the "aura" of the sacred.[6] For Gadamer too tradition is the locus of experience, including the experience of art.[7] Gadamer likewise emphasizes the personal challenge of authentic art, which, he says, demands that we look at the injustice of our lives and change ourselves.[8] He is equally critical of Enlightenment humanism and the aesthetic consciousness that champions the autonomy of art.[9] For Gadamer no less than for the Traditionalists, art is an event of truth that reflects and creates community founded in religious life. Art for Gadamer, as for the Traditionalists, connects the human and the divine by way of order and the beautiful.[10]

However, the differences between the Traditionalist religious school and philosophical hermeneutics are ultimately fundamental, and their very similitude makes those differences exemplary for thought. The school's traditionalism is not simply traditional. It features a conceptual doctrine of true or authentic religion and a sweeping critique of the spiritual history of the West. The Traditionalists believe that the connection of the human and the divine that prevailed in traditional cultures and becomes visible in traditional art has been lost, decisively in the West and now, with globalization, potentially worldwide.[11] There has been a degeneration of art, they argue, which constitutes both symptom and partial cause of that loss. These aspects of the school reflect a kind of intellectualism that, I shall argue, is itself infected with the modernist disease against which they rail.

Gadamer too sees a crisis of modernity in which the loss of the sacred certainly plays a part. He has addressed this problem in terms of the balance between the spiritual and the scientific, which has tipped dangerously to the latter.[12] This is entirely sympathetic to the Traditionalist perspective. However, art to Gadamer, far from being symptom and partial cause, is reaction and partial cure – a recurring reconnection, in fact, with the divine. To Gadamer there is no progressive decay in the history of Western art, but rather a constant renewal of meaning in response to differing historical circumstances. Far from wholesale rejection of that art as nontraditional and naturalistic, Gadamer's more inclusive perspective embraces its accomplishments. The modern artistic revolt against traditional art itself forms a genuine tradition of a different kind but with equally religious implications, marked by opposition to scientism and industry-driven forms of social organization, and shaped by the loss of over-arching community. The new tradition is more ephemeral, including the shifting and multiple communities formed around the work of art of those who learn from it as they learn from it, for to Gadamer the "being" of the artwork is not thing but activity. Perhaps one could describe it more broadly as the community of liberal education in a largely uncomprehending world. Where the Traditionalists find only ills of globalizing modernity, Gadamer finds spiritual renewal that counters those ills. This difference is hugely important to the way we view the world and to our prescriptions, both personal and social, for its betterment. The former urges us to struggle against the forces of globalization in an attempt to return to historical forms of traditional life; the latter, to work within those forces to hold fast in a new world to the endless search for the good, the beautiful, and the true.

The Traditionalists trace the origins of the loss of the sacred to the ancient Greeks, who in their view created a "naturalistic" art based on imitation of the exterior looks of things.[13] In a complex historical path, this Greek abandonment of traditional art took root in Medieval Christianity. The Medieval Christians developed an overly rationalized theology, partly to oppose the rationalizations of Greek philosophy but partly too to incorporate the power of Greek thought into Christian revelation, and their art developed a concomitant "realism."[14] Under this pressure the authentic Christian tradition faltered. In art this authentic spiritual tradition had consisted of icon painting, the crafts, and some liturgical music.[15] The Gothic cathedral, representing the peak of Christian traditional craftsmanship, is the most perfect expression of this tradition. But the movement away from iconographic medieval painting to Renaissance realism reflects an essential degeneration, which continued through the Reformation and Enlightenment into modernity, where the perversion of the spirit has reached its final conclusion in the subjectivist emptiness of surrealism and other movements of modern art. This is not to say that all in modernity is worthless, for the Spirit will emerge in the darkest of circumstances from time to time; but with the authentic tradition eclipsed, the fate of the West is nihilism and art is its clearest expression.[16] As Frithjof Schuon puts it, all of Western art starting with the Greeks descends to sentimentality.[17] From this point of view, one might consider the artistic tradition or traditions of the West generally as glorified kitsch. This degeneracy can only be countered by the revitalization of traditions of the East that have not yet fully succumbed to the blandishments of globalization. And for that hope, it is crucial to distinguish authentic or traditional art from naturalistic and modernist depredations.

PLATO AND THE GREEKS

Are the Traditionalists not simply repeating and updating Plato's argument against Greek art? Such, at any rate, is their claim. Traditionalist writings are peppered with references to Plato, which provide a kind of scholarly anchor to their concept of art and with it, their critical history of the West.[18] Gadamer shares the Traditionalist respect for Plato, who may be said similarly to anchor his own work. Their divergent interpretations of Plato's critique of Greek art are therefore paradigmatic for understanding their differences.

The so-called "Greek miracle" consisted, as Frithjof Schuon puts it, of "anthropolatry" and "cosmolatry." This perversion of the "exteriorized and exteriorizing intelligence" is the real meaning of Greek rationalism and source of the profane and superficial imitation of nature in Greek art.[19] In his critique of the poets, Plato disparaged it for being the worthless activity that it was.[20] At issue for the Traditionalists is not the question of imitation, but rather what is imitated. In Plato's view, as Nasr says following Coomaraswamy, it should be archetypes; in Plato's critique it was nature.[21] Archetypes, which transcend mundane form, cannot be rendered naturalistically. Even though the perfect human form is the perfect work of God's art,[22] to imitate it as such is to reduce it to the human level. Traditional art is always stylized.

Coomaraswamy's seminal essay "A Figure of Speech or a Figure of Thought?" provides a sustained interpretation of Plato's philosophy of art. Coomaraswamy weaves together points taken from various Platonic dialogues and other traditions, creating a unified narrative to show that "the true philosophy of art is always and everywhere the same."[23] Plato, more than any other, has coherently articulated that philosophy. "It is the basis of Plato's criticism of naturalistic poets and painters," says Coomaraswamy, "that they know nothing of the reality but only the appearances of things, for which their vision is overkeen; their imitations are not of the divine originals, but are only copies of copies."[24] Plato's description of artistic imitation in the *Republic* is of particular importance to Coomaraswamy and the Traditionalists. The school takes at face value Socrates' argument that artistic imitation is thrice removed from the original. Using examples of "naturalistic" imitation, Socrates goes from the one form of any given thing to its many exemplars in reality, to the imitation of those exemplars in painting whose greatest success is therefore the greatest deception. The artist, the imitator at furthest remove, may paint such things as bits and reins but does not have the horseman's understanding of the use of bits and reins (601b–c). The artist is ignorant of the meaning of what he paints.

The case of Homer, however, does not belong to the representative arts. In Coomaraswamy's recounting, Homer is singled out for opprobrium because he reduced the gods and heroes to human dimensions and wrote of their passions so as to arouse our own. "The *katharsis* of Plato's City," Coomwaraswamy concludes, "is to be effected not by such exhibitions as this, but by the banishment of artists who allow themselves to imitate all sorts of things, however shameful."[25]

Coomaraswamy's interpretation is narrowly true to the text. It relies on taking Socrates' or the Athenian Stranger's (or, conflating them, Plato's) literal statements from various places and integrating them with remarks on the meaning of art from other traditions. It creates a persuasive argument insofar as the same doctrine of art is gleaned from a variety of religious traditions from a variety of times. Coomaraswamy, and the Traditionalists after him, are communicating not the hesitancy of "love of wisdom" but the certainty of wisdom itself, a fact that is emphasized by Nasr.[26] The school's "philosophy" turns out to be a doctrine of sacred knowledge. In contemporary discourse "philosophy" has become so conflated with "doctrine" that some might not even find this description problematic. Nasr, for one, does not. He does distinguish between "dogma" and "doctrine," the former being a "fixation" of the latter, with the latter taken "metaphysically" as a kind of "signpost" toward "ineffable" sacred knowledge.[27] This strikes me, however, as a distinction without a difference. It is clear neither why a signpost pointing in the right direction should not be fixed nor how the school's signposts have avoided fixation. Gadamer, for whom no statement is conclusive, would certainly object to identifying philosophy with doctrine.

Like Coomaraswamy, Gadamer finds the epitome of Plato's critique of art in his critique of the tragic poets in general and Homer in particular. However, Gadamer goes in a very different direction. He agrees that naturalistic imitation reflects degenerate art, but he considers that theory of imitation, common though it is, too trivial to warrant refutation. At no time, he argues, has "imitation" carried so simplistic a meaning in serious thinking about art.[28] Even kitsch, a phenomenon of interest to Gadamer, does not descend to that level. Kitsch, Gadamer says, reflects the loss of the traditional world and, like the Traditionalist aesthetic, expresses the desire to return.[29] Certainly for Gadamer the concept of naturalistic imitation does not begin to describe ancient Greek aesthetics, as the Traditionalists claim.[30]

Homer's poetry to Gadamer is not reducible to imitation of the passions and the question of the artistic depth of tragedy cannot easily be shunted aside. Outside the context of the *Republic* it is difficult to sustain the argument that Homer is shallow, although to this day he has proven difficult of discursive access and on the surface, with a literalist interpretation, there may seem to be nothing but passion and divine deceit. The profound but unselfconscious poetic formulations of Homeric times were, Gadamer argues, "defenseless" against the "arbitrary perversions" of the spirit of sophistic education. The problem, then, was not the art itself, but the morality (or lack thereof) that could be twisted from the art.[31] The *Republic* responds to this degeneracy with a picture in words of its total opposite. As such it is itself a deceptive imitation of the true.[32] Gadamer points out that Socrates actually rejects Homer on the grounds that his influence was not practical – he did not found a state, make ingenious discoveries, create a school, or even achieve the success of the sophists (600a ff.). Who today would demand such practical standards for poetic meaning, or indeed for anything of intellectual significance?[33] Only in the context of the *Republic*'s founding of a radically new city does this denigration of artistic accomplishment make sense. Socrates complains about Homer's depiction of the gods' passions not because they inaccurately imitate the divine but rather because of their deleterious influence on education. Thus in context of the dialogue the reasons for rejecting Homer are practical and pedagogical rather than religious.[34]

In the *Republic* this argument about Homer is described in terms of the "quarrel between poetry and philosophy" (607b). To Gadamer this famous "quarrel," rather than reflecting Coomaraswamy's one-sided argument about superficial imitation versus *Sophia perennis*, is ultimately a lover's quarrel over which could achieve greater proximity to the divine.[35] This lover's quarrel does not directly concern the thrice-removed imitation of representative arts, for the poet is himself the tool of his art whose object is not things but rather the experience of life. In Gadamer's interpretation, the point of the argument about naturalistic imitation in the plastic arts is to shed light on the poetic imitation of the soul. Painting that represents external appearance without understanding is analogous with the poet's representation of deeper truth about which, in his inspiration or *enthousiasmos* ("being filled with the gods"),[36] he has no conscious comprehension. Thus the "quarrel between poetry and philosophy" turns on the question of knowledge rather than superficiality.

Moreover, Gadamer points out that the critique of Homer is not a synecdoche for Greek art as such; the tragic poets (and early Greek philosophers as well) purified Homer's myths of their traditional form in search of new ways to uncover their truth. Gadamer has stressed in a number of studies the plastic nature of myth, which could be taken up ever again in new ways and reshaped for poetic purposes, which effectively combined the seriousness of religious truth with the life of play. Myth, says Gadamer, always concerns religious truth.[37] Yet as an art form it evades doctrinal rigidity, This rigidity, which descends to a kind of literalism, actually represents the error of "naturalism" in the interpretation of poetry. The impulse of Greek art is only naturalistic in the sense that to the Greeks the divine is everywhere in the cosmos, which does not imply pantheism or even polytheism as normally understood.[38] Their entire effort in portraying the human form is to reach toward the ideal, the divine idea, of perfect human beauty.[39]

CANONS OF ART

Rather than attempt to deny the greatness of Greek art, Gadamer tries to understand Plato's critique in its dramatic context. In the *Republic* Socrates is creating an entirely new city in speech in which the education of the guardians requires art to be subordinated to moral education. This has a polemical purpose. The real target of Socrates' critique, in Gadamer's view, is not art as such but rather its role in contemporary Athenian morality. In this period traditional education, which was largely a matter of inculcating virtue or excellence (*arête*) by habit and example, had given way to the first teachers of Greece, the sophists. The sophists claimed to be able to teach excellence as a conscious matter. This effectively tore to shreds the Greek *ethos*, the spirit of Greek culture and the Greek relation to the divine, which in all traditional cultures is unconscious and hence vulnerable.[40]

This takes us to the overriding consideration of the *Republic*, the need for conscious awareness in the absence of the traditional communitarian *ethos* upon which the Traditionalist school insists. "Socratic philosophy is not to be circumvented," Gadamer concludes.[41] The widening of consciousness through what Gadamer has called "the conversation of mankind" is, in fact, the only effective response to the worldwide loss of traditional *ethos* that the school has rightly perceived. Philosophy shares with sophistry bringing knowledge to consciousness but does not descend to sophistic cynicism. Gadamer speaks of the "content and dogma" of the new sophistic *ethos* of Greece in which a cosmic sense of justice gives way to self-interest narrowly defined.[42] This prefigures what the Traditionalists correctly fear to be a worldwide trend in modernity.

In bringing to discourse that which had previously been inculcated habitually and unconsciously, the sophists undermined the greatness of Greek artistic vision by reversing the order of the human and the divine, hastening a process of degeneration that was already leading Greek religion in the direction of what we today understand as polytheism. By seeking the ideal within the real, Greek art was vulnerable to this reversal. Anthropomorphism, in religion and in the poetry at the base of Greek religion, necessitated a philosophical opposition that called for a thoroughgoing purification, says Gadamer, of "the whole of the tradition."[43] This opposition, sharpened to an extreme in the educational utopia of the *Republic*, results in "an arbitrarily and radically purified poetry, which is no longer a reflection of human life but the language of an intentionally beautified lie."[44] The lie is the possibility of creating political unity on the natural basis of the soul by moral education. Since to Gadamer the human can only be understood in light of the divine,[45] the artistic reflection of human life is deeply religious; but real art cannot be morally purified with legal sanctions. The Traditionalists, in the intensity of their desire to preserve and recover tradition, fail to appreciate this limitation.

Problems with the Traditionalist interpretation of Plato are indicative of problems with the Traditionalist philosophy of art, which are in turn indicative of problems with Traditionalist religious doctrine in general. The Traditionalists provide specific criteria for authentic or true art. Schuon, for example, argues that sculpture in stone must be static in order to conform not only with the timeless unity of God, but also with the properties of the material – a different kind of "naturalism," to be sure, which denigrates much of Western sculpture from the Greeks through Rodin and beyond.[46]

Movement has no place in this stylization, for movement denotes temporality and the whole purpose of art is to help us transcend the temporal. This seems counter-intuitive insofar as artistic beauty is thought to capture graceful movement as opposed to stolid immobility. The Traditionalists find Plato's support particularly helpful to overcome contemporary prejudices about what constitutes beauty in art. This is clearly brought out in the difference between Pharaonic and Greek art. Coomaraswamy, Schuon, and Nasr all claim that Plato preferred Egyptian to Greek art because its frozen anti-naturalistic stylization is closer to the timelessness of essential unity.[47] Nasr and Schuon merely assert this point but Coomaraswamy, the earliest of the group, provides three relevant references to Plato's *Laws*. Two of these refer back to a single passage in Book II (657a–b), where the Athenian Stranger speaks of Egyptian success for "ten thousand years" in enacting by law songs that "possess a natural correctness" that are "permanently consecrated" so that they are no more or less beautiful at the end than at the beginning.[48] This accomplishment would "be the task of a god or godlike man," in this case attributed to the female goddess Isis.[49]

The Stranger praises the Egyptian injunction against change in artistic style in context of a discussion of the role of pleasure and pain in raising children. Coomaraswamy argues that law (or censorship, as he says) is useful for overcoming the fluctuations of pleasure and pain.[50] And clearly the law must appeal to the sacred to last for long periods of time. But this does not address the instantiation of the beautiful. The problem is not only that pleasure and pain are ever-changing but also that the life of the beautiful, far from being frozen, is ever constituted anew. As Gadamer points out, the beautiful, and with it the true in art, is encountered as singular event or work "whose particularity cannot be surpassed."[51] This means that the truth manifested in the beautiful is different in each work, even though the true itself may ultimately be one. As the Athenian Stranger puts it, "it is the duty of every man and child – bond and free, male and female – and the duty of the whole State, to charm themselves unceasingly with the chants we have described, constantly changing them and securing variety in every way possible, so as to inspire the singers with an insatiable appetite for the hymns and with pleasure therein" (665c). To Coomeraswamy this is not a problem insofar as the new creations are within a preset stylistic tradition, for the particularity of the beautiful must be circumscribed by changeless canons of form. "New songs, yes," he says, "but never new kinds of music, for these may destroy our whole civilization."[52] The question is whether such a changeless artistic tradition can retain permanent vitality. The Egyptian example is not entirely reassuring in this regard, and to apply it to contemporary conditions would be absurd. To hold the sacred at such a remove from the movement intrinsic to human experience is to calcify the life of art by insisting on a single style of manifestation of the infinite divine for any given culture, a single artistic impulse, so to speak, for all time. There is a fundamental paradox of "the one and the many" here that cannot be ignored. For all that truth connotes changeless unity, the beautiful representation of truth is many, infinitely or indefinitely many, just as the beautiful traditions, even according to the Traditionalist scheme of things, are indefinitely many. In a globalized world, can the movement of the beautiful be held back in the name of the immobility of the true?

What happens when law denies change? On the one hand, resisting change is the very goal of law and source of social stability. On the other hand, we may glean from the dialogue's context that this goal is ultimately impossible to achieve and perverse in its effects. Law has its limitations. Pleasure and pain are not only impediments to spiritual ecstasy, but also basic components of human experience, which is why their proper control is so important to a child's education. The ultimate result of such legal codifying of divine manifestation, insofar as it is successful, is human self-hatred in light of the gap between human life and the eternal rather than joy in light of the divine. That self-hatred creates a dour spiritual disposition that easily turns to self-regarding pride and vice. The forced overcoming of pleasure and pain meant to raise the human spirit redounds perversely upon the soul.[53]

The Traditionalists' focus on the unity of the divine has led them to neglect the central issue of this passage in context of the dialogue, which is the nature of law. Their failure to take into account the context of the

Athenian Stranger's comments about Egypt commits a literalist error. One might speculate that the Traditionalists seize upon the Athenian Stranger's remark about Egyptian art because it provides convenient support for their own thesis. To the Traditionalists as a school scholarship often seems not so much a matter of dispassionate study as of supportive citation. It is perilous to cite Plato without taking dialogical and dramatic context into account. Indeed, Gadamer goes so far as to assert that "any interpretation is wrong from the start which neglects this context and seeks to pass judgment on isolated statements which Plato makes."[54] That is exactly what the Traditionalists do.

The Traditionalist demand for static stylization goes to the heart of their doctrine of how art must be directed toward the timeless unity of God. Gadamer agrees that the significance of the beautiful work of art reaches beyond its immediacy.[55] However, for him this is a matter not of reference, but of instantiation. Since the work of art is a singular instantiation of the beautiful, its aura of the sacred is self-contained. This aura is indeed evoked in the static art of religious traditions, including the archaic period of Greek art, but the great art of the West also reflects the presence of the divine in the beauty of life.[56] In this sense the art that the Traditionalists dismiss as "naturalism" arguably reflects a kind of irreversible aesthetic progress: the Greeks went beyond their Archaic style without turning back, but Classical art set a standard to which human beings have constantly returned. In Gadamer's view the great work of art reflects "the inseparable belonging together of the real and the ideal, as it is conceived in the principle of intellectual intuition."[57]

Here it becomes necessary to distinguish the static from the moment. The real and the ideal are most beautifully united in the moment. As Goethe puts it, in the plastic arts the greatest emotional impact depicts the moment of "sudden transition" suspended, as it were, between before and after.[58] In this moment time is transcended in the sense that motion comes to rest and rest, to motion. The sudden, says Gadamer, opens a new dimension of "eschatological" time that opens onto changeless being.[59] Nasr too recognizes the importance of the moment in intellectual intuition, but he restricts its artistic relevance to the act of meditation on his school's canon of "traditional" static works and, as far as I have been able to determine, ignores the possibility that it could be instantiated in the work of art itself.[60] Yet that possibility, I would argue, quite precisely determines the "greatness" of the great art of the Western tradition. In his hurry to dispose of modernism and its antecedents, Nasr misses a central phenomenon of Western artistic representation.

THE QUESTION OF THE CONCEPT

In its artistic instantiation the moment cannot simply be conceptualized. All art, says Gadamer, carries an "excess of meaning" that cannot be captured in discourse.[61] The constriction of Traditionalist philosophy of art is a reflection of their commitment to a particular way of discussing religious matters that defines them as a school. Part of the school's conceptual argument is their exposition of the inner nature of religious truth as esoteric knowledge that cannot be told but must be found through traditional purifying practices such as ritual, prayer, and meditation. Thus the doctrine combines conceptual clarity with substantive mystery, intellectual understanding of the divine with the much deeper, life-long task of understanding, as Nasr puts it, the divine "Intellect." Traditional art, says Nasr, expresses this esoteric dimension.[62] In this sense, then, art for the Traditionalists also transcends the conceptual. Yet there is a crucial difference. Nasr's assertion constitutes a statement of religious faith rather than an observation about art. To "see" the esoteric in art we must compare it with an intimation of the transcendent in religion. To "see" Gadamer's transcendence of the conceptual, we need only reflect on the experience of art itself.

Nasr's statement of faith is in conceptual language. It is a doctrine of the meaning of art. The basic divide between Gadamer and the Traditionalist school concerns the purpose of conceptual thought and its relation to art. In a globalized world the post-Homeric "language of concepts" initiated by the sophists and then the philosophers has become the universal mode of discourse.[63] That is certainly the language of the Traditionalist school, as opposed to most of the traditions of which they speak. The Traditionalist school, however, is engaging in this "conversation of mankind" naively in the sense that their writings simply state matters of religious import as though they were matters of fact captured in declarative sentences. This is a major reason for the school's appeal: they believe that they are telling the truth the way it is, and the reader has no difficulty following their unambiguous manner of expression, amplified with reference to the art, both poetic and plastic, of the great religious traditions worldwide. That manner of expression is no less unambiguous when it speaks of transcendent ambiguities such as esoteric religion. It provides satisfying categories in which we can believe without the actual experiences to which it refers.

Philosophical hermeneutics, on the other hand, is not a "school." Gadamer always prided himself on his students' independence of thought, a value that seems entirely absent among the Traditionalists. At issue is whether philosophy can transcend doctrine, which is a question that comes to the fore in the philosophy of art. For Gadamer the life of tradition is its renewal in every person and every generation, which means not only constancy but change, whereas the Traditionalists seek the unchanging truth *simpliciter* in which renewal is recognition without difference. The Truth, capitalized, stands unambiguously to be said as such. The Traditionalists, with their exclusive focus on the unity of truth, understand the same truth the same way, which lends itself to the common doctrines of a "school." We are thrown back to the problem of "the one and the many." For Gadamer all understanding means to understand differently and he goes so far as to describe understanding as "the mystery of individuality."[64] For the Traditionalists "individuality" is mere subjectivity and hence does not enter into their discourse. They have confused the problematic unity of truth with adherence to unproblematic doctrinal unity.

This has major consequences for their philosophy of art. Both see the beautiful in art as key to truth, but Gadamer understands the beautiful not from a pre-given standard of unity but rather as positing its own self-contained standard, at the same time that the beautiful always displays itself in unity, as a whole. In that sense even the work reflecting an artist's rebellion against the beautiful displays beauty in the ordered whole by virtue of which we call it a "work." Thus Gadamer can find beauty, and hence truth, in modernist phenomena such as Picasso's cubism, which Nasr dismisses as "below" form.[65] To the Traditionalists, form serves transcendence toward the divine only as long as it reflects their canon of religious traditions.

Once the traditional world of the Traditionalists is broken, it cannot be repaired by a return to the tradition, however devout, however sophisticated, or however forceful that attempt to return may be. This is clearly reflected in art. In modernity art requires constant renewal, not within an irremediably fragmented overarching tradition, but rather of tradition itself. The broken community of traditional art must give way to a re-creation of community in art by individual artists. This requires an alienation from the degenerated common world "to the borders of the non-understandable," as Gadamer puts it, in order to effect a fresh homecoming in living truth.[66] That is why attempts to create in a past manner, such as contemporary works in impressionist style, seem invariably trivial and derivative.

This living truth, ineffable and ephemeral though it is as such, is for all people and for all times. Gadamer, with the Traditionalists, argues that "historicism" as a form of relativism is quite wrong. Truth is not relative to temporal change. For him art provides the decisive evidence.[67] The collapse of temporal distance in art allows us to engage the works of any tradition from any time. We can call this phenomenon the "contemporaneousness" of all great works of art – a contemporaneousness that the Traditionalists also accept, but only selectively, according to a stylistic grouping by religion that excludes traditions deemed insufficiently religious and individual accomplishment deemed insufficiently traditional.

"Historicism," however, is different from "historicity." The limitation of Traditionalist doctrine is a failure to take historicity, which means human finitude, seriously. For them monotheistic belief obviates all difficulties of explanation. Their perspective is from the eternal, which to Gadamer is precisely the error of metaphysics, and the Traditionalists indeed describe their enterprise as Metaphysics, unaware or unconcerned that the term might be considered problematic. All answers are answers to questions, and the way we frame the questions and hence understand the answers, Gadamer has stressed, changes from person to person, generation to generation, and historical epoch to historical epoch, whether we know it or not. This is no less true when the questions concern timeless issues of life, death, and meaning. In matters of Quranic exegesis, Traditionalist scholarship, drawing on a millennium and more of careful thought, is unobjectionable.[68] The problems arise with their shared doctrines of religious truth, art, and history. For all their critique of modernity, such writings are conceptually modern, deeply, albeit unconsciously, influenced by ideals of systematization and univocal expression in Western science. Thus they fall prey to a kind of scientism or modernism precisely when they think they are demonstrating the limits of science and modernity. Eternal truth brought to language is temporal, and the price of ignoring that paradox is a way of thinking that simplifies. Such simplification has a name: ideology.

ART AND IDEOLOGY

The Traditionalists seem to take their bearings from an axiomatic belief in unity and judge human life in terms of its proximity to that unity. Scholarship becomes a means to support this pre-given schema. In this sense their conceptualization is formulaic. In philosophical terms the Traditionalist simplification concerns what Gadamer calls the "dialectical" nature of unity or, to repeat the terms of Greek philosophy, the relation of "the one and the many."[69] The difference here between Traditionalism and philosophical hermeneutics goes to the heart of conceptual expression. Briefly put, for Gadamer unity is "dialectical" because starting from experience, as responsible thinking must,[70] the "one" is always found among many "ones," which means that unity is constituted by difference. Thus the unity of any work of art, by virtue of which we can call it a "work," does not "refer" to the unity of the cosmic whole, but instantiates that unity by virtue of its own self-contained wholeness. This constitutes the difference between art and concept. That is why the work of art overcomes the dichotomy of the profane and sacred so that art, as Gadamer says, "always has something sacred about it."[71] The Traditionalists do not delve into the complexities of the concepts of unity, wholes, and the whole in the Western philosophical and artistic traditions, which limits both the expressive potential of their formulations and the possibility of fruitful dialogue outside the discourse of the school. Moreover, the school also does not occupy itself with respectfully engaging other ways of thinking. It is hardly surprising that such dialogue has not taken place.

"Ideology" means, in part at least, that the perception of reality is filtered through a set of fixed ideas. One consequence in terms of art is a tendency to explain poetry conceptually as though all poems had a "message" that can thereby be captured or grasped. This becomes visible in Nasr's work. Two poems that Nasr quotes, one at the end of Chapter One of his major work *Knowledge and the Sacred*, and the other at the beginning of Chapter Two, nicely illustrate the problem.[72] The first, by Dylan Thomas, concerns an old man who has lost his way.

> Too proud to die, broken and blind he died
> The darkest way, and did not turn away,
> A cold kind man brave in his narrow pride
> Being innocent, he dreaded that he died
> Hating his God, but what he was plain.
> An old kind man brave in his burning pride.

Nasr uses this poem to illustrate the loss of spirituality in Western modernity. The second, not surprisingly by Frithjof Schuon, is about finding the Truth despite Western modernity in return to primordial tradition.

> For though all seem lost, yet All is found
> In the last who is the First. Faithful pageant
> Not amiss is thy mime, for manifest in thee
> Omega is an archway where Alpha stands framed
> The First who comes Last, for likewise art thou
> The season of seeds, O season of fruits.

According to Gadamer's understanding of art, however, the real or poetic meaning is reversed. The first, in the beauty of its freestanding expression, itself constitutes an affirmation of the divine in the very "message" of its loss. The second, with its verbal flaccidity, leaves the respondent with no more than the ideas Nasr takes from it. He has appropriated the literal statements of the poems univocally while ignoring their poetry. Art has been reduced to ideology. Nasr, who explicitly criticizes Western culture for its increasing inability to understand poetry, hoists himself on his own petard.[73]

Nasr's own poetry, which has received the highest praise from a member of his school, also nicely illustrates the same problem.[74] Here is an example:

> Thou are the One, Absolute, Infinite
> The Goodness that is the source of all goodness.
> Above all that we can think or imagine,
> Thou art above thou-ness itself,
> Thy Essence beyond all that is, nay beyond the light of Being.
> We bear witness to Thy Oneness.
> Yet, who are we this Oneness to behold?
> It is Thou within us who testifies,
> Who bears witness to Thy Oneness pure.
> Our selves are but veils hiding Thee from Thee,
> But Thy Self resides in our heart of hearts
> And our substance, pure and primordial,
> Reflects now and forever the glory of Thy Oneness.[75]

I submit that this betrays a kind of aesthetic "tin ear." It is doctrinally identical to his prose but poetically trivial. Art is not only subordinated but subsumed by ideological conceptual standards. This is not intended as a blanket indictment. For traditional art that conforms to Traditionalist concepts, the school's work in general and Nasr's in particular can be insightful. It does, however, indicate a limitation that is, I submit, not merely personal but ideological.

Part of the attraction of the Traditionalist aesthetic ideology is that it provides a sense of explanatory completeness, of doing justice to all artistic phenomena. Thus classical music, which might seem to provide a counterexample to the progressive spiritual deterioration of the West, is greatly appreciated as an exception to that progression. There are indefinitely many such exceptions among individual artists due to the fact that "the Spirit shows through," even in the most degenerate of places. If one objects that this renders the thesis irrefutable, so much the better – the truth, after all, permits of such complexities.

I have suggested that a central problem with the Traditionalist school is that it reverses the dependence of conclusion on inquiry, which is fundamental to genuine scholarship. *Philosophia perennis*, as a way that human beings approach *Sophia perennis*, must always proceed differently in reaching toward the same. The Traditionalists start from the unity of the "same," whereas philosophical hermeneutics embraces the more modest "reaching."[76] The term *perennis* in *Sophia perennis* connotes the eternal. In *philosophia perennis*, the *perennis* to Gadamer would merely refer to the fact that the conversation of philosophy is endlessly renewed. Philosophy is only "on the way" and the failure to appreciate that incompleteness lends itself to the *idées fixes* and erroneous judgmentalism that characterize ideological thinking. The experience of human finitude, Gadamer maintains, "can remove the false claim of gnostic self-certainty from the self-understanding of faith."[77] This observation was made in the context of Christian theological debate but it applies, *mutatis mutandis*, to the Traditionalist version of Islam. Such certainty is theologically suspect, conceptually flawed, and aesthetically fatal. Ambiguity in the "meaning" of art that demands the observer's active participation to constitute it as a work creates the spiritual community of modern art. For Gadamer it is the prime virtue of modernism; for the Traditionalists, such art dissolves into meaninglessness. Their faith-based judgmentalism limits their historical options. By denigrating Western art, the school closes itself to any new *ethos* of viable communities of the human spirit in contemporary life.

Nevertheless, in contemporary political circumstances one can only be sympathetic with the value of the Traditionalist argument because by hearkening back to the great traditions of the great religions, especially Islam, it provides much-needed opposition to modern fundamentalisms. The Traditionalists offer one of the most articulate voices of moderate Islam. By drawing us back to the wisdom of the ages, the effect or contemporary *Wirkungsgeschichte* of the Traditionalist school may arguably be *counter*-ideological, pitting a benign ideology that melts into responsible practice against more virulent forms of the same modernist deformity. This may be the *only* effective counter on a mass level in the present *Zeitgeist*. Perhaps this is what philosophy should look like when it seeks to become edifying. However, it also illustrates the fact that philosophy that seeks to become edifying is no longer philosophy. It sinks below with the claim of soaring above.

Gadamer explains ideology as a result of the "soaring desire" of the human heart for certainty.[78] The ideologue has all the answers that really matter. Like ancient sophistry, ideology is a matter of conscious knowledge or at least of the claim to conscious knowledge. It is systematized sophistry for the modern age, characterized more by self-deception than indoctrination by malicious or impersonal forces. Ideologues view their knowledge as both comprehensive and transparently coherent. According to Gadamer, the phenomenon of sophistry is so close to philosophy as to constitute a "nascent impurity" within philosophy itself.[79] In the juxtaposition of the Traditionalist school and Gadamer's hermeneutical phenomenology we find a perfect illustration of that impurity, so close – and yet so far, infinitely far away. Art makes that chasm visible.

The American University in Cairo

NOTES

[1] The school itself favors the "Traditionalist" label. It is rightly termed a "school" both because its members share premises and worldview and because they make constant reference to each others' works. While deeply appreciative of all great religions, especially Christianity and Hinduism but including Native American, with the seminal figures of René Guénon and Frithjof Schuon and the contemporary influence of Seyyed Hossein Nasr the school has coalesced around Islam. My study has begun with Nasr and broadened to members of the school to whose work he provides relevant references. These include Schuon, Titus Burkhardt, and, most prominently for this study, Ananda Coomaraswamy. Without pretending to have read the entire school, I believe that these are sufficient to supply a fair overview.

[2] See for example, Seyyed Hossein Nasr, *Islamic Art and Spirituality* (Oxford: Oxford University Press, 1990), pp. 7, 13; Titus Burkhardt, "The Universality of Sacred Art," in William Stoddart (ed.), *The Essential Titus Bruckhardt* (Bloomington: World Wisdom, Inc., 2003), p. 91; and elsewhere. To the Traditionalist school this is a truism.

[3] See for example Nasr, *Knowledge and the Sacred* (Albany: SUNY Press, 1989), p. 262.

[4] "An art worthy of the name is beautiful because it is true." Burkhardt, "The Universality of Sacred Art," *op. cit.*, p. 88. See also Nasr, *Knowledge and the Sacred, op. cit.*, p. 252: "Sacred art which lies at the heart of traditional art has a sacramental function and is, like religion itself, at once truth and presence ..." This is another truism of the school.

[5] Nasr, *Knowledge and the Sacred, op. cit.*, p. 253. Nasr adds, "... because it is based upon a science of the cosmic which is of a sacred and inward character and in turn is the vehicle for the transmission of a knowledge which is of a sacred nature." As Schuon puts it: "... sacred art has this particularity, that its essential content is a revelation ..." Frithjof Schuon, "The Degrees of Art," *Studies in Comparative Religion* 10: (Autumn 1976): 195.

[6] Hans-Georg Gadamer, "Wort und Bild – so wahr, so seiend," in *Gesammelte Werke*, Vol. 8 (Tübingen: J.C.B. Mohr, 1993), p. 378.

[7] "Tradition is what is to be experienced." Gadamer, *Truth and Method*, Second Revised Edition, trans. Joel Weinsheimer and Donald G. Marshall (New York: Continuum, 1993), p. 358.

[8] See Gadamer, "Die Aufgabe der Philosophie," in *Das Erbe Europas: Beiträge* (Frankfurt am Main: Suhrkamp Verlag, 1989), p. 168.

[9] This critique comprises Part I of *Truth and Method.*

[10] "The "meaning" of art ... does not seem to me to be tied to special social conditions .... On the contrary, the experience of the beautiful, and particularly the beautiful in art, is the invocation of a potentially whole and holy order of things, wherever it may be found." Gadamer, "The Relevance of the Beautiful," in Robert Bernasconi (ed.), trans. Nicholas Walker, *The Relevance of the Beautiful and Other Essays* (Cambridge: Cambridge University Press, 1986), p. 32.

[11] For a general discussion of the religious decline of the West, see Nasr, *Knowledge and the Sacred, op. cit.*, p. 34 ff.

[12] Gadamer observes rather caustically in this regard, "It seems to me that European civilization has been derelict in its duty to the law of balance [between science and spirit, logic and art] for the last three hundred years. It has in an admirable manner brought the culture of science and its technical and organizational application to total development." Gadamer, "Vom Wort zum Begriff. Dire Aufgabe der Hermeneutik als Philosophie," in Bernd Klüser (ed.), *Die Moderne und die Grenze der Vergegenständlichung* (München/Maising: Anderland Verlagsgesellschaft mbH), p. 39. (My translation.)

[13] See for example Nasr, *Knowledge and the Sacred, op. cit.*, p. 257 f. For the Traditionalist critique of naturalism, see for example Ananda K. Coomaraswamy, "Why Exhibit Works of

Art?", in Rama P. Coomaraswamy (ed.), *The Essential Ananda K. Coomaraswamy* (Bloomington, Indiana: World Wisdom, Inc., 2004), p. 114: "In other words, a real art is one of symbolic and significant representation; a representation of things that cannot be seen except by the intellect."

[14] Nasr, *Knowledge and the Sacred, op. cit.*, pp. 34–36.

[15] Burckhardt, "Christian Art," in *The Essential Titus Burckhardt, op. cit.*, p. 112.

[16] "To be sure, there were artists who sought desperately for some spiritual significance in their work and life and reflected elements of quality in their paintings such as we find among a number of impressionists, but often they reached a nihilism which in a number of cases even resulted in suicide, such as we see in the tragic life of a painter as gifted as Van Gogh or in the equally tragic life of Rothko .... Instead of surrendering the ego to the Self or realizing the reality of the Void in its metaphysical sense ... such artists finally sought to annihilate themselves through external destruction of their earthly lives, as if one could destroy a sacred text by simply throwing it into the fire." Nasr, "Reply to Eliot Deutsch," in Lewis Edwin Hahn, Randall E. Auxier, and Lucian W. Stone, Jr. (eds.), *The Philosophy of Seyyed Hossein Nasr* (Chicago: Open Court, 2001), pp. 386–387.

[17] Frithjof Schuon, *The Transcendent Unity of Religions*, ed. Huston Smith (Wheaton, IL: The Theosophical Publishing House, 1993), p. 63.

[18] There is, however, a difference between casual references and scholarly citations. Coomaraswamy is the only Traditionalist to provide extensive Platonic citations, although Nasr provides extensive Coomaraswamy citations (such being the nature of a school).

[19] Schuon, "The Degrees of Art," *op. cit.*, p. 200. Nasr quotes from Schuon's book *Le Soufisme voile et quintessence*: "The 'Greek miracle' is in effect the substitution of reason for the Intellect, of fact for principle, of phenomenon for Idea, of form for essence, of man for God, and that in art as well as in thought." My translation; Nasr does not translate quotations. Nasr, *Knowledge and the Sacred, op. cit.*, p. 60, note 110.

[20] Schuon, "Foundations of an Integral Aesthetics," in *Studies in Comparative Religion* 10: 3 (Summer 1976): 132.

[21] Nasr, *Knowledge and the Sacred, op. cit.*, p. 263 and note 18, pp. 276–277.

[22] Nasr, *Ibid*, pp. 256–257.

[23] Coomaraswamy, "A Figure of Speech or a Figure of Thought?", in *The Essential Ananda K. Coomaraswamy, op. cit.*, p. 37.

[24] *Ibid*, p. 29. Coomaraswamy's Platonic reference here is to *The Republic*, 601, where Socrates uses everyday or "naturalistic" examples of imitation of reins and bit in horsemanship to set forth his thesis that artistic imitation is imitation of imitation of the true reality. However, this whole discussion is in context of the critique of tragedy in general and Homer in particular.

[25] *Ibid*, p. 25. Socrates puts Coomaraswamy's point somewhat differently at 607a: "... Homer is the most poetic and first of the tragic poets; but you must know that only so much of poetry as is hymns to gods or celebration of good men should be admitted into a city." Elsewhere in this essay endorsing Plato's critique of Homer and the tragic poets, Coomaraswamy proposes the thesis that the purpose of tragedy is to purge the passions and release the immortal soul, which purgation he finds also in Indian texts. But does that not elevate Greek tragedy to high art in his terms? There seems to be some confusion here. See *Ibid*, pp. 25–26.

[26] On the certitude of religious knowledge, see for example Nasr, *Knowledge and the Sacred, op. cit.*, p. 109: "With Schuon's writings the full-fledged revival of tradition as related to the rediscovery of the sacred in the heart of all traditions and by virtue and through the aid of tradition in the heart of virgin nature, sacred art, and the very substance of the human being has taken place, making it possible amidst a world suffocating from the poisonous atmosphere of nihilism and doubt for those who 'are called' to gain access to knowledge of the highest order

rooted in the sacred and therefore inseparable from the joy and light of certitude." Also: "Beauty bestows upon intelligence that highest gift which is certitude." Nasr, *Knowledge and the Sacred, op. cit.*, p. 269.

27 Nasr, "Reply to Eliot Deutsch," in *The Philosophy of Seyyed Hossein Nasr, op. cit.*, p. 383.

28 Gadamer, "Art and Imitation," in *The Relevance of the Beautiful and Other Essays, op. cit.*, p. 94.

29 Gadamer, "Ende der Kunst?" *Gesammelte Werke*, Vol. 8, *op. cit.*, p. 212. While I would not want to strain the comparison, this does suggest that kitsch in art may exhibit a structural similarity to ideology in thinking. Both rest on simplification born of nostalgic desire for a kind of unity and coherence no longer attainable in the conditions of modern life. This is no less true when the doctrine of the ideology is future-utopian.

30 This is, of course, not only Gadamer's view. As Goethe puts it, "The artists of antiquity were not laboring under our present-day misconception that a work of art must appear to be a work of nature; rather, they identified their works of art as such by a conscious arrangement of components, employed symmetry to clarify the relationship among these components, and so made a work of art comprehensible. Through slight variations in symmetry and positioning the most effective contrasts become possible." Johann Wolfgang von Goethe, "Ancient Art," in John Gearey (ed.), trans. Ellen van Nardroff and Ernest H. von Nardroff (trans.), *Essays on Art and Literature* (Princeton: Princeton University Press, 1986), p. 16.

31 Gadamer, "Plato and the Poets," in P. Christopher Smith, (ed.), trans. *Dialogue and Dialectic* (New Haven: Yale University Press, 1980), p. 61.

32 "Plato's critique of the poets is thus to be interpreted in terms of the two faces which the *Republic* presents: on the one hand, the strict utopian constitution of the state and, on the other, a satirical criticism of existing states. The very immoderation of this critique of the poets gives us tangible evidence of the purpose which Plato has in mind. It is his aim to bring about the possible, i.e., the actual, education of the political human being by providing a picture of the impossible, i.e., and organized paideia whose unlimited capability derives entirely from itself and in no way from a given ethos." Gadamer, "Plato and the Poets," *op. cit.*, p. 53.

33 *Ibid*, p. 41.

34 Gadamer points this out, with emphasis, in *Ibid*, p. 44.

35 "The great drama of Greek literature displays poetry and philosophy wrestling from early on like two great contenders for the prize of depicting and representing the genuine religious experience of the Greek world." Gadamer, "Thinking as Redemption," trans. Joel Weinsheimer (New Haven: Yale University Press, 1999), p. 81. See also Gadamer, "Religion and Religiosity in Socrates," in John J. Cleary (ed.), trans. Richard Velkley, *Proceedings of the Boston Area Colloquium in Ancient Philosophy*, Vol. 1 (Lanham/London: University Press of America, 1986), p. 63.

36 Gadamer's emphasis on this literal meaning is pointed out in a Translator's Note to the English edition. Gadamer, "Plato and the Poets," *op. cit.*, p. 42.

37 Gadamer, "Myth in the Age of Science," in *Hermeneutics, Religion, and Ethics, op. cit.*, p. 92.

38 Gadamer cites Walter F. Otto's study of the Greek gods, which concludes that each of the Olympic gods provides a perspective on the whole of Being, except Zeus, the figure that unifies that whole. See *Ibid*, p. 100.

39 One simple indication of this is the fact that classical coins portray gods and mythical heroes rather than actual individuals, as opposed to the portraits of emperors on Roman coins.

40 "Plato's Socratic insight was that a binding political ethos, which would assure the proper application and interpretation of poetry, no longer existed once sophism had come to define the spirit of education." "Political," taken here in its broadest sense, means culture and culture, religion. Gadamer, "Plato and the Poets," *op. cit.*, p. 50.

41 *Ibid.*, p. 41.
42 *Ibid.*
43 *Ibid.*, p. 48.
44 *Ibid.*, p. 58.
45 Gadamer, "Friendship and Self-Knowledge," in *Hermeneutics, Religion, and Ethics, op. cit.*, p. 140.
46 "In the case of sculpture ... it [is] necessary to respect the immobility of matter by suppressing movement or by reducing it to an essential, balanced and quasi-static type ..." Schuon, "The Degrees of Art," *op. cit.*, p. 201.
47 See Nasr, *Knowledge and the Sacred, op. cit.*, p. 258; Schuon, "Foundations of an Integral Aesthetics," *op. cit.*, p. 132; "The Degrees of Art," *op. cit.*, p. 200, and *The Transcendental Unity of Religions, op. cit.*, p. 72; and Coomaraswamy, "A Figure of Speech or a Figure of Thought?", p. 28 and "Why Exhibit Works of Art?", p. 114, in *The Essential Ananda K. Coomaraswamy, op. cit.*
48 The other two references are to 665c and 700c. Coomeraswamy, "A Figure of Speech or a Figure of Thought?", in *The Essential Ananda K. Coomaraswamy, op. cit.*, p. 45, Note 37. The statement for this reference is as follows: "Now we know that Plato ... is always praising what is ancient and deprecating innovations ... and that he ranks the formal and canonical arts of Egypt far above the humanistic Greek art that he saw coming into fashion" (p. 28). See also his unattributed quotation from the same section of the *Laws* in "Why Exhibit Works of Art?", in *The Essential Ananda K. Coomaraswamy, op. cit.*, p. 114: "It was anything but 'the Greek miracle' in art that Plato admired; what he praised was the canonical art of Egypt in which 'these modes (of representation) that are by nature correct had been held for ever sacred.' "
49 Here I follow the Loeb Library's R. G. Bury translation, which is closer to the language of Coomaraswamy's quotation than later translations.
50 Coomaraswamy, "A Figure of Speech or a Figure of Thought?", in *The Essential Ananda K. Coomaraswamy, op. cit.*, pp. 24–25.
51 Gadamer, "The Relevance of the Beautiful," in *The Relevance of the Beautiful and Other Essays, op. cit.*, p. 37.
52 Coomaraswamy, "Why Exhibit Works of Art?", in *The Essential Ananada K. Coomaraswamy, op. cit.*, 114.
53 See 716a4–b3 with 747b6–d1. I am drawing on Seth Benardete's reading in *Plato's* Laws: *The Discovery of Being* (Chicago and London: The University of Chicago Press, 2000), p. 66f. This phenomenon is visible today in the draconian spiritual and legal codes of *salafi* Islam.
54 Gadamer, "Plato and the Poets," in *Dialogue and Dialectic, op. cit.*, p. 48.
55 Gadamer, "The Relevance of the Beautiful," in *The Relevance of the Beautiful and Other Essays, op. cit.*, p. 32.
56 "In Archaic Greek art the genre of particular things had outweighed their specific, individual qualities in artistic representation. Hence abstraction, expressed through the geometricization of natural forms, dominated Archaic art. In the fourth century ... it is possible to detect the first indications of a taste ... for the representation of specifics without any emphatic suggestion of the genre or form (in the Platonic sense) from which they were derived. Realism, in short, began to undermine the long-standing role of abstraction in Greek art. In the art of the High Classical period ... these two poles of artistic thinking – the absolute and the relative – seem to have been magically balanced." J.J. Pollitt, *Art and Experience in Classical Greece* (Cambridge: Cambridge University Press, 1972), p. 96.
57 Gadamer, "Goethe and Philosophy," trans. Robert H. Paslick, *Literature and Philosophy in Dialogue: Essays in German Literary Theory* (Albany: SUNY Press, 1994), p. 13. Gadamer

shares this appreciation of intellectual intuition with the Traditionalists, as opposed to much of the philosophical tradition since Kant. See discussion in Ernest Wolf-Gazo, "Nasr and the Quest for the Sacred," in *The Philosophy of Seyyed Hossein Nasr*, *op. cit.*, 289 ff.

58 Goethe, "Ancient Art," in *Essays on Art and Literature*, *op. cit.*, pp. 17–18, 20. Such emotional impact is, of course, radically different from the emotion in art that the Traditionalists denigrate as "sentimentality."

59 Gadamer, "Articulating Transcendence," in Fred Lawrence (ed.), *The Beginning and the Beyond: Papers from the Gadamer and Voegelin Conferences* (Chico: Scholars Press, 1984), pp. 7–8, with "Plato's *Parmenides* and its Influence," trans. Margaret Kirby, in *Dionysus* 7 (December 1983): 16.

60 Nasr, *Knowledge and the Sacred*, *op. cit.*, p. 271.

61 Gadamer, "Aesthetics and Hermeneutics," David E. Linge (ed.), trans., *Philosophical Hermeneutics* (Berkeley: University of California Press, 1976), p. 102.

62 Nasr, *Knowledge and the Sacred*, *op. cit.*, p. 268.

63 As modern children of the West, Gadamer observes, "we are compelled to speak the language of concepts." Gadamer, "Letter to Dallmayr," in Diane P. Michelfelder and Richard E. Palmer (eds.), *Dialogue and Deconstruction* (Albany: SUNY Press, 1989), p. 101.

64 Gadamer, "The Hermeneutics of Suspicion," in Gary Shapiro and Alan Sica (eds.), *Hermeneutics: Questions and Prospects* (Amherst: The University of Massachusetts Press, 1984), p. 57.

65 Nasr, "Reply to Eliot Deutsch," in *The Philosophy of Seyyed Hossein Nasr*, *op. cit.*, p. 386. See also Nasr, *Knowledge and the Sacred*, *op. cit.*, p. 256.

66 Gadamer, "Ende der Kunst?", in *Gesammelte Werke*, Vol. 8, *op. cit.*, p. 220.

67 *Ibid*, p. 209.

68 See for example *Islam, Fundamentalism, and the Betrayal of Tradition: Essays by Western Muslim Scholars*, Joseph E.B. Lumbard (ed.) (Bloomington, Indiana: World Wisdom Inc., 2004).

69 See for example Gadamer, *The Idea of the Good in Platonic-Aristotelian Philosophy*, trans. P. Christopher Smith (New Haven: Yale University Press, 1986), p. 137. This fundamental issue can only be touched upon here.

70 Gadamer, "The Phenomenological Movement," in *Philosophical Hermeneutics*, *op. cit.*, p. 172.

71 Gadamer, *Truth and Method*, *op. cit.*, p. 150.

72 Nasr, *Knowledge and the Sacred*, *op. cit.*, p. 48 and p. 65.

73 Nasr, "Reply to Eliot Deutsch," in *The Philosophy of Seyyed Hossein Nasr*, *op. cit.*, p. 384.

74 "[In his poetry] Nasr has chosen to share his divine gift and to sing in 'the language of the birds' for the first time in his life. His book of poetry could only have been authored by a mystic attuned to otherworldly sapiental experience. *Poems of the Way* culminates the scholar's philosophical arguments in a moving admission of direct experience: Nasr has evolved from lecturing about *Knowledge and the Sacred* to celebrating his having attained knowledge *of* the sacred." Luce López-Baralt, "Knowledge *of* the Sacred: The Mystical Poetry of Seyyed Hossein Nasr," in *The Philosophy of Seyyed Hossein Nasr*, *op. cit.*, p. 401. It should be noted that Nasr's poetry is written in English.

75 Nasr, *Poems of the Way* (Oakton, VA: The Foundation for Traditional Studies, 1999), p. 14.

76 Nasr quotes Schuon: '"*Philosophia perennis*' is generally understood as referring to that metaphysical truth which has no beginning, and which remains the same in all expressions of wisdom. Perhaps it would here be better or more prudent to speak of a '*Sophia perennis*'..." Nasr, *Knowledge and the Sacred*, *op. cit.*, p. 88, note 18.

[77] Gadamer, "On the Problem of Self-Understanding," in *Philosophical Hermeneutics*, *op. cit.*, p. 49.
[78] Gadamer, "Emilio Betti," in *Gesammelte Werke*, Vol. 10, *op. cit.*, p. 435.
[79] Gadamer, "Dialectic and Sophism in Plato's Seventh Letter," in *Dialogue and Dialectic*, *op. cit.*, p. 122.

LJUDMILA MOLODKINA

AESTHETIC VIRTUALITY OF THE ARCHITECTURAL–NATURAL LANDSCAPE IN MODERN COMMUNICATIONS

… Any landscape is an ideal material for expressing thoughts of certain character.

F. Novalis

In the global system of modern communications there are plenty of options that integrate the reasoning with diversity and spontaneity of life, which makes every cultural space unique and inimitable, an inherent project that may be attributable exclusively to such reasoning. To speak about superiority or primacy of one culture would mean to play down the importance of others'. It would be much more rational and more effective to search for "cultural universalism" consolidating people in their vital (true-life) world both in its routine and more socialized aspects. Here we need to behold a new conception of reasoning, to explore fundamentals of rationality in the bowels of life itself. According to A-T. Tymieniecka, the human status, having been generated during an onto-poetic process of life self-individualization, in a capacity of midmost virtuality, includes creative imagination as a specifically human pathway of life.

Alongside with intellectual and moral evaluative-semantic factors of creative virtuality, A-T. Tymieniecka distinguishes a poetically-aesthetic feeling, which is not isolated in the framework of psychic experience, but proliferates far beyond the boundaries of survival functioning.[1] Aesthetic virtuality endows novelty of meaningfulness to the vital, empirically significant natural environment. In the search of "cultural universalism" throughout modern communications continuum, the nature and architecture in their symbiosis are the most "allied" and comprehensible attributes for the human status. On the background of natural–architectural reification through conscious and unconscious perception and aesthetic articulation, the human contribution to the process of universal life becomes the most valuable and explicit.

A-T. Tymieniecka (ed.), Analecta Husserliana XCIII, 423–430.

First of all, introduction of the "architectural–natural landscape" category is focused on to its aesthetic–phenomenological structuring in human consciousness. Without any repudiation, au contraire, by way of cross-fertilization of philosophic ideas, which were drawn from different conceptual complexes, we will try to concentrate our attention on phenomenological aesthetics of landscape and its perception by a contemporary recipient (spectator) in a wide communicative space. However, I believe it is appropriate to depict briefly the loop of conceptual measurements of landscape and to note that in this "conflict of interpretations" (let me use Ricker's term) the concept of "landscape" is quite complicated and diversified.[2]

An architectural–natural landscape appears as an optically delineated and pronounced natural space with inscribed homogenetic elements – architectural structures, which in many cases become imaginative dominants mostly due to their historical features. Theodore Adorno in its famous *Aesthetic Theory* presented landscape culture as a synthesis of nature and architectural buildings. "Quite often historical buildings together with their geographic environment, which supplied stone they were built of, are perceived as the very beauty."[3] Architectural–natural landscapes are undoubtedly linked with history; the history reflects in them and "feeds" them semantically, saturates them with various notional "strata." Continuity of historic development literally embodies in the architectural–natural form, thus "dynamically integrating" the landscapes, "as it usually happens in the works of art."[4]

In the époque of romanticism due to the cult of ruins, the architectural–natural landscape, which carries the deep "traces of the past," is digested by individual and collective consciousness as a phenomenon, and widely crops out as an aesthetic stratum, although subsequently falls into disrepair, turning into an advertising pad for concerts and recreation activities or amusement events, a kind of "asylum" from odious reality. In an architectural–natural landscape the historic newsworthiness adopts its aesthetic shape and simultaneously preserves "the traces of former true-life crucifixion." Some cultural landscapes with inherent ruins and partially survived buildings narrate the stories of "the langsyne grievance instinct with anguish of body and mind, bereft of sound long ago … There is no beauty without historic memory."[5]

The process of empathizing with such landscapes generates a phenomenological integrity of aesthetic perception and contemplation. Nature and architecture are constituted as intentions in the spectator's consciousness at the junction of imagination of the past and the empathy of the present, and are perceived as certain "situations." Casting a "glance" on such a landscape "situation," the interpreter yields its meaning, vanquishes structures already generated in his/her consciousness, produces new ones, rejects the targeted structures and focuses on the feasible ones. The architectural–natural landscape, turned into a correspondent of consciousness in its different modus as a "phenomenon of the existential" builds up significance, a "network of notional intentions" (quoting Merleau-Ponty). An opportunity to construe the intentional life of consciousness in a modern socio-cultural situation, so multivalent and contradictory, open for a dialogue with the world, and which shapes the character and mechanisms of "vital communications" between consciousness, human behavior and material reality, to my mind, appears to be the most realistic and typical, if we fall back to the analysis of the aesthetic–phenomenological perception of the architectural–natural landscape by a traveler or a tourist.

Actualization of various forms of modern cultural communications such as "travel," "tourism" or "museum," is initiated by the people's desire to cognize something unforeknowable. Conventionally, let us exclude cases when an architectural–natural cultural object becomes subject to vain entertaining digestibility, and serves as a kind of "attenuating stuffer" in addition to overall touristic "relaxation." Let us not count for downsides of the so called "organized tourism," which "distorts intrinsic content of natural experience," involved in the system of metathesis. In the framework of the "travel industry" the "unaffected nature's experience," according to T. Adorno, is not binding, it is neutral and apologetic. "Nature became a sanctuary, a reservation and an alibi sui generis."[6] Sometimes a human enjoys a "morally-narcissistic contentment" (say, having a gust of feeling nature!), or a true sense of the beautiful is easily substituted by a contemplation of bridal processions out in the country. Specifically this is the case with historical architectural–natural territories such as artistic memorial manors, historic castles, villas etc. This is a "shriveled" nature's experience, not true and not original; because the "living nature wishes for silence, inspiring the allocution of those who may adopt its experience – and such words free them from monadological captivity for a moment."[7]

Nevertheless, no matter how skeptical some philosophers could be in their attitude to traveling, at present the exploration of the global cultural, artistic and historical heritage is mainly due to world-wide touristic communications, which involves the Reasoning, Feelings and Emotions of a contemporary human being. Nowadays the philosophic aesthetics of tourism is quite latitudinous and diversified and, from my point of view, may represent a broad variety of challenging issues: "tourist – museum – communications," "tourism and historic memory," "landscape – architecture – museum," "nature – history – memory," "ecotourism as topical modus in cultural communicative space" etc. Perhaps the "traveling syndrome" and "traveling therapy" phenomena can be explained by the fact, that an individual experiences "the happiness of his/her association with nature" only when he/she realizes, that being an imaginary subject of certain dominating and infinite quantitative substance, he/she has fallen away from nature, has dissociated from environment, and then he/she is trying to throw his/her image on nature, feeling a kind of intimacy to it. A human is trying figuratively to "escape" from the "secondary," humanized and socialized nature to the nature of the "origin," genuine habitat that gave birth to him.

Travel appears as a kind of "path," a "procession to the past" by the instrumentality of the present. This, in its own way, is an "immersion" of the reasoning and feelings to Another, the Other, "not mine," but, at the same time, attractive and alluring enough, even if such Another is alien for various reasons: religious, confessionary, ethical and other considerations. "Touristic space" and "space of a traveler" today has become an arena of the most intensive development of modern communications, which ensures a possibility of direct contact with foretime realities. Transcendental Ego immanently includes transcendental Alter-Ego, in which the Other has been intentionally infelt. In case of necessity, transcendental subjectivity of an individual tourist intergrades up to the level of inter-subjectivity or transcendental sociality in a general context of communications.

A spectator-tourist first imagines and anticipates his/her "rendezvous with the past," then perceives, empathizes, and contemplates the observed. Quite a few modus step in the "work" of consciousness. Vladislav Tatarkevich strongly believes that the aesthetic empathy of the architectural–natural landscape starts with concentration and observation. "In order to perceive the beauty of wild nature or the art, you need to concentrate a glance,"[8] believes the famous Polish aesthete. As though in a due with aesthetic assertions of Roman Ingarden pertaining to "introductory emotions," which are characterized by excitement, the "existential basis" of an architectural monument, about its "substantiation" and "initiation into a temple,"[9] V. Tatarkevich identifies some other instances of aesthetic empathy: expectation, excitement, admiration, "unemotional humility," association, predilection, pleasure, rapture, day-dreaming etc,[10] which undoubtedly have great significance for constituting a phenomenological architectural–natural image.

Material perception of a landscape as a "vital world," which started with natural reduction aimed at its architectural and natural components, is gradually substituted with eidetic reduction, which allows to endow a surrounding object with philosophic–aesthetic significance, rhetorical meaning, imagining them as "pure phenomena," which "live" (remain) in consciousness even in cases when immediateness of the perceptional contact disappears and another modus of consciousness – memory – starts functioning. Architectural–natural landscape, in its Noema (material sense) remains unchanged, but constantly varies in its Noesis (modus of intentional consciousness) and the way of visual perception of a tourist-spectator. This phenomenological process expands through integration of two plotlines in an architectural–natural "text": natural "movement" (the beauty of mountains, trees, valleys, seas, lakes etc.) merges with "architectural theme" (castles, villages, monasteries and other developments).

Imagination plays a great role in the phenomenological process of landscape memorialization. Architectural natural space, perceived as memorial or memorial monument, witnesses the virginity of nature: this is a memory of "antediluvian," ancient places. In the eyes of a tourist the entire architectural–natural space is museified. Landscape "themes" are read by way of excursion (to the seaside, mountains etc.). There is a ground for surprise: "all of a sudden ...." Greater fineness to such sensations is attributed by architectural impregnations to natural intrinsic context (for instance, Rupit village high in the mountains of Catalonia). "Touristic" consciousness is constituted by a number of emotions, which we call "touristic motives". In our consciousness a motive of the sea or ocean emotionally produces an intention of endless watery waste, as eternity and permanency. This is a self-sufficient concept that never changes. It is absolutely beautiful and absolutely unpredictable. There is a sensation of anonymity, submergence into absolute "nowhere"; the waves fascinate, intimidate, calm and thrill memorial feelings. Because this abyss in our sensations may be perceived as a resting place for shipwrecks, treasures, drowned cities, and, what is more thrilling and exciting for humans, the "deep-six" of sailors and single civilizations.

Emotional "touristic" motives of mountains, flatlands, trees, flowers, snow-capped hill, castle, monastery etc. – all these are given or served up to the consciousness of a tourist, his sensations and thoughts in the process of traveling: either by overview through excursion as an integral part of travel or through a museum method. In this case the entire accumulated experience of a traveler is of great importance (that is: read about it, friends told me, saw it in the catalogue, heard it over the radio, watched it on TV, searched it on the internet etc.). This sensuous empiric element is consistently regenerating, thrills our imagination, and intensifies our perception at the moment of "seeing live." We start empathizing objects as though broadcasted from the past, and which are offered through museum-excursion methods to be heard or observed "de visu" in a modern context. These objects, as a result of reduction change, are conventionally memorialized in the tourist's consciousness, thus obtaining intentionality, and producing "extended intentional threads" (Merleau-Ponty).

A tourist (traveler, pilgrim) receptively and perceptively submerges into the observed, recalls the things that were visited earlier, sensuously experienced, imagines and represents his/her intrinsic fusion with nature. Important are his/her bodily presence, a "touch," an involvement in the architectural–natural space. We need to consider that a tourist or a traveler is not only a spectator, a recipient, but also an "associate" of coexistence on the level of imagination; he/she broadcasts the past to the present, "then" transfers to "now." He/she lives through "then" "now," the "then" projects to "today," an image of a by-gone event, in which he/she "directly" participates, is constituted in his/her consciousness. All these result in a feeling of undoubted pride of such a "rendezvous" with realities which existed in the past and which may become a subject of his own perception.

A motive of rendezvous in a traveler's space is closely connected with a motive of a road. A path made by a tourist is essential and significant for the structuring of intentional objects. The immersion into Another, Alien, but very attractive because of to its yet unknown and distant nature, starts from the scene from the window of a bus, plane or train. However, the direct intercourse starts already now, it foretastes the beginning of a real visual contact, a touch (for instance, a road to Genoa, Sorrento, Catalonia, Blenheim, to Russian monasteries or manors, to Italian villas or English castles). It is the road, the path that anticipates a rendezvous in reality; it is the road, the path that speaks about recognition and delight or, vice versa, irrecognition and disappointment.

What is the moment when a sense of memory arises in our consciousness, when an intentional object becomes a kind of memorial monument? There are certain landscapes which are literally museified, such as Italian villas, English historical castles or Russian memorial manors. These cultural objects have already been memorialized by way of museification. However, there are landscapes which are only conventionally museified (partially or piecewise). We speak about Italy as a country-museum, or we speak about Greece as an antique country-museum, the cradle land of our civilization. In these cases the museified are only some fragments of natural and architectural spaces, but it looks like the entire surrounding landscape with inscribed homogenetic elements is a unified museum continuum. Here comes up a phenomenon of integrity of architectural–natural landscape perception. Here is the unity of intentional objects and methods of their presentation. If we speak about the literally museified architectural–natural landscape, then there is a "tourist-museum" communication scheme: a tourist in compliance with his/her travel schedule visits a museum, or takes an excursion.

In the second case a spectator-tourist phenomenologically "museifies" in his/her consciousness, that is, to a certain extent, memorializes in his/her perception, imagination and empathy the entire visual accessible natural surrounding with all homogenous impregnations. Especially this is the case with visual perception in the dynamics of touristic communications of huge landscapes with various infelt architectural objects (temples, castles, monasteries, medieval villages). From the point of view of phenomenology these are imagined and perceived as a unified intrinsic space-museum, the reservoirs of which perpetually store the inviolable, historically natural exhibits, majestically conciliative "eyewitnesses" of by-gone époques, civilizations, or intact fragments of wildlife and, moreover, which attract our keen interest (let me recall once again Catalonia, Italian scenery, Egypt, Scotland, or world-famous places for pilgrims – Montserrat in Spain, Crete's Acrotiri, the Solovki monastery, Trinity Sergiy Lavra in Russia etc).

The "museum – excursion – pilgrimage" concept was shaped up under the influence of the tourist philosophy as a topical means of communications. As I noted earlier, the methods of presentation or "museum servicing" of architectural–natural landscape are first of all connected with its visual perception and institutionalization as an intentional object.

State University of Land Use Planning, Moscow

NOTES

[1] A-T. Tymieniecka, "Theme: Gardens and the Passion for the Infinite //Gardens and the Passion for the Infinite," in *Analecta Husserliana* LXXVIII (Dordrecht: Kluwer Academic Publishers, 2003) pp. 1–4.

[2] In the philosophic literature a landscape is analyzed as a concept of geographic and artistic tradition, the landscape is structured as a frame, which constitutes a comprehension of cosmic order and chaos. The externalized "metaphysics of landscape" orients the philosophy on to comprehension of multi-dimensionality of topological structures of existence and human reasoning. There is also a concept of "visual" landscape, which is localized physically, historically and biologically. It is a kind of an image of a supra-individual "universe" that inhabits creative boutiques of philosophers, artists, writers etc. The concept of "verbal" landscape allows for considering the geographic space as a landscape, which has "lost physics" and which has acquired a meaning of aestheticized "rhetoric." A chamber landscape, as though being inserted into an alien context, is described, interpreted and infelt in the framework of verbal image categories, such as "seascapes" of Kierkegaard, subterranean space elements of Nietzsche, "mountain space" of Heidegger and "subjectivation zone" and metaphoric spheres of Delez. There are also other types of landscape such as "corporeal" landscape, which gradually becomes viewless and deformed, connected with psychomotor effects that stimulate the creation of a masterpiece. "Corporeal" landscape literally directs the reasoning into a right tideway of creative research. For instance, we may recall the "ascension to the depths" as a "dionysiac dancing line" of Nietzsche or Heidegger's "up-thrusting" as a "crease edge" etc. A.A. Gritsanov "Landscape," in *Contemporary Dictionary on Philosophy* (Minsk, 2003), p. 542.

[3] Theodor W. Adorno. *Aesthetic Theory* (Moscow, 2001), p. 96.
[4] *Ibid.*, p. 96.
[5] *Ibid.*, p. 97.
[6] *Ibid.*, p. 102.
[7] *Ibid.*, p. 102–103.
[8] W. Tatarkewic, *The History of the Six Notions* (Moscow, 2002), p. 332.
[9] R. Ingarden, *Researches on Aesthetics* (Moscow, 1962), pp. 203–260.
[10] Tatarkewic, *The History of the Six Notions* (Moscow, 2002), pp. 329–359.

MARTIN NKAFU NKEMNKIA

VITALOGICAL AESTHETICS. THE IDEA OF BEAUTY IN AFRICAN CULTURE, ART AND PHILOSOPHY

INTRODUCTION

I presented in my book *African Vitalogy – A Step forward in African Thinking*[1] in paragraph 8.3, transcendental properties of life in the following order: the truth or the true of the truth, unity, the good and the beauty of everything. I would like to invert the order of these properties because I earlier said in the same text that each and everyone of them could be treated as the first, because of their indispensability for life to manifest itself. This is why I am going to centre this exposition and reflection on the Good, the goodness and the beautifulness of everything in the universe. Let me start with the following preposition:

> The good as such does not exist but we have people who are good and do good, beauty as such does not exist in itself but rather things and objects present themselves as beautiful, nicely looking and therefore desirable.

The world and all that exists can therefore be presented under a series of principles, summarised in: *there is harmony in the rhythm of the universe.*

A: UNIVERSAL PRINCIPLES OF AFRICAN VITALOGICAL AESTHETICS. FIRST ARTICULATION

1. There is a perfect organisation of the planets and their movements set up by the Creator of life, independent from the desire and will of the creatures, especially from that of human beings.
2. All that is beautiful is necessary for the survival of human beings and fundamental for the way that all that exists manifests itself in the concert of the living.
3. In order to be perceived, every existence has to be qualified and the fundamental *attribute of all that exists is to be found in the colour and perfume of the universe*. Therefore colours and perfume are the course of every beauty.

A-T. Tymieniecka (ed.), Analecta Husserliana XCIII, 431–442.

4. All that manifests itself in one or another form as beautiful, is necessarily desirable, and all that is desirable is obviously good, since that which is good, imposes itself as a fundamental substance (aliment) for the survival of humanity. This means that beauty and the art of beauty are at the origin of happiness.
5. Thus, everything that presents or manifest itself to the rest of the existing is a necessary and an indispensable aliment for the survival of other existing realities. Therefore, the first form of beauty in nature is found in *colour*, to be considered as universal attributes, as the quality of the quantity of every reality. By the way, the intensity of the desire expressed by any individual towards objects and persons are proportional to the capacity of reception of the knowing subject.
6. All that is beautiful and good, and that which is tasteful, produces a divine reaction in human beings. It generates joy and happiness in them as well as harmony among people, family, friends, groups and the community.

B: NECESSARY CONSEQUENCES FROM WHAT HAS JUST BEEN SAID

(a) This is why all human actions presuppose an aim. *Beautifulness, goodness, joy and happiness* consequently provoke similar sentiments in people. This is why I earlier declared that beauty and the art of beauty are at the origins of happiness in living beings, in human beings in particular, so much so that all that exists, exercises an eternal attraction between all and everything, influenced by a fragrance that fills the soul, a fragrance that has never been produced by any person on earth before.

(b) This is why we can say that beautiful and good things are properties of nature and constitute its basic attributes. In this case the art of being good is the human person's natural tendency and it is what promotes or generates harmony in the universe as well as the reciprocity of beauty and the happiness of all and everything that exists.

(c) Beauty and goodness, considered in this light, would be the basic nourishment of the inner and supernatural life of human beings. It is what makes one desire eternity for oneself and for others, eternity for all that exists because whatever exists and lives, is "willed" by the good that the Creator "willed" for all living beings, for the entire creation.

(d) The attempt to reproduce nature in its various expressions is certainly what can therefore be express as *art*. The attempt to concretely reproduce the various aspects of nature, making it the criterion of beauty, is what we want to consider as *aesthetics* and moreover, *African vitalogical aesthetics*, the science of beauty and of beautiful things, the science of goodness.

(e) Since beauty and goodness are now known as the basic attributes of nature itself, in the study of African culture as a vitalogical enquiry, as a science of the totality, African art is an attempt to represent the divine, the source and cause of all that is beautiful and goodness.

(f) Every attempt to reproduce aspects of nature which generate sentiments of the divine in human beings, awakens in them the desire for immortality. Therefore, one can easily affirm that the African vision of beauty, that is, aesthetics, as seen from the African perspective, is the sum of every goodness that exists in each and everything, expressed particularly in "art" which, in turn, includes figures, colours, sounds and melodies, objects of speculation and reason for philosophical enquiry.

One can therefore continue to articulate the principles already mentioned in the first part of this paper.

C: UNIVERSAL PRINCIPLES OF AFRICAN VITALOGICAL AESTHETICS. SECOND ARTICULATION

7. For Africans the idea of Beauty is first an interior act (subjective), secondly it is an exterior act that influences subjects, conditioning them to act or behave according to its command. In fact, subjects are moved, attracted by the impact of what is being perceived from objects.

   The influence of these irradiations from objects has the same capacity of stimulating and satisfying the fantasy of he who perceives, so much so that one is invited to contemplate nature as such.
8. African society consists mostly of rhythm. Planets have their rhythm, nature has its time: seasons and all alike are manifested as an eternal beauty without end, as are the flowers, waterfalls, trees, mountains, seas.
9. Human beings are artists by vocation. That is why it has been possible to build houses, construct towns, learn how to reproduce and imitate nature through painting and drawing, carving and mouldering statues in bronze and clay, weaving clothes.

   Nature is always feasting: the birds on the land, seasons of the year, blossoming of flowers, the movements of the planets, the colour of the dresses. Briefly, one can say that everything seems to be ordered to generate happiness and joy in the living, especially in human beings. Each day offers an occasion for man to celebrate life. Man makes of each day a hymn to life. When people celebrate an anniversary, a birthday, one is perpetually celebrating life.

10. Many phenomena transmit beauty to living beings: the blow of the winds, the fresh air of morning, of late evening, sunrise and sunset, sounds, melodies and music, songs of birds and of people, steps of dances, the perfumes of the world which generate desires and intensifies the will for living and the conquest of life.
11. In order to contemplate the good and the beauty, one has to be serene.
12. Human behaviour presupposes a certain end: reaching happiness and a permanent state of well-being. Beautiful things and objects as well as the good (attractive things) provoke in human beings sentiments which are substantially similar in all of them.
13. In everyday life we realise that beautiful and "tasteful" (good) things and objects are intrinsic properties of nature, constituting its profound attributes. Human creativity, the art of being good (behaving well), of exercising goodness is the natural habit and attitude of human beings and it is that which enables one to perceive and reach the harmony of the universe, to experience reciprocity, to contemplate the beauty hidden in everything and in everyone.

D: NECESSARY CONSEQUENCES RESULTING FROM THE FOLLOWING PRINCIPLES

(a) Therefore, that which is beautiful and attractive is at the origin of happiness and joy, expressed by every human being and nature itself as a hymn to life. This is why everything and everyone desires one another and no one desires living alone. This attraction is caused by the perfume that fills the heart and souls, perfume that has never been produced by any human being since it can only be scented by them.

(b) For this reason again, the beautiful and the good that generate happiness and joy in human beings are fundamental substances (aliments) of the interior and supernatural life in them. It is that which enables one to desire eternal life for oneself and for others as well as everlasting life for the entire creation. This is also why beauty as such can only be perceived, experienced and desired as the the Creator wanted it for his creatures.

E: UNIVERSAL PRINCIPLES OF AFRICAN VITALOGICAL AESTHETICS. THIRD ARTICULATION

14. The effort of representing all aspects of nature is the criterion of individuation and evaluation of beauty as such and beautifulness and it is what is

being considered as African vitalogical aesthetics, the science of beauty, of beautiful things and objects in life.

15. Since the beautiful and the good are constitutive attributes of nature itself, in African culture and thinking as vitalogy and as science of the all (totality), what is being considered as art is an attempt to represent the divine nature of all that exists, since it is said to be that which enables everything and everyone to be attractive, to be beautiful and handsome, nicely looking, desirable and "tasteful" at the same time.

    Certainly, for this reason, in an attempt to reproduce nature while imitating it at the same time, one does not create nature, making it better than what it is but rather symbolising and subliming its divine character. It is enough to see the forms of African masks, statures in their almost indefinite forms, always incarnating the figures they are representing.

    This is also because nature alone is the master of all that is really beautiful without end, nature alone is the most complex architecture existing, nature alone is the mother of all that is nice, of all beauty.

16. Every attempt to reproduce aspects of nature with the aim of generating sentiments of the divine in the souls of the living, generates in the artist the desire of immortality, for nature in itself is immortal. This is why it is also said to be divine, since all that is divine is also eternal, and eternity is equivalent to immortality.

    This is the main reason why we affirmed that the African vision of the beautiful, or rather of vitalogical aesthetics, is the sum of all that is good in everything, expressed in *works of arts* which furthermore is articulated in figures, colours, sounds and melody. In this we observe the harmony of the rhythm of the universe, a divine and eternal one.

17. In general, what is considered to be good (nice), is that which is visible and therefore desirable as already said in the above. In the hierarchy of goodness and of beautiful things and objects, mostly it is life that imposes itself as goodness and likeness, as beauty and divine. That which renders something to be indispensable is in fact the goodness it bears and expresses as will, caused by the beauty containing in them. In fact, people are admirable because they transmit both beauty and joy, happiness and wisdom, life through generations and hopes for an ever better future.

18. All that is beautiful generates in the souls of people a sentiment of admiration and of pleasure, due to the capacity of attraction and donation. All that is good is beautiful and contagious, is necessarily shared. This is the case of the world in its totality, such are objects, things and persons generating knowledge and experiences, such is life in its sublimity. Only that which is beautiful is lifesome, can be wanted, can be deserved.

This is why it becomes obvious to affirm that all that is beautiful (nice) is necessarily good (tasteful), though not everything has the same capacity of diffusing its beauty, not possessing the same intensity of beauty as well as taste to offer to everything and everyone.

19. Material objects or things always present a certain grade of beauty and, at the same time, change the forms and the modality of manifesting themselves in public: the "become something else", transforming themselves from one form of being to another. In doing so, they also change and modify; their beauty. For example, a flower is beautiful and diffuses its perfume as it remains alive but as soon as it loses its colour and perfume, it no longer attracts anyone for it does not generate happiness any longer in anyone. Instead the flower sometimes, causes unhappiness, sorrow as well as sadness in people.

20. Spiritual beauty exists. Such beauty, which in any case is a progressive effect from material objects and things, sanctifies the desire of the souls and it is that which generates the sentiment of love in the hearts of human beings. Therefore, beauty derived from objects of desire is considered as life originating from something else or someone else and as such does not remain in the subjects themselves. The force of beauty of the objects is that which "defends" them from death, making them eternally desirable and always offering a better and beautiful aspect never seen or perceived before.

    This is what creates in people the tension towards sanctity: the desire of the beautiful and the good. In fact, for African people, the only fear is that, one day, one may loose the capacity of living, not being anymore attractive. Not wanted, one may run the risk of dying alone somewhere and be forgotten by the living.

21. Good and beautiful things, beauty and beautifulness as predicates and hymn to life and to nature, remain the only attribute among all the known "ones" capable of attraction.

    Good manners, behaviour, way of dressing or clothing oneself, the choice of colours and of the type of dresses to wear, the way of combing one's hair, the way of walking on the road, the sound of one's voice, are all means through which each one tries to render his or herself desirable to others. One renders himself/herself desirable in order to be wanted and to be loved by others.

22. All that is beautiful is therefore considered as an instrument of relation and of unity between and among everything and everyone. This is why the "Good" has been seen in two ways: (a) as objective good (which is reached or acquired through the movement of the "Will" towards determinated things and (b) as subjective – supernatural – good (that which

remains in the spiritual sphere of human existence after the knowledge of the objects desired, which are sensitive goods).

23. Of the two goods, everyone can reach the first (objective good) because the material character does not completely satisfy the capacity of desiring, proper of the human will and likeness. As for the second (subjective good), a personal effort is demanded to conserve what is acquired. Such a good is desired, not yet only by instinct but especially through the intellect and the soul (spiritual good).

Spiritual goods are superior to the material ones though more fragile at the same time. It is, in fact, that which provokes love in the hearts of human beings towards nature, towards his pairs, towards the absolute. It is that which provokes sentiments of faults and guiltiness for wrong doings in persons. It is that which reveals to the human intelligence the existence of immortality, of the divine nature.

In all that God created, there is a seed of the "good," of the divine, of immortality which is, by the way, the characteristic of God's life.

F: NECESSARY CONSEQUENCES RESULTING FROM THE FOLLOWING PRINCIPLES

(a) All this said, it is good to know the reality in its substance, to be acquainted with other human beings as ones self, in order to discover that the goodness of all that exists is life itself which is being communicated to everything and to everyone alike.

The "will" and "tension" towards one another is desire, is affection, is reciprocal attraction, is inclination and above all, is love which is being disseminated everywhere. Goodness is the enjoyment of the desired, it is that which keeps one in the condition to attain beatitudes and contemplation of the beauty and of the good in the universe.

(b) The good is again said to be that which is really wanted and hoped by everyone, by everybody. Wanting life, desiring life and goodness is really tending towards perfection because only that which is perfect satisfies the immortal soul.

(c) For this reason, goodness is the permanence of life in all living, in all aspects and states in which things and reality present or manifest themselves.

(d) If one could say in a key sentence what these attributes of life (goodness, beauty, joyfulness) are, it is convenient to say that they are aspects of the life of the Creator impressed on the creation and the creatures.

(e) At this point, it becomes necessary to affirm that each of these properties is transcendental and plays an indispensable role in the life of the creatures for the determination of everything in its gender and kind.
(f) For this reason, there is no preference in the order of presenting them. One could start form anywhere to investigate on reality.
(g) Such are, in fact, the truth (or the true of the truth), oneness, unity or united aspects of life (things in relation to each other), such is beauty (or beautiful, nice and handsome things and people), such is good (or goodness of thing and peoples, tasteful things etc.).

CONCLUSION

Let us conclude with this last preposition: That which is beautiful or is related to beauty or again to the idea of beauty is intrinsically rooted in the DNA of every human being, constituting to focus points for desiring life for ever. Life is the only everlasting beauty every living being possesses and will not for any reason lose it.

Pontifical Lateran University, Rome, Vatican City

NOTE

1 Martin Nkafu Nkemnkia, *Il pensare africano come "vitalogia"* (Roma: Città Nuova, 1995).

SELECTED BIBLIOGRAPHY

Albert Ott, "Akan Gold Weights," *Transactions of the Historical Society of Ghana* 9 (1968).
P.A. Allison, *African Stone Sculpture* (London, 1968).
P.A. Allison, *Cross River Monoliths* (Lagos, 1968).
P.A. Allison, "Mask Makers and Myth in Western Liberia" in Anthony Forge (ed.), *Primitive Art and Society* (London/New York, 1973).
Anthony Forge (ed.), *Primitive Art and Society* (London/New York, 1973).
Audrey I. Richards, *Chisungu: a Girls' Initiation Ceremony among the Bemba of Northern Rhodesia* (1956).
Ba Amadou Hampaté and Germaine Dieterlen, *Koumen, texte initiatique des pasteurs peuls* (Paris/The Hague, 1961).
G.T. Basden, *Niger, Ibos* (London, 1938).
U.H. Beier, *African Mud Sculpture* (Cambridge, 1963).
Beltramo Ceppi Zevi Caudia *Arte Africana*, Quando Dio abitava Ife (catalogo mostra Firenze 2005).
Bernard Fagg, "Recent Work in West Africa: New Light on the Nok Culture," *World Archaeology* 1, 1969.

Berthold Söderberg, "Antelope Horn Whistles with Sculptures from the Lower Congo," *Ethnos* (Stockholm) 1(4) (1966/7).
Bohumil Holas, *Animaux dans l'art ivoirien* (Paris, 1969).
J.S. Boston, "Some Northern Ibo Masquerades," *Journal of the Royal Anthropological Institute*, Part 1, 1960.
Brigitte Menzel, *Goldgewichte aus Ghana* (Berlin, 1968).
Bruce Lincoln, "The Religious Significance of Women's Scarification among the Tiv," *Africa* xlv (3) (1975).
Bruzzichelli Pia (a cura di) *Arte, Africa e Cristo* (Assisi: Pro Civitate Cristiana, 1963).
Burchard Brentjes, *African Rock Art* (London, 1969).
K.L. Carroll, "Three Generations of Yoruba Carvers," *Ibadan* (1961): 12.
K.L. Carroll, *Yoruba Religious Carving* (London, 1967).
T.J.H. Chappell, "The Yoruba Cult of Twins in Historical Perspective," *Africa*, Ixiv, 3, 1974.
Charles Monteil, *Les Bambara du Ségou et du Kaarta* (Paris 1924).
Charles de Brosses, *Du Culte de Dieux Fétiches*, 1700.
Clara Odugbesan, "Femininity in Yoruba Religious Art," in M.M. Douglas and P.M. Kaberry (eds.), *Man in Africa* (London/New York, 1969).
Claude Tardits, "Panneaux sculptes bamoun," *Objets et Mondes* 2 (1962).
C.K. Cooke, *Rock Art of South Africa* (Cape Town, 1969).
J. Cornet Art *de l'Afrique au pays du fleuve Zaire* (Bruxelles, 1972).
Daniel P. Biebuyck, "Function of a Lega Mask," *International Archives of Ethnology*, xlvii, 1, Leiden, 1954.
Daniel P. Biebuyck, *Tradition and Creativity in Tribal Art* (Los Angeles/London, 1969).
Daniel P. Biebuyck, *Lega Culture, Art, initiation and Moral Philosophy among a Central African People* (Los Angeles/London, 1973).
Daniel F. McCall and Edna G. Bay (eds.), *African Images. Essays in African Iconology* (New York and London, 1975).
Derrick Stenning, *Savannah Nomads*, London, 1959.
Dominique Zahan, *Sociétés d'initiation bambara, le Korè* (Paris/The Hague, 1960).
Domenique Zahan, *White, Red and Black Colour Symbolism in Black Africa.* (Eranos Jahrbuch, 1972), pp. 365–395.
M.M. Douglas and P.M. Kaberry, (eds.), *Man in Africa* (London/New York, 1969).
Douglas Fraser, (ed.), *The Many Faces of Primitive Art: a Critical survey* (Englewood Cliffs, 1966).
Douglas Fraser and Herbert M. Cole (eds.), *African Art and Leadership* (Madison, 1972).
Edictiones poligrafa S.A. Barcellona, *Arte dell' Africa negra* (Espana, 1976).
C. Edward Hopen, *The Pastoral Fulbe Family in Gwandu* (London/Ibadan, 1958).
Elizabeth Marshall Thomas, *The Harmless People* (London, 1959).
Elliot Picket, "The Animal horn in African Art," *African Arts* 4 (1971).
Eugène Roosens, *Images africaines de la mère et l'enfant* (Louvain/Paris, 1967).
E.E. Evans-Pritchard, *Witchcraft, Oracles and Magic among the Azande* (Oxford, 1937).
Frank Willett, "A Hunter's Shrine in Yorubaland, Western Nigeria," *Man* 334 (1959).
Frank Willett, "A Further Shrine for a Yoruba Hunter," *Man* 66 (1965).
Frank Willett, *African Art* (London, 1971).
Frank Willett and J. Picton, "On the Identification of Individual Carvers: A Study of Ancestor Shrine Carvings from Owo, Nigeria," *Man* 1 (1967).
Franz M. Olbrechts, *Les arts du Congo belge* (Bruxelles, 1959).
Gabriel Mande, (a cura di) *Capire l'Arte africana* (Bergamo: Luchetti, 1987).

A. Gerbrands, "Art as an Element of Culture, especially in Negro Africa," Mededeelingen van het Rijksmuseum voor Volkenkunde 12, Leiden (1957).
Germaine Dieterlen, *Essai sur la religion bambara* (Paris, 1951).
George W. Harley, "Masks as Agents of Social Control in Northeast Liberia," *Peabody Museum Papers*, 32 (2) (1950).
J. Girard, "Dynamique de la Société Ouobé," *Mémoires de l'Institut Fondamental d'Afrique Noire* 78 (Dakar, 1967).
E. Haafe and J. Zwenemann, "Krankheitsdarstellungen an Afrikanischen Masken," *Tribus*, 20 (November, 1971).
Hamo Sassoon, "Cave Paintings recently discovered near Bauchi, Northern Nigeria," *Man* 70 (1960).
Hans Cory, *Wall Paintings by Snake-charmers in Tanganyika* (London, 1953).
Hans Cory, *African Figurines* (London, 1956).
H. Hartwig, "Wooden Dolls for Unmarried Girls and Childless Women," *Baessler Archiv* (1969).
Henri Brandt, *Nomades du Soleil* (Lausanne, 1956).
Henri Koch, *Magie et chasse dans la forêt camerounaise* (Paris, 1968).
Henri Lhote, *The Search for the Tassili Frescoes: The Rock Paintings of the Sahara* (London, 1959).
Herbert Cole, "Art as Verb in Iboland," *African Arts* (Autumn), 1969.
H. Herskovits, "The Art of Dahomey," *American Magazine of Art* (1934).
W.R. Horton, *The Gods as Guests* (Lagos, 1960).
W.R. Horton, "The Kalabari World View," *Africa* (October, 1962).
W.R. Horton, "The Kalabari Ekine Society," *Africa* (April, 1963).
W.R. Horton, *Kalabari Sculpture* (Lagos 1965).
Ivan Bargna, *Arte Africana* (Milan: Jaca Book, 2003).
Jacqueline Delange, *Arts et peuples de l'Afrique noire* (Paris, 1967).
Jalmar Rudner and Ione Rudner, *The Hunter and his Art* (Cape Town, 1970).
James C. Faris, *Nuba Personal Art* (London, 1972).
James H. Vaughan, "Rock Paintings and Rock Gongs among the Marghi," *Man* 63 (1962).
Jean Gabus, *Au Sahara II Arts et symboles* (Neuchâtel, 1958).
Jean Gabus, *Art nègre* (Neuchâtel, 1957).
Jean Laude, *Arts anciens du pays dogon* (Paris, 1959).
Jean Laude, *Irons of the Dogon* (New York, 1964).
Jean Laude, *The Arts of Black Africa* (Los Angeles/London, 1971).
G.I. Jones, *The Trading States of the Oil Rivers* (London, 1963).
G.I. Jones, "Sculpture of the Umuahia Area of Nigeria," *African Arts* 4 (1971).
Joseph Henry, *L'âme d'un peuple africain: les Bambara* (Münster, 1910).
Julius E. Lips, *The Savage hits back or the Whiteman through Native Eyes* (London, 1937).
C. Kjersmeier, *Ashanti Weights* (Copenhagen, 1948).
C. Kjersmeier, *Centres de style de la sculpture nègre africaine*, 4 vols. (Paris/Copenhagen, 1935–1938).
A. Kyerematen, "The Royal Stools of the Ashanti," *Africa* xxxix (1) (1969).
Ladislas Segy, "Shango Sculptures," *Acta Tropica* 12 (1955).
Ladislas Segy, "The Yoruba Ibeji statue," *Acta Tropica* 27 (1970).
J.P. Lebeuf, *L'art ancien du Tchad: bronzes et céramiques* (Paris: Grand Palais, 1962).
R. Lecoq, *Les Bamiléké* (Paris, 1953).
Louis Perrois, *La statuaire fang, Gabon* (Paris, 1972).
Louis Perrois, "La statuaire des Fang du Gabon," *Arts d'Afrique* 7 (automne, 1973).

Louis Tauxier, *La religion bambara* (Paris, 1927).
Marcel Griaule, *Masques dogons* (Paris, 1938).
Margaret Trowell, *African Design* (London, 1960).
Margaret Trowell, *Classical African Sculpture* (London, 1964).
Marguerite Dupire, *Peuls nomades* (Paris: Institut d'ethnologie, 1962).
Marian Wenzel, *House Decoration in Nubia* (London, 1972).
Marie-Claude Dupré, "Le système des forces *nkisi* chez les Kongo d'après le troisième volume de K. Laman," *Africa*, Ixv, 1, 1975.
Marie-Louise Bastin, *L'Art décoratif Tshokwe* (Lisbon, 1961), 2 vols.
Marie-Louise Bastin, "L'art d'un peuple d'Angola," *African Arts* (Spring), 1969.
Marylin Houlberg, "Yoruba Twin Sculpture and Ritual," unpublished MPhil thesis, University of London, 1968.
R. Mauny, "Gravures, peintures et inscriptions rupestres de l'ouest africain," (1954).
R. Mauny, *Masques mende de la société bundu (Sierra Leone)* (Dakar, 1959).
Mazonowica, "Prehistoric Rock Painting at Tassili," *African Arts* 1 (1968).
M.D. McLeod, "Gold weights of Asante," *African Arts* (1, Autumn 1971).
C. C. Meek, *Law and Authority in a Nigerian Tribe* (London, 1937).
Michael Leiris and Jacqueline Delange, *African Art* (London, 1968).
Misquitela Lima, *Fonctions sociologiques des figurines du culte hamba* (Luanda, 1971).
O. Nuoffer, *Afrikanische Plastik in der Gestaltung von Mutter und Kind* (Dresden, 1927).
Pascal James Imperato, "Wool blankets of the Fulani of Mali," *African Arts* (1973).
Patricia Vinnicombe, "Myth, Motive and Selection in Southern African Rock Art," *Africa* xlii (3) (1972).
Paula Ben-Amos, "Professionals and Amateurs in Benin Court Carving" in Daniel F. McCall and Edna G. Bay (eds.), *African Images. Essays in African Iconology* (New York/London, 1975).
D. Paulme, *Women in Black Africa* (London, 1963).
Foss Perkins, "Festival of Ohworu at Eywreni," *African Arts* 4 (1973).
Peter J. Ucko and Andrée Rosenfeld, *Palaeolithic Cave Art* (London, 1967).
Peter M. Weil, "The Masked Figure and Social Control: the Mandinka Case," *Africa* xli (4) (1971).
Philip E. Smith, "Problems and Possibilities of the Prehistoric Rock Art of Northern Africa," *African Historical Studies* i(1) (1968).
Philip J.C. Dark, *An Introduction to Benin Art and Technology* (Oxford, 1973).
Pierre Harter, "Four Bamileke Masks," *Man* 4 (1969).
Pierre Harter, "Les masques 'dits' Batcham," *Arts d'Afrique* 3 (1972).
R. S. Rattray, *Ashanti* (Oxford, 1923).
R. S. Rattray, *Religion and Art in Ashanti* (Oxford, 1927).
Robert Brainì and Adam Pollock, *Bangwa Funerary Art* (London, 1972).
Robert Brain, *Arts and Society in Africa.* (Hong Kong: Longman, 1980).
C. Salvadori and A. Fedders, *The Maasai* (London, 1973).
M.W. Smith (ed.), *The Artist in Tribal Society* (London, 1961).
Solange De Ganay, "On a Form of Cicatrisation among the Bambara," *Man* (1965), 49.
F. Starkweather, *Traditional Igbo Art* (Ann Arbor, Michigan, 1966).
A. Steinmann, *Maske und Krankheit* (Basle: CIBA Zeitschrift 89, 1943).
Suzanne Lafon, "La parure chez les femmes peules du Bas Sénégal," *Notes africaines*, IFAN, 46 (avril. 1950).
Suzanne Rudy, "Royal sculpture in the Cameroons Grasslands," in Douglas Fraser and Herbert M. Cole (eds.), *African Art and Leadership* (Madison, 1972).

P.A. Talbot, *The Peoples of Southern Nigeria*, 4 Vols (London, 1921).
P.A. Talbot, *Some Nigerian Fertility Cults* (London, 1927).
P.A. Talbot, *Tribes of the Niger Delta, Their Religion and Customs* (London, 1932).
Théodore Delachaux, "Méthodes et instruments de divination en Angola," *Acta tropica* iii, 2, 1946.
Thurstan Shaw, *Igbo-Ukwu: an Account of Archaeological Discoveries in Eastern Nigeria* (Ibadan, 1970).
V.W. Turner, "A Lunda Love Story ... Human Problems in British Central Africa," *Rhodes Livingstone Journal* 9 (1955).
P.J.L. Vandenhoute, *Classification Stylistique du masque Dan et Gnéré de la Côte d'lvoire occidentale*, Mededeelingen van het Rijksmuseum voor Volkenkande, Leiden 4.
R. Verly, *Les Mintadi* (Louvain, 1955).
Victor C. Uchendu, *The Igbo of Southeastern Nigeria* (New York, 1965).
Virginia Coulon, "Niominka pirogue ornaments," *African Arts* (Spring), 1973.
Viviana Pagnes, *Les Bambara* (Paris, 1954).
Warren L. d'Azevedo, *The Traditional Artist in African Societies* (London: Bloomington, 1973).
William B. Fagg, *Afro-Portuguese Ivories* (London, 1959).
William B. Fagg, *Nigerian Images* (New York, 1963).
William B. Fagg, *Tribes and Forms in African Art* (London, 1965).
William B. Fagg, *Divine Kinship in Africa* (London: British Museum, 1970).
G. Williams, *African Designs from Traditional Sources* (New York, 1971).
H.C. Woodhouse, "Rock Paintings of Southern Africa," *African Arts* (Spring, 1969).
Yves Urvoy, "L'art dans le territoire du Niger," *Etudes nigériennes II*, IFAN, Centre IFAN, (1955).
Zdenka Volavkova, "Nkisi Figurines of the Lower Congo," *African Arts* 2 (1972).
R. Zeller, "Die Goldgewichte von Asante," *Baesler Archiv Beiheft III* (Leipzig/Berlin, 1942).
J. Zwernemann, "Eine aussergewöhnliche Aufsatzmaske von den Bangwa, West Kamerun," *Tribus* (1972).

NAME INDEX

Analecta Husserliana

The Yearbook of Phenomenological Research

1. Tymieniecka, A-T. (ed.), *Volume 1 of Analecta Husserliana.* 1971 ISBN 90-277-0171-7
2. Tymieniecka, A-T. (ed.), *The Later Husserl and the Idea of Phenomenology.* Idealism – Realism, Historicity and Nature. 1972 ISBN 90-277-0223-3
3. Tymieniecka, A-T. (ed.), *The Phenomenological Realism of the Possible Worlds.* The "A Priori", Activity and Passivity of Consciousness, Phenomenology and Nature. 1974 ISBN 90-277-0426-0
4. Tymieniecka, A-T. (ed.), *Ingardeniana.* A Spectrum of Specialised Studies Establishing the Field of Research. 1976 ISBN 90-277-0628-X
5. Tymieniecka, A-T. (ed.), *The Crisis of Culture.* Steps to Reopen the Phenomenological Investigation of Man. 1976 ISBN 90-277-0632-8
6. Tymieniecka, A-T. (ed.), *The Self and the Other.* The Irreducible Element in Man, Part I. 1977 ISBN 90-277-0759-6
7. Tymieniecka, A-T. (ed.), *The Human Being in Action.* The Irreducible Element in Man, Part II. 1978 ISBN 90-277-0884-3
8. Nitta, Y. and Hirotaka Tatematsu (eds.), *Japanese Phenomenology.* Phenomenology as the Trans-cultural Philosophical Approach. 1979 ISBN 90-277-0924-6
9. Tymieniecka, A-T. (ed.), *The Teleologies in Husserlian Phenomenology.* The Irreducible Element in Man, Part III. 1979 ISBN 90-277-0981-5
10. Wojtyła, K., *The Acting Person.* Translated from Polish by A. Potocki. 1979 ISBN Hb 90-277-0969-6; Pb 90-277-0985-8
11. Ales Bello, A. (ed.), *The Great Chain of Being* and *Italian Phenomenology.* 1981 ISBN 90-277-1071-6
12. Tymieniecka, A-T. (ed.), *The Philosophical Reflection of Man in Literature.* Selected Papers from Several Conferences held by the International Society for Phenomenology and Literature in Cambridge, Massachusetts. Includes the essay by A-T. Tymieniecka, *Poetica Nova.* 1982 ISBN 90-277-1312-X
13. Kaelin, E. F., *The Unhappy Consciousness.* The Poetic Plight of Samuel Beckett. An Inquiry at the Intersection of Phenomenology and literature. 1981 ISBN 90-277-1313-8
14. Tymieniecka, A-T. (ed.), *The Phenomenology of Man and of the Human Condition.* Individualisation of Nature and the Human Being. (Part I:) Plotting the Territory for Interdisciplinary Communication. 1983 *Part II* see below under Volume 21. ISBN 90-277-1447-9

Analecta Husserliana

15. Tymieniecka, A-T. and Calvin O. Schrag (eds.), *Foundations of Morality, Human Rights, and the Human Sciences.* Phenomenology in a Foundational Dialogue with Human Sciences. 1983 ISBN 90-277-1453-3
16. Tymieniecka, A-T. (ed.), *Soul and Body in Husserlian Phenomenology.* Man and Nature. 1983 ISBN 90-277-1518-1
17. Tymieniecka, A-T. (ed.), *Phenomenology of Life in a Dialogue Between Chinese and Occidental Philosophy.* 1984 ISBN 90-277-1620-X
18. Tymieniecka, A-T. (ed.), *The Existential Coordinates of the Human Condition: Poetic – Epic – Tragic.* The Literary Genre. 1984 ISBN 90-277-1702-8
19. Tymieniecka, A-T. (ed.), *Poetics of the Elements in the Human Condition.* (Part 1:) The Sea. From Elemental Stirrings to Symbolic Inspiration, Language, and Life-Significance in Literary Interpretation and Theory. 1985
For Part 2 and 3 *see below* under Volumes 23 and 28. ISBN 90-277-1906-3
20. Tymieniecka, A-T. (ed.), *The Moral Sense in the Communal Significance of Life.* Investigations in Phenomenological Praxeology: Psychiatric Therapeutics, Medical Ethics and Social Praxis within the Life- and Communal World. 1986 ISBN 90-277-2085-1
21. Tymieniecka, A-T. (ed.), *The Phenomenology of Man and of the Human Condition.* Part II: The Meeting Point Between Occidental and Oriental Philosophies. 1986 ISBN 90-277-2185-8
22. Tymieniecka, A-T. (ed.), *Morality within the Life- and Social World.* Interdisciplinary Phenomenology of the Authentic Life in the "Moral Sense". 1987
Sequel to Volumes 15 and 20. ISBN 90-277-2411-3
23. Tymieniecka, A-T. (ed.), *Poetics of the Elements in the Human Condition.* Part 2: The Airy Elements in Poetic Imagination. Breath, Breeze, Wind, Tempest, Thunder, Snow, Flame, Fire, Volcano... 1988 ISBN 90-277-2569-1
24. Tymieniecka, A-T., *Logos and Life.* Book I: Creative Experience and the Critique of Reason. 1988 ISBN Hb 90-277-2539-X; Pb 90-277-2540-3
25. Tymieniecka, A-T., *Logos and Life.* Book II: The Three Movements of the Soul. 1988 ISBN Hb 90-277-2556-X; Pb 90-277-2557-8
26. Kaelin, E. F. and Calvin O. Schrag (eds.), *American Phenomenology.* Origins and Developments. 1989 ISBN 90-277-2690-6
27. Tymieniecka, A-T. (ed.), *Man within his Life-World.* Contributions to Phenomenology by Scholars from East-Central Europe. 1989 ISBN 90-277-2767-8
28. Tymieniecka, A-T. (ed.), *The Elemental Passions of the Soul.* Poetics of the Elements in the Human Condition, Part 3. 1990 ISBN 0-7923-0180-3
29. Tymieniecka, A-T. (ed.), *Man's Self-Interpretation-in-Existence.* Phenomenology and Philosophy of Life. – Introducing the Spanish Perspective. 1990 ISBN 0-7923-0324-5
30. Rudnick, H. H. (ed.), *Ingardeniana II.* New Studies in the Philosophy of Roman Ingarden. With a New International Ingarden Bibliography. 1990 ISBN 0-7923-0627-9

Analecta Husserliana

31. Tymieniecka, A-T. (ed.), *The Moral Sense and Its Foundational Significance: Self, Person, Historicity, Community*. Phenomenological Praxeology and Psychiatry. 1990 ISBN 0-7923-0678-3
32. Kronegger, M. (ed.), *Phenomenology and Aesthetics*. Approaches to Comparative Literature and Other Arts. Homages to A-T. Tymieniecka. 1991 ISBN 0-7923-0738-0
33. Tymieniecka, A-T. (ed.), *Ingardeniana III*. Roman Ingarden's Aesthetics in a New Key and the Independent Approaches of Others: The Performing Arts, the Fine Arts, and Literature. 1991
Sequel to Volumes 4 and 30 ISBN 0-7923-1014-4
34. Tymieniecka, A-T. (ed.), *The Turning Points of the New Phenomenological Era*. Husserl Research – Drawing upon the Full Extent of His Development. 1991 ISBN 0-7923-1134-5
35. Tymieniecka, A-T. (ed.), *Husserlian Phenomenology in a New Key*. Intersubjectivity, Ethos, the Societal Sphere, Human Encounter, Pathos. 1991 ISBN 0-7923-1146-9
36. Tymieniecka, A-T. (ed.), *Husserl's Legacy in Phenomenological Philosophies*. New Approaches to Reason, Language, Hermeneutics, the Human Condition. 1991 ISBN 0-7923-1178-7
37. Tymieniecka, A-T. (ed.), *New Queries in Aesthetics and Metaphysics*. Time, Historicity, Art, Culture, Metaphysics, the Transnatural. 1991 ISBN 0-7923-1195-7
38. Tymieniecka, A-T. (ed.), *The Elemental Dialectic of Light and Darkness*. The Passions of the Soul in the Onto-Poiesis of Life. 1992 ISBN 0-7923-1601-0
39. Tymieniecka, A-T. (ed.), *Reason, Life, Culture, Part I*. Phenomenology in the Baltics. 1993 ISBN 0-7923-1902-8
40. Tymieniecka, A-T. (ed.), *Manifestations of Reason: Life, Historicity, Culture*. Reason, Life, Culture, Part II. Phenomenology in the Adriatic Countries. 1993 ISBN 0-7923-2215-0
41. Tymieniecka, A-T. (ed.), *Allegory Revisited*. Ideals of Mankind. 1994 ISBN 0-7923-2312-2
42. Kronegger, M. and Tymieniecka, A-T. (eds.), *Allegory Old and New*. In Literature, the Fine Arts, Music and Theatre, and Its Continuity in Culture. 1994 ISBN 0-7923-2348-3
43. Tymieniecka, A-T. (ed.): *From the Sacred to the Divine*. A New Phenomenological Approach. 1994 ISBN 0-7923-2690-3
44. Tymieniecka, A-T. (ed.): *The Elemental Passion for Place in the Ontopoiesis of Life*. Passions of the Soul in the *Imaginatio Creatrix*. 1995 ISBN 0-7923-2749-7
45. Zhai, Z.: *The Radical Choice and Moral Theory*. Through Communicative Argumentation to Phenomenological Subjectivity. 1994 ISBN 0-7923-2891-4
46. Tymieniecka, A-T. (ed.): *The Logic of the Living Present*. Experience, Ordering, Onto-Poiesis of Culture. 1995 ISBN 0-7923-2930-9

Analecta Husserliana

47. Tymieniecka, A-T. (ed.): *Heaven, Earth, and In-Between in the Harmony of Life*. Phenomenology in the Continuing Oriental/Occidental Dialogue. 1995 ISBN 0-7923-3373-X
48. Tymieniecka, A-T. (ed.): *Life. In the Glory of its Radiating Manifestations*. 25th Anniversary Publication. Book I. 1996 ISBN 0-7923-3825-1
49. Kronegger, M. and Tymieniecka, A-T. (eds.): *Life. The Human Quest for an Ideal*. 25th Anniversary Publication. Book II. 1996 ISBN 0-7923-3826-X
50. Tymieniecka, A-T. (ed.): *Life. Phenomenology of Life as the Starting Point of Philosophy*. 25th Anniversary Publication. Book III. 1997 ISBN 0-7923-4126-0
51. Tymieniecka, A-T. (ed.): *Passion for Place. Part II*. Between the Vital Spacing and the Creative Horizons of Fulfilment. 1997 ISBN 0-7923-4146-5
52. Tymieniecka, A-T. (ed.): *Phenomenology of Life and the Human Creative Condition*. Laying Down the Cornerstones of the Field. Book I. 1997 ISBN 0-7923-4445-6
53. Tymieniecka, A-T. (ed.): *The Reincarnating Mind, or the Ontopoietic Outburst in Creative Virtualities*. Harmonisations and Attunement in Cognition, the Fine Arts, Literature. Phenomenology of Life and the Human Creative Condition. Book II. 1997 ISBN 0-7923-4461-8
54. Tymieniecka, A-T. (ed.): *Ontopoietic Expansion in Human Self-Interpretation-in-Existence*. The I and the Other in their Creative Spacing of the Societal Circuits of Life. Phenomenology of Life and the Creative Condition. Book III. 1997 ISBN 0-7923-4462-6
55. Tymieniecka, A-T. (ed.): *Creative Virtualities in Human Self-Interpretation-in-Culture*. Phenomenology of Life and the Human Creative Condition. Book IV. 1997 ISBN 0-7923-4545-2
56. Tymieniecka, A-T. (ed.): *Enjoyment*. From Laughter to Delight in Philosophy, Literature, the Fine Arts and Aesthetics. 1998 ISBN 0-7923-4677-7
57. Kronegger, M. and Tymieniecka, A-T. (eds.): *Life. Differentiation and Harmony ...Vegetal, Animal, Human*. 1998 ISBN 0-7923-4887-7
58. Tymieniecka, A-T. and Matsuba, S. (eds.): *Immersing in the Concrete*. Maurice Merleau-Ponty in the Japanese Perspective. 1998 ISBN 0-7923-5093-6
59. Tymieniecka, A-T. (ed.): *Life – Scientific Philosophy/Phenomenology of Life and the Sciences of Life*. Ontopoiesis of Life and the Human Creative Condition. 1998 ISBN 0-7923-5141-X
60. Tymieniecka, A-T. (eds.): *Life – The Outburst of Life in the Human Sphere*. Scientific Philosophy/Phenomenology of Life and the Sciences of Life. Book II. 1998 ISBN 0-7923-5142-8
61. Tymieniecka, A-T. (ed.): *The Aesthetic Discourse of the Arts*. Breaking the Barriers. 2000 ISBN 0-7923-6006-0
62. Tymieniecka, A-T. (ed.): *Creative Mimesis of Emotion*. From Sorrow to Elation; Elegiac Virtuosity in Literature. 2000 ISBN 0-7923-6007-9

Analecta Husserliana

63. Kronegger, M. (ed).: *The Orchestration of The Arts – A Creative Symbiosis of Existential Powers.* The Vibrating Interplay of Sound, Color, Image, Gesture, Movement, Rhythm, Fragrance, Word, Touch. 2000 ISBN 0-7923-6008-7
64. Tymieniecka, A-T. and Z. Zalewski (eds.): *Life – The Human Being Between Life and Death.* A Dialogue Between Medicine and Philosophy, Recurrent Issues and New Approaches. 2000 ISBN 0-7923-5962-3
65. Kronegger, M. and Tymieniecka, A-T. (eds.): *The Aesthetics of Enchantment in the Fine Arts.* 2000 ISBN 0-7923-6183-0
66. Tymieniecka, A-T. (ed.): *The Origins of Life, Volume I: The Primogenital Matrix of Life and Its Context.* 2000 ISBN 0-7923-6246-2; Set ISBN 0-7923-6446-5
67. Tymieniecka, A-T. (ed.): *The Origins of Life, Volume II: The Origins of the Existential Sharing-in-Life.* 2000 ISBN 0-7923-6276-4; Set ISBN 0-7923-6446-5
68. Tymieniecka, A-T. (ed.): *PAIDEIA.* Philosophy/Phenomenology of Life Inspiring Education of our Times. 2000 ISBN 0-7923-6319-1
69. Tymieniecka, A-T. (ed.): *The Poetry of Life in Literature.* 2000 ISBN 0-7923-6408-2
70. Tymieniecka, A-T. (ed.): *Impetus and Equipoise in the Life-Strategies of Reason.* Logos and Life, volume 4. 2000 ISBN 0-7923-6731-6; HB 0-7923-6730-8
71. Tymieniecka, A-T. (ed.): *Passions of the Earth in Human Existence, Creativity, and Literature.* 2001 ISBN 0-7923-6675-1
72. Tymieniecka, A-T. and E. Agazzi (eds.): *Life – Interpretation and the Sense of Illness within the Human Condition.* Medicine and Philosophy in a Dialogue. 2001 ISBN Hb 0-7923-6983-1; Pb 0-7923-6984-X
73. Tymieniecka, A-T. (ed.): *Life – The Play of Life on the Stage of the World in Fine Arts, Stage-Play, and Literature.* 2001 ISBN 0-7923-7032-5
74. Tymieniecka, A-T. (ed.): *Life-Energies, Forces and the Shaping of Life: Vital, Existential.* Book I. 2002 ISBN 1-4020-0627-6
75. Tymieniecka, A-T. (ed.): *The Visible and the Invisible in the Interplay between Philosophy, Literature and Reality.* 2002 ISBN 1-4020-0070-7
76. Tymieniecka, A-T. (ed.): *Life – Truth in its Various Perspectives.* Cognition, Self-Knowledge, Creativity, Scientific Research, Sharing-in-Life, Economics...... 2002 ISBN 1-4020-0071-5
77. Tymieniecka, A-T. (ed.): *The Creative Matrix of the Origins.* Dynamisms, Forces and the Shaping of Life. 2003 ISBN 1-4020-0789-2
78. Tymieniecka, A-T. (ed.): *Gardens and the Passion for the Infinite.* 2003 ISBN 1-4020-0858-9
79. Tymieniecka, A-T. (ed.): *Does the World Exist?* Plurisignificant Ciphering of Reality. 2003 ISBN 1-4020-1517-8

Analecta Husserliana

80. Tymieniecka, A-T. (ed.): *Phenomenology World-Wide.* Foundations – Expanding Dynamics – Life-engagements. A Guide for Research and Study. 2002
ISBN 1-4020-0066-9
81. Tymieniecka, A-T. (ed.): *Metamorphosis.* Creative Imagination in Fine Arts, Life-Projects and Human Aesthetic Aspirations. 2004 ISBN 1-4020-1709-X
82. Tymieniecka, A-T. (ed.): *Mystery in its Passions.* Literary Explorations. 2004
ISBN 1-4020-1705-7
83. Tymieniecka, A-T. (ed.): *Imaginatio Creatrix.* The Pivotal Force of the Genesis/Ontopoiesis of Human Life and Reality. 2004. ISBN 1-4020-2244-1
84. Tymieniecka, A-T. (ed.): *Phenomenology of Life. Meeting the Challenges of the Present-Day World.* 2005. ISBN 1-4020-2463-0
85. Tymieniecka, A-T. (ed.): *The Enigma of Good and Evil: The Moral Sentiment in Literature.* 2005. ISBN 1-4020-3575-6
86. Tymieniecka, A-T. (ed.): *Temporality in Life as Seen Throught Literature: Contributions to Phenomenology of Life.* 2007. ISBN 978-1-4020-5330-6
87. Tymieniecka, A-T. (ed.): *Human Creation Between Reality and Illusion.* 2005.
ISBN 1-4020-3577-2
88. Tymieniecka, A-T. (ed.): *Logos of Phenomenology and Phenomenology of the Logos. Book One: Phenomenology as the Critique of Reason in Contemporary Criticism and Interpretation.* 2005. ISBN 1-4020-3678-7
89. Tymieniecka, A-T. (ed.): *Logos of Phenomenology and Phenomenology of the* Logos. Book Two: The Human Condition in-the-unity-of-everything-there-is-alive. *Individuation, Self, Person, Self-determination, Freedom, Necessity.* 2005.
ISBN 1-4020-3706-6
90. Tymieniecka, A-T. (ed.): *Logos of Phenomenology and Phenomenology of the Logos. Book Three: Logos of History - Logos of Life. Historicity, Time, Nature, Communication, Consciousness, Alterity, Culture.* 2005. ISBN 1-4020-3717-1
91. Tymieniecka, A-T. (ed.): *Logos of Phenomenology and Phenomenology of the Logos. Book Four: The Logos of Scientific Interrogation. Participating in Nature – Life – Sharing in Life.* 2005. ISBN 1-4020-3736-8
92. Tymieniecka, A-T. (ed.): *Logos of Phenomenology and Phenomenology of the Logos. Book Five: The Creative Logos. Aesthetic Ciphering in Fine Arts, Literature and Aesthetics.* 2005. ISBN 1-4020-3743-0
93. Tymieniecka, A-T. (ed.): *Phenomenology of Life - From the Animal Soul to the Human Mind. Book One: In Search of Experience.* 2007. ISBN 978-1-4020-5191-3

MIX
Papier aus verantwortungsvollen Quellen
Paper from responsible sources
FSC® C105338

If you have any concerns about our products,
you can contact us on
ProductSafety@springernature.com

In case Publisher is established outside the EU,
the EU authorized representative is:
Springer Nature Customer Service Center GmbH
Europaplatz 3, 69115 Heidelberg, Germany

Printed by Libri Plureos GmbH
in Hamburg, Germany